Animal Magnetism, Early Hypnotism, and Psychical Research, 1766 – 1925

An Annotated Bibliography

Bibliographies

in the

History of Psychology and Psychiatry

A Series

Robert H. Wozniak, General Editor

Animal Magnetism, Early Hypnotism, and Psychical Research, 1766–1925

An Annotated Bibliography

Adam Crabtree

KRAUS INTERNATIONAL PUBLICATIONS
White Plains, New York
A Division of Kraus-Thomson Organization Limited

First Printing 1988

Printed in the United States of America

This book is printed on acid-free paper. It is Smyth sewn
and casebound in F-grade Library Buckram, which contains
no synthetic fibers. The paper and binding of this book
meet the guidelines for permanence and durability of the
Committee on Production Guidelines for Book Longevity
of the Council on Library Resources.

Library of Congress Cataloging in Publication Data

Crabtree, Adam.
Animal magnetism, early hypnotism and psychical research, 1766–1925.

(Bibliographies in the history of psychology and psychiatry)
Includes indexes.
1. Animal magnetism—Bibliography. 2. Hypnotism—Bibliography.
3. Psychical research—Bibliography.
I. Title. II. Series.
Z6878.A54C73 1988 [BF1141] 048.1547 87-29746
ISBN 0-527-20006-9 (alk. paper)

Contents

Historical Introduction vii
Format of the Bibliography xvii
Acknowledgments xxi
Glossary of Terms xxiii
References xxix
Animal Magnetism, Early Hypnotism and Psychical Research
Early Works: Pre–1800 1
1800–1809. 55
1810–1819. 63
1820–1829. 81
1830–1839. 95
1840–1849 115
1850–1859 153
1860–1869 199
1870–1879 225
1880–1889 245
1890–1899 295
1900–1909 339
1910–1919 379
1920–1925 407
Name Index 437
Title Index 455
Subject Index 517

Historical Introduction

Animal magnetism is little known today. Most historical scholars would probably be hard pressed to write more than a brief paragraph about Franz Anton Mesmer (1734–1815) and his discovery. Yet, for approximately seventy-five years from its beginnings in 1779, animal magnetism flourished as a medical and psychological specialty, and for another fifty years it continued to be a system of some influence.

When one examines the history of animal magnetism and its offshoots, it seems incredible that this once powerful system is now almost completely forgotten. That animal magnetism is no longer practiced is hardly surprising. The theory of animal magnetism in its original form would be difficult for most moderns to accept. What is puzzling is that the story of animal magnetism is so neglected.

Animal magnetism is not comparable to certain medical fads which flourished for a time and then died out. Such crazes did not significantly shape medical or psychological theory and practice, nor did they significantly affect the evolution of those disciplines. Animal magnetism, on the other hand, had a profound impact on medicine, psychology, and psychical research (today called parapsychology), as a brief examination of its history will show.

Franz Anton Mesmer and Animal Magnetism

The seeds of thought that gave birth to animal magnetism may be found in Mesmer's thesis *Dissertatio physico-medica de planetarum influxu* of 1766, which he wrote for a doctorate in medicine at the University of Vienna. In this treatise Mesmer developed the notion of "animal gravitation," a force which he considered to be both the cause of universal gravitation and the foundation for all bodily properties, and which he believed to affect organisms in the most intimate way. Mesmer believed that animal gravitation connected living things to the stars and was the basis for healthy functioning, since it harmonized the body in a fashion comparable to the tuning of a musical instrument.

Mesmer's interest in invisible forces found concrete expression in his early medical practice, where he experimented with using iron magnets to treat illness. Spurred on by success, Mesmer enthusiastically turned his attention to revising his theory of "animal gravitation." Retaining his central idea of a universal force that is the foundation for health and disease, he renamed that force "animal magnetism," finding it to possess many of the characteristics associated with mineral magnetism.

The more Mesmer experimented, the more he became disenchanted with using iron magnets to heal. He came to believe that the physician himself is a magnet of a very special kind, capable of channeling the invisible "magnetic fluid" that pervades the universe into the body of the sick person and bringing about the magnetic balance necessary for a cure.

The basic principles of Mesmer's theory of animal magnetism were articulated in twenty-seven propositions in his *Mémoire sur la découverte du magnétisme animal* of 1779. Among the more informative are the following:

> 1) There exists a mutual influence between the celestial bodies, the earth and animate bodies; 2) The means of this influence is a fluid that is universally distributed and continuous . . . and which, by its nature, is capable of receiving, propagating and communicating all impressions of movement; 3) This reciprocal action is governed by mechanical laws as yet unknown. . . .; . . . 8) The animal body experiences the alternative effects of this agent which insinuates them into the nerves and affects them immediately; 9) It particularly manifests itself in the human body by properties analogous to the magnet. . . .; 10) Because the property of the animal body which makes it susceptible to the influence of heavenly bodies and to the reciprocal action of those around it is analogous to that of the magnet, I decided to call it "animal magnetism." . . . .; . . . 23) The facts will show, following the practical rules that I will establish, that this principle can heal disorders of the nerves immediately, and other disorders mediately.

Mesmer eventually discontinued the use of iron magnets entirely, relying instead on the application of newly evolved animal–magnetic techniques. These techniques involved "magnetic passes" or sweeping movements of the hands to direct magnetic fluid to diseased parts of the patient's body. Using these methods, Mesmer performed some remarkable, if controversial, cures in Austria and Germany and attempted to gain acceptance for his theory of animal magnetism from the medical establishment of Vienna. He was not successful in this endeavor and in 1778 decided to go to Paris, where, he believed, new ideas were more favorably considered.

In Paris, Mesmer set up two treatment clinics, one for the rich and the other for the poor. The sick flocked to him and he treated them by the hundreds over the next few years. During that time there were many among both rich and poor who testified to being cured by animal magnetism, in some cases of long-term chronic illnesses.

During this period, Mesmer made attempts to get the medical establishment of Paris to approve his theory of animal magnetism, but try as he might, he could not gain a sympathetic hearing. On the contrary, the medical faculty at Paris became alarmed at the popularity of Mesmer's clinics and moved to suppress them. In 1784 two commissions were constituted to investigate animal magnetism, both appointed by the king of France. One was made up of members of the Royal Academy of Sciences and the Faculty of Medicine; it included some of the country's most eminent scientists and functioned under the chairmanship of Benjamin Franklin (1700–1790), then the American ambassador to France. The second commission was composed of physicians of the Royal Society of Medicine. The resulting investigations were carried out in the face of Mesmer's objections and without his cooperation. Both commissions filed reports unfavorable to animal magnetism, although a member of the second commission wrote a dissenting opinion recommending further investigation. The first commission also drew up a secret report for the king on potential dangers to morals through the misuse of magnetic techniques.

The publication of the reports in 1784 was followed by a flood of treatises in response, many written by physicians. Some supported the conclusions of the commissions; others were strongly critical. In the latter category were the protests of dozens of medical practitioners who had themselves been using animal magnetism, in their opinion very successfully. They criticized the commissioners for both their attitude and the technique of investigation.

Meanwhile Mesmer had become embroiled in a controversy about how his theory and technique were to be taught. Mesmer desired to secure his financial condition by charging a fee to those who wanted to be trained in animal magnetism. A scheme was worked out on Mesmer's behalf by a banker, Guillaume Kornmann (b. ca. 1840) and a lawyer and freethinker, Nicholas Bergasse (1750–1832). It involved the founding of Societies of Harmony—which were to be considered the official organs for teaching animal magnetism, membership in the Society being gained through the subscription of a considerable sum of money. From the parent Society of Harmony in Paris, dozens more were established throughout France. Mesmer and Bergasse, who had become the chief spokesman of the Society of Harmony in Paris, eventually had a falling out and a split resulted.

Mesmer grew more and more disillusioned with Paris and undertook a series of trips away from that city. Eventually, he settled in Germany and lived in comparative seclusion. Although he continued periodically to write on animal magnetism after 1790, he was not very actively involved in its affairs. In 1812, the Berlin Academy of Science, surprised to discover that Mesmer was still alive, sent Karl Christian Wolfart (1778–1832) to find out about animal magnetism directly from its discoverer. Wolfart remained with Mesmer for two years, putting together what would be the master's last treatise on animal magnetism. It was published under the title *Mesmerismus. Oder System der Wechselwirkungen* in 1814. Mesmer died in 1815.

Magnetic Sleep

Although Mesmer's personal fortunes and fame waned after 1790, animal magnetism (also called "mesmerism") flourished. This was due in no small part to the work of the Marquis de Puységur (1751 – 1825), one of Mesmer's most loyal and enthusiastic pupils. Puységur discovered that some individuals fell into a kind of trance when animal magnetism was applied to them. Although appearing to be asleep, they were still conscious and could reply to questions and convey information. In this state of "magnetic sleep," as Puységur called it, the patient was very suggestible, taking for reality any fantasy the magnetizer might depict. Upon awakening from magnetic sleep, the patient would remember nothing that had taken place while asleep.

Puységur was fascinated by this unusual state of consciousness, so different from ordinary waking consciousness. He discovered that many in this state could apparently diagnose their own illnesses and those of others, and even prescribe effective remedies for the conditions they perceived. He also noticed that although magnetized subjects had no memory in the waking state for occurrences in the state of magnetic sleep, they did retain a continuous memory from sleep state to sleep state. Noting these two separate chains of memory that accompanied the two distinct states of consciousness, Puységur came to view magnetic sleep and the waking state as "two different existences." From this seed, the notion of a seemingly separate mind or self operating covertly within the human psyche took root. It came to fruition some one hundred years later with the work of Pierre Janet and his concept of the "subconscious," as described below.

Puységur noted the similarity between "magnetic sleep" and the natural phenomenon of "sleepwalking" or "somnambulism," the only difference between the two states being that in magnetic sleep the subject is in a special connection or "rapport" with the magnetizer, whereas in sleepwalking the sleeper is in rapport with no one. Because of the similarity, Puységur called the newly discovered state "magnetic somnambulism." Another term that eventually came into use was "artificial somnambulism."

Hypnotism

Puységur's work had a powerful influence on the practitioners of animal magnetism. Mesmer, working from a markedly mechanistic model of the human organism, had emphasized the physical action involved in magnetic healing. Puységur's orientation was much more psychological. From his experiments with magnetic sleep, he developed the rudiments of a psychotherapy based upon the investigation of somnambulistic consciousness. He evolved a theory of mental disturbance as a state of "disorderly somnambulism" in which the individual moves in and out of a condition of disturbed somnambulism in a chaotic manner. Puységur's psychological orientation is also demonstrated by the importance he placed on the role of human will when magnetizing and the need for the magnetizer to exercise "good will" in order to be effective. Although Puységur's views differed in these significant ways from those of Mesmer, he nevertheless retained Mesmer's notion of a "magnetic fluid" that passes between magnetizer and patient.

Puységur became a very influential figure in the history of animal magnetism. His psychological concerns were taken up by many investigators and this eventually led to a new formulation of the theory of animal-magnetic phenomena. That formulation was first hinted at in the writings of the Abbé Faria (1755–1819) and Alexandre Bertrand (1795–1831) and reached its culmination in the work of the Manchester physician James Braid (1795–1860). In 1842 Braid coined the term "hypnotism" or "nervous sleep" to replace "animal magnetism," intending to do away with any notion of a physical agent such as "magnetic fluid" that passes between magnetizer and subject and produces the phenomena of somnambulism. Braid described hypnotism as a psycho-physiological state that needs no operator and can be self-induced. He also emphasized the role of suggestion both in producing the hypnotic state and in bringing about the healing effects associated with it.

Eventually Braid's view became the dominant one and his terminology the accepted nomenclature. This took some time, however, and animal magnetism in its traditional form remained a force to be reckoned with for another sixty years.

Three Streams Flowing from Animal Magnetism

It is possible to trace three distinct currents of thought flowing directly from the discovery of animal magnetism. These three streams may be identified as 1) psychological, 2) medical, and 3) parapsychological.

Psychological Stream

The most important of these three currents, from an historical point of view, is the psychological stream. Justly it can be said that Mesmer's discovery of animal magnetism was a pivotal moment in the evolution of modern psychology and psychotherapy. It led to Puységur's investigation of the consciousness manifested in magnetic sleep and the eventual discovery of a subconscious realm of mental activity. It also led to Braid's teaching about hypnotism as a psychological phenomenon and the resulting exploration of the psychotherapeutic power of suggestion.

The magnetic tradition of Puységur and the hypnotic tradition of Braid were both very much in evidence in mid–nineteenth-century France, particularly in experimentation with somnambulism and its effects. Braid's writings were "discovered" in France around 1860, and by the 1870s and 1880s, men trained in psychological observation, such as Charles Richet (1850–1935), Henri Beaunis (1830–1921), and Joseph Delboeuf (1831–1896), began to become involved in work on hypnotic phenomena.

In the 1860s, Ambroise Liébeault (1823–1901), a provincial physician, had undertaken some special observations of his own. He used hypnotism to treat the illnesses of some of his clients with great success. Liébeault believed that hypnotism was based on suggestion and that its healing effects were due to the power of suggestion. Hippolyte Bernheim (1840–1919), professor of medicine at Nancy, was impressed by Liébeault's results, and the two initiated what came to be called the Nancy School of hypnotism.

Meanwhile, the highly respected neurologist Jean-Martin Charcot (1825–1893) was developing his own ideas about the nature of hypnotism based on his work with hysterical patients at the Salpêtrière Hospital. Because Charcot worked from a more physicalist model of hypnotism, his ideas came into conflict with those of Liébeault and Bernheim. The resulting competition between the Nancy School and the School of the Salpêtrière continued for many years, resulting in extremely valuable experimental studies of hypnotism, suggestion, and hysteria.

Among the associates of Charcot, although not an adherent of his school, was Pierre Janet (1859–1947). Janet had a particular interest in hysteria and the automatisms associated with that condition. From his observations, he developed the notion of the "subconscious," a realm of mental activity in which emotional disorders originate. His ground-breaking work in this area made possible the development of all modern psychotherapies that accept the reality of an unconscious realm of mental and emotional activity influencing ordinary human life. Janet's discoveries had a strong impact on the subsequent work of Max Dessoir (1867–1947), Morton Prince (1854–1929), Boris Sidis (1867–1923), and William James (1842–1910), among others.

Multiple personality was one of the disorders that was most closely studied by Janet and others who were interested in fathoming the mechanism of the subconscious. It was considered to be a special form of hysteria in which somnambulistic consciousness had taken the form of well-defined, distinct personalities. Janet, Alfred Binet (1857–1911), Eugène Azam (1822–1899), and others pioneered work in this area and used their findings to throw light not only on hysterical disorders but also on the nature of hypnotic consciousness in the normal individual.

These insights into the subconscious carried implications for possible treatment methods for the emotionally disturbed. Sigmund Freud (1856–1939) was influenced by these developments, and his earlier works clearly reflect the magnetic–hypnotic tradition with its gradual unveiling of an unconscious mental life.

Medical Stream

In its origins, animal magnetism was a healing system. It was based on a view of the human organism as a self-healing entity requiring the proper balance of a universal "magnetic fluid" that affects the ebb and flow of the life force. The techniques of animal magnetism were geared to restoring that balance in persons suffering from illness.

Although Puységur's work diverted the attention of many magnetizers to psychological pursuits, there remained a powerful current of interest in the healing work Mesmer originally envisioned. In the literature it is not always easy to differentiate between those interested in the investigation of somnambulistic consciousness and those interested in magnetic healing, since more often than not practitioners were involved with both. However, the two concerns were distinguishable in practice. This is reflected by the fact that many of the thousands of books written on animal magnetism before 1925 have one section dealing with the treatment of disease and another dealing with somnambulistic phenomena.

Puységur himself had pointed out that many individuals, when put into a state of magnetic sleep, would spontaneously diagnose their own illnesses and those of others. He also described instances in which the somnambulist prescribed treatment by specific medicines or medical procedures. He considered this to be one of the great benefits of magnetic sleep, claiming that somnambulists were almost always correct in their diagnosis and that their prescribed treatments were often successful.

In 1826 there appeared a hefty treatise of nearly twelve hundred pages compiled by Simon Mialle (b. ca. 1790) entitled *Exposé par ordre alphabetique des cures opérés en France par le magnétisme animal*. This work gives some idea of the vast extent of the tradition of magnetic healing in the decades following Mesmer. Here Mialle details cases of cure through the application of animal magnetism between 1774 and 1826. In each case there is a description of the disease treated, the animal-magnetic procedure employed, and the results produced. Each instance is documented by source, and the reader cannot help but be impressed by the sheer volume of work of this kind being done in those early years.

The healing tradition of animal magnetism continued well beyond the year 1826, extending even into the twentieth century. After 1880, books on animal-magnetic healing often incorporated chapters on healing by suggestion, taking a page from the successful medical use of hypnotism by Liébeault and his followers.

Besides healing, another medical use of animal magnetism was as an anesthetic for surgery. The first well-documented surgical operation on an individual in a state of magnetic somnambulism was performed in Paris on April 16, 1829. The mesmerist was Pierre Jean Chapelain and the surgeon was Jules Cloquet (1790–1883), later famous for his works on anatomy. The surgery was for the removal of a cancerous breast from a sixty-four-year-old woman, a Madame Plantin. The earliest use of animal magnetism as an anesthetic in the United States seems to have been a painless tooth extraction performed by the mesmerist Bugard in 1836. It seems that this new use of animal magnetism did not really come into its own until the early 1840s. In 1842 a Dr. Ward successfully performed the amputation of a leg at the thigh upon a mesmerized patient in London. His influential colleague, John Elliotson (1791–1868), immediately took up the cause for this medical use of animal magnetism, and in India James Esdaile (1808–1859), carried out dozens of serious operations on magnetized patients in the mid-1840s. At about the same time a series of surgical operations were performed under Dr. Loisel in Cherbourg, France.

All this promising activity involving animal magnetism as an anesthetic soon faded, however, with the introduction in Britain of ether as an analgesic in 1847. Although animal magnetism or hypnotism did not become widely used as an anesthetic, some surgical operations under its agency continued to be performed long after chemicals were well established in that role. In fact, the rise of interest in hypnotism in France around 1860 was strongly associated with its successful use in surgery.

Parapsychological Stream: Psychical Research

Psychical research, the scientific study of the paranormal, may be said to have had its official beginning in 1882 with the establishment of the Society for Psychical Research in Britain. Psychical research was the direct result of certain developments arising from animal magnetism. These were: 1) the occult medico-philosophical tradition in Germany that adopted animal-magnetic theory; 2) the development of "magnetic magic" in France; and 3) the rise of spiritualism in the United States.

In Germany animal magnetism developed a strong early following among those influenced by romantic philosophy, so prominent at the end of the eighteenth century. Literary men such as Jean Paul Richter (1763–1825) and E.T.A. Hoffman (1776–1822), physicians including Johannes Kaspar Lavater (1741–1801) and Friederich Hufeland (1774–1839), and religious philosopher Johannes Heinrich Jung-Stilling (1740–1817) are examples of thinkers who found the notion of a universal magnetic agent that connected all beings and was the source of life and health a most congenial concept. The spiritual philosophy of Emanuel Swedenborg (1688–1772) had made inroads of its own in late eighteenth-century Germany; magnetic somnambulists began to have Swedenborgian style visions, communicating with the world of spirits in mesmeric ecstasy. Paranormal-type phenomena, such as clairvoyance and precognition, were common in these circles, astounding the curious observer. While there were many who were content to account for these marvels through a romantic, occult-oriented philosophy, some felt the need for a more scientific approach that could examine the facts systematically and evaluate their credibility, a need that would not be met until the rise of psychical research in the latter part of the nineteenth century.

The French developed their own particular melding of animal magnetism and occult tradition. The most influential magnetizer of this kind was the Baron Du Potet De Sennevoy (1796–1881). He developed a system called "magnetic magic" that revised animal magnetism's traditional doctrine of a universal magnetic fluid by incorporating it within the older notion of a universal spiritual power, which serves as the basis for "natural magic." This concept, so different from the mechanical view of Mesmer's, considered magnetism to be the bond between spirit and matter, or body and soul. In Du Potet's view, mesmerizers who recognized the true nature of magnetism could work "magic," producing marvelous cures and various paranormal phenomena.

Animal magnetism began making significant inroads in the United States from the mid–1830s on. Lectures by Charles Poyen St. Sauveur (d. 1844) on animal magnetism excited the imagination of the country and led to the emergence of magnetic practitioners of a peculiarly American type. Itinerant magnetizers wandered the countryside with professional somnambulists at their sides, stopping in the local towns to give medical clairvoyant readings. For a fee, the somnambulist would diagnose an illness and prescribe remedies. The visionary Andrew Jackson Davis (1826–1910) began his career as such an itinerant somnambulist and eventually became an author of great popularity, using the magnetic trance to dictate his spiritual treatises. All this magnetic activity prepared the way for the rise of Spiritualism, initiated by "spirit rapping" in the home of John Fox in 1848. This spirit activity centered on the daughters of the household, and news of the purportedly paranormal activity of the Fox sisters spread rapidly throughout the United States, reaching England, France, and Germany within a few years. Spiritualist "mediums" appeared who claimed to be able to communicate with the departed on the "other side." Typically, the medium would go into a self-induced trance and produce paranormal phenomena of the mental or psychic type (clairvoyance, telepathy, precognition, etc.) or the physical type (levitation of objects, materialization of forms, production of mysterious lights, etc.).

When Spiritualism spread to England in the early 1850s, it found a very receptive home. Within a few years spiritualistic mediums could be found in great abundance throughout the country. Spiritualist churches were established; spiritualist alliances were formed; and spiritualist books and newspapers came into print. There was such a proliferation of apparently paranormal spiritualistic phenomena that serious-minded people voiced the need for a careful scientific investigation to discover whether these things were real or illusory.

The successful spread of Spiritualism was to a large extent due to the popularity of a fad that grew out of spiritualistic circles and emigrated to Great Britain and Europe in 1853. This was the practice of "table tipping," "table turning," or "table tapping," as it was commonly called. A group of people would gather around the parlour table, rest their hands in a circle on its surface, and wait for spontaneous movement to occur. Sometimes the table would rotate; at other times it would rise and fall on one side, tapping a leg on the floor. The tapping would be read as an alphabetical code, and a message would be deciphered. Many explained the phenomenon in terms of the action of spirits of the dead communicating with the living; others attributed the movements and messages to the action of animal magnetic fluid emanating from the participants; still others believed the participants were simply deluding themselves, the movement being produced by their own unconscious physical exertions.

In Germany, France, England, and the United States, the association between magnetic somnambulism and paranormal phenomena of the spiritualistic type was very strong. Many of the books and articles that appeared wove their way back and forth between the two areas, giving the impression that it was impossible to discuss one without dealing with the other. It is not surprising, then, that when the Society for Psychical Research was formed in England in 1882, it undertook to investigate not only the validity of spiritualistic phenomena, but also the nature of animal magnetism and hypnotism.

In the thirty years preceding the foundation of the Society for Psychical Research, there had been a number of notable attempts to investigate the phenomena of Spiritualism scientifically. Some were carried out by scientists, others by individuals untrained in the procedures of systematic investigation. The results were uneven and inconclusive. So when a group of academics, most of them associated with Cambridge, decided to set up a society that would undertake a study employing stringent scientific criteria, there was enthusiasm for the idea both from intellectuals and the Spiritualists themselves. The Society was fortunate to have the nearly full-time involvement of a number of highly gifted investigators and within a few years began publishing its *Proceedings* and a journal. This activity generated a great many similar studies of the paranormal by some of the brightest minds of the day. The result was the publication of a mass of material on psychical research that continued well into the twentieth century.

Conclusion

The three streams flowing from the discovery of animal magnetism often merged. Writers in the psychological stream such as William James and Charles Richet sometimes dealt with issues of psychical research. On the other hand, the literature of psychical research was rich in psychological writings of real significance. An example of this is F.W.H. Myers's (1843–1901) classical work, *Human Personality and Its Survival of Bodily Death* (1903), which is generally considered to be a significant contribution to the investigation of the subconscious.

The crossover among the three streams is also illustrated by the fact that both those interested in the psychological stream and those drawn to psychical research often investigated the healing and medical aspect of animal magnetism. Among the former, for example, were Hippolyte Bernheim, Jean-Martin Charcot, and Pierre Janet, who explored the relationship between the healing effects of animal magnetism and psychological factors, such as suggestion; among the latter were the members of the Society for Psychical Research whose studies focused on the nature of magnetic healing.

The histories of animal magnetism, hypnotism, and psychical research are inextricably intertwined. As will be evident from the annotated entries in the bibliography, the literature of any one of these areas cannot but include the literature of the other two.

Format of the Bibliography

This bibliography is intended to include the literature of animal magnetism and those streams of thought that can be identified as flowing directly from it. Conceived in the broadest possible terms, that literature is vast, far exceeding the scope of this work.

This bibliography has been consciously circumscribed for two reasons. The first is the desire to include only those works that are immediately connected with animal magnetism and the themes that arise directly from it. The second is the intention to produce a work focused on the most significant writings within the historical tradition arising from animal magnetism.

The effect of including only those works that are closely connected with the themes arising from animal magnetism is most easily illustrated by describing the categories of literature that have *not* been included. Omitted are works that deal exclusively with occultism, possession, or witchcraft; theosophy, anthroposophy, Christian Science, or other spiritual philosophies; theology or religious thought; and conjuring or stage magic. While spiritualist writings have true importance for the history of animal magnetism and its offshoots, only those works have been included here that depict its development from mesmeric influences or that play a significant role in the rise of psychical research. This means that the bibliography does *not* include stories about clairvoyants, seers, or prophets; books relating communications from spirits; spiritualistic speculations about the afterlife or related matters; and collections of ghost-lore.

The second reason for circumscribing the bibliography, the wish to focus on the most significant writings arising within the animal magnetic tradition, necessitated the omission of works which, while legitimately part of the history of animal magnetism and its offshoots, are relatively minor. While it is difficult to define the criteria by which such a selection was made, a few words can be said about the thinking that was involved. Rating low on the list of works to be included were writings that simply summarize the work of others, collections of cases that have been dealt with in previous works, and popularizations intended to simplify and condense more serious treatises. In the field of hypnotism, this tended to exclude books on stage hypnotism, handbooks of hypnotic practice, pamphlets meant for home study of hypnotism, treatises on personal magnetism, and writings on auto-hypnosis and auto-suggestion. There are, of course, some works in these categories that were influential or otherwise significant, and they have been incorporated as appropriate into the bibliography.

Time Frame

The bibliography begins with the year 1766, the date of publication of Mesmer's medical thesis *Dissertatio physico-medica de planetarum influxu,* which contains the first seeds of animal magnetism. The bibliography ends with the year 1925. This date was chosen for a number of reasons. First, by 1925 publication of works on animal magnetism had almost completely ceased. Second, shortly after 1925 the study of hypnotism entered a new phase, in which researchers such as Clark Hull, M. M. White, and others adopted innovative laboratory and statistical methods to explore its nature and effects. Third, by 1925 most of the classical works of psychical research had been published, and psychical research too was about to enter a new stage of development. Like that of hypnotism, this new stage, initiated by the work of Joseph Banks Rhine at Duke University in the late 1920s, involved the introduction of laboratory and statistical methods into the study of paranormal phenomena. Indeed, what had been "psychical research" became popularly referred to as "parapsychology."

Annotations

Annotations are intended to provide information about the content of the work and thereby indicate its place in the history of the field. The length of the annotation depends to some extent upon the significance of the entry in that history. In some cases, due to inaccessibility of a given work, annotations were compiled without benefit of direct inspection.

Annotations are given for approximately one-third of the entries. The intention is to provide annotations for the most important items and a sufficient variety of less significant works to convey to the reader a sense of the evolution of the literature.

Form of the Entries

Entries are listed by year, from 1766 to 1925. Within each year, works are listed alphabetically. Each entry contains full bibliographic information, including author, title and publication data. Many entries contain annotations.

FORMAT OF THE BIBLIOGRAPHY

Publication information is given in English: cities are cited in their commonly accepted English form (e.g., "Munich" rather than "München," and "Rome" rather than "Roma"); multiple publishers are joined by English conjunctives (e.g., "Bailliére and Dentu" rather than "Bailliére et Dentu"); and "The Author" is used instead of "chez l'Auteur."

Included with every item is the designation [H] or [P], and in some instances [H & P]. These initials stand for Hypnotism, Psychical Research, and both.

Every effort has been made to provide information about the first edition of each book. In the few cases in which that could not be obtained, information about a later edition is given. For books in languages other than English, English translations known to exist are listed. It is intended ordinarily to give the earliest English translation.

Finally, undated works have been assigned the most accurate date that can be ascertained. In some cases, because of inaccessibility of the works, entries lack information about publisher or pages.

Acknowledgments

I would like to acknowledge a number of people who helped make books available for my research. Particular thanks go to Bill Williams, O. C., a man with an abiding interest in the working principles of the mind, who opened to me his fine private library of works on psychical research. I also would like to express my appreciation for the assistance given me by Jane Lynch, Senior Interlibrary Loan Technician for the Robarts Library and her staff at the University of Toronto.

I am deeply grateful for the hospitality, advice, and assistance given me in Freiburg by Eberhard Bauer of the Psychologisches Institut at the University of Freiburg, who is editor of the *Zeitschrift für Parapsychologie und Grenzgebiete der Psychologie.* In addition, I am particularly indebted to both Eberhard Bauer and Professor Hans Bender for making available to me the rare works contained in the Fanny Moser Library in Freiburg. My thanks also to Professor Heinz Schott at the Institut für Geschichte der Medizin at the University of Freibrug.

I would also like to acknowledge with thanks a timely grant for this project provided by Therafields Foundation of Toronto.

Dr. Joel Whitton's assistance in my work has been most important, as has been the information and inspiration given me by my colleague, John Gach. Valuable research for this project was carried out by Matthew O'Sullivan, along with Tom Snyders, Erin Clark, and Laurel Paluck.

My special thanks to my wife Josephine for her research work, editing, and overall support for this project.

Finally, I would particularly like to express my appreciation to Professor Rob Wozniak, editor of this bibliographic series, for his suggestion that I undertake this work, for his encouragement while I was working on the project, and for his invaluable input into each step of the process.

Glossary of Terms

amnesia: Absence of memory; sometimes experienced after coming out of a state of trance.

animal magnetism: Healing system devised by Franz Anton Mesmer; it posited the existence of a universal magnetic fluid that is central in the restoration and maintenance of health. Its use sometimes produced a trance state in the patient; this aspect eventually became known as hypnotism.

artificial somnambulism: A trance state brought about through the application of animal magnetic or hypnotic techniques.

automatism: An action produced by an individual without conscious knowledge.

baquet: The magnetic baquet was an invention of Franz Anton Mesmer designed to store and distribute animal magnetic fluid. It was a wooden tub partially filled with bottles of magnetized water seated on powdered glass and iron filings. The tub had a wooden cover with iron rods extending upward through the cover and then bent at right angles to be accessible to those using the device.

braidism: Another term for hypnotism (q.v.).

clairvoyance: The ability to be aware of objects, people, or events through means other than the five senses.

conjuring: The use of trick and illusion to produce striking effects.

control: In the context of spiritualism, a spirit who possesses the body of a medium and takes charge of a séance.

dissociation: The separation of any group of mental processes from the rest of the psyche.

divided consciousness: A way of referring to two distinct consciousnesses: the waking consciousness and the consciousness operative in artificial somnambulism.

double consciousness: Another term for divided consciousness (q.v.).

double memory: The state of having two apparently distinct memory chains: that of the waking state and that of the somnambulistic state.

doubling of the personality: The production of an apparent second personality or second self present in the subconscious and in some cases operative in the world.

dual personality: A disorder which involves the functioning of two distinct personalities in the life of one individual.

ectoplasm: A substance produced by mediums which is the basis for materializations (q.v.).

electro-biology: A doctrine originating in the United States that holds that the will of one individual can modify the physical or mental state of another; it is an alternate explanation for animal magnetism.

fascination: A hypnotic technique that involves engaging the eyes of the subject in an intense way; it produces a strong impulse to imitate the hypnotist.

higher phenomena of mesmerism or somnambulism: Unusual phenomena produced by some individuals in the trance state, including: physical rapport in which the subject experiences the sensations of the mesmerizer; mental rapport with the ability to read the mesmerizer's thoughts; clairvoyance or awareness of things at a distance in space or time; and ecstasy or an elevated state of consciousness in which the subject has an awareness of spiritual things.

hypnotism: The term coined by James Braid (1795–1860) to replace "animal magnetism." Its complete form is "neuro-hypnotism" and means "nervous sleep."

hysteria: An emotional disturbance that manifests in a variety of physical symptoms, such as blindness, anesthesia, or paralysis. These symptoms are produced by subconscious functions that are dissociated from normal awareness.

lower phenomena of mesmerism or somnambulism: Less extraordinary phenomena produced by some individuals in the trance state, including: a sleep-waking kind of consciousness, divided consciousness (q.v.), loss of sense of identity, suggestibility, heightened memory, deadening of the senses and insensibility to pain, and rapport or a special connection with the mesmerizer.

lucid somnambulism: The state of somnambulism (q.v.) accompanied by clairvoyance (q.v.).

magic: In the context of this bibliography, this term is used to refer to the occult traditions of the western world.

magnetic crisis: A critical point reached when someone is treated by animal magnetism; it may involve anything from convulsions to sleep.

magnetic fluid: A universal, infinitely fine substance that pervades the universe and is characterized by an ebb and flow; it was believed to have certain properties usually associated with magnets, such as attraction, repulsion, and polarity.

magnetic medicine: A medical tradition beginning with Paracelsus that was based upon the notions of sympathy, antipathy, and a universal magnetism.

magnetic passes: Repeated regular movements of the hands (usually in a downward direction) made by a practitioner of animal magnetism to cure an illness; these movements were usually made at a slight distance from the body of the sick person.

magnetic sleep: Another term for artificial somnambulism (q.v.).

magnetic somnambulism: Another term for artificial somnambulism (q.v.).

magnetism: In the context of this bibliography, this term is usually used as the equivalent of animal magnetism (q.v.).

magnetization: Applying animal magnetism to an individual, usually using magnetic passes.

magnetization at a distance: Applying animal magnetism to an individual who is not in the presence of the magnetizer.

magnetizer: A person who applies animal magnetism.

materialization: The mysterious appearance of temporary forms composed of ectoplasm (q.v.) that possess human physical characteristics (e.g., hands, faces, or full figures).

medical clairvoyance: Clairvoyant diagnosis of disease, sometimes accompanied by prescription for treatment.

medium: A person who serves as a link between this world and the spirit world, or in a more general sense, one in whose presence paranormal phenomena can be observed.

mental healing: Healing illnesses through the use of the mind and will.

mental or psychic phenomena of spiritualism: Extraordinary phenomena of a non-physical kind associated with spiritualism, including: clairvoyance, telepathy, precognition, retrocognition, and astral travel.

mesmerism: Used as the equivalent of animal magnetism (q.v.).

mesmerizer: Another term for magnetizer (q.v.).

metalotherapy: A technique of treating disease by the direct application of various metals and compounds.

multiple personality: A disorder which involves the functioning of two or more distinct personalities in the daily life of an individual.

paranormal phenomena: Those phenomena that transcend the limits of what is usually considered to be physically possible. Equivalent of "supernormal phenomena."

perkinism: A healing technique employing metallic tractors invented by Elisha Perkins (1741–1799) in the late eighteenth century United States; it has certain aspects in common with animal magnetism.

phenomena of mediumship: Those paranormal occurrences that happen in connection with spiritualistic séances; they include mental phenomena such as clairvoyance and precognition, and physical phenomena such as the movement of objects without the use of physical force and materializations of human form.

phenomena of spiritualism: See phenomena of mediumship.

phrenology: An approach developed by Franz Joseph Gall (1758–1828) at the end of the eighteenth century that claimed that character and personality could be analyzed by examining the shape and size of various parts of the skull; it was an attempt to relate personality traits to brain development.

physical phenomena of spiritualism: Extraordinary phenomena of a physical kind associated with the practices of spiritualism; they include the movement of objects without the application of physical force, materializations of the human form, the production of sounds without any apparent physical cause, and the manifestation of lights for which there seems to be no normal explanation.

planchette: A heart-shaped piece of wood mounted on casters with a pencil pointed downwards, designed to use to communicate with spirits; the hands of the operator were placed on top of the instrument and it wrote on paper.

precognition: Paranormal knowledge of future events.

preno-magnetism: A technique that combined animal magnetism and phrenology.

psychic: *Adjective:* endowed with extraordinary mental powers, such as clairvoyance or precognition. *Noun:* a person reputed to possess psychic abilities.

psychical research: The scientific investigation of supernormal phenomena (q.v.); later called parapsychology.

psychometry: The paranormal ability to sense the history of an object by touching or holding that object.

retrocognition: Paranormal knowledge of past events.

scrying: Divination carried out by gazing at crystalline or shiny objects.

somnambulism: A state of consciousness which has characteristics of both sleep and waking; as a spontaneous phenomenon it is called sleepwalking or sleeptalking; induced deliberately through animal magnetism or hypnotism, it is called artificial somnambulism.

spiritism: The belief that human beings survive death and may communicate with the living.

spiritualism: A modern religio-philosophical movement that began in the United States in 1848 and embodies the beliefs of spiritism (q.v.).

subconscious: A part of the human psyche normally outside conscious awareness, which is the arena of mental and emotional activity that may affect a person's thoughts and behavior; the term was coined by Pierre Janet (1859–1947), who was one of the most important investigators of subconscious phenomena.

subliminal consciousness: A term coined by F.W.H. Myers (1843–1901) to designate the realm of human activity that is "below the threshhold" *(limen)* of awareness. It is the source of instinctual impulses and subconscious complexes, and the arena of human paranormal faculties. The subliminal consciousness is the counterpart of supraliminal ("above the threshhold") consciousness, the ordinary self of daily life.

supernormal phenomena: Unusual phenomena for which there seems to be no explanation by the known laws of science; they include such things as telepathy, clairvoyance, apparitions, telekinesis, and materializations.

table tapping: A phenomenon in which a number of individuals are seated around a table, usually with hands joined, and a rising and falling of one side of the table with a tapping of one of its legs on the floor occurs; often the taps spell out messages by alphabetical code.

table tipping: The same as "table tapping" (q.v.).

table turning: A phenomenon in which a number of individuals are seated or stand around a table, usually with hands joined, eventually bringing about the rotation of the table; often practiced in connection with table tapping (q.v.).

talking tables: See table tapping.

telekinesis: The movement of objects apparently without the application of physical force, considered to be accomplished by the power of the mind or a "psychic force."

telepathy: The communication of information from one mind to another apparently without using the recognized channels of sense; also called thought transference.

unconscious: That part of an individual's mind that produces actions or mental processes without that individual's conscious participation.

References

Amadou, Robert (ed.). *Le magnétisme animal.* Paris: Payot, 1971.

Artelt, Walter. *Der Mesmerismus in Berlin.* Mayence: Akademie der Wissenschaft und der Literatur, 1966.

Barrucand, Dominique. *Histoire de l'hypnose en France.* Paris: Presses universitaires de France, 1967.

Benz, Ernst. *Franz Anton Mesmer (1734–1815) und seine Ausstrahlung in Europa und Amerika.* Munich: Wilhelm Fink, 1976.

Benz, Ernst. *Franz Anton Mesmer und die philosophischen Grundlagen des "animalischen Magnetismus."* Mainz: Akademie der Wissenschaften und der Literatur, 1977.

Bibliotheca Esoterica. Cataloque annoté et illustré de 6707 ouvrages anciens et modernes qui traitent des sciences occultes. . . . Brueil-en Vexin: Yvelines, 1975. [Originally published ca. 1912].

Blake, John B. (ed.). *A Short title Catalogue of Eighteenth Century Printed Books in the National Library of Medicine.* Bethesda, Maryland: National Institute of Health, 1979.

Bloch, George (ed.). *Mesmerism: A Translation of the Original Scientific and Medical Writings of F. A. Mesmer.* Los Altos, California: William Kaufmann, 1980.

Bousfield, Wendy (ed.). *Catalog of the Maurice M. and Jean H. Tinterow Collection of Works on Mesmerism, Animal Magnetism, and Hypnotism.* Wichita, Kansas: Wichita State University, 1983.

Bramwell, J. Milne. *Hypnotism: Its History, Practice and Theory.* New York: Julian Press, 1956. (The first edition was published in 1903.)

Brown, Slater. *The Heyday of Spiritualism.* New York: Pocket Books, 1972.

Bunn, Walter von. "Die Anfange der hypnotischen Anasthesie." *Deutsche medizinische Wochenschrift* 79(1954): 336–340.

Buranelli, Vincent. *The Wizard from Vienna.* New York: Coward, McCann & Geoghegan, 1975.

Caillet, Albert Louis. *Manuel bibliographique des sciences psychiques ou occultes*. 3 vols. Paris: Lucien Borbon, 1912.

Carlson, Eric T. "Charles Poyen Brings Mesmerism to America." *Journal of the History of Medicine and Allied Sciences* 15(1960):121–132.

Carlson, Eric and Simpson, Meribeth. "Perkinism Vs. Mesmerism." *Journal of the History of the Behavioral Sciences* 6(1970):16–24.

Catalogue of the Library of the Society for Psychical Research. Boston: G. K. Hall, 1976.

Cerullo, John J. *The Secularization of the Soul*. Philadelphia: Institute for the Study of Human Issues, 1982.

Chertok, Leon. *Le non-savoir des psy. L'hypnose entre la psychanalyse et la biologie*. Paris: Payot, 1979.

Chertok, Leon and De Saussure, Raymond. *The Therapeutic Revolution from Mesmer to Freud*. New York: Brunner/Mazel, 1979.

Crabtree, Adam. "Mesmerism, Divided Consciousness and Multiple Personality." In *Franz Anton Mesmer und die Geschichte des Mesmerismus*, edited by Heinz Schott. Stuttgart: Franz Steiner, 1985.

Crabtree, Adam. "Explanations of Dissociation in the First Half of the Twentieth Century." In *Split Minds and Split Brains*, edited by Jacques Quen. New York: New York University Press, 1986.

Darnton, Robert. *Mesmerism and the End of the Enlightenment in France*. Cambridge, Massachusetts: Harvard University Press, 1968.

Dessoir, Max. *Bibliographie des modernen Hypnotismus*. Berlin: Carl Duncker, 1888.

Dessoir Max. *Erster Nachtrag zur Bibliographie des modernen Hypnotismus*. Berlin: Carl Duncker, 1890.

Dessoir, Max (ed.). *Der Okkultismus in Urkunden*. 2 vols. Berlin: Ullstein, 1925.

Dingwall, Eric J. *Abnormal Hypnotic Phenomena: A Survey of Nineteenth-Century Cases*. 4 Vols. New York: Barnes & Noble, 1967–1968.

Dureau, Alexis. *Histoire de la médecine et des sciences occultes. Notes bibliographiques pour servir à l'histoire du magnétisme animal. Analyse de tous les livres, brochures, articles de journaux publiés sur le magnétisme animal, en France et à l'étranger, à partir de 1766 jusqu'au 31 décembre 1868*. Paris: The Author and Joubert, 1869.

Edmonston, William E. *The Induction of Hypnosis*. New York: John Wiley & Sons, 1986.

Ellenberger, Henri. *The Discovery of the Unconscious: The History and Evolution of Dynamic Psychiatry*. New York: Basic Books, 1970.

Figuier, Louis. *Histoire du merveilleux dans les temps modernes. Vol. 4: Le magnétisme animal*. 2 ed. Paris: L. Hachette, 1860.

Frankau, Gilbert. *Mesmerism by Doctor Mesmer*. London: MacDonald, 1948.

Fuller, Robert C. *Mesmerism and the American Cure of Souls*. Philadelphia: University of Pennsylvania Press, 1982.

Gallini, Clara. *La sonnambula meravigliosa. Magnetismo e ipnotismo nell 'Ottocento italiano*. Milan: Giangiacomo Feltrinelli, 1983.

Gartrell, Ellen G. *Electricity, Magnetism, and Animal Magnetism. A Checklist of*

Printed Sources: 1600–1850. Wilmington, Delaware: Scholarly Resources Inc., 1975.

Gauld, Alan. *The Founders of Psychical Research.* New York: Schocken, 1968.

Goldsmith, Margaret. *Franz Anton Mesmer. The History of an Idea.* London: Arthur Barker, 1934.

Grattan-Guinness, Ivor. *Psychical Research. A Guide to Its History, Principles and Practices.* Wellingborough, Northamptonshire: Aquarian Press, 1982.

Haynes, Renée. *The Society for Psychical Research: 1882–1982.* London: Macdonald, 1982.

Ince, R. B. *Franz Anton Mesmer. His Life and Teaching.* London: William Rider, 1920.

Inglis, Brian. *Natural and Supernatural. A History of the Paranormal from Earliest Times to 1914.* London: Hodder and Stoughton, 1977.

Inglis, Brian. *Science and Parascience. A History of the Paranormal, 1914–1939.* London: Hodder and Stoughton, 1984.

Jensen, Ann and Watkins, Mary Lou. *Franz Anton Mesmer: Physician Extraordinaire.* New York: Helix Press, 1967.

Jervey, Edward. "La Roy Sunderland: 'Prince of the Sons of Mesmer'." *Journal of Popular Culture* 9(1976): 1010–1026.

Kaplan, Fred. " 'The Mesmeric Mania': The Early Victorians and Animal Magnetism." *Journal of the History of Ideas* 35(1974): 691–702.

Kerner, Justinus. *Franz Anton Mesmer aus Schwaben; Entdecker des thierischen Magnetismus. Erinnerungen an denselben, nebst Nachrichten von den letzten Jahren seines Lebens zu Meersburg am Bodensee.* Frankfurt: Literarische Anstalt, 1856.

Kiesewetter, Carl. *Geschichte des neueren Occultismus. Geheimwissenschaftliche Systeme von Agrippa von Nettesheym bis zu Carl du Prel.* Leipzig: Wilhelm Friedrich, (1891).

Kiesewetter, Carl. *Franz Anton Mesmer's Leben und Lehre. Nebst einer Vorgeschichte des Mesmerismus, Hypnotismus und Somnambulismus.* Leipzig: Max Spohr, 1893.

Leibrand, Werner. *Romantische Medizin.* Hamburg and Leipzig: H. Goverts Verlag, 1937.

Leibrand, Werner. *Die spekulative Medizin der Romantik.* Hamburg: Claassen, 1956.

Ludwig, August Friedr. *Geschichte der okkultistischen (metapsychichen) Forschung von der Antike bis zur Gegenwart. I Teil: Von der Antike bis zur Mitte des 19. Jahrhunderts.* Pfullingen: Johannes Baum, 1922.

McGuire, Gregory R. *La Marginalisation de la parapsychologie: étude historique de l'orthodoxie et du contrôle dans une communauté scientifique.* (Privately published paper), 1982.

McGuire, Gregory R. *Presentism and the Role of Parapsychology in the History of Psychology.* Paper presented at the 15th annual meeting of CHEIRON: The International Society for the History of Behavioral and Social Sciences held at Gledon College, York University, Toronto (June 15–18, 1983).

McGuire, Gregory R. *The Collective Subconscious: Psychical Research in French*

Psychology (1880–1920). Paper presented at a symposium entitled: Controversies in Psychology During France's Belle Epoque, conducted at the 92nd Annual Meeting of the American Psychological Association, Toronto. (August 25–28, 1984).

Mialle, Simon. *Exposé par ordre alphabetique des cures opérées en France par le magnétisme animal, depuis Mesmer jusqu'a nos jours (1774–1826).* 2 vols. Paris: J. G. Dentu, 1826.

Milt, Bernhard. *Franz Anton Mesmer und seine Beziehungen zur Schweiz. Magie und Heilkunde zu Lavaters Zeit.* Zurich: Leemann, 1953.

Moore, R. Laurence. *In Search of White Crows. Spiritualism, Parapsychology, and American Culture.* New York: Oxford University Press, 1977.

Moser, Fanny. *Der Okkultismus, Taüschungen und Tatsachen.* 2 vols. Zurich: Orell Fussli, 1935.

Mottelay, Paul Fleury (ed.). *Bibliographical History of Electricity & Magnetism Chronologically Arranged.* London: Charles Griffin, 1922.

Myers, Frederick W. H. *Human Personality and Its Survival of Bodily Death.* 2 vols. London: Longmans, Green, and Co. 1903.

National Laboratory of Psychical Research. *Proceedings.* 1 Part 2, 1929: *Short-title Catalogue of Works on Psychical Research, Spiritualism, Magic, Psychology, Legerdemain and Other Methods of Deception, Charlatanism, Witchcraft, and Technical Works for the Scientific Investigation of Alleged Abnormal Phenomena from circa 1450 A.D. to 1929 A.D.*

Palfreman, Jon. "Mesmerism and the English Medical Profession: A Study of Conflict." *Ethics in Science and Medicine* 4(1977): 51–66.

Parssinen, Terry. "Professional Deviants and the History of Medicine: Medical Mesmerists in Victorian Britain." *Sociological Review Monographs,* No. 27, 1979.

Pattie, Frank. "Mesmer's Medical Dissertation and Its Debt to Mead's *De imperio solis ac luna.*" *Journal of the History of Medicine and Allied Sciences* 11(1956): 275–287.

Pleasants, Helene. *Biographical Dictionary of Parapsychology, with Directory and Glossary.* New York: Helix Press, 1964.

Podmore, Frank. *Modern Spiritualism, a History and a Criticism.* 2 vols. London: Methuen, 1902.

Podmore, Frank. *Mesmerism and Christian Science: A Short History of Mental Healing.* London: Methuen, 1909.

Podmore, Frank. *The Newer Spiritualism.* London: Fisher Unwin, 1910.

Quen, Jacques. "Case Studies in Nineteenth Century Scientific Rejection: Mesmerism, Perkinism, and Acupuncture." *Journal of the History of the Behavioral Sciences* 11(1975): 149–156.

Quen, Jacques. "Mesmerism, Medicine, and Professional Prejudice." *New York State Journal of Medicine* 76(1976): 2218–2222.

Rausky, Franklin. *Mesmer ou la révolution thérapeutique.* Paris; Payot, 1977.

Rosen George. "Mesmerism and Surgery: A Strange Chapter in the History of Anesthesia." *Journal of the History of Medicine* 1(1946): 527–550.

Schneider, Emil. *Der animale Magnetismus. Seine Geschichte und seine Beziehungen zur Heilkunst.* Zurich: Konrad Lampert, 1950.

Schott, Heinz. "Die Mitteilung des Lebensfeuers. Zum therapeutischen Konzept von Franz Anton Mesmer (1734–1815)." *Medizin-Historisches Journal* 17(1982): 195–214.

Schott, Heinz (ed.). *Franz Anton Mesmer und die Geschichte des Mesmerismus.* Stuttgart: Franz Steiner, 1985.

Schroeder, H. R. Paul. *Geschichte des Lebensmagnetismus und des Hypnotismus. Vom Uranfang bis auf den heutigen Tag.* Leipzig: Arwed Strauch, 1899,

Shepard, Leslie A. (ed.). *Encyclopedia of Occultism & Parapsychology.* 2 vols. New York: Avon, 1978.

The Society for Psychical Research. Proceedings. 1927–1934: Vols. 37, 38, 39, 40, and 42.

Tatar, Maria M. *Spellbound: Studies on Mesmerism and Literature.* Princeton, New Jersey: Princeton University Press, 1978.

Tinterow, Maurice. *Foundations of Hypnosis From Mesmer to Freud.* Springfield, Illinois: Charles C. Thomas, 1970.

Tischner, Rudolf. *Geschichte der okkultistischen (metapsychischen) Forschung von der Antike bis zur Gegenwart. II. Teil: Von der Mitte des 19. Jahrhunderts bis zur Gegenwart.* Pfullingen: Johannes Baum, 1924.

Tischner, Rudolf. *Franz Anton Mesmer. Leben, Werk und Wirkungen.* Munich: Münchner Drucke, 1928.

Tischner, Rudolf and Bittel, Karl. *Mesmer und sein Problem: Magnetismus—Suggestion—Hypnose.* Stuttgart: Hippokrates-Verlag Marquardt & Cie., 1941.

Usteri, Paul. *Specimen bibliothecae criticae magnetismi sic dicti animalis.* Gottingen: Joannes Christ. Dieterich, 1788.

Vinchon, Jean. *Mesmer et son secret.* Paris: A. Legrand, (1936).

Walmsley, D. M. *Anton Mesmer.* London: Robert Hale, 1967.

Wyckoff, James. *Franz Anton Mesmer. Between God and Devil.* Englewood Cliffs, New Jersey: Prentice-Hall, 1975.

Wydenbruck, Nora. *Doctor Mesmer: An Historical Study.* London: John Westhouse, 1947.

Wygrant, Larry J. (ed.). *The Truman G. Blocker, Jr. History of Medicine Collections: Books and Manuscripts.* Galveston: University of Texas Medical Branch, 1986.

Early Works: Pre-1800

1766

1. ***Mesmer, Franz Anton.***
Dissertatio physico-medica de planetarum influxu. Vienna: Ghelen, 1766, 48 pp. ENGLISH: "Physical-Medical Treatise on the Influence of the Planets," in *Mesmerism.* Translated and edited by George Bloch. Los Altos, California: William Kaufmann, 1980.

The first published writing of Franz Anton Mesmer is a dissertation presented to the University of Vienna medical school for the degree of doctor of medicine. While the title page of the dissertation carries the initials of a doctorate in liberal arts and philosophy after Mesmer's name, there is a serious doubt that this degree was ever conferred. Although there is no record of what Mesmer studied in the years from 1755 to 1759, the rest of his education is known, and there is little reason to think he attained a degree during that hiatus. Mesmer came to Vienna in 1759 to study at the university. After one year in law, he began a six-year program in the medical school, finishing with this dissertation. At the very beginning of the thesis Mesmer states that he is attempting to continue the work of Richard Mead (1673–1754) who wrote about the influence of the stars on men. Mesmer emphasizes that he is not talking about an astrological understanding of that influence, but a purely physical, scientific one. After a general discussion of the laws of planetary motion, centrifugal force, and gravitation, he writes of his notion that there must be tides in the atmosphere just as there are in the ocean. Frank Pattie of the University of Kentucky, in his study of the influence of Mead on Mesmer, points out that this idea is central in Mead's *De imperio solis ac lunae* (1704), and that Mesmer sometimes reproduces

Mead's own words on the matter, only slightly altered, without giving him credit. Mesmer then presents his own original ideas. He says that just as there are tides in the sea and the atmosphere, so also there are tides in the human body. There is, he asserts, a universal gravitation by which our bodies are affected. Through this influence emanating from the stars, our bodies are caused to resonate in a harmonious fashion. This fact, says Mesmer, must be taken seriously by medical practitioners, for if human bodies are violently shaken by the action of celestial bodies, then understanding the nature of that influence is of utmost importance. This generalized influence is labeled by Mesmer "animal gravitation." Some years later, this concept will reappear, somewhat modified by his experience with magnets, as "animal magnetism." [H]

1775

2. *Hell, Maximillian.*

Unpartheyischer Bericht der allhier gemachten Entdeckungen der sonderbaren Würkung der kunstlichen Stahlmagneten in verschiedenen Nervenkrankheiten. Vienna: n.p., 1775.

[H]

3. *Mesmer, Franz Anton.*

Gedruckte Antwort des Herrn Dr Mesmer vom 19. Januar 1775. N.p., n.p., 1775.

Mesmer's third published writing. Apparently the treatise first appeared in a Viennese periodical and then as a separate pamphlet. Now it can only be found in the *Sammlung der neuesten gedruckten und geschriebenen Nachrichten* (*see* entry number 9). It is a response to an article of Maximillian Hell, an expert on the construction of magnets, who wrote an article critical of Mesmer's *Schreiben über die Magnetcur* (entry number 5). Hell claimed that the cures ascribed by Mesmer to animal magnetism were really due to the action of magnets. In his response, Mesmer not only reasserts that animal magnetism was the cause, he also denies that magnets are of any use in treating illness. [H]

4. *Mesmer, Franz Anton.*

Herrn Dr Mesmers Schreiben an die Frankfurter vom 10. Mai 1775. N.p., n.p., 1775.

A letter on magnetism addressed to the inhabitants of Frankfort. It can be found today only in the *Sammlung der neuesten gedruckten und geschriebenen Nachrichten* (*see* entry number 9). [H]

5. *Mesmer, Franz Anton.*
Schreiben über die Magnetkur von Herrn. A. Mesmer, Doktor der Arzneygelahrtheit, an einen auswartigen Arzt. (Vienna): Joseph Kurzbock, 1775, (1) + 14 pp. ENGLISH: "Letter from M. Mesmer, Doctor of Medicine at Vienna to A. M. Unzer, Doctor of Medicine, on the Medicinal Usage of the Magnet," in *Mesmerism.* Translated and edited by George Bloch. Los Altos, California: William Kaufmann, 1980.

Written on January 5, 1775, to Doctor Johann Christoph Unzer of Altona, this treatise appeared as a pamphlet and was then immediately published in the *Neuer gelehrter Mercurius* (edited by Unzer). In this attempt to explain his theory of cure through magnetism, Mesmer first uses the term "animal magnetism" in print. This term he now equates with the "animal gravitation" of his *Dissertation* (*see* entry number 1). He distinguishes animal magnetism from mineral magnetism, but at the same time shows the analogy between the two. Mesmer points out that the animal magnetic fluid penetrates everything and can be stored up and concentrated, like "electric fluid." Like mineral magnetism, animal magnetism can operate at a distance. An edition of this work published in 1776 has an important addition: a section titled *Anhang von einigen Briefen und Nachrichten.* This appendix contains excerpts from letters describing cures performed by Mesmer and ascribed to animal magnetism. [H]

6. *Unzer, Johann Christoph.*
Beschreibung eines mit dem kunstlichen Magneten angestellten medicinischen Versuchs. Hamburg: Herold, 1775, 144 pp.

Having read of Mesmer's work, Unzer, a physician and editor of the periodical *Neuer gelehrter Mercurius*, experimented with magnets in the treatment of his own patients. The results were good, and he wrote this favorable opinion of the medical use of magnets. [H]

1776

7. *Klinkosch, Joseph Thaddaus.*
Schreiben den Thier. Magnetismus u. die sich selbst wieder ersetzende Kraft Betreffend. Prague: n.p., 1776.

Klinkosch included the alleged cures through exorcism performed by Gassner and those carried out by Mesmer using animal magnetism in the same category. He considered them to be false and delusory and argued that if any such cures occurred, it must be through electricity rather than magnetism. [H]

1778

8. *Mesmer, Franz Anton.*
Discours sur le magnétisme. N.p.: n.p., (1778?).

A small treatise found today only in *L'antimagnétisme* by Paulet (*see* entry number 93). There are questions about both authorship and date. Amadou believes it a genuine piece by Mesmer, while Frank Pattie has his doubts. There is no date given by Paulet who took his version from the *Recueil des effets salutaires de l'aimant* (*see* entry number 22). Amadou places the treatise somewhere between Mesmer's controversy with Hell and the publication of his *Mémoire* (*see* entry number 10). In the treatise, Mesmer describes how he first became aware of a "magnetic quality" in his own person that had effects upon the bodies of the sick, analogous to that produced by mineral magnetism. [H]

9. *Mesmer, Franz Anton.*
Sammlung der neuesten gedruckten und geschriebenen Nachrichten von Magnet-Curen, vorzüglich der Mesmerischen. Leipzig: Hilscher, 1778, 4 + 194 pp.

A collection of journal articles and polemical pamphlets written by Mesmer, Hell and others about the nature and efficacy of magnetic healing. For some of those writings this is the only remaining source. Mesmer himself acknowledged the accuracy of the reproductions in this collection. [H]

1779

10. *Mesmer, Franz Anton.*
Mémoire sur la découverte du magnétisme animal. Geneva and Paris: Didot le jeune, 1779, vi + 85 pp. ENGLISH: "Dissertation on the Discovery of Animal Magnetism," in *Mesmerism.* Translated and edited by George Bloch. Los Altos, California: William Kaufmann, 1980.

A foundational work in the history of modern psychology. In this, Mesmer's first and most influential public presentation of his theory of animal magnetism, he describes the context of his discovery of animal magnetism, depicts the first cures performed through its application, and sets forth twenty-seven propositions which delineate its nature and effects. Having experimented with iron magnets to treat illnesses, Mesmer came to the conclusion that the human body itself is a magnet and that the physician, using his own body magnetically, can produce the most effective cures. To promote his discovery, Mesmer moved in 1778 to Paris, which was the intellectual center of the Europe of his day. His Parisian clinic claimed many remarkable cures. This dramatic success and the *Mémoire* of 1779 caused his fame to spread quickly throughout France. Mesmer considered himself above all a physician and a scientist. In the foreword to the *Mémoire*, he states his basic belief that "nature affords a universal means of healing and preserving men," and throughout the text he emphasizes his conviction that he has discovered a natural principle whose laws would eventually be revealed by critical observation of the facts. According to Mesmer, there exists in nature a universal agent which, through as yet unknown laws, produces a mutual influence among the heavenly bodies, the earth, and living things. When this agent is observed operating in living organisms, it is seen to have properties of attraction similar to those of the magnet, even exhibiting polarity. Because of this similarity to mineral magnetism, Mesmer names this universal agent "animal magnetism." He claims that it acts upon the nerves of living things and that its discovery makes available a powerful means of curing illness and preserving health. Animal magnetism operates in the organism by means of an extremely fine "fluid" (the term common among scientists of the time to denote any subtle substance or influence), which Mesmer calls "magnetic fluid." Cure of disease is brought about by the direct intervention of the physician himself. He uses the magnetic power of his own body to influence the ebb and flow of magnetic fluid in that of his patient, restoring the natural balance of animal magnetic currents and thus aiding nature in the cure of disease. Mesmer states that in this way animal magnetism can cure nervous disorders directly and other disorders indirectly. Although Mesmer lived and wrote until 1815, he never significantly altered the outline of his theory as it is presented in the *Mémoire* of 1779. It contains all the basic principles which were to be applied to treatment of the sick by "magnetizers" for decades to come. [H]

1780

11. *D'Eslon, Charles.*

Discours prononcé en l'assemblée de la Faculté de Médecine de Paris le 18 septembre 1780. N.p., n.p., (1780?).

D'Eslon was physician to the Comte d'Artois, the regent of the Faculty of Medicine of Paris. Mesmer's first important associate in Paris, D'Eslon was educated by him in the theory and practice of animal magnetism. This well-constructed explanation and defence of animal magnetism was directed by D'Eslon to his medical colleagues in Paris. It resulted in their demand that he give up his involvement with the practice of animal magnetism. It can be found in Mesmer's *Précis historique* (1781), pp. 173 ff. (*see* entry number 17). [H]

12. *D'Eslon, Charles.*
Observations sur le magnétisme animal. London and Paris: Didot, 1780, (4) + 151 pp.

Because of his standing in the medical world, D'Eslon gave Mesmer credibility among the intelligentsia of Paris. This book was his major opus on animal magnetism in which he describes his first exposure to animal magnetism and how he became convinced of its efficacy. He adheres to all of Mesmer's teachings about the nature of the phenomenon, although he does not emphasize the doctrine of a magnetic fluid. D'Eslon stresses the importance of the fact that animal magnetism is effective as a treatment for illness. He knew this from his own experience, having been cured by Mesmer of a life-long ailment. Showing little concern about the niceties of theory, D'Eslon's appreciation of the practical efficacy of animal magnetism marks him as a sincere promoter of what he thought to be a great benefit to mankind. [H]

13. [*Horne, de, ———*].
Réponse d'un médecin de Paris à un médecin de province, sur le prétendu magnétisme animal de M. Mesmer. Vienna and Paris: L. A. Delalain le jeune, (1780), 16 pp.

De Horne was physician to the Comtesse d'Artois and the Duke of Orleans. In this booklet he criticizes the animal magnetic fluid of Mesmer, claiming it is simply an electro-magnetic influence. [H]

14. [*Paulet, Jean Jacques*].
Les miracles de Mesmer. N.p.: n.p. (1780), 23 pp.

Paulet, a botanist and physician who was strongly opposed to animal magnetism, was editor of the *Gazette de Santé*. This pamphlet is a reprint of two book reviews from that journal: the first, a review of D'Eslon's *Observations sur le magnétisme animal* (entry number 12); the second, a review of de Horne's *Réponses d'un médecin de province* (entry number 13). In the reviews, Paulet makes some extravagant claims against Mesmer and animal magnetism. [H]

1781

15. *Bergasse, Nicolas.*
Lettre d'un médecin de la Faculté de Paris à un médecin du College de Londres; ouvrage dans lequel on prouve contre M. Mesmer que le magnétisme animal n'existe pas. The Hague: n.p., 1781, 70 pp.

Bergasse was a lawyer, philosopher, and political theorist from Lyons. In 1781 he was successfully treated by Mesmer and became his devoted follower. In this letter he declares his belief in the efficacy of the cures performed by Mesmer and condemns the closed attitude of orthodox medicine. [H]

16. [*Fournier-Michel,* ———].

Lettre à M. Mesmer, et autre pièces concernant la maladie de mademoiselle de Berlancourt de Beauvais. Beauvais: P. Desjardins, 1781, 15 pp.

An important early testimony of a cure by Mesmer with independent witnesses. The condition and cure (paralysis of part of the body of a young woman) are described in some detail. [H]

17. *Mesmer, Franz Anton.*

Précis historique des faits relatifs au magnétisme animal jusques en Avril 1781. Tr. de l'allemand. London: n.p., 1781, (8) + 229 + (2) pp.

This collection of documents and comments was translated into French from an outline written by Mesmer in German. The original outline was later destroyed and the German version of this work published in 1783 was a translation from the French edition. The translator and editor was apparently D'Eslon. The work attempts to give a history of animal magnetism to date by reproducing and commenting on important relevant documents. The history of animal magnetism is divided into five time periods: 1) dealings with the Faculty of Medicine at Vienna, 2) dealings with the Academy of Sciences at Paris, 3) dealings with the Royal Society of Medicine at Paris, 4) various activities in the two years following, and 5) dealings with the Faculty of Medicine at Paris. Mesmer uses the documentation format to reiterate his views and emphasize his side in the various disputes in which he had been involved. [H]

18. [*Thouvenel, Pierre*].

Mémoire physique et médicinal, montrant des rapports évidens entre les phénomènes de la baguette divinatoire, du magnétisme et de l'électricité, avec des éclaircissements sur d'autres objects non moins importants, qui y sont relatifs. Paris and London: Didot, 1781, (3) + 304 pp.

A study of the use of the divining rod and its effectiveness in discovering hidden sources of water. Thouvenel recalls the tradition of magnetic medicine and the theories of earlier writers concerning a universal magnetic force which accounts for such mysterious powers. He writes about "animal electricity" and "animal magnetism" as derived from that tradition, not from the writings of Mesmer. The similarities between these ideas of Thouvenel and those of Mesmer are, however, striking. [H]

1782

19. *Bourzeis, Jacques Aimée de.*
Observation très-importante sur les effets du magnétisme animal. Paris: P. F. Gueffier, 1782, 28 pp.

A little work complaining about the treatment Mesmer gave to the author's patient. [H]

20. *D'Eslon, Charles.*
Lettre de M. d'Eslon, docteur régent de la Faculté de Paris, et médecine ordinaire de Monseigneur le comte d'Artois, à M. Philip, docteur en médecine, doyen de la Faculté. The Hague: n.p., 1782, 144 pp.

One of a number of attempts by D'Eslon to present his views on animal magnetism to the medical establishment at Paris. [H]

21. [*D'Eslon, Charles*].
*Lettre de M. le Marquis de***, à un médecine de province.* N.p.: n.p., (1782), 46 + (2) pp.

A collection of letters on animal magnetism, including a letter by Mesmer protesting D'Eslon's claim to represent the interests of animal magnetism (see *Lettre sur un fait relatif à l'histoire* . . . , entry number 23) and correspondence between Mesmer and D'Eslon on the matter. The collection is ascribed to D'Eslon. [H]

22. [*Harsu, Jacques de*].
Recueil des effets salutaires de l'aimant dans les maladies. Geneva: B. Chirol and E. Didier, 1782, 60 + 276 pp.

An extremely rare book which contains, among other things, the original of a small treatise by Mesmer entitled: *Discourse sur le magnétisme.* This treatise was later published in *L'Antimagnétisme* by Paulet (*see* entry number 94). [H]

23. [*Mesmer, Franz Anton*].
Lettre sur un fait relatif à l'histoire du magnétisme animal adressée à M. Philip, doyen de la Faculté de Médecine de Paris. London and Aix-la-Chapelle: n.p., 15 pp.

This letter was addressed by Mesmer to the dean of the Faculty of Medicine at Paris and written to protest D'Eslon's claim to be in possession of Mesmer's teaching. According to Mesmer, D'Eslon said that Mesmer would not be returning to Paris and for that reason the commission being appointed by the Faculty to investigate animal magnetism should examine his (D'Eslon's) work. Mesmer objects, saying that D'Eslon cannot possibly claim to present a complete picture of the theory and practice of animal magnetism. [H]

24. ***Puységur, Antoine Hyacinte Anne de Chastenet, comte de.***
*Lettre de M. le C** de C**P** a M. le P**E** de S***. N.p.: n.p., 1782, 59 pp.

[H]

25. ***Retz, Noel de Rochefort.***
Lettre sur le secret de M. Mesmer ou réponse d'un médecin à un autre, qui avait demandé des eclaircissements à ce sujet. Paris: Méquignon, 1782, 22 pp.

Retz, "physician ordinaire" to the King of France, rejects the theory of animal magnetism. However, he does admit that cures have been brought about through its application. His explanation is that the cures were accomplished through "imagination." [H]

26. ***Thouret, Michel Augustin and Andry, Charles Louis François.***
Observations et recherches sur l'usage de l'aimant en médecine; ou Mémoire sur le magnétisme médicinal. Paris: L'imprimerie de monsieur, 1782, 168 pp.

This work is an extract from the *Mémoires* of the Société royale de médecine for the year 1779. The authors describe the medical uses of magnets, beginning with a history of the subject. They then take up contemporary practitioners who use magnets to heal, including Franz Anton Mesmer. They concentrate on Mesmer's use of the mineral magnet, but they note his *Mémoire* of 1779 in their footnote and refer to his original "discoveries." Generally, the authors convey a positive attitude towards Mesmer's work. In 1784, however, Thouret revised his view, becoming very critical of animal magnetism in his *Recherches et doutes* (1784) (entry number 116). [H]

1783

27. ***Andry, Charles Louis François and Thouret, Michel Augustin.***
Rapport sur les aimans presentés par M. l'abbé Le Noble; lu dans la séance tenus au Louvre, le mardi premier avril 1783. Paris: P. D. Pierres, (1783), 266 pp. [H]

28. ***Bacher, Alexandre André Philippe Frédéric.***
Grande belle découverte du magnétisme animal. N.p.: n.p., 1783, 15 pp.

This pamphlet consists of a letter from Mesmer to a Doctor Philip (1782) with comments added by Bacher. [H]

29. *Court de Gébelin, Antoine.*
Lettre de l'auteur de monde primitif à messieurs ses souscripteurs sur le magnétisme animal. Paris: Valleyre l'aîné, 1783, 47 pp.

Court de Gébelin was one of the most highly respected intellectuals of his day. His monumental nine-volume *Monde primitif* (1773–1784) was a virtual encyclopedia of studies in comparative linguistics. In this letter he tells of how he was cured of a serious illness by Mesmer's animal magnetism. He passionately appeals for a serious study of that system and indicates the directions that study might take. Court de Gébelin had long sought to unearth the outlines of a primitive science present in the great cultures of the west. In animal magnetism he believed he had discovered the true basis for that science. [H]

1784

30. *Bailly, Jean Sylvain.*
Exposé des expériences qui one été faites pour l'examen du magnétisme animal. Lu à l'Académie des sciences, par M. Bailly en son nom & aux nom de Mrs. Franklin, Le Roy, de Bory, et Lavoisier, le 4 Septembre 1784. Paris: Imprimerie Royale, 1784, 15 pp.

Presented as a courtesy to the members of the Academy of Sciences which had been commissioned by the king to investigate animal magnetism. This brief report was intended to give them an account of some of the experiences which the commissioners had during their investigation. It is considerably shorter than the official report (*see* entry number 31). [H]

31. *Bailly, Jean Sylvain, ed.*
Rapport des commissaires chargés par le roi de l'examen du magnétisme animal. Paris: Imprimerie Royale, 1784, 66 pp.

In the spring of 1784 the King of France appointed this commission made up of members of the Academy of Sciences to investigate the claims of animal magnetism. He chose some of the most eminent men of science of his day. The chairman was Benjamin Franklin (1706–1790), a founding father of the United States of America, ambassador of that country to France, and a person highly knowledgeable in electricity and terrestrial magnetism. The commission's president was Antoine Laurent Lavoisier (1743–1794), a follower of Condillac and one of the most important chemists of the age. The secretary of the commission and editor of its report was the famous astronomer Jean Sylvain Bailly (1736–1793). The commission also included the director of the Academy of Sciences, Jean Baptiste Leroy (1724–1800), an investigator of electricity of some note. The fifth member of the commission was the physician de Bory, about which nothing is known today. The commission began its investigations on March 12, 1784, and published its report in August of that year. Both this commission and one made up of members of the Royal Society of Medicine, appointed by the King at the same time, investigated animal magnetism as practiced by Charles D'Eslon, a disciple of Mesmer. D'Eslon wanted this official inquiry, while Mesmer strongly opposed it. By cooperating with the commissions, D'Eslon effectively removed himself from his teacher's fold. Although D'Eslon's theory of animal magnetism, as presented to the commission, was somewhat different from that of Mesmer, the commissioners did not seem to be bothered by that fact. They contended that theory made no difference to their mandate, which was to decide about the existence and utility of animal magnetism. Their conclusion was that they found no evidence for the existence of an animal magnetic fluid. They ascribed any cures or improvement of health that might occur through the application of animal magnetism to the action of "imagination." The report was very influential and became a center of a vigorous controversy which raged for a number of years, with pamphlets and books being written for and against its conclusions. [H]

32. [*Bailly, Jean Sylvain*].

Rapport secret présenté au ministre et signé par la commission précédente. N.p.: n.p., 1784, 10 pp.

This secret report by the Franklin commission was not published at the time of the public report. It was presented privately to the King of France and appeared for the first time in print in 1800 (*see* entry number 213.). [H]

33. [*Bailly, Jean Sylvain*].

Supplément au deux rapports de MM. les commissaires de l'Académie et de la Faculté de médecine et de la Société royale de médecine. Amsterdam and Paris: Gueffier, 1784, 80 pp.

A summary of cures through animal magnetism attested to by the magnetizer that performed them or by the patient who had been cured. The cures were in many cases not described with the detail needed to judge the effectiveness of the treatment. [H]

34. [*Barbeguière, J. B.*]

La maçonnerie mesmérienne, ou leçons prononcées par Fr. Mocet, Riala, Themola, Seca et Célaphon, de l'Ordre des F. de l'harmonie, en Loge mesmérienne de Bordeaux, l'an des influences 5.784, et le premier du mesmérisme. Amsterdam: n.p., 1784, 83 + (1) pp.

The pronouncement of a group of Bordeaux freemasons concerning their views on mesmerism. They were at first orthodox members of the Bordeaux Lodge of Harmony founded to teach animal magnetism. However, in 1784 they broke away for a brief period, but were soon returned to the fold. [H]

35. [*Barré, P. Y. and Radet, J. B.*]

Les docteurs modernes, comédie-parade en un acte et en vaudeville, suivie du Banquet de santé, divertissement analogue mêlé de couplets représentée pour la première fois à Paris par les comédiens italiens ordinaires du Roy, le mardi 16 Novembre 1784. Paris: Brunet, 1784, 59 pp.

Two plays staged in Paris in 1784. *Les docteurs modernes* satirized Mesmer and D'Eslon. *Banquet de santé* continues the theme. It takes place in a healing salon equipped with a magnetic "baquet." The plays represent Mesmer and D'Eslon as charlatans who make fun of their dupes. The play occasioned an impassioned reaction from a supporter of Mesmer named Duval d'Eprémesnil (*see* entry 57). [H]

36. *Bergasse, Nicolas.*

Considérations sur le magnétisme animal, ou sur la théorie du monde et des êtres organisés, d'après les principes de M. Mesmer, par M. Bergasse avec des pensées sur le mouvement, par M. le Marquis de Chastellux, de l'Académie française. The Hague: n.p., 1784, 149 pp.

Bergasse, along with Kornmann, helped Mesmer found the Society of Harmony of Paris, the first of many which would take on members for a fee to teach them the doctrine and techniques of animal magnetism. Bergasse delivered lectures to the members of the Paris Society, and these lectures were distilled into the content of the present book. Bergasse and Mesmer had their disagreements and although in *Considérations* Bergasse vigorously defends Mesmer against all attacks, divergences of doctrine do nevertheless appear. Within a year after the publication of the book, Mesmer and Bergasse publicly ended their association. [H]

37. [*Bergasse, Nicolas*].

Dialogue entre un docteur de toutes les universités et académies du monde connu, notamment de la faculté de médecine fondée à Paris dans la rue de la Bucherie, l'an de notre salut 1472 et un homme de bon sens, ancien malade du docteur. Paris: Gastellier, 1784, 24 pp.

A satirical work written by Bergasse and published anonymously. It is aimed at those physicians who condemned animal magnetism. [H]

38. [*Bergasse, Nicolas*].

Théorie du monde et des êtres organisés suivant les principes de M. . . . Paris: n.p., 1784, 15 + 21 + 16 pp.

Produced by Bergasse, at a time when he was still a loyal and fanatical disciple of Mesmer, for the "adepts" of animal magnetism — those initiated into its secrets through membership in the lodge of Harmony. Composed in a format that reminds one of certain alchemical texts, the work presents text and symbols according to a code that only the initiated could understand. Beginning with the basic postulates of Mesmer's theory, Bergasse moves on to examine in detail the properties of matter and motion, and concludes with a section on the nature of man. [H]

39. [*Bombay*, ———].
Procédés du magnétisme animal. (Paris?): n.p., 1784, 53 pp.

The work of an experienced magnetizer who was influenced by the work of Puységur. Attributed to the physician Bombay, the book contains descriptions of the use of physical objects as repositories of magnetism which can be used to treat the ill. In this category are trees, tubs of water, and the *baquet* as devised by Mesmer. The author tells how to impart magnetic fluid to these objects. [H]

40. *Bonnefoy, Jean Baptiste.*
Analyse raisonée des rapports des commissaires chargés par le roi de l'examen du magnétisme animal. Lyon and Paris: Prault, 1784, (4) + 98 pp.

The preeminent magnetizer of Lyon at the time of the report of the Franklin Commission, the physician Bonnefoy strongly criticized the report using, among other things, arguments drawn from the electrical science of the day. [H]

41. *Bormes, baron de*, ———.
Lettres de M.l.B.d.B. à M.P.L.G.H.D.L.S., à Marseille, sur l'existence du magnétisme animal et l'agent universel de la nature dont le Dr Mesmer se sert pour opérer ses guérisons . . . avec le moyen de se bien porter sans le secours du médecin. Geneva and Paris: Couturier, 1784, 87 pp. [H]

42. [*Bouvier, Marie André Joseph*].
Lettres sur le magnétisme animal; où l'on discute l'ouvrage de M. Thouret, intitulé: Doutes et recherches . . . et le rapport des commissaires sur l'existence. . . . Brussels: n.p., 1784, 103 + (3) p.

Written by the physician Bouvier of Versailles, the work defends the reality of animal magnetism and its effectiveness as a cure. Bouvier himself used animal magnetism successfully in his medical practice. [H]

43. [*Brack*, ———].
Histoire du magnétisme en France, de son régime et de son influence, pour servir à developper l'idée qu'on doit avoir de la médecine universelle. Vienna and Paris: Royez, 1784, 32 pp.

Brack was a physician who wrote a number of pamphlets against Mesmer and animal magnetism. In this work he provides interesting information about the foundation of magnetic Societies of Harmony in Paris and the provinces. [H]

44. [*Brack*, ———].
Lettre de Figaro au Comte Almaviva sur la crise du magnétisme animal, avec des détails propres à fixer enfin l'opinion sur l'inutilité de cette découverte;

nouvelle édition précédée et suivie des réflexions qui ont rapport aux circonstances présentes, traduites de l'espangnol. (Madrid and) Paris: n.p., 1784, 45 pp.

One of a number of pamphlets written by the physician Brack against animal magnetism. The first edition of this work is not known to be extant and reference to a Spanish original seems to be a literary fiction. [H]

45. [*Brissot de Warville, Jacques Pierre*].

Un mot à l'oreille des académiciens de Paris. N.p.: n.p., 1784, 24 pp.

Brissot, a French revolutionary, social critic and theoretician, believed that animal magnetism could serve as a means of achieving social reform. In this work, among other things, he writes of how animal magnetism could make the rich more human and concerned about the poor. [H]

46. [*Cambry, Jacques*].

Traces du magnétisme. The Hague: n.p., 1784, 48 pp.

An erudite study of the "traces" of animal magnetic phenomena in antiquity. Written by an antiquarian and supporter of Mesmer. [H]

47. [*Cloquet, ———*].

Détail des cures opérées à Buzancy, près Soissons par le magnétisme animal. Soissons: n.p., 1784, 42 pp.

An important and very rare document in the history of animal magnetism that contains, among other items, a letter written by a M. Cloquet describing in detail the induction of artificial somnambulism by the Marquis de Puységur. This letter is the first mention in print of "magnetic sleep," as Puységur would call it in his *Mémoires* (see entry number 105), written a few months later. Puységur's historic discovery of what would eventually be termed "hypnotic trance" was a turning point in the history of modern psychology and psychotherapy. The treatise also contains a letter written by Puységur to Bergasse, a letter written by the Rev. Gérard, Superior General of the Order of Charity, a brief description by Puységur of 62 cures performed through animal magnetism at Buzancy, and a description of a cure performed by Mesmer on the son of the banker Kornmann. [H]

48. *Confession d'un médicin académicien et commissaire d'un rapport sur le magnétisme animal avec les remontrances et avis de son directeur.*

(Paris): n.p. (1784), 70 pp.

[H]

49. [*Dampierre, Antoine Esmonin Marquis de*].

Réflexions impartiales sur le magnétisme animal, faites après la publication du rapport des commissaires chargés par le roi de l'examen de cette décou-

verte. Geneva and Paris: Barthélemy Chirol (Geneva) and Périsse le jeune (Paris), 1784, 50 pp.

Dampierre was a theologian, magistrate, and president of the parliament of Bourgogne. A member of the mystical Lyons school of freemasons, he developed a philosophy of animal magnetism that viewed it as an aid to the healing and social evolution taking place according to hidden laws of nature. Dampierre delineates four different types of "magnetic crisis" that can be experienced by an individual suffering from illness. The first is the necessary and salutary crisis that an ill person may experience before magnetic treatment—certain gestures and actions that demonstrate nature's attempt to heal. The second type of crisis is that produced by suggestive individuals, when being magnetized, through imitation and the action of imagination—this being a useless and even harmful type of crisis. The third type is the crisis produced through fear upon seeing another person in the throes of a violent crisis—this also being a harmful type of crisis. The fourth type of crisis is that produced by the action of animal magnetism in susceptible persons who have a strong desire to remain in the state of crisis—this type being dangerous to the patient. Since none of these crises leads, with the possible exception of the first, to a fruitful conclusion, Dampierre and his colleagues at Lyons sought an alternate, positive healing crisis. Dampierre believed that the crises most often produced by animal magnetism as practiced by those who used the techniques of Mesmer were of the harmful type described. He considered these crises to be embarrassing and obscene for the patient and narcissistically flattering for the magnetizer. Dampierre's solution was the magnetic technique developed by his fellow Freemason at Lyons, the Chevalier Barberin. In contrast to that of Mesmer which relied on magnetic "passes" that involved physical contact, or were applied at the most a few inches from the body, the magnetizing of Barberin was done at a distance, sometimes a distance of miles. This approach could produce magnetic somnambulism with the attendant apparently paranormal phenomena first noted by the Marquis de Puységur (q.v.). This Lyons brand of Freemason animal magnetism was as such strongly oriented towards the occult worldview of the magical tradition of the West. [H]

50. ***Les débris du baquet ou Lettres critiques de la requête de Mesmer.***
Paris: n.p., 1784, 23 pp.

[H]

51. ***Décret de la faculté de médecine de Paris, du 24 août 1784 par lequel est adopté le Rapport des commissaires (Français et Latin).***
Paris: Marchand, 1784, 23 pp.

[H]

52. *D'Eslon, Charles.*
Observations sur les deux rapports de MM. les commissaires nommés par sa majesté pour l'examen du magnétisme animal. Philadelphia and Paris: Clousier, 1784, (1) + 31 pp.

A critique of the report of the commissioners charged with examining animal magnetism. D'Eslon's criticism is based on direct knowledge of their investigations. He condemns their prohibition of the practice of animal magnetism and says it is unenforceable. [H]

53. *D'Eslon, Charles.*
Supplément aux deux rapports de MM. les commissaires de l'Academie & de la Faculté de Médecine, & de la Société Royale de Médecine. Amsterdam: Gueffier, 1784, 80 pp.

[H]

54. *Devillers, Charles Joseph.*
Le colosse aux pieds d'Argille. (Paris): n.p., 1784, iv + 176 pp.

Devillers was a member of the very active lodge of Freemasons of Lyon. When its members took sides on the issue of animal magnetism in 1784, he supported the opposition. In this book he admits that some of the phenomena of animal magnetism are genuine, but denies that they are due to the action of Mesmer's magnetic fluid. He attempts to show that the power of imagination is sufficient explanation. Charles Joseph Devillers should not be confused with Charles de Villers (1765–1815) who wrote a novel in 1787 entitled *Le magnétiseur amoureux* (*see* entry number 175). [H]

55. *Doppet, François Amédée.*
Traité théorique et pratique du magnétisme animal. Turin: Jean Michel Briolo, 1784, 80 pp.

Doppet was a pupil of D'Eslon and practitioner of animal magnetism. He held a somewhat unusual attitude toward magnetism: he believed magnetism was an important medical tool and used it himself, but was not sure that it was all that the theory claimed to be. [H]

56. [*Dufau, Julien*].
Remarques sur la conduite du sieur Mesmer et de son commis le P. Hervier, et de ses autres adhérents; où l'on tache de venger la médecine de leurs outrages. (Paris): n.p., 1784, 30 pp.

The purpose of this pamphlet is to destroy the credibility of Hervier and his cure by Mesmer through animal magnetism. [H]

57. [*Éprémesnil, Jean Jacques Duval d'*].
Réflexions préliminaires à l'occasion de la pièce intitulée Les docteurs modernes jouée sur le Théatre Italien, le seize, Novembre 1784. (Paris): n.p., 1784, 3 pp.

In this leaflet, Éprémesnil, a vehement supporter of Mesmer at this time, expresses his indignation about the play *Les docteurs moderne* (*see* entry number 35) which satirized Mesmer, D'Eslon, and the practice of animal magnetism. Éprémesnil scattered the leaflet into the audience during a performance. [H]

58. [*Éprémesnil, Jean Jacques Duval d'*].

Suite des réflexions préliminaires à l'occasion des Docteurs modernes. (Paris): n.p., 1784, 8 pp.

A second statement on the part of Éprémesnil supporting Mesmer in face of the jibes of the play *Les docteurs modernes* (*see* entry number 35). Like the first leaflet, this one was tossed to the audience during a performance. [H]

59. *Fontette Sommery, comte de.*

Lettre à M. d'Eslon, médecin ordinaire de Monseigneur le Comte d'Artois. Glasgow and Paris: Prault, 1784, 27 pp. [H]

60. *Foughet, Par.*

Lettre d'un médecin de la faculté de Paris à M. Court de Gébelin, en réponse à celle que ce savant a adressée à ses souscripteurs et dans laquelle il fait un éloge triomphant du magnétisme animal. Bordeaux: Bergeret, 1784, 69 pp.

A letter addressed to Court de Gébelin about his supposed illness and cure through animal magnetism. The author, a physician of Paris, states that it was very likely that Court de Gébelin had not been ill at all, but if he had been ill, then it was the healing power of nature itself, not animal magnetism, that had cured him. The letter also contains a number of criticisms of the way animal magnetism was being practiced in Paris. [H]

61. *Gallert de Montjoie, Cristophe Félix Louis.*

Essai sur la découverte du magnétisme animal. N.p.: n.p., 1784, 9 pp.

[H]

62. *Gallert de Montjoie, Cristophe Félix Louis.*

Lettre sur le magnétisme animal, où l'on examine la conformité des opinions des peuples anciens & modernes, des sçavans & notament de M. Bailly avec celles de M. Mesmer: et où l'on compare ces mêmes opinions au rapport des commissaires chargés par le roi de l'examen du magnétisme animal adressé à Monsieur Bailly de l'Académie des Sciences etc. Paris and Philadelphia: Pierre J. Duplain, 1784, (4) + v–viii + 136 pp.

In the first part of this defence of Mesmer and animal magnetism, Gallert de Montjoye tries to find points of rapprochement between Mesmer and the astronomer Bailly. He then compares the ideas of Mesmer to those of Descartes and Newton, siding with Newton against Bailly in his view of matter and motion. The author explains Mesmer's view of the ebb and flow of magnetic fluid and attempts to show how it is in agreement with the best contemporary views of physics. In the second part, Gallert de Montjoie takes up the report of the Franklin commission, devoting considerable space to the issue of the place of the imagination in the action of animal magnetism. He examines the place of the will in the action of the magnetic fluid, stating that it is principally by the will that the fluid is directed and that it is involved in magnetization at a distance. [H]

63. [*Gardane, Joseph Jacques de*].

Eclaircissements sur le magnétisme actuel. London: n.p., 1784, 36 pp.

A treatise on animal magnetism by a man who experimented with electricity as a medical aid. He accepted the reality of the effects of animal magnetism, but believed that imagination was probably the principal cause. [H]

64. *Gérardin,* ———.

Lettre d'un Anglais à un Français sur la découverte du magnétisme animal et observations sur cette lettre. Paris: Bouillon, 1784, 24 pp.

A letter in favor of animal magnetism. Some scholars attribute this pamphlet to Sébastien Gérardin. [H]

65. *Gilbert,* ———.

Mémoire en réponse au rapport de MM. les Commissaires chargés par le roi de l'examen du magnétisme animal. Morlaix: n.p., 1784, 26 pp.

Critical of the findings of the Franklin commission. [H]

66. *Gilbert, Jean Emmanuel.*

Aperçu sur le magnétisme animal ou résultats des observations faites à Lyon sur ce nouvel agent. Geneva: n.p., 1784, 76 pp.

Gilbert was a highly reputable physician and professor of medicine. Among other things he was known for his opposition to all forms of quackery in medicine, having written extensively on the subject. In this treatise he attests to the genuineness of the healing effects of animal magnetism, which he had repeatedly witnessed with his own eyes. Gilbert does not say that he agrees with Mesmer's theories of magnetic fluid. However, as a strong believer in the healing powers of nature, he is sympathetic with the practice of animal magnetism. [H]

67. [*Girardin,* ———].

Observations adressées à Mrs. les commissaires chargés par le Roi de l'examen du magnétisme animal; sur la manière dont ils y ont procédé, & sur leur rapport. Par un médecin de province. London and Paris: Royez, 1784, 36 pp.

Comments addressed to the king's two commissions that investigated animal magnetism, offered by a man who was present at the experiments the commissions carried out with the aid of D'Eslon. The author's comments are meant to correct certain errors in the report and further explain the nature of the treatment. [H]

68. [*Girardin, ———*].

Observations adressées à messieurs les commissaires de la Société royale de médecine, nommés par le Roi . . . Sur la manière dont ils on procédé, et sur le rapport qu'ils en ont fait. Par un médecin de P—. Pour servir de suite à celles qui ont été adressées sur le même objet à MM. les commissaires tirés de la Faculté de médecine & de l'Académie royale des sciences de Paris. London and Paris: Royez, 1784, (1) + 17 pp.

A second set of comments continuing observations started in the first (*see* entry number 67). [H]

69. [*Guigoud-Pigale, Pierre*].

Le baquet magnétique, comédie en vers et en deux actes. London and Paris: Gastellier, 1784, 126 pp.

A farce in two acts about animal magnetism. [H]

70. *Hervier, Charles.*

Lettre du père Hervier aux habitants de Bordeaux. Paris: n.p., 1784, 4 pp.

A letter extolling animal magnetism. [H]

71. *Hervier, Charles.*

Lettre sur la découverte du magnétisme animal à M. Court de Gébelin. (Peking and) Paris: Couturier, 1784, viii + 48 pp.

Hervier expresses his enthusiasm for animal magnetism, the result of the healing that it had accomplished for him personally. Like Court de Gébelin, to whom he addresses the letter, Hervier was cured of an illness through the application of Mesmer's techniques. This letter provides a detailed description of the illness and treatment. [H]

72. [*Jussieu, Antoine Laurent de*].

Rapport de l'un des commissaires chargés par le Roi, de l'examen du magnétisme animal. Paris: Veuve Harissart, 1784, 51 pp.

Jussieu strongly disagreed with the conclusions of the principal report which followed the investigation of animal magnetism by the Société Royale de Médecine in 1784. He stated his own views in this treatise. The commission had seen the demonstrations of animal magnetism given by D'Eslon and Lafisse (Mesmer having refused to take part in the investigation). He distinguished four different kinds of facts observed by the commissioners concerning animal magnetism: the first were those general positive effects about which it was not possible to come to any conclusions as to cause; the second were those which were negative, showing only the non-action of the alleged magnetic fluid; the third were effects, either positive or negative, which could be attributed to the work of the imagination; and the fourth were those positive effects that could only be explained through the action of some unknown agent. Jussieu concluded that although the existence of a magnetic fluid had not been proven, there were enough effects of the fourth kind to justify the continued use of animal magnetism and further investigations of the exact nature of those effects. [H]

73. [*La Grezie, Bertrand de*].

Le magnétisme animal dévoilé par un zélé citoyen français. Geneva: n.p., 1784, 36 pp.

A denunciation of Mesmer and animal magnetism by a man Mesmer failed to cure. La Grezie states his belief that magnetizers simply harm their clients. He also presents his own theory of the possibility of communicating electrical fluid from one person to another. [H]

74. [*Landresse, C. de*].

Le cri de la nature, ou le magnétisme au jour; ouvrage curieux et utile pour les personnes qui cherchent à étudier les causes physiques du magnétisme ainsi que les phénomènes qui s'y rapportent. London and Paris: n.p., 1784, 40 pp.

The author himself had been cured of an illness through animal magnetism by D'Eslon. Because of his appreciation for the technique, he wants to make it known to the public. [H]

75. *Lettre a un magistrat de province, sur l'existence du magnétisme*.

N.p.: n.p., (1784), 32 pp.

[H]

76. *Lettre de M.A.***à M.B.***, sur le livre intitulé: Recherches & doutes sur le magnétisme animal de M. Thouret*.

Brussels: n.p., 1784, (1) + 42 pp.

[H]

77. *Lettre sur la mort de Court de Gébelin*.

Paris: n.p. 1784, 6 pp.

[H]

78. [*Mahon, Paul Augustin Olivier*].

Examen sérieux et impartial du magnétisme animal. London and Paris: Royez, 1784, (1) + 43 pp.

A booklet attributed to the physician Mahon. He writes in favor of animal magnetism, and states that if his fellow physicians look into the matter seriously, they will find much of value there for their medical practice. [H]

79. [*Mahon, Paul Augustin Olivier*].

Lettre de l'auteur de l'Examen sérieux et impartial du magnétisme animal à M. Judel, médecin membre de la Société de l'Harmonie, où, en répondant à la critique qu'en a faite ce docteur, et qu'il a insérée dans les affiches du pays chartrain, on fait voir que les disciples de d'Eslon peuvent être aussi instruits de la doctrine du magnétisme animal, que ceux de M. Mesmer et quelquefois mieux. Paris and Philadelphia: n.p., 1784, 16 pp.

Mahon, a physician and supporter of animal magnetism, comments on the opinion of Judel that the commissioners who investigated animal magnetism should not have used D'Eslon as their practitioner, since he did not know the art as well as Mesmer. Mahon did not agree with Judel's position. [H]

80. *Marat, Jean Paul.*

Mémoire sur l'électricité médicale. Paris: N. T. Méquignon, 1784, 111 pp.

A treatise in which the author mentions animal magnetism in the context of a discussion of medical applications of electricity. [H]

81. *Mesmer, Franz Anton.*

Apologie de M. Mesmer; ou, Réponse à la brochure intitulée: Mémoire pour servir à l'histoire de la jonglerie dans lequel on démontre les phénomènes du mesmérisme. (Paris?): n.p., 1784, 8 pp.

[H]

82. *Mesmer, Franz Anton.*

*Lettre de M. Mesmer à M. Le Cte de C*** (31 août 1784)*. N.p.: Imprimerie royale, (1784), 11 pp.

The letter includes the "Copie de la requête à nos seigneurs de Parlement en la grand' Chambre" in which Mesmer objects to the procedure used by the commission appointed by the king to examine animal magnetism. Mesmer asserts that they were not justified in carrying out their investigation with the aid D'Eslon, who did not have the true doctrine of animal magnetism. [H]

83. *Mesmer, Franz Anton.*

*Lettre de M. Mesmer à M**** (*Vicq-d'Azyr*), *Paris, 16 aout 1784*. N.p.: n.p., 1784, 6 pp.

Originally appearing in the journals of the day, this letter is found in the *Recueil des pièces les plus intéressantes sur le magnétisme animal* (*see* entry number 86). It contains Mesmer's criticisms of Thouret's *Recherches et doutes sur le magnétisme animal* (*see* entry number 116). Thouret had pointed to Mesmer's dependency on sixteenth- and seventeenth-century writers. Mesmer denies such dependency, stating that Thouret builds his case largely on the writings of Maxwell, whom Mesmer had not even read and whose doctrine is in any case very different from that of Mesmer. [H]

84. *Mesmer, Franz Anton.*

Lettres de M. Mesmer à messieurs les auteurs du Journal de Paris, et à M. Franklin. N.p.: n.p., (1784), 14 pp.

Letters to the editors of the *Journal de Paris* and to Benjamin Franklin, head of the commission appointed by the king to investigate animal magnetism. Mesmer complains that the commission carried out its investigation with D'Eslon, whose work Mesmer had rejected. He asserts that history will be the judge of the worth of his discovery. [H]

85. *Mesmer, Franz Anton.*

Lettres de M. Mesmer, à M. Vicq-d'Azyr, et à messieurs les auteurs du Journal de Paris. Brussels: n.p., 1784, 30 pp.

A collection of letters written by Mesmer to the editors of the *Journal de Paris* and a letter written by him in response to a critical article by Vicq-d'Azyr published in that journal. [H]

86. [*Mesmer, Franz Anton*].

Lettre d'un médecin de Paris à un médecin de province. (Paris): n.p., 1784, 16 pp.

Addressed to a "physician of the Commission" appointed by the king to investigate animal magnetism, this letter accuses D'Eslon of contravening an agreement made between him and Mesmer concerning the control of the propagation of animal magnetism. The text of the agreement is given. [H]

87. *Mesmer, Franz Anton et al.*

Recueil des pièces les plus interessantes sur le magnétisme animal. N.p.: n.p., 1784, (4) + (6) + 7–468 pp.

An important collection of works written on the subject of animal magnetism by Mesmer and others. It includes Mesmer's *Mémoire* of 1779 (entry number 10); the *Lettre sur la mort de M. Court de Gébelin* (entry number 77); *Détail des cures operées à Buzancy près Soissons par le magnétisme animal* (entry number 47); and many others. [H]

88. *Mesmer guéri ou Lettre d'un provincial au R.P.N. . . . , en réponse à sa lettre intitulée Mesmer blessé.*

London and Paris: n.p., 1784, 13 pp.

This letter supports the views of Hervier expressed in his *Lettre sur la découverte du magnétisme animal* (*see* entry number 71). [H]

89. *Moulinié, Charles.*

Lettre sur le magnétisme animal adressée à M. Perdriau, pasteur et professeur de l'église et de l'académie de Genève. Paris: n.p., 1784, 25 pp.

An unusual document written by a Christian magnetizer. Moulinié was the pastor of a church and an extremely effective practitioner of animal magnetism. His mere presence seemed to be enough to heal some people, and the poor were often the beneficiaries of his magnetic power. He was extremely enthusiastic about the great good he believed Mesmer's discovery had bestowed upon mankind. [H]

90. [*Muletier, ———*].

Réflexions sur le magnétisme animal, d'après lesquelles on cherche à établir le degré de croyance que peut mériter justqu'ici le système de M. Mesmer. Brussels and Paris: Couturier, 1784, (1) + 43 pp.

The author believes that Mesmer's remarkable success was due only to the fascination with the marvelous that was popular at the time. He mentions use of the technique of fixation of the eyes to produce a convulsion and closure of the eyelids. [H]

91. *Observations sur le rapport des commissaires chargés par le Roi de l'examen du magnétisme animal. Par M.G.C. Membre de diverses académies.*

Vienna: n.p., 1784, 20 pp.

[H]

92. *Orelut, Pierre.*

Détail des cures opérées à Lyon par le magnétisme animal, selon les principes de M. Mesmer. Précédé d'une lettre à M. Mesmer. Lyon: Faucheux, 1784, 27 pp.

Orelut begins with a letter to Mesmer telling him that when he (Orelut) arrived in Lyon he found the city to be in a state of excitement about animal magnetism and the cures being attributed to it. He states that his purpose in writing this account of various well-attested cures is to help make known Mesmer's marvelous discovery. The author then describes in some detail the nature of the cases treated and the positive effects produced. [H]

93. *O'Ryan, ———.*

Discours sur le magnétisme animal lu dans une assemblée du College des médecins de Lyon le 15 septembre 1784. Dublin: n.p., 1784, 31 pp.

A well-written treatise arguing against animal magnetism. A professor of medicine at the University of Montpellier, O'Ryan bases his opposition on cases of animal magnetic practice that he had actually observed. [H]

94. [*Paulet, Jean Jacques*].

L'antimagnétisme, ou Origine, progrès décadence, renouvellement et réfutation du magnétisme animal. London: n.p., 1784, (2) + 252 pp.

Paulet's longest and most thorough critique of Mesmer and his theory of animal magnetism. The author begins with an examination of the "traces of magnetism" to be found in the authors of the sixteenth and seventeenth centuries, such as Paracelsus, Van Helmont, and Fludd. This is followed by a discussion of Mesmer's propositions on animal magnetism and certain writings of the followers of Mesmer. Paulet also examines the work of the "stroking doctors" such as Greatrakes and the "miracles" of the exorcist Gassner, viewing them as the product of the imagination. This is one of the most important and intelligent of the early critiques of animal magnetism. It includes a famous engraved frontispiece depicting Mesmer drawing magnetic fluid from the heavens and conferring it upon the ill. [H]

95. [*Paulet, Jean Jacques*].

Mesmer blessé ou réponse à la lettre du P. Hervier sur le magnétisme animal. London and Paris: Couturier, 1784, 34 pp.

Paulet sets out to convince Hervier that his cure by Mesmer was not remarkable, because he (Hervier) had not been truly ill in the first place. [H]

96. [*Paulet, Jean Jacques*].

Mesmer justifié. Constance and Paris: n.p., 1784, (2) + 46 pp.

A farcical piece which makes fun of Mesmer and his practices by pretending to give serious instructions about how to use animal magnetism, while really mocking the whole procedure. It prescribes, for instance, a "blissful disposition" on the part of the operator and "blind submission" on the part of the patient. It also goes into great detail about the necessity of knowledge of the "poles and equators" to properly use animal magnetism. [H]

97. [*Peaumerelle, C. J. de B. de*].

La philosophie des vapeurs ou Correspondance d'une jolie femme, nouvelle édition, augmentée d'un petit traité des crises magnétiques à l'usage des mesmériennes. Paphos and Paris: Royez, 1784, xxii + 168 pp.

[H]

98. [*Pétiau, abbé*].

Autres reveries sur le magnétisme animal à un académicien de province. Brussels: n.p., 1784, 48 pp.

The Abbé Petiot was a secretary of the Paris Society of Harmony and a friend of Bergasse. In this pamphlet he defends Mesmer against D'Eslon and voices his disagreement with the findings of the king's commission on animal magnetism. [H]

99. [*Pétiau, abbé*].

Lettre de M. l'abbé P. . . de l'Académie de la Rochelle, à M. . . de la meme Académie sur le magnétisme animal. N.p.: n.p., 1784, 7 pp. [H]

100. [*Philip,* ———].

La Mesmériade, ou le triomphe du magnétisme animal. Poème en 3 chants dédié à la lune. Geneva and Paris: Couturier, 1784, 15 pp.

This satirical piece is attributed to Doctor Philip, dean of the faculty of medicine at Paris at the time when it carried out its investigation of animal magnetism. [H]

101. *Poissonnier, Pierre Isaac, Caille, Claude Antoine, Mauduyt de Varenne, Pierre Jean Claude, and Andry, Charles Louis François.*

Rapport des commissaires de la Société royale de médecine nommés par le Roi pour fair l'examen du magnétisme animal, imprimé par ordre du Roi. Paris: Imprimerie royale, 1784, 39 pp.

This report was compiled by a commission of the Royal Society of Medicine set up by the king to investigate the claims of animal magnetism. This commission was constituted at approximately the same time as a second commission (The Franklin Commission), also appointed by the king, made up of nine eminent scientists from the Academy of Sciences. The commission of the Royal Society of Medicine began its investigations on April 5, 1784. The findings of this commission condemned animal magnetism. Hampered by a lack of scientific method and a surfeit of theoretical dogmatism, however, its report proved to be far less significant than that of the Franklin commission. One of the members of the Royal Society's commission strongly disagreed with its published findings and wrote his own report (*see* Jussieu, entry number 72). [H]

102. *Pressavin, Jean Baptist.*

Lettres sur le magnétisme. (Lyon): n.p., (1784), 16 pp.

[H]

103. *Pressavin, Jean Baptiste.*

Suite de la correspondance de Monsieur Pressavin gradué etc., avec les magnétiseurs de la mème ville. N.p.: n.p., 1784, 15 pp.

[H]

104. *Prony, Gaspard Clair François Marie Riche.*

Nouvelles cures opérées par le magnétisme animal. Paris: n.p., 1784, 64 pp.

[H]

105. ***Puységur, Armand Marie Jacques de Chastenet, Marquis de.*** *Mémoires pour servir à l'histoire et à l'établissement du magnétisme animal.* Paris: Dentu, 1784, (7) + 8–232 pp.

A work of great significance for the history of modern psychology. Puységur was an artillery officer in the army, a colonel of the regiment of Strasbourg. A member of an old and distinguished family, he had inherited a large property in Buzancy near Soissons and spent most of his time there looking after his land and occasionally carrying out experiments with electricity. Having heard about animal magnetism and its marvelous curative powers, he went to Paris to learn from Mesmer. Returning to his estate at Buzancy, he began to use animal magnetism to alleviate the ills of local residents. Among the first he treated was a peasant named Victor Race who was suffering from a fever and congestion of the lungs. When applying the magnetic passes to the man, Puységur noticed that Victor had fallen asleep. This sleep was not, however, a normal one, for he could still communicate with Puységur. While in this state Victor showed himself to be extremely suggestible and even seemed, in Puységur's estimation, to be able to read his magnetizer's thoughts. When returning to his normal state of consciousness, Victor remembered nothing of what had happened. Puységur also noted that there seemed to be a specially close relationship between himself and Victor, a relationship that he would later (*Suite de Mémoires*, entry number 148) call an "intimate rapport." He also was struck by the dramatic change in personality that Victor underwent between the state of magnetic sleep and his normal state: in the latter he was of rather ordinary or even slow wit, while in the former he became extremely bright, perceptive, and articulate. Puységur termed this newly discovered condition "magnetic sleep" and "magnetic somnambulism," since he immediately noticed the similarity between this state and that of natural "somnambulism" or "sleep-walking." Puységur went on to further investigate this state through experimentation with other patients. He noted that they all showed the same characteristics as the ones that Victor had demonstrated. His work with Victor and subsequent investigations are described in the *Mémoires*. Puységur's discovery of artificial somnambulism started a whole new trend in the practice of animal magnetism, shifting the emphasis from the physical to the psychological. The alteration in consciousness between the state of magnetic sleep and the normal waking state, with its attendant amnesia, revealed, within human beings, a double or divided consciousness with two memory chains. This revelation opened up a new line of investigation that would eventually lead to the psychological concepts of the "subconscious," the "subliminal self," and the "unconscious" in the latter part of the nineteenth century. The second edition (1809) and the third and most complete edition (1820) are supplemented by the *Suite de Mémoires* (entry number 148) which further develops the discoveries of the *Mémoires*. [H]

106. ***Puységur, Jacques Maxine Paul de Chastenet, Comte de.*** *Rapport des cures opérées à Bayonne par le magnétisme animal, adressé à M. l'abbé de Poulouzat, conseiller clerc au Parlement de Bordeaux, par le comte* *de Puységur, avec notes de M. Duval d'Espremenil, conseiller au Parlement de Paris.* Bayonne and Paris: Prault, 1784, 72 pp. ENGLISH: "Report of cures by animal magnetism occurring at Bayonne with verifications," in Maurice Tinterow, *Foundations of Hypnosis.* Springfield, Illinois: Charles C. Thomas, 1970.

The Comte de Puységur, brother of the Marquis de Puységur, was second in command of the Regiment of Languedoc when the occurrences described in this pamphlet took place. He had learned the techniques of animal magnetism and found occasion to use them at this posting. Here he mentions some sixty cures accomplished through animal magnetism. One of the most curious was that of a dog which had been injured by an angry soldier. The Comte de Puységur approached the animal, which seemed to be barely alive, and applied magnetic passes. The dog was restored to good health in the space of a few minutes. This seems to be the first example in the literature of animal magnetism of the application of magnetic healing to an animal. [H]

107. ***Rapport de la Société royale de médecine sur l'ouvrage*** *intitulée Recherches et doutes sur le magnétisme animal, etc.*
Paris: n.p., 1784, 22 pp.

[H]

108. ***Rapport du rapport de MM. les Commissaires nommés par le*** *Roi, etc. par un amateur de la vérité excité par l'imagination, l'attouchement et magnétisé par le bon sens et la raison. Adressé à M. Caritides, fils de cet illustre savant qui avait conçu l'ingénieux projet de mettre toutes les côtes du royaume en port de mer, actuellement résident au Monomotapa.*
Peking and Paris: Couturier, 1784, 34 pp.

[H]

109. ***[Retz, Noel de Rochefort].***
Mémoire pour servir à l'histoire de la jonglerie, dans lequel on démontre les phénomènes du mesmérisme. Nouvelle édition précédée d'une lettre sur le secret de M. Mesmer. . . . On y a joint une réponse au Mémoire qui paroit ici pour la première fois. London and Paris: Méquignon, 1784, (1) + 47 + (1) + 8 pp.

Contains a reprint of Retz's earlier *Lettre* (entry number 25) with the assertion that it had anticipated the conclusions of the report of the Franklin commission of 1784. In the *Mémoire* Retz deals with the healing "impostures" that have been perpetrated over the ages. He mentions Gassner's exorcisms, the magnetic medicine of the seventeenth century, the "powder of sympathy," and other "impostures." Although the title seems to indicate the work is to be a defense of Mesmer, it in fact rejects animal magnetism as the latest of the "impostures." This second edition seems to be the only one still extant. [H]

110. *Salaville, Jean Baptiste.*
Le moraliste mesmérien ou lettres philosophiques sur l' influence du magnétisme. London and Paris: Berlin, 1784, 132 pp. [H]

111. [*Servan, Joseph Michel Antoine*].
Doutes d'un provincial, proposés à messiers le médecins-commissaires chargés par le roi de l'examen du magnétisme animal. Lyon and Paris: Prault, 1784, (4) + 136 pp.

Servan was a distinguished French lawyer and a correspondant of Voltaire and d'Alembert. He had been cured by a mesmerist when traditional medicine had failed to help him. In this treatise, Servan defends Mesmer's views on animal magnetism in the wake of the negative reports of the two commissions. He addresses some well-considered questions to the members of the commission of the Société Royale de Médecine. These "doubts" question: why they expected quick cures through animal magnetism and dismissed its efficacy on the basis of brief trials; why they did not choose optimal conditions for success but arbitrarily set up conditions they preferred; and why with their limited experiençe they felt justified in drawing such far-reaching conclusions about the non-existence of animal magnetism. Servan's treatise is one of the most thoughtful contemporary criticisms of the findings of the commission. [H]

112. [*Servan, Joseph Michel Antoine*].
Questions du jeune Docteur Rhubarbini de Purgandis, adressés à Messieurs les docteurs-regents, de toutes les facultés de médecine de l'universe, au sujet de M. Mesmer & du magnétisme animal. Padua: n.p., 1784, xii + 50 pp.

[H]

113. *Sousselier de la Tour, comte.*
L'ami de la nature, ou Manière de traiter les maladies par le prétendu magnétisme animal. Dijon: Capel, 1784, xiii + 176 pp.

[H]

114. *Swinden, Han Hendrik van.*
Recueil de mémoires sur l'analogie de l'électricité et du magnétisme: couronnés & publiés par l'Académie de Baviere; traduits du Latin & de L'Allemand, augmentés de notes, & de quelques dissertations nouvelles. 3 vols. Haye: Libraires Associés, 1784.

Van Swinden was an eminent physicist and first president of the Royal Institute of the Netherlands. This collection contains a section called *Reflexions sur le magnétisme animal, et sur le système de M. Mesmer.* [H]

115. ***Le systéme de la rose magnétique.***
(Paris): n.p., 1784, 18 pp.

The dating of this work is uncertain, but it is believed to be 1784–1789. [H]

116. ***Thouret, Michel Augustin.***
Recherches et doutes sur le magnétisme animal. Paris: Prault, 1784, xxxv + (1) + 251 pp.

Thouret was a member of the Royal Society of Medicine in Paris and one of the leading spokesmen of the opposition of that society to animal magnetism and the teachings of Mesmer. In this work Thouret claims that his main concern is not to examine the details of cures being performed by animal magnetism, but to trace the history of the theory and practice of animal magnetism. He nevertheless clearly sides with those who reject animal magnetism as an illusion. Admitting that many persons of stature accept animal magnetism as an effective cure, Thouret uses his considerable erudition to show that such cures are not new and that Mesmer was simply the most recent of a long tradition of thinkers who posited a hidden power of nature that produces healing effects. He cites Paracelsus, Kircher, Maxwell, and Fludd as examples of men who held views similar to those of Mesmer. He also points out that there have been many healers over the ages who have accomplished cures resembling Mesmer's, mentioning the exorcist Gassner and the "stroking doctor" Greatrakes as examples. Thouret's learned critique was extremely influential and served as a starting point for much of the discussion at the time about the originality and effectiveness of animal magnetism. [H]

117. **[*Thouvenel, Pierre*].**
Second mémoire physique et médicinal, montrant des rapports évidents entre les phénomènes de la baguette divinatoire, du magnétisme, et de l'électricité, avec des éclaircissements sur d'autres objets non moins importans, qui y sont relatifs. London and Paris: Didot le jeune, 1784, (2) + 268 pp.

After presenting summaries of some of the responses to his first *Mémoire* (*see* entry number 18), Thouvenal describes subsequent attempts to use the divining rod to seek water. He points out that he has been charged by the king to investigate the mineral and medicinal waters of the realm. Once again Thouvenal brings in "animal magnetism," which he defines as the flux and reflux to which the "electrical matter" in all living things is subject. As in the first *Mémoire*, Thouvenel makes no mention of Mesmer and his own brand of animal magnetism, which by this time must have been known to him. [H]

118. **[*Tissart de Rouvres, Jacques Louis Noel, marquis de*].**
Nouvelles cures opérées par le magnétisme animal. (Paris): n.p., (1784), 64 pp.

The work describes successful treatment of various illnesses by animal magnetism. [H]

119. ***La vision contenant l'explication de l'écrit intitulé: Traces du magnétisme, et la théorie des vrais sages.***
Memphis and Paris: Couturier, 1784, iv + 31 pp.

[H]

1785

120. **[*Archibold,* ———, ed.]**
Recueil d'observations et de faits relatifs au magnétisme animal, présenté à l'auteur de cette découverte, et publié par la société de l'harmonie de Guienne. Paris and Bordeaux: Pallandre jeune, 1785, (2) + 168 pp.

[H]

121. ***Bell, John.***
Animal Electricity and Magnetism, &c. Demonstrated after the Laws of Nature; with New Ideas upon Matter and Motion. In Two Parts. (London): The Author, 1785, 36 + vi + 7–44 pp.

John Bell, trained in the theory and practice of animal magnetism in Paris, witnessed Puységur's experiments with magnetic somnambulism. In this, the least memorable of his works on animal magnetism, Bell presents a confused physics of magnetism, animal magnetism, magnetic fluid, etc. He does have some interesting things to say with regard to the place of "idea" and "will" in directing the action of the magnetic fluid, and his description of the use of the hands and movement in the application of animal magnetism is also informative. [H]

122. **[*Bergasse, Nicolas*].**
Confession d'un médecin académicien et commissaire d'un rapport sur le magnétisme animal, avec les remontrances et avis de son directeur. N.p.: n.p., 1785, 70 pp.

A satirical confession of wrongs by a fictitious member of the commission that condemned animal magnetism. Attributed to Bergasse. [H]

123. ***Bergasse, Nicolas.***
Observations de M. Bergasse sur un écrit du docteur Mesmer, ayant pour titre: Lettre de l'inventeur du magnétisme animal à l'auteur des Reflexions préliminaires. London: n.p., 1785, (2) + 101 pp.

Here Bergasse announces the split that had opened between Mesmer and himself. It is valuable for the information it provides about contemporary events concerning the fortunes of animal magnetism. [H]

124. *Bergasse, Nicolas.*

Supplément aux Observations de M. Bergasse, ou Règlemens des sociétés de l'harmonie universelle, adoptés par la société de l'harmonie de France dans l'assemblée générale tenue à Paris, le 12 Mai 1785; avec des notes pour servir à l'intelligence du texte. N.p.: n.p., (1785), 32 pp.

The constitution for the Societies of Harmony which were to be set up all over France. The parent society in Paris was to have special privileges, being designated the "Society of Harmony of France." The rules lay out the conditions under which an individual may become a member and the fees for membership. In the rules Mesmer was given the title of "Perpetual President." Bergasse was principal speaker of the Society in Paris and the chief architect of these rules. [H]

125. *Bonnefoy, Jean Baptiste.*

Examen du Compte rendu par M. Thouret, sous le titre de Correspondance de la Société royale de médecins, relativement au magnétisme animal. Lyon: n.p., 1785, 59 pp.

[H]

126. [*Brack, ———*].

Testament politique de M. Mesmer, ou la précaution d'un sage, avec le dénombrement des adeptes; le tout traduit de l'Allemand par un Bostonien. Leipzig and Paris: n.p., 1785, 50 pp.

A pamphlet against Mesmer. The reference to a German original seems to be a literary fiction. [H]

127. *Bruno, de, ———.*

Recherches sur la direction du fluide magnétique. Amsterdam and Paris: Gouffier, 1785, viii + 206 pp.

De Bruno developed a theory of magnetic fluid that was similar to that of Mesmer whom he cites. He posits one universal magnetic fluid, rather than many, which explains all physical phenomena. [H]

128. *Carra, Jean Louis.*

Examen physique de magnétisme animal; analyse des éloges & des critiques qu'on en a faits jusqu'à présent; et développement des véritables rapports, sous lesquels on doit en considérer le principe, la théorie, la pratique & le secret. London and Paris: E. Oufroy, 1785, 98 pp.

Carra was a prolific writer in many fields, including that of physics. In this work he seems to disagree with Mesmer's theory of magnetism, but he agrees that the techniques devised by Mesmer were efficacious for healing illness. Carra gives his own somewhat convoluted physical and philosophical explanations about why this is so. [H]

129. [*Caullet de Veumorel, ———, ed.*].

Aphorismes de M. Mesmer, dictés à l'assemblée de ses élèves, & dans lesquels ou trouve ses principes, sa théorie & les moyens de magnétiser; le tout forant un corps de doctrine, developpé en trois cents quarant-quatre paragraphes, pour faciliter l'application des commentaires au magnétisme animal. Ouvrage mis au jour par M. C. de V. Paris: M. Quinquet, 1785, (2) + xxiv + 172 + (4) pp.

A compilation of "class notes" taken down from talks given by Mesmer to those he was training in the theory and practice of animal magnetism. Edited and published by Caullet de Veaumorel, a disciple of D'Eslon, the book was rejected by Mesmer. It is nonetheless believed to faithfully reproduce Mesmer's teaching. The *Aphorismes* was a very popular book and went through many editions. [H]

130. [*Delandine, Antoine François*].

De la philosophie corpusculaire, ou des connaissances et les procédés magnétiques chez les divers peuples. Paris: Cuchet, 1785, iii + 200 + (3) pp.

A wide ranging study of phenomena that are analogous to animal magnetism. The author covers everything from electricity and magnetism in the human body to the curative effects of music. His speculations on the nature of sympathetic cures are particularly interesting. [H]

131. *Delandine, Antoine François.*

Notice historique sur les systèmes et les écrits anciens qui se rapportent au magnétisme animal. Paris: n.p., (1785), 16 pp.

An attempt to trace the historical antecedents of animal magnetism. Delandine works along the same lines as those pursued in his *De la philosophie corpusculaire* (entry number 130). [H]

132. *D'Eslon, Charles.*

Lettre adressée par M. d'Eslon aux auteurs du Journal de Paris et voluntairement refusée par eux, concernant l'extrait de la correspondance de la Société royale relativement au magnétisme animal, rédigé par M. Thouret et imprimé au Louvre. N.p.: n.p., (1785), 7 p.

A criticism of the views of Thouret on the nature of animal magnetism. [H]

133. *Devillers, Charles Joseph.*

L'antimagnétisme martiniste ou barbériniste; observations trouvées manuscrites sur la marge d'une brochure intitulée: Réflexions impartielles sur le magnétisme animal, faites après la publication du Rapport des commissaires, &c. Lyon: n.p., 1785, 43 pp.

A work opposing the brand of animal magnetism being practiced in Lyon by a number of practitioners associated with Freemasonry, particularly those under the leadership of the Chevalier de Barberin. [H]

134. [*Doppet, François Amédée*].
Oraison funèbre du célèbre Mesmer, auteur du magnétisme animal et président de la Loge de l'Harmonie. Grenoble: n.p., 1785, 39 pp.

A satirical piece written on the supposed death of Mesmer. [H]

135. *Elie de la Poterie, Jean Antoine.*
Examen de la doctrine d'Hippocrate, pour servir à l'histoire du magnétisme animal. Bordeaux: n.p., 1785, 87 pp.

[H]

136. [*Éprémesnil, Jean Jacques Duval d'*].
Sommes versées entre les mains de monsieur Mesmer pour acquérir le droit de publier sa découverte. Paris: n.p., 1785, 8 pp.

A pamphlet, written by a former supporter of Mesmer, objecting to Mesmer's claim to hold the exclusive right to propagate the doctrine of animal magnetism. Éprémesnil contends that since he and others had paid Mesmer a good deal of money (listed in detail in the pamphlet) to teach them animal magnetism, they should now be able to teach it to the public. [H]

137. [*Favrye, Mme. de la*].
Les rêves d'une femme de province sur le magnétisme animal, ou Essai théorique & pratique sur la doctrine à la mode. London and Paris: n.p., 1785, (1) + 42 pp.

[H]

138. [*Fortia de Piles, Alphonse, Jourgniac de St-Médard, François de, and de Boisgelin, Louis*].
Correspondance de M. M. . . . sur les nouvelles découvertes du baquet octogne, de l'homme-baquet et du baquet moral, pour servir de suite aux aphorismes. Recueillie et publiée par MM. de F. . . . ; J. . . . et B. . . . Libourne and Paris: n.p., 1785, 163 pp. [H]

139. [*Fournel, Jean François*].
Essai sur les probabilités du somnambulism magnétique: pour servir à l'histoire du magnétisme animal. Paris: Gastelier, 1785, (2) + 70 pp.

After the discovery of "magnetic somnambulism" by Puységur in 1784, Fournel was the first person to attempt to theorize about the nature of this new phenomenon. He sees magnetic somnambulism as a state midway between waking and sleep, a state essentially the same as natural somnambulism, which had been widely recognized as a reality. Fournel points out that the seemingly extraordinary phenomena associated with magnetic somnambulism, such as suggestibility and clairvoyance, have been noted for centuries in connection with natural somnambulism. Speaking of the sudden rise to popularity of magnetic somnambulism, he estimates the number of somnambulists in Paris and the provinces to be in the neighborhood of six thousand. Fournel makes a strong case for accepting magnetic somnambulism as a genuine phenomenon which deserves further study. [H]

140. ***Fournel, Jean François.***
Mémoire pour M. Charles Louis Varnier . . . appellant d'un décret de la Faculté; contre les doyen et docteurs de ladite Faculté, intimés. Paris: Herissant, 1785, 68 pp.

[H]

141. **[*Fournel, Jean François*].**
Remonstrances des malades aux médecins de la Faculté de Paris. Amsterdam: n.p., 1785, (1) + 103 pp.

[H]

142. ***Histoire véritable du magnétisme animal, ou nouvelles preuves de la réalité de cet agent tirées de l'ancien ouvrage d'un vieux docteur.***
The Hague: n.p., 1785, 16 pp.

A satirical treatise written in opposition to animal magnetism. The "old doctor" is François Rabelais (1490–1553). [H]

143. **[*Laugier, Esprit Michel*].**
Parallèle entre le magnétisme animal, l'électricité et les bains médicinaux par distillation, &c. appliqués aux maladies rebelles. On a joint à ce précis l'art de conserver la santé, & de guérir les maladies le plus rebelles. . . . Paris: Morin, 1785, 12 + 91 pp.

The author uses the popular interest in animal magnetism to advertize the use of medicinal baths and other approaches such as exercise and music to treat illnesses. However, there is very little about animal magnetism in the pamphlet. Its mention in the title was obviously just to arouse the curiosity of the reader. [H]

144. ***Mesmer, Franz Anton.***
Lettre de l'auteur de la découverte du magnétisme animal à l'auteur des Réflexions préliminaires. Pour servir de réponse à un imprimé ayant pour titre: Sommes versées entre les mains de M. Mesmer pour acquérir le droit de publier sa découverte. (Paris): n.p., (1785), 26 pp.

Written by Mesmer to defend himself against accusations leveled at him by a former supporter, Jean Duval d'Éprémesnil, in a work entitled: *Sommes versées entre les mains de monsieur Mesmer* . . . (entry number 136), Éprémesnil had objected to Mesmer's claim that those who had paid Mesmer to teach them had no right to make the doctrine known to the public. Mesmer counters that it was explicitly stated in their agreement with him that the doctrine of animal magnetism remains his property and that only he can determine how it is to be propagated. [H]

145. *Mullatera, Giovanni Thommaso.*
Del magnetismo animale, e degli effetti ad esso attribuiti nella cura delle umane infermita. Biella: Antonio Cajani, 1785, (6) + 7–60 pp.

Apparently the earliest Italian book on animal magnetism, and there are no references to it in any of the bibliographical sources for animal magnetism. It was published with the "imprimatur" of the Roman Catholic Church. Mullatera dedicates the book to Innocenzo Laneri, Professor of medicine at the University of Torino. He begins with something of an apology for writing a book on the subject of animal magnetism, a subject which is of questionable merit. He points out that some French commissions had already dismissed it as a matter of imagination. But since there are people in Italy, at Piedmont, who are nonetheless practicing it, something needs to be written in response. Mullatera examines the background of magnetic medicine in sixteenth- and seventeenth-century Europe, pointing out the similarity between the teachings of Mesmer and those of Paracelsus, Van Helmont and Fludd. For his contemporary sources he uses principally the reports of the commissions (including that of Jussieu) and the propositions of Mesmer. He finds animal magnetism to be of no particular value as a method of cure and places it in the category of useless, fantastic medical treatments. [H]

146. *Nouvelle découverte sur le magnétisme animal; ou, Lettre adressée à un ami de province, par un partisan zélé de la verité.*
N.p.: n.p., 1785, 64 pp.

[H]

147. [*Paulet, Jean Jacques*].
Réponse à l'auteur des Doutes d'un provincial, proposés à MM. les médecins commissaires chargés par le Roi de l'examen du magnétisme animal. London: n.p., 1785, 70 pp.

Paulet takes Servan, author of the *Doutes d'un provincial* (*see* entry number 111) to task for his criticism of the medical establishment and its rejection of animal magnetism. [H]

148. [*Puységur, Armand Marie Jacques de Chastenet, marquis de*].
Suite des mémoires pour servir à l'histoire et à l'établissement du magnétisme animal. Paris and London: n.p., 1785, 256 pp.

Through this work Puységur meant to supplement the findings of his monumental *Mémoires* of 1784 (entry number 105). Here he further develops his notion of the importance of the "will" in the action of animal magnetism and presents more information about the phenomena of magnetic somnambulism. Also, for the first time, he uses the word "rapport" to describe the special connection between magnetizer and somnambulist mentioned in the *Mémoires.* [H]

149. *Recit de l'avocat-général de —, aux chambres assemblées du* *public, sur le magnétisme animal.*
Philadelphia and Paris: P. J. Duplain, 1785, 39 pp.

[H]

150. *Requête burlesque et arrêt de la cour du parlement,* *concernant la suppression du magnétisme animal.*
N.p.: n.p., 1785, 21 pp.

[H]

151. *Société de l'harmonie de Guienne.*
Recueil d'observations et des faits relatifs au magnétisme animal présenté à l'auteur de cette découverte et publié par la Société de l'harmonie de Guienne. Paris: n.p., 1785, 168 pp.

[H]

152. [*Tardy de Montravel, A. A.*].
Essai sur la théorie du somnambulisme magnétique. London: n.p., 1785, 108 pp.

The first treatise to attempt to present a comprehensive theory of magnetic somnambulism. It was published in 1785 shortly after the essay of Fournel (*see* entry number 138), which Tardy de Montravel knew and appreciated. Like Fournel, he notes that since its modest beginnings at Buzancy in the previous year, the phenomena of magnetic somnambulism could now be found in Paris, Strasbourg, and throughout all the provinces of France. Tardy de Montravel had observed many somnambulists, but he bases his newly formulated theory of magnetic somnambulism chiefly on experiments he conducted with a certain Mademoiselle N., the first person he seriously attempted to magnetize. He found that she seemed to possess certain extraordinary abilities already noted in magnetic somnambulists by Puységur and others. These qualities included the ability to diagnose her own illness and those of others and the ability to perceive clairvoyantly. He also states that his Mademoiselle N. could directly see the magnetic fluid and describe its colors. Tardy de Montravel believed the somnambulist could exercise these powers because of a "sixth sense" that is activated in the somnambulistic state. He held that this sixth sense proceeded from the stomach area and that somnambulists could see and hear with their stomachs. In this he anticipated the writings of Petetin who made the same discovery in his work with hysterics (*see* Petetin, *Mémoire,* entry number 171). Tardy de Montravel describes the special connection between magnetizer and somnambulist as a "harmonic rapport," and compares it to "platonic love." He emphasizes, however, that the relationship between them is completely moral, and that, should the magnetizer attempt anything improper, the somnambulist would immediately awaken. This work is one of the most important and influential early writings on magnetic somnambulism, being cited in nearly all treatises on the subject written before 1800. [H]

153. ***Thomas d'Onglée, François Louis.***
Rapport au public de quelques abus en médecine; avec des réflexions & notes historiques, critiques & médicales. Paris: Herissant, 1785, (3) + 169 pp.

[H]

154. ***Thouret, Michel Augustin.***
Extrait de la correspondance de la Société royale de médicine, relativement au magnétisme animal. Paris: Imprimerie royale, 1785, 74 pp.

[H]

155. ***Valleton de Boissière, ———.***
Lettre de M. Valleton de Boissière, médecin à Bergerac, à M. Thouret, médecin à Paris, pour servir de réfutation à l'Extrait de la correspondance de la Société Royale de Médecine, relativement au magnétisme animal. Philadelphia: n.p., 1785, 240 pp.

[H]

1786

156. ***Annales de la société harmonique des amis réunis de** Strasbourg, ou Cures des membres de cette société ont opérées par le magnétisme animal.*
Vols. 1–3; 1786–1789.

[H]

157. ***Archiv für Magnetismus und Somnambulismus.***
Vols. 1–2; 1786–1788.

This journal originated in Strasbourg and was edited by Johann Lorenz Böckmann. [H]

158. [*Barberin, Chevalier de*].

Système raisonné du magnétisme universel. D'après les principes de M. Mesmer, ouvrage auquel on a joint l'explication des procédés du magnétisme animal accomodés au cures des différentes maladies, tant par M. Mesmer que par M. le Chevalier de Barberin et par M. de Puységur, relativement au somnambulisme ainsi qu'une notice de la constitution des Sociétés dites de l'harmonie . . . Par la Société de l'harmonie d'Ostende. Ostende: n.p., 1786, v + (3) + 133 pp.

A member of the very active Lyon branch of Freemasonry and a friend of the celebrated philosopher Louis Claude de Saint-Martin (1743–1803), the Chevalier de Barberin developed a mystical and spiritualist type of animal magnetism that quickly influenced many practitioners in France. The notion of a physical magnetic fluid was de-emphasized and a more psychological—even magical—view of animal magnetic action took its place. The magnetic passes were made without touching the body and there was much emphasis placed on magnetizing at a distance—even at very great distances. The importance of the will was emphasized, and the magnetizer was expected to be in tune with the patient in order sympathetically to diagnose and then heal the person. Barberin rejected as harmful the convulsive crisis so often connected with magnetic cures, placing importance on the gentle crisis, the magnetic somnambulism of Puységur. He also truly believed in the pronouncements of his magnetic somnambulists, both for their usefulness in the healing process and for their spiritual messages. [H]

159. [*La Breteniere, de,* ———].

*Extrait du journal de ce qui s'est passé concernant le somnambulisme magnétique de Mme ***.* (Paris?): n.p., (1786), 28 pp.

A description of cures and other phenomena associated with the somnambulist Madame de La Breteniere. [H]

160. [*Lutzelbourg, comte de*].

*Cures faites par M. Le Cte. de L******. Sindic de la Société de Bienfaisance établie à Strasbourg. . . . Avec des notes sur les crises magnétiques appellées improprement somnambulisme.* (Strasbourg): Lorence & Schouler, 1786, (7) + 8–92 pp.

This work is better known by the title of its second expanded edition: *Extrait des journaux d'un magnétiseur,* (1786). The Comte de Lutzelbourg learned to magnetize directly from Puységur and in this work the similarities in approach are apparent. In this work Lutzelbourg distinguishes four degrees of "magnetic crisis," the fourth being characterized by an inability to feel pain and a completely reliable clairvoyance. Like Puységur, Lutzelbourg shows himself to be very interested in the magnetic subject's emotional state and how that affects the cure. [H]

161. [*Tardy de Montravel, A. A.*].

Journal du traitement magnétique de la demoiselle N. Lequel a servi de base à l'Essai sur la théorie du somnambulisme magnétique. London: n.p., 1786, xxxii + 255 pp.

A detailed description of Tardy de Montravel's first experiences with magnetic somnambulism, which were held with Mademoiselle N., a particularly good somnambulistic subject. The author develops at length his notion of a sixth sense which is brought into operation in the magnetic state. Tardy de Montravel used these experiences as the starting point for his *Essai sur la théorie du somnambulisme magnétique* (entry number 152). [H]

162. [*Tardy de Montravel, A. A.*].

Suite du traitement magnétique de la demoiselle N., lequel a servi de base à l'Essai sur la théorie du somnambulisme magnétique. London: n.p., 1786, 206 pp. [H]

1787

163. ***Birnstiel, F. H.***

Gesammelte Acten-Stücke zu Aufdeckung des Geheimnisses des sogenannten thierischen Magnetismus in einigen freundschaftlichen Briefen dem Herrn Ernst Gottfried Baldinger mitgetheilet. Marburg: Neue Academ. Buchhandlung, 1787, 96 pp.

Birnstiel was a well-known professor of medicine at Marburg; Baldinger was a physician. This collection of letters is one of the earliest German works that critically examines the nature of animal magnetism. [H]

164. ***Favrye, Mme. de la.***

Idées de physique ou résumé d'une conversation sur la cause des sensations avec la composition de la poudre de sympathie, ouvrage dédié aux dames de Paris. Paris: Gastellier, 1787, 111 pp.

In support of animal magnetism and one of the few works written by a woman in the early years of its history. [H]

165. ***Gmelin, Eberhard.***

Über thierischen Magnetismus. In einem Brief an Herrn Geheimer Rath Hoffmann in Mainz. 2 parts in one vol. Tübingen: Heerbrandt, x + 134 + 247 pp.

Gmelin, a physician, was a member of a distinguished family of intellectuals at Tübingen. Intrigued by a paper on animal magnetism written by

Hoffman, Gmelin decided to experiment with this potential source of healing on his own patients. In this work he presents detailed case histories of his magnetic treatments and draws preliminary conclusions about the nature of animal magnetism. [H]

166. ***Lettre à Madame la Comtesse de L . . . contenant une*** *observation magnétique faite par une somnambule sur un enfant de six mois.*
Besançon: n.p., 1787, v + (7)–16 pp.

[H]

167. ***Lutzelbourg, Comte de.***
Extrait du journal d'une cure magnétique. Traduit de l'allemand. Rastadt: J. W. Dorner, 1787, 14 pp.

[H]

168. ***Magnetische Magazin für Nieder-Deutschland.***
Vols. 1–8; 1787–1790.

Published in Bremen and edited by Arnold Wienholt. [H]

169. **[*Meltier,* ———].**
Lettre adressée à M. le Marquis de Puységur sur une observation faite à la lune, précédée d'un système nouveau sur le mécanisme de la vue. Amsterdam: n.p., 1787, 84 pp.

The letter is written as a farce. [H]

170. **[*Mouilleseaux, de* ———].**
Appel au public sur le magnétisme animal, ou Projet d'un journal pour le seul avantage du public, et dont il serait le coopérateur. Strasbourg: n.p., 1787, 100 pp.

Mouilleseaux proposes the establishment of a journal that will publish articles on the systematic and scientific study of animal magnetism (a journal that never came into being). In the process of making his proposal, the author gives an informative picture of the present state of affairs with regard to animal magnetism. He has a note on the phenomena of magnetic somnambulism that is one of the best summaries of that subject of the time. In this note Mouilleseaux makes what seems to be the very first published reference to the phenomenon that would much later be called "post-hypnotic suggestion." He writes that somnambulists are sometimes observed "to execute, in their natural [waking] state, the will of their magnetizer; they do this through an irresistible compulsion without being able to give a reason for their action and without realizing that the intimation of this will was made while they were in crisis [magnetic somnambulism]—and the effect of this will might occur right away or many days later" (p. 83). [H]

171. *Petetin, Jacques Henri Desiré.*
Mémoire sur la découverte des phénomènes que présentent la catalepsie et le somnambulisme, symptomes de l'affection hystérique essentielle, avec des recherches sur la cause physique des ces phénomènes. Premiere partie. Mémoire sur la découverte des phénomènes de l'affection hystérique essentielle, et sur la méthode curative de cette maladie. Second partie. (Lyon?): n.p., 1787, 62 + 126 + (1) pp.

Petetin describes a number of cases of hysteria that he treated through the induction of magnetic somnambulism. He believed that certain hysterics spontaneously enter somnambulistic states and that magnetic somnambulism could be better understood through the experiences of these patients. He, like Tardy de Montravel before him, believed that somnambulists could see and hear from the stomach area. [H]

172. [*Tardy de Montravel, A. A.*].
Journal du traitement magnétique de Madame B. . . . pour servir de suite au Journal du traitement magnétique de la Dlle. N. . . . & de preuve à la théorie de l'Essai. Strasbourg: Librairie Académique, 1787, 279 pp.

[H]

173. [*Tardy de Montravel, A. A.*].
Lettres pour servir de suite à l'essai sur la théorie du somnambulisme magnétique. London: n.p., 1787, (4) + 65 + (1) pp.

Two years after the publication of his *Essai sur la théorie du somnambulisme magnétique* (entry number 152), Tardy de Montravel wrote these letters in answer to criticisms of that earlier work. The letters also contain comments on certain issues raised in his *Journal de la traitement magnétique de Demoiselle N.* (entry number 161) and *Journal de traitement magnétique de Madame B.* (entry number 172). These letters have become rather rare. [H]

174. [*Ulrich, A.*].
Der Beobachter des thierischen Magnetismus und des Somnambulismus. Strasbourg: Lorenz and Schuler, 1787, (12) + 243 pp.

The author writes with the purpose of bringing a balance to the controversy for and against animal magnetism. He says that he has himself observed errors of judgment and unfounded conclusions reached by the supporters of animal magnetism, but this should not serve to lead to the condemnation of that doctrine. What is needed, he says, is a balanced and careful investigation of the facts, not a wholesale dismissal of the phenomenon because of admitted shortcomings in some of its supporters. [H]

175. [*Villers, Charles de.*]
Le magnétiseur amoureux, par un membre de la société harmonique du régiment de Metz. Geneva (Besançon): n.p., 1787, viii + 229 pp.

Villers was a friend and aide-de-camp to the Marquis de Puységur and was strongly influenced by Puységur's approach to animal magnetism. Villers was a member of the society of harmony of the Metz artillery regiment associated with that of the Marquis at Strasbourg. His highly philosophical theory of animal magnetism was clearly influenced by that of the Lyon school of the Chevalier de Barberin. This work is a novel, but is as much a theoretical treatise on animal magnetism as it is a work of fiction. An important writing in the history of animal magnetism and psychotherapy, it develops a definite psychology of the relationship between magnetizer and magnetized. The title has a significant nuance of meaning, referring both to the magnetizer-hero, Valcourt, who is in love with the daughter of the household in which the novel is set, and to the central theme of the book, which discusses the affection and "cordiality" that the magnetizer must exercise toward his patient if he wants to cure effectively. Villers does not seem to believe in the existence of a magnetic fluid—at least he does not place any importance upon a physical agent in the action of animal magnetism. Rather he sees animal magnetism as the work of the soul, a spiritual entity, which makes use of the will to bring about the desired curative effects. Villers mentions two different kinds of suggestion at work in the relationship between magnetizer and patient: the first being a kind of "identification" that is established between their souls, and the second an "inspiration" that follows from expectations about the effects which are supposed to take place. Villers, more than any other writer of the age, emphasized the psychological interaction of the magnetic relationship and insisted that there must be a "moral affection" on the part of the magnetizer in order to heal. He believed that patients must put their trust in the magnetizer, opening themselves completely to his influence, and he in turn must exercise a familial benevolence towards them. This emphasis on the emotional interaction of successful magnetic treatment was an important step towards establishing principles of a magnetic psychotherapy which only much later (in the work of Liébeault, Janet, Brever, Freud, etc.) would come to fruition.

This work has a curious publication history. Today only one copy of the first edition is extant. Villers had sent some copies to his friends, including the Marquis de Puységur, but the rest were destroyed by the minister of police. In 1825 Puységur, presumably using the copy given him by Villers, decided to issue a new edition. As he states in his preface, he made some changes, but left the work principally intact. The new version was published in two-volume form (Paris: Dentu, (4) + 296; (4) + 281 pp.). However, Puységur, for some unknown reason (perhaps he feared the title could give animal magnetism a bad name), ordered this entire new edition to be destroyed. In any case the second edition is extremely rare. [H]

176. *Wienholt, Arnold.*

Beitrag zu den Erfahrungen über den thierischen Magnetismus. Hamburg: n.p., 1787, 80 pp.

[H]

177. *Würtz, Georg Christophe.*

Prospectus d'un nouveau cours théorique et pratique de magnétisme animal, réduit à des principes simple de physique, de chymie, et de médecine. Dans lequel on démontrera le système de M. Mesmer, et ses procédés; on rectifiera quelques unes de ses erreurs; on analysera la cause et le mécanisme par le quel les differents effets magnétiques sont produits; on prouvera enfin l'analogie qu'ils ont avec beaucoup d'autres effets naturels, et pourquoi ils ne présentent rien d'opposé aux connaissances que nous avions jusqu'ici de l'économie animale. Strasbourg: Treuttel, 1787, 54 pp.

Würtz is a strong supporter of orthodox animal magnetism as taught by Mesmer. He compares Mesmer to Galileo and Harvey whose important discoveries were rejected by their contemporaries. Würtz says that Mesmer, recognizing a power in nature noted by some of the ancients but not understood by them, was able to lift this healing technique out of the mire of superstition and into the realm of science. He criticizes the views of D'Eslon and the condemnation of the Franklin commission, stating that if a proper investigation had been made, Mesmer would have been vindicated. He then sets out a proposal for a course of study of animal magnetism, highlighted by an investigation of the physics, chemistry and metaphysics of magnetism. The proposed course would also deal with the practical application of animal magnetism to healing and with procedures to be used for different cases. Würtz insists that, for the most effective use of magnetic healing, the practitioner must combine a grasp of these techniques with a knowledge of conventional medicine. Within this framework he suggests an examination of the effects of magnetic somnambulism, particularly questioning the accuracy of the medical pronouncements of somnambulists. In discussing this issue, Würtz states that Mesmer knew of magnetic somnambulism in his practice and recognized the reality of a special sixth sense that can operate in a magnetized person, but Mesmer and his followers were very cautious about this faculty. [H]

1788

178. *Bell, John.*

An Essay on Somnambulism, or Sleep-walking, produced by Animal Electricity and Magnetism. As well as by Sympathy, &c. Dublin: The Author, 1788, 38 pp.

In the preface to this work, Bell describes his initiation into the study of animal magnetism by "Father Harvier" at the Augustin Convent in Paris. Soon after that, Bell met the Marquis de Puységur and learned about magnetic somnambulism. Bell states that he then developed his own practice of animal magnetism, and it is from these experiences that he writes the present treatise. Bell finds magnetic somnambulism the most interesting aspect of animal magnetism because the phenomena it produces are so striking. He writes of the sixth sense that comes into play in the somnambulist and the rapport or "analogy" that exists between magnetist and magnetized. Bell's treatment of the subject is fairly thorough, a much more readable essay than his first endeavor in the field (see *Animal Electricity and Magnetism*, entry number 120). This is the first lengthy discussion of magnetic somnambulism written in English. [H]

179. *Inchbald, Elizabeth Simpson.*

Animal Magnetism: A Farce, in Three Acts, as Performed at the Theatre Royal, Covent-Garden. Dublin: P. Byron, (1788), 36 pp.

This play was also published in New York in 1809 and performed there in the New York Theatre. [H]

180. *Levade, Louis, Reynier, Jean François and Berchem, Jacob Pierre Berthout van.*

Rapport fait à la Société des Sciences Physiques de Lausanne, sur un somnambule naturel. Lausanne: Henri Vincent, 1788, 61 pp.

[H]

181. *Lo-Looz, Robert de.*

Recherches physiques et métaphysiques sur les influences célestes, sur le magnétisme universel et sur le magnétisme animal dont on trouve la pratique de temps immémorial chez les Chinois. London and Paris: Couturier, 1788, 148 pp.

Lo-Looz was a Belgian physician who was also a spiritual philosopher. This book devotes about sixty pages to animal magnetism, a phenomenon which the author claims to have known about before Mesmer. Lo-Looz also finds hints in the writings of the Chinese that they had long known about animal magnetism. [H]

182. *Lutzelbourg, comte de.*

Dieu, l'homme et la nature. Tableau philosophique d'une somnambule. (London): n.p., 1788, 36 pp.

A treatise containing pronouncements on spiritual matters by a magnetic somnambulist, along with prescriptions for how to apply animal magnetism. [H]

183. *Lutzelbourg, comte de.*

Nouveaux extraits des journaux d'un magnétiseur depuis 1786 jusqu'au mois d'avril 1788. Strasbourg: n.p., 1788, 99 pp.

The continuation of Lutzelbourg's unsigned *Cures faites par M. Le Cte. de L . . .* of 1786 (*see* entry number 160). [H]

184. *Meiners, Christoph.*

Über den thierischen Magnetismus. Lemgo: Werner, 1788, (8) + 340 + (4) pp.

One of the earliest German treatises on animal magnetism. Acknowledging that more often than not animal magnetism has been a subject of ridicule, Meiners undertakes to present enough information about it to convince the reader that it is a respectable subject of inquiry. He concentrates on animal magnetism as a healing art and describes the method to be used in magnetizing and the marks that characterize the magnetized state, emphasizing that this state has a remarkable power to bring about healing in the physical organism. In the process of examining this healing power, Meiners describes in detail a number of interesting case histories. [H]

185. *Rosenmüller, Johann Georg.*

Briefe über die Phänomene des thierischen Magnetismus und Somnambulismus. Leipzig: G. J. Goschen, 1788, 106 pp.

This treatise, better known in the French translation (*Lettre à la Société exéqétique et philanthropique de Stockholm . . .*), was written by the author at the request of the Exegetic and Philanthropic Society of Stockholm. That society had issued a publication in the form of a letter entitled *Lettre sur la seule explication satisfaisante des phénomènes du magnétisme animal et du somnambulisme . . .* in the same year addressed to the Société des amis réunis de Strasbourg, and had sent a copy to Rosenmüller. His comments are highly critical of the position taken by members of the Swedish society. He objects to the basically religious orientation of their explanation of the phenomena of animal magnetism, their belief that supernatural and spirit forces are at work. Rosenmüller believes that a purely natural explanation for the phenomena is sufficient and best. Where the society attributes all sickness to the action of evil spirits and all cure to the removal of those spirits, Rosenmüller says the modern physicians must look for natural and physical causes for ill health. This even-handed and well-written work is one of the best discussions of the problems of occult interpretations of magnetic phenomena to appear before 1800. [H]

186. *Société Exégétique & Philantropique, Stockholm.*

Lettre sur la seule explication satisfaisante des phénomènes du magnétisme animal et du somnambulisme déduite des vrais principes fondés dans la connaissance du créateur, de l'homme, et de la nature, et confirmée par l'expérience. Stockholm: L'imprimerie Royal, 1788, 40 pp.

The Exegetic and Philanthropic Society of Stockholm was founded to study and promote the teachings of Emmanuel Swedenborg (1688–1772), a visionary and intellectual of great influence. The theory of animal magnetism, and especially the experiences connected with magnetic somnambulism, were very attractive to this society and it incorporated them into its world view. This treatise, addressed to the Société des amis réunis de Strasbourg, attempts to formulate the results of this amalgamation. It is not surprising that the Swedish society found the writings of the Strasbourg society, oriented as they were to the psychological and moral aspects of magnetic somnambulism, congenial to its philosophical framework. The Stockholm society believed that supernatural and spirit forces were at work in the creation of disease and so must be involved in the cure. The Stockholm Society attempted to show that this Swedenborgian view of illness was the only reasonable way to explain the phenomena of animal magnetism and magnetic somnambulism. This letter was commented on by Rosenmüller in the previous entry (*see* entry number 185). [H]

187. ***Usteri, Paulus.***

Specimen bibliothecae criticae magnetismi sic dicti animalis. Göttingen: Joann. Christ. Dieterich, 1788, 44 pp.

Written to obtain a degree in medicine and surgery, this is an interesting but very abbreviated list of works on animal magnetism. It cites the title of the work, place of publication, and number of pages. Its greatest value is that it lists periodical articles on animal magnetism and also gives locations of periodical reviews of the books mentioned. [H]

1789

188. ***Edwards, D.***

Treatise on Animal Magnetism; Discovering the Method of Making the Said Magnets, for the Cure of Most Diseases Incident to the Human Body. From the Writing of Paracelsus, Tentzelius, Fludd, Boulton, &c. London: Wagstaff, 1789, 17 pp.

[H]

189. ***Pratt, Mary.***

A List of a Few Cures Performed by Mr. and Mrs. De Loutherbourg of Hammersmith Terrace, Without Medicine. London: J. P. Cooke, for The Author, 1789, 9 pp.

[H]

190. ***Rahn, Johann Heinrich.***

Über Sympathie und Magnetismus. Aus den Lateinischen übersazt und mit anmerkungen begleitet von Heinrich Tabor. Heidelberg: F. L. Pfahler, 1789, 272 pp.

A doctoral thesis presented in 1786, with a translator's preface written in 1788. "Sympathy" in this treatise basically refers to the connection between body and soul. Rahn states that he has studied the writings on animal magnetism and has not discovered anything essentially new. Rather he has found a revision of old opinions about a universal world spirit that also exists in the human body. This universal world spirit not only produces the general mutual influence between all bodies, heavenly and earthly, but also produces the special sympathy between one man and another, and is, in the last analysis, the bond between body and soul. Rahn notes there is a remarkable connection between individuals who are magnetized at the same time and he relates this connection to the old, well-known notion of natural sympathy. But he says that people must be willing to learn what new things the magnetists have to teach them, and, with the idea of an animal magnetic material, perhaps something novel has been added to traditional knowledge of the phenomena. [H]

1790

191. ***Martin, John.***
Animal Magnetism Examined: in a Letter to a Country Gentleman. London: Stockdale, 1790, 70 pp.

One of the earliest British works on animal magnetism. Martin believed magnetizers to be mere hustlers drumming up business with empty promises of cure. [H]

192. **[*Pearson, John*].**
A Plain and Rational Account of the Nature and Effects of Animal Magnetism: in a Series of Letters. With Notes and Appendix by the Editor. London: W. and J. Stratford, 1790, 51 pp.

The author casts a skeptical eye on the purported effectiveness of animal magnetism in curing illness. [H]

193. ***A Practical Display of the Philosophical System called*** *Animal Magnetism, in Which is Explained Different Modes of Treating Diseases. . . .*
London: n.p., 1790, 16 p.

This early British treatise reflects the views of Puységur, stating that the effects of animal magnetism can be produced through both the use of the hands and the exercise of the will. The author emphasizes the importance of a good will for effective treatment. [H]

194. ***Sauviac, Joseph Alexandre Betbezé Larue de.***
Recherches physiques sur le magnétisme; insérées dans le Journal des Savans, en l'année 1790. N.p.: n.p., 1790, 26 pp.

An attempt to look at the questions of animal magnetism and animal electricity from a physiological point of view. The author tends to downplay the importance of the effects of animal magnetism, although he admits it does exist, and to emphasize the importance of animal electricity. [H]

1791

195. *Absonus, Valentine.*
Animal Magnetism. A Ballad, with Explanatory Notes and Observations: Containing Several Curious Anecdotes of Animal Magnetisers, Ancient as well as Modern. London: n.p., 1791, 44 pp.

[H]

196. *Eckhartshausen, Karl von.*
Verschiedenes zum Unterricht und zur Unterhaltung für Liebhaber der Gaukeltasche, des Magnetismus, und anderer Seltenheiten. Gesammelt und herausgegeben von dem Hofrath von Eckhartshausen. Munich: Joseph Lindauer, 1791, (xvi) + xxxvi + 345 pp.

[H]

197. *Marat, Jean Paul.*
Les charlatans modernes; ou, Lettres sur le charlatanisme académique. (Paris): Marat, 1791, 40 pp.

Mesmer is included among Marat's charlatans. [H]

198. *Stearns, Samuel.*
The Mystery of Animal Magnetism Revealed to the World, Containing Philosophical Reflections on the Publication of a Pamphlet Entitled, A True and Genuine Discovery of Animal Electricity and Magnetism: also, an Exhibition of the Advantages and Disadvantages that may Arise in Consequence of Said Publication. London: M. R. Parsons, 1791, (2) + 58 pp.

[H]

199. *Voltelen, Floris Jacobus.*
Florentii Jacobi Voltelen Oratio de magnetismo animali: publice habita Lugduni Batavorum die VIII. Februarii a. CDDCCLXXXXI. quum magistratum academicum solemniter deponeret. Lugduni Batavorum (?): n.p., 1791, (10) + 45 pp.

[H]

200. ***Wonders and Mysteries of Animal Magnetism Displayed,*** *or the History, Art, Practice, and Progress of that Useful Science, from Its First Rise in the City of Paris, to the Present Time. With Several Curious Cases and New Anecdotes of the Principal Professors.*
London: J. Sudbury, 1791, 35 pp.

[H]

1792

201. ***Bell, John.***
The General and Particular Principles of Animal Electricity and Magnetism, &c. in Which Are Found Dr. Bell's Secrets and Practice, As Delivered to His Pupils in Paris, London, Dublin, Bristol, Glocester, Worcester, Birmingham, Wolverhampton, Shrewsbury, Chester, Liverpool, Manchester, &c. &c. Shewing How To Magnetise and Cure Different Diseases; to Produce Crises, as well as Somnambulism, or Sleepwalking; and in That State of Sleep to Make a Person Eat, Drink, Walk, Sing and Play Upon Any Instruments They Are Used To, &c. To Make Apparatus and Other Accessaries To Produce Magnetical Facts; Also To Magnetise Rivers, Rooms, Trees . . . (London): The Author, 1792, vi + (7)–80 pp.

The most influential of the early British works on animal magnetism. Bell was trained in animal magnetism at the Paris Society of Harmony and at the beginning of the book includes a reproduction of a certificate of fellowship signed by Bergasse, Kornman, and others. It is clear that Bell was also strongly influenced by the work of Puységur, for he devotes a great deal of space to the subject of magnetic sleep, something that would not have been touched on in Paris. Bell uses his own terminology to describe the phases of animal magnetism. He distinguishes between the "strong crisis," which involves convulsions, and the "gentle crisis," which involves sleep. Like Puységur, Bell speaks of a "sixth sense" by which the somnambulist may diagnose disease and predict its course. He also emphasizes the importance of the will in the process of magnetization and even mentions the ability to will an absent person into the magnetic state. Bell used this book as a sort of written manual to supplement lectures on the subject of animal magnetism that he delivered throughout Britain. [H]

202. ***Litta Biumi Resta, Carlo Matteo.***
Riflessioni sul magnetismo animale; fatte ad oggetto di illuminare i suio cittadini aveudolo trovato salutare in molti mali. (Milan?): Constante Cordialita, 1792, 234 pp.

[H]

1795

203. *Sibly, Ebenezer.*
A Key to Physic, and the Occult Sciences. (London): The Author, 1795, (4) + 395 + (1) pp.

A handbook for health practice that combines occult procedures with medical electricity and animal magnetism. The book contains some remarkable plates depicting these practices. [H]

1796

204. *Perkins, Benjamin Douglas.*
The Influence of Metallic Tractors on the Human Body, in Removing Various Painful Inflamatory Diseases, etc., by which a New Field of Enquiry is Opened in the Modern Science of Galvanism or Animal Electricity. London: J. Johnson and Ogilvy and Son, 1796, xi + 99 + (1) pp.

Benjamin Perkins was the son of the American doctor Elisha Perkins (1741 – 1799) who developed a healing technique involving the use of metallic tractors. The approach in some way resembled that used by the practitioners of animal magnetism; partly because of that, it received a hostile reception from the American medical establishment. Benjamin took up the cause, promoting the tractors in the United States and England. He received a warmer reception in England than in America, and so he published most of his writings on "perkinism" in Great Britain. [H]

205. *Perkins, Elisha.*
Certificates of the Efficacy of Dr. Perkins's Patent Metallic Instruments. New London, Connecticut: S. Green's Press, 1796, 16 pp.

Elisha Perkins was an American physician who developed a healing technique involving the use of metallic tractors. The approach in some way resembled that used by the practitioners of animal magnetism. "Perkinism" received a hostile reception from the American medical establishment, who condemned this "stroking" technique as useless and therefore dangerous. Animal magnetism itself had made little headway in the United States at this time, having been condemned in France by Benjamin Franklin's commission and spoken poorly of by Jefferson. Only George Washington, because of the request of his friend Lafayette, gave it any kind of a hearing through a brief correspondance with Mesmer. Perkins too seems to have gotten Washington's ear, since he sold him a set of his tractors. These devices were made of dissimilar metals, gold and silver, and were approximately three inches in length. Held in the doctor's hands, they were used to gently stroke the body in the direction of the heart. Although Mesmerism was known to the American medical world at the time when Perkins developed his tractors, it is not clear to what extent he was influenced by Mesmer's teaching, for he says little about the theory of their operation. His son Benjamin was more vocal about theoretical issues (*see* his *The Influence of Metallic Tractors* . . . , entry number 204). This collection of certificates simply testifies to the efficacy of the tractors. [H]

1797

206. ***Perkins, Elisha.***
Evidences of the Efficacy of Doctor Perkins's Patent Metallic Instruments. New London, Connecticut: S. Green's Press, 1797, 32 pp.

[H]
See Elisha Perkins, *Certificates of The Efficacy* . . . (entry number 205).

207. ***Vaughan, John.***
Observations of Animal Electricity. In Explanation of the Metallic Operation of Dr. Perkins. Wilmington: W. C. Smyth, 1797, 32 pp.

An attempt to explain the efficacy of the metallic tractors of Elisha Perkins (*see* the works of Elisha and Benjamin Perkins) in terms of animal electricity. [H]

1798

208. ***De Mainauduc, John Benoit.***
The Lectures of J. B. de Mainauduc, M. D. Part the First. London: Printed for the Executrix, 1798, xii + (4) + 230 pp.

De Mainauduc received his medical training and set up his medical practice in London. In 1782 he moved to Paris to become "Quarterly Physician to the King of France." There he studied for a time with Charles D'Eslon. In 1785 he returned to London and soon began to teach a "Science" of healing without the use of conventional medical means. In the *Lectures*, he says that this Science of healing is not new, but has been practiced by physicians for two hundred fifty years. But, he insists, it had not been recognized for the science it is because it was "enveloped with ridiculous nostrums or machinery" (p. viii). So De Mainauduc claims for himself "new discoveries" about this Science, consisting of his particular philosophy and techniques. His "Science" is, in fact, a type of spiritual healing (based upon the action of the spirit, mind, and will) presented in the framework of a rather simplistic atomistic physiology: "Thus it appears that every human Being possesses the power of striking other forms with the particles which are flying off from his own body; or, to state it in its proper light, we must say, that it is the prerogative of Spirit in many, by vigorous exertions, to propel the atoms of its own body against and through the pores of any other form in nature" (p. 111). De Mainauduc writes about "invisible fingers" that can manipulate the interior of the patient's body: "The Operator's own emanations, when duly influenced, become for the Operator, invisible fingers, which penetrate the pores, and may be truly considered as the natural and only ingredients which are or can be adapted to the removal of nervous, or of any other affections of the body" (pp. 120–121). Making liberal use of quotations from the Bible, De Mainauduc paints his discovery of the principles of the new Science as God's gift to medicine. He bemoans the false philosophies that have over the ages become associated with this healing power. Chief among those philosophies was that of Mesmer, who "pillaged the subject from Sir Robert Fludd, and found to a certainty the existence of the power; undisposed to attend to our Saviour's information, he preferred loadstones and magnetic ideas to the service of the Great Author, and after performing several accidental cures, his magnetism and his errors shared the fate of his predecessors" (pp. 224–225). [H]

209. ***Perkins, Benjamin Douglas.***
The Influence of Metallic Tractors on the Human Body, in Removing Various Painful Inflammatory Diseases . . . Lately Discovered by Dr. Perkins . . . and Demonstrated in a Series of Experiments and Observations, by Professors Meigs, Woodward, Rogers, &c, &c. by which the Importance of the Discovery is Fully Ascertained, and a New Field of Enquiry Opened in the Modern Science of Galvanism, or, Animal Electricity. . . . London: Printed for J. Johnson, and Ogilvy and Son., 1798, xi + 99 + (1) pp.

See earlier works by Elisha Perkins and his son (Entries 204–206). [H]

1799

210. ***Herholdt, Johan Daniel.***
Experiments with the Metallic Tractors in Rheumatic and Gouty Affections, Inflammations, and Various Topical Diseases, as Pub. by Surgeons Herholdt and Rafn . . . tr. into German by Professor Tode . . . thence into the English Language by Mr. Charles Kampfmuller: also Reports of about One Hundred and Fifty Cases in England . . . By medical and Other Respectable Charac-

ters. Ed. by Benjamin Douglas Perkins. London: Printed by L. Hanfard, for J. Johnson, 1799, (1) + (vii)–xxiv + 355 + (3) pp.

The editor is the son of Elisha Perkins who invented the metallic tractors described here. (*See* entries 204–206). [H]

211. *Mesmer, Franz Anton.*

Mémoire de F. A. Mesmer, docteur en médecine, sur ses découvertes. Paris: Fuchs, 1799, xii + 110 pp. ENGLISH: "Dissertation by F. A. Mesmer, Doctor of Medicine, on His Discoveries," in *Mesmerism.* Translated and edited by George Bloch. Los Altos, California: William Kaufmann, 1980.

One of Mesmer's most important works. Writing twenty years after his history-making *Mémoire* of 1779 (entry number 10), Mesmer complains that his discovery, animal magnetism, has had great difficulty in attaining acceptance in the medical world. He hopes that the present treatise will correct mistaken notions and clarify the relationship of animal magnetism to the principles of physics. He also undertakes to explain the phenomenon of somnambulism and its place in the practice of animal magnetism. Although Mesmer never mentions it, the reader is aware that the work of the Marquis de Puységur is the backdrop against which Mesmer's comments on this subject are made. With Puységur's discovery of "magnetic somnambulism," Mesmer was confronted with a problem: how could he acknowledge the reality of this phenomenon without giving it too prominent a place in his system? One of the main goals of this *Mémoire* was to attempt to solve that problem. In his description of the physics of animal magnetism, Mesmer reasserts his mechanical view of what takes place. If there had been any question that Mesmer's theory was occultist in orientation, that is clearly answered in the negative here. Everything is explained in terms of matter and motion. The magnetic fluid is composed of infinitesimally small particles that move in streams. Health and disease are the result of the free or blocked flow of those streams. In the process of explaining this system, Mesmer makes a point of distinguishing it from "animal electricity," a distinction that a number of authors had failed to make. Taking up the subject of somnambulism, Mesmer states that it is a phenomenon known throughout history. He gives it his own particular interpretation, however, calling it "critical sleep" ("critical" referring to "crisis," a special physical state of the body as it attempts to deal with illness). He says that all the phenomena connected with "critical sleep," such as knowledge of things at a distance and the ability to diagnose disease (see the works of Puységur) are to be explained in terms of the actions of the magnetic fluid, and therefore in terms of a purely mechanical physical system. He insists that traditional occult interpretations of these things are mistaken and disclaims any connection with those of his followers who have gone in that direction. Mesmer admits that somnambulism or "critical sleep" is quite likely to occur during the application of animal magnetism, and he explains that fact in terms of the "workings of all the machine's systems" ("machine" meaning the body). In this *Mémoire*, Mesmer accomplishes his chief aim: to expunge from the interpretation of his teaching any hint of occultist views. [H]

212. ***Perkins, Benjamin Douglas.***

Reports of about One Hundred and Fifty Cases in England Demonstrating the Efficacy of the Metallic Practice in a Variety of Complaints Both upon the Human Body and on Horses. London: J. Johnson, 1799.

Another of the younger Perkins's books on the metallic tractors of Elisha Perkins. [H]

1800–1809

1800

213. ***Bailly, Jean Sylvain.***
"Rapport secret sur le mesmérisme." In *Le conservateur . . . de N. François (de Neufchateau)* 1: 146–155.

The first public appearance of the secret report drawn up by the Franklin commission on animal magnetism in 1784 (*Rapport secret présenté au ministre et signé par la commission précédente*, entry number 32). It was meant for the king's eyes only, due to the delicate nature of the subject matter. In it the commissioners express their concern about potential misuses of animal magnetism, particularly the possible sexual arousal of female subjects by male magnetizers. Although the report contains some false information (that it is always men who magnetize women), it does give an accurate picture of the procedures: the contact between the knees of the magnetizer and those of the patient; the placement of the hand on the stomach or abdominal area; the reaching of one hand behind the patient's body; and the general physical proximity and attunement involved in the process of magnetization. The commission states that the patient does not simply regard the magnetizer as a physican; he is, after all, still a man. The report also points out that often the female subject experiences an ecstasy of sorts when in the magnetic crisis, a buildup of emotions which is followed by a languor and a kind of sleep of the senses. The emphasis of the commission is on not only the danger of overt sexual acts performed by the magnetizer, but also the fact that the process may well awaken sexual passions latent in the female patient which she will then seek to fulfill in fornication or adultery. [H]

214. ***Haygarth, John.***

Of the Imagination, as a Cause and as a Cure of Disorders of the Body: Exemplified by Fictitious Tractors, and Epidemic Convulsions. Bath: R. Cruttwell, 1800, iii + 43 pp.

Haygarth, a physician of Bath, and a friend, William Falconer, undertook to test the efficacy of the metallic tractors of Elisha Perkins (*see* entry numbers 204–206) in treating ill patients. They used wooden tractors painted to look like the metallic ones and discovered that the wooden ones were just as effective as the metallic ones. The experiments are described in this book. Haygarth's conclusion is that the real healing force at work is the imagination of the patient, and that this remarkable power of the imagination needs further serious study. [H]

215. ***Heineken, J.***

Ideen und Beobachtungen den thierischen Magnetismus und dessen Anwendung betreffend. Bremen: Friedrich Wilmans, 1800, x + (2) + 231 + (2) pp.

[H]

216. ***Mesmer, Franz Anton.***

Lettre de F. A. Mesmer, docteur en médecine, sur l'origine de la petite vérole et le moyen de la faire cesser, suivie d'une autre lettre du même adressé aux auteurs du Journal de Paris, contenant diverses opinions relatives au système de l'auteur sur le magnétisme animal. Paris: Impr. des Sciences et Arts, 1800, 17 pp.

[H]

217. ***Perkins, Benjamin Douglas.***

The Efficacy of Perkins' Patent Metallic Tractors, in Topical Diseases of the Human Body and Animals; Exemplified by 250 Cases from the First Literary Characters in Europe and America. To Which is Prefixed A Preliminary Discourse in Which the Fallacious Attempts of Dr Haygarth to Detract from the Merits of the Tractors, are Detected, and Fully Confuted. London, Bath and Edinburgh: J. Johnson, Cadell and Davies, etc., 1800, vi + 135 pp.

A description and defense of the use of the metallic tractors of Elisha Perkins (1741–1799) written by his son. [H]

1801

218. ***Perkins, Benjamin Douglas.***

Cases of Successful Practice with Perkins's Patent Metallic Tractors: Communicated since Jan. 1800, the Date of the Former Publication, by Many Scien-

tific Characters. To Which are Prefixed, Prefatory Remarks. . . . London: Cooke, 1801, xvi + 92 pp.

[H]

219. *Wagener, Samuel Christoph.*

Neue Gespenster kurze Erzählungen aus dem Reiche der Wahrheit. 2 vols. Berlin: Friedrich Mauer, 1801, (3) + lxiv + 303; (3) + lxxii + (8) + 288 pp.

A description of supernatural occurrences in the eighteenth century. The work is an important attempt to assemble facts related to paranormal events in an effort to explain them. [P]

220. *Winter, George.*

Animal Magnetism. History of its Origin, Progress and Present State; its Principles and Secrets Displayed, as Delivered by the Late Dr Demainauduc. To which is Added, Dissertations on the Dropsy; Spasms; Epileptic Fits . . . with Upwards of One Hundred Cures and Cases. Also Advice to Those who Visit the Sick . . . a Definition of Sympathy; Antipathy; The Effects of the Imagination on Pregnant Women; Nature; History; and on the Resurrection of the Body. Bristol: Routh, 1801, (vii) + 223 pp.

Winter did not believe in the efficacy of animal magnetism to cure disease; he had tried it himself with no success. He did believe that imagination could play a part in producing results. A section written by De Mainauduc is entitled: "Principles of the Science in Animal Magnetism." [H]

1802

221. *Petetin, Jacques Henri Désiré.*

Nouveau mécanisme de l'électricité fondé sur les lois de l'équilibre et du mouvement, démontré par des expériences qui renversent le système de l'électricité positive et négative et qui etablissent ses rapports avec le mécanisme caché de l'aimant et l'heureuse influence du fluid électrique dans les affections nerveuses. Lyon: Bruyset ainé, iv, xxvii, 1802, 300 pp.

[H]

222. *Wienholt, Arnold.*

Heilkraft des thierischen Magnetismus nach eigenen Beobachtungen. 3 vols. Lemgo: Meyer, 1802–1806, (18) + 3–504; xvi + 637 + (1); (2) + 436 pp.

A massive collection of writings on animal magnetism by one of Germany's earliest practitioners. Wienholt was a physician from Bremen who began his magnetic practice in 1786 when, along with a Dr. Olbers, he healed a nineteen-year-old girl of a severe nervous condition. Wienholt went on to found the *Magnetische Magazin für Niederteutschland*, one of the more important early German periodicals in the field, in 1787. His practice of animal magnetism lasted for more than fifteen years, and this master work spans that whole period. After outlining a brief history of animal magnetism in Germany and describing the details of his own magnetizing process, Wienholt unfolds a panorama of his healing career from 1787. Many of the cases he presents are nervous disorders, ranging from epilepsy to hysterical loss of voice. The third volume, posthumously published, contains a collection of seven lectures on the nature of somnambulism, and seven more lectures on the physics of the living body. The lectures on somnambulism were translated by J. C. Colquhoun (*Seven Lectures on Somnambulism*, Edinburgh: Adam and Charles Black, 1845; *see* entry number 226). [H]

1803

223. *Fiard, Jean Baptiste.*

La France trompée par les magniciens, les démonolâtres et les magnétiseurs du dix-huitième siècle. Paris: Gregoire, 1803, iv + 200 pp.

A work claiming that the practitioners of animal magnetism were subject to demoniacal and occult influences. [H]

1804

224. *Perkinean Society.*

Transactions of the Perkinean Society, Consisting of a Report on the Practice with the Metallic Tractors, at the Institution in Frith-Street, and Experiments Communicated by Several Correspondents. London: A. Topping. (4) + 51 + (2) pp.

Case histories of individuals treated by the metallic tractors of the American doctor Elisha Perkins (*see* entries 204–206). Perkins developed a unique theory of healing that to some degree resembled that of Mesmer, although Perkins's ideas evolved independently. [H]

1805

225. *Petetin, Jacques Henri Désiré.*

Electricité animal, prouvée par la découverte des phénomènes physiques et moraux de la catalepsie hystérique, et de ses variétés; et par les bons effets de l'électricité artificielle dans le traitment de ces maladies. Lyon: Bruyset et Buynand, 1805, xii + 156 pp.

In this work, Petetin supplements the observations made in his first studies of hysteria (*Mémoire sur la découverte des phénomènes que présentent la catalepsie et le somnambulisme* . . . (entry number 171) and describes his treatment of hysteria with electricity. There is a second edition, in two volumes, that appears in 1808. It contains, in addition to the present material, a very long note on Petetin's life and writings, a letter written to Petetin in 1808, and a lecture delivered by Petetin in 1807. [H]

226. *Wienholt, Arnold.*

D. Arnold Wienholt's psychologische Vorlesungen über den naturlichen Somnambulism. Aus den literarischen Nachlass des Verfassers besonders abgedruckt. Lemgo: Mener, 1805, (2) + 144 pp. ENGLISH: *Seven Lectures on Somnambulism.* Translated by J. C. Colquhoun. Edinburgh: Adam and Charles Black, 1845.

Extracted from Wienholt's *Heilkraft des thierischen Magnetismus* of 1802–1804 (*see* entry number 222). [H]

1806

227. [*Vélye, abbé de*].

Du fluid-universel, de son activité et de l'utilité de ses modifications par les substances animales dans le traitement des maladies. Aux étudiants qui suivent les cours de toutes les parties de la physique. Paris: Delance, 1806, xv + 218 + (1) pp.

After two chapters on electrical machines and galvanism, the author deals with the subject of magnetic somnambulism and its power to cure illnesses. He presents cases of his own in which this treatment has been successful. There seems to have been a much shorter version of this work published the same year. [H]

1807

228. ***Puységur, Armand Marie Jacques de Chastenet, marquis de.***
Du magnétisme animal, considéré dans ses rapports avec diverses branches de la physique générale. Paris: Desenne, 1807, ix + 478 pp.

Puységur's most theoretically oriented work on animal magnetism. After stating how he believes one becomes convinced of the reality of animal magnetism, he discusses the "physics" of animal magnetism, placing it in the context of the known phenomena of heat, fire, electricity, light, mineral magnetism, etc. He then outlines a brief history of animal magnetism in France, from Mesmer's time in Paris through his (Puységur's) foundation of the Society of Harmony in Strasbourg, appending a section on the magnetic systems of the "spiritualists" of Lyon. Puységur then describes how to magnetize and states his own views on the nature of magnetic action and somnambulism, including a discussion of clairvoyance and the role of imagination. The final section of the book is a collection of letters written to Puységur by various people on the subject of animal magnetism. Among them is a most interesting letter from the brother of the Marquis, Maxime, who describes procedures to be used when many magnetizers are operating at one time under the leadership of a chief magnetizer who sets the tone for the whole group. He also mentions the necessity for a magnetizer to keep control of individuals in magnetic crisis through the use of his will. [H]

229. **[*Puységur, Armand Marie Jacques de Chastenet, marquis de*].**
Procès verbal du traitement par l'actions magnétique d'une femme malade (par la rupture d'un vaisseau dans la poitrine) de près Soissons. N.p.: n.p., 1807, 39 pp.

[H]

1808

230. ***Judel, ———.***
Considérations sur l'origine, la cause et les effets de la fièvre, sur l'électricité médicale, et sur le magnétisme animal. Paris: Treuttel et Wurtz, 1808, 149 pp.

Judel devotes a part of this work to animal magnetism, which he sees as a useful medical tool. [H]

231. ***Jung-Stilling, Johann Heinrich.***
Theorie der Geister-Kunde, in einer Natur-, Vernunft- und Bibelmässigen Beantwortung der Frage: Was von Ahnungen, Geschichten und Geisterer-

scheinungen geglaubt und nicht geglaubt werden müsse. Nurnberg: Raw, 1808, xxviii + 380 pp.

Born in Westphalia, Germany, Jung-Stilling was raised in a pious Christian home and dreamed of becoming a preacher. Instead he became a physician and eventually professor at the universities of Heidelburg and Marburg. The important *Theorie* attempts to unify the data of animal magnetism and somnambulism within a far reaching spiritualistic philosophy. The author takes up issues such as the state of the soul after death, visions and apparitions, presentiments and prophecy, and other spiritistic matters. He discusses the higher states of somnambulism which make the world of the spirits accessible to those still in the body and thereby anticipates a whole body of literature that would combine these various concerns some forty years later. Jung-Stilling was strongly influenced by Swedenborg but also conducted his own experiments with somnambulism. [H & P]

1809

232. ***Jung-Stilling, Johann Heinrich.***
Apologie der Theorie der Geisterkunde. Nurnberg: Raw, 1809, 77 pp.

A sequel to Jung-Stilling's *Theorie* (entry number 231). Here he writes about the notion of Hades or Scheol. Otherwise, the work contains nothing basically new. [H & P]

1810–1819

1811

233. ***Asklepieion: Allgemeines medicinish-chirurgisches Wochenblatt.***
Vols. 1–2; 1811–1812.

This periodical was published in Berlin and edited by Karl Christian Wolfart (1778–1832). [H]

234. ***Hufeland, Friedrich.***
Über Sympathie. Weimar: Landes-Industrie-Comptoirs, 1811, x + (2) + 228 pp.

An important treatise on the immediate connection or bond, called "sympathy," existing between things in nature. Hufeland analyses the nature of "magnetic rapport" between magnetizer and magnetized in terms of this universal sympathy. He states that magnetizer and somnambulist are so closely bound in a sympathetic relationship that the two form "one individual" (p. 117), and that the soul of the magnetizer is to some extent also the soul of the somnambulist. Sympathy, Hufeland says, causes thoughts and feelings to be held in common by the two, and even muscular action is communicated to the somnambulist by the muscular movement of the magnetizer. [H]

235. ***Kluge, Karl Alexander Ferdinand.***
Versuch einer Darstellung des animalischen Magnetismus, als Heilmittel. Berlin: C. Salfeld, 1811, (5) + iv–xiv + 612 + (2) pp.

One of the most researched and widely read early German works on animal magnetism. In some way Kluge's book could be seen as a bibliographical essay on the subject, with numerous references to writings in the area and related fields. The assistant of Christoph von Hufeland (entry number 234), Kluge carried out magnetic treatments on the patients of his mentor, but his book deals mainly with experiments and theories of other magnetizers. [H]

236. *Poli, Giuseppe Saverio.*
Breve saggio sulla calamita e sulle sue virtù medicinale. Palermo: Reale stamperia, 1811, vi + 73 pp.

A treatise on the medical use of magnets with favorable mention of animal magnetism. [H]

237. *Puységur, Armand Marie Jacques de Chastenet, marquis de.*
Recherches, expériences et observations physiologiques: sur l'homme dans l'état de somnambulisme naturel, et dans le somnambulisme provoqué par l'acte magnétique. Paris: J. Dentu and The Author, 1811, (10) + xii + (13)–430 + (1) pp.

A significant study of natural and magnetic somnambulism by the discoverer of the latter. Puységur believed that one of the main differences between natural and artificial somnambulism was in the matter of "rapport." He states that with magnetic somnambulism, the somnambulist is in "rapport" (or a state of special connection) with the magnetizer. But with natural somnambulism, the somnambulist is in "rapport" with no one. The other difference between the two conditions, Puységur says, is in the way they are brought about. Ordinarily natural somnambulism follows upon sleep. Artificial somnambulism, on the other hand, is produced through the application of the techniques of animal magnetism. This book also examines the faculties exhibited by individuals in the somnambulistic state, providing many case histories as illustrations. Puységur concludes with a look at the dangers that must be guarded against in the use of magnetic somnambulism. [H]

1812

238. *Mesmer, Franz Anton.*
Allgemeine Erläuterungen über den Magnetismus und den Somnambulismus. Als vorläufige Einleitung in das Natursystem. Aus dem Askläpieion abgedruckt. Halle and Berlin: Hallischen Waisenhauses, 1812, 78 pp.

A compilation of articles that appeared in the *Askläpieion* in 1812. Here Mesmer makes his second attempt to tackle the issue of the nature of somnambulism and its relationship to animal magnetism, which he had first developed in the *Mémoire* of 1799 (*see* entry number 211). Again, his intention is to remove any superstitious or religious elements in the explanation of somnambulism and to explain it in terms of his mechanistic theory of animal-magnetic fluid. He first gives a physical explanation for the nature of animal magnetism itself, describing it as an "invisible fire" and distinguishing it from "animal electricity," a subject of great interest at the time. Moving on to somnambulism, he refers to it in the same terms as he had in the *Mémoire*, that is as "critical sleep," pointing out that often the healing crisis of animal magnetism takes place in connection with sleep and in that state the unusual phenomena associated with somnambulism may develop. He states that these phenomena have been recognized in all ages. They include: foreseeing an illness and predicting its course; prescribing effective medicines for a disease without the benefit of medical knowledge; seeing and sensing distant objects; receiving impressions from the will of another; and other phenomena. Mesmer explains how these things are possible with a purely mechanical theory of physiology and the action of the magnetic fluid over a distance. From this he further elaborates his notion of an "inner sense" which leads to knowledge of things at a distance and an instinct about health. He also explains "rapport" in the same terms. [H]

239. ***Montegre, Antoine François Jenin de.***
Du magnétisme animal et de ses partisans; ou, Recueil de pièces importantes sur cet objet, précédé des observations récemment publiées. Paris: D. Colas, 1812, (4) + 139 pp.

An important collection of articles on animal magnetism accompanied by the official reports of the two French commissions of 1784, including the secret report concerning potential moral abuses of animal magnetism. The work also contains a letter highly critical of Puységur's views of animal magnetism. [H]

240. ***Puységur, Armand Marie Jacques de Chastenet, marquis de.***
Continuation du traitement magnétique du jeune Hébert (mois de Septembre). Paris: Dentu, 1812, (2) + xxvi + 27–107 + (1) pp.

See Puységur's *Les fous, les insensés* . . . (entry number 241). [H]

241. ***Puységur, Armand Marie Jacques de Chastenet, marquis de.***
Les fous, les insensés, les maniaques et les frénétiques ne seraient-ils que des somnambules désordonnés? Paris: J. G. Dentu, 1812, (4) + vii + (8)–91 pp.

The publication history of this and subsequent related works deserves a special note. This first description of Puységur's treatment of Alexandre Hébert was followed by the publication in the same year of *Continuation du traitement magnétique du jeune Hébert* . . . (entry number 240). Then in 1813 there appeared *Appel aux savans observateurs du dix-neuvième siècle, de la décisions portée par leurs prédécesseurs contre le magnétisme animal, et fin du traitement du jeune Hébert* (entry number 245). Later in 1813 this latter work was republished in one volume with *Les fous* . . . and *Continuation du traitement magnétique* . . . (see *Appel aux savans observateurs* . . . , 1813). The reason for this somewhat confusing set of related publications is that together they constitute a description of the ongoing treatment of Alexandre Hébert that concluded in 1813. *Les fous* . . . and the subsequent writings are little known but extremely significant for the history of modern psychotherapy. In his description of the treatment of Alexandre Hébert, Puységur shows that he was in the process of evolving a magnetic psychotherapy that embodied insights that were ahead of his time. Animal magnetism had from its beginnings made use of a kind of "empathy" between magnetizer and patient. Puységur, with his concept of "intimate rapport" (*Suite de mémoires* . . . , 1785) had taken that notion a step further. With *Les fous* . . . , he shows an awareness of the healing dynamic between magnetizer and patient that is very close to the modern concepts of transference and countertransference. Alexandre Hébert was a boy of twelve and a half who suffered from paroxysms of rage in which he was a danger for both himself and those around him. He experienced severe headaches and would fall into fits of weeping and moaning while hitting his head against the wall, sometimes even attempting to throw himself from windows. Puységur undertook to treat his condition by using animal magnetism. The boy was a good subject, immediately falling into states of somnambulism. Eventually, Puységur took the boy into his own home and kept him by his side for months. The magnetic treatment became a veritable psychotherapy, involving long conversations, revelations of secrets in the state of magnetic somnambulism, and even the analysis of dreams. Alexandre, who had been troubled from age four, eventually underwent a relatively complete cure, with only certain memory problems remaining. Through this experience, and an earlier one with an artillery soldier, Puységur developed a theory of mental disturbances. He came to believe that "most insanity is nothing but disordered somnambulism" (*Les fous* . . . , p. 54). He held that the disturbed person was in a state of disorder precisely because he or she was caught in a magnetic rapport with someone no longer present (Alexandre, for instance, maintained a hidden rapport with his mother). Puységur believed that the cure was to replace this disturbing rapport by a healthy rapport with the magnetizer. This "transference" oriented theory of psychotherapy seems to be unique for the time. [H]

1813

242. ***Boin, ———.***

Coup d'oeil sur le magnétisme et examen d'un écrit qui a paru sous ce titre: Lettre sur le magnétisme à M. . . . à Paris par M. Morisson de Bourges. Bourges: n.p., 1813, 29 pp.

A note on animal magnetism written in the form of a critique of a letter by Morisson (*see* entry number 244). [H]

243. *Deleuze, Joseph Philippe François.*
Histoire critique du magnétisme animal. 2 vols. Paris: Mame, 1813, (4) + 298; (4) + 340 pp.

Deleuze is a central figure in the history of animal magnetism. After serving as a lieutenant in the French infantry, he decided to devote himself to the study of the natural sciences, eventually becoming assistant naturalist of the Garden of Plants in Paris and secretary to the association of the Museum of Natural History. Deleuze was highly respected by his contemporaries as a great scholar with a balanced approach to scientific issues. In 1785 Deleuze heard about the work of Puységur, but found it hard to believe that he had been able to do what was claimed. However, when a respected friend of Deleuze went to see Mesmer and was then able to induce somnambulism, Deleuze decided to visit his friend and find out for himself. He was impressed with the demonstration and began to pursue his own study of animal magnetism. He was influenced most strongly by the ideas of Puységur, and it is clear from reading Deleuze's writings that he was like Puységur in temperament and attitude towards the people he worked with. The *Histoire* is Deleuze's first work on animal magnetism and it is one of the most important ever written on the subject. In the process of depicting the history, Deleuze also conveys a great deal of information about the theory and practice. In his presentation, he pays a great deal of attention to detail and strives to be objectively fair. Although a partisan of animal magnetism, Deleuze does not close his eyes to legitimate criticism. The *Histoire* is about as balanced a treatment as one could find from a man who was engaged in a daily practice of that art. [H]

244. *Morisson, ———.*
Lettre sur le magnétisme animal adressée à M. . . . à Paris. Bourges: n.p., 1813, 13 pp. [H]

245. *Puységur, Armand Marie Jacques de Chastenet, marquis de.*
Appel aux savans observateurs du dix-neuvième siècle, de la décision portée par leurs prédécesseurs contre le magnétisme animal, et fin du traitement du jeune Hébert. Paris: Dentu, 11 + 127 pp.

This work was republished in the same year in one volume with *Continuation du journal du traitement du jeune Hébert* (September, October, and November). Paris: Dentu, (2) + 11 + (1) + vii + (1) + (8)–91 + (1) + xxvi + (27)–109 + (1) + 127 pp. (*see* entry number 241). [H]

246. *Strombeck, Friedrich Karl von.*
Geschichte eines allein durch die Natur hervorgebrachten animalischen Magnetismus und der durch denselben bewirkten Genesung; von dem Augenzeugen dieses Phänomens. Braunschweig: F. Vieweg, 1813, xxxii + 216 p.

Strombeck writes about a seventeen-year-old girl whose attacks of hysterical trance were cured through the use of magnetic sleep. The girl was a member of Strombeck's household and he spent a great deal of time applying animal magnetic treatments and writing down the resulting conversations with her while she was magnetized. This book is an important contribution to the history of psychotherapy. [H]

247. [*Velye, abbé de*].

Somnambulisme ou supplément aux journaux dans lesquels il a été question de ces phénomènes physiologiques. Paris: Brébault, (1813), 84 pp.

[H]

1814

248. *Annales du magnétisme animal.*

Nos. 1–48: 1814–1816.

Continued as: *Bibliothèque du magnétisme animal*, vols. 1–8, 1817–1819. (Entry number 272.) This in turn was continued as: *Archives du magnétisme animal*, vols. 1–8, 1820–1823. (Entry number 297.) [H]

249. *Mais, Charles.*

The Surprising Case of Rachel Baker, Who Prays and Preaches in her Sleep: with Specimens of her Extraordinary Performances Taken Down Accurately in Short Hand at the Time; and Showing the Unparalleled Powers She Possesses to Pray, Exhort, and Answer Questions, During Her Unconscious State. The Whole Authenticated by the Most Respectable Testimony of Living Witnesses. New York: S. Marks, 1814, 34 pp.

An account of a "sleep-talker," Rachel Baker, who did just what the title says. She is depicted as a "hale country lass of nineteen," quite taciturn, who speaks with a heavy southern drawl. But when asleep she would deliver exhortations and prayers with a "clear, harmonious voice." The book describes her condition and gives an example of her preaching. [H]

250. *Mesmer, Franz Anton.*

Mesmerismus. Oder System der Wechselwirkungen, Theorie und Anwendung des thierischen Magnetismus als die allgemeine Heilkunde zur Erhaltung des Menschen. Herausgegeben von Dr. Karl Christian Wolfart. Berlin: Nikola, 1814, lxxiv + 356 pp.

Mesmer's final work on animal magnetism, written with the assistance of Karl Christian Wolfart, who was sent to see Mesmer by the Berlin Academy of Science. Strangely, the original was written in French by Mesmer, and then translated into German, Mesmer's native tongue, by Wolfart. According to Tischner (*Franz Anton Mesmer*, p. 101), Wolfart's translation was, unfortunately, often stiff and bungling. Wolfart's carelessness extended even to his giving Mesmer's name incorrectly as "Friedrich" Anton Mesmer. This error has somehow been perpetuated by many modern libraries. *Mesmerismus* presents Mesmer's overall view of animal magnetism and its implications for human life. It is divided into two parts: "Physik" and "Moral." The first part discusses the physical nature of animal magnetism and its relationship to such phenomena as fire and electricity. In his explanations, Mesmer is consistent with the mechanistic approach he held from the beginning. "There is one uncreated primary essence—God. In the universe there are two primary essences—matter and motion. All possibilities unfold from the action of motion upon matter" (p. 33). Mesmer also takes up the issue of somnambulism and the "inner sense," repeating much that can be found in his *Erläuterungen* of 1812 (entry number 238). The second part examines the implications of animal magnetism for the moral life. Based on the central notion of universal magnetic harmony, much expounded by Bergasse, Mesmer draws conclusions about the true nature of morality, freedom, social life and religious worship. [H]

251. ***Puységur, Armand Marie Jacques de Chastenet, marquis de.***
Les verités cheminent, tot ou tard elles arrivent. Paris: Dentu, 1814, (2) + 14 pp.

[H]

252. ***Stieglitz, Johann.***
Über den thierischen Magnetismus. Hannover: Hahn, 1814, xvi + 671 pp.

Stieglitz's approach to animal magnetism was that of a skeptic. He did not go along with the common theory of animal magnetism, and he was slow to accept its genuineness. However, he took this position not as one who had experimented in the field, but as a critic of the literature. This book is his major work in the field of animal magnetism. While revealing his skepticism, it also shows that he was not entirely closed to the data. He did believe that Mesmer had discovered an important natural force, but he thought that the nature of the phenomena of animal magnetism could not yet be decided. [H]

1815

253. ***Bruining, Gerbrand.***
Schediasma, de Mesmerismo ante Mesmerum. Groeningen: Van Boekeren, 1815, 88 pp.

The author takes up the question of whether animal magnetism was known before Mesmer, discussing writings of the Greeks, Romans, and Egyptians. [H]

254. ***Devotional Somnium; or A Collection of Prayers and Exhortations, Uttered by Miss Rachel Baker, in the City of New York, in the Winter of 1815, During her Abstracted and Unconscious State; To Which Pious and Unprecedented Exercises is Prefixed, An Account of Her Life, with the Manner in Which She Became Powerful in Praise of God and Addresses to Man; Together with a View of That Faculty of the Human Mind which is Intermediate between Sleeping and Waking. The Facts, Attested by the Most Respectable Divines, Physicians, and Literary Gentlemen; and the Discourses, Correctly Noted by Clerical Stenographers. By Several Medical Gentlemen.***

New York: S. Marks, 1815, (1) + 298 pp.

This unusual account is significant for a number of reasons. Although not the earliest (*see* Charles Mais, entry number 249), it is the most detailed account available of Rachel Baker, the "sleeping preacher." She was nineteen-years old when the book was written and was becoming fairly well known to the public, having recently been brought to New York for observation. The young woman had gone through a series of stages before the preaching phase began. In 1811 she had become extremely depressed and disturbed because she believed herself to be the world's greatest sinner. Then in January of 1812 she went through a night of terror followed by a feeling of great peace. From then she began to give sermons about God, his love, and the evil of sin. Awake she remembered nothing of these talks. In her ordinary state she was not considered to be a good thinker or "sensible." But while preaching in her sleep, she struck those who heard her as extremely cultivated and wise, speaking of religious matters with great passion and conviction. The contrast between the young woman's waking state and somnambulistic states is very similar to that reported of Victor Race by Puységur (1784). In addition to presenting a vivid account of Rachel Baker and her phenomena, the book has a number of small treatises by various authors on sleep walking and related matters. Particularly noteworthy is one by Samuel L. Mitchill on "somnium" or "that condition of the human faculties which is intermediate between sleeping and waking." The book also contains a description of an apparent case of multiple personality, the Rev. Dr. Tennent, which is one of the earliest ever published. [H]

255. **[*Fustier, abbé*].**

Le mystère des magnétiseurs et des somnambules, dévoilé aux âmes droites et vertueuses. Par un homme du monde. Paris: Legrand, 1815, 55 pp.

An attack on animal magnetism by a theologian. [H]

256. *Wolfart, Karl Christian.*
Erläuterungen zum Mesmerismus. Berlin: Nikola, 1815, xvi + 296 pp.

Wolfart, Professor at the University of Berlin, had assisted Mesmer in the production of Mesmer's *Mesmerismus* . . . (entry number 250.) Dedicated to Mesmer (still living when the book was published), *Erläuterungen zum Mesmerismus* was meant to serve as a running commentary on *Mesmerismus*. . . . [H]

1816

257. *Arndt, W.*
Beytrage zu den durch animalischen Magnetismus zeither bewirkten Erscheinungen. Aus eigner Erfahrung. Breslau and Leipzig: C. Cnobloch, 1816, vi + 430 pp.

[H]

258. *Eschenmayer, Carl Adolph von.*
Versuch die scheinbare Magie des thierischen Magnetismus aus physiologischen und psychischen Gesetzen zu erklären. Stuttgart and Tübingen: J. G. Cotta, 1816, (2) + 180 pp.

Eschenmayer was a German physician who was particularly interested in philosophy and mysticism. He began to investigate animal magnetism shortly after 1800 and became co-editor of *Archiv für den thierischen Magnetismus* (entry number 269) with Kieser, Nasse, and Nees von Esenbeck. He eventually taught at the University of Tübingen, remaining in that post until 1836, when he returned to his medical practice. Influenced by the nature-philosophy of Schelling, Eschenmayer had a special interest in the ancient occult traditions. Here, as well as in later works, he seeks out the parallels between those traditions and the contemporary phenomena of animal magnetism. [H & P]

259. [*Hénin de Cuvillers, Etienne Félix, baron d'*].
Réponse aux articles du Journal des Débats, contre le magnétisme animal. Paris: Dentu, 1816, 24 pp.

The editor of the *Archives du magnétisme* here defends the genuineness of animal magnetism. [H]

260. *Hufeland, Christoph Wilhelm.*
Auszug und Anzeig der Schrift des Hernn Leibmedikus Stieglitz über den thierischen Magnetismus, nebst Zusätzen. Berlin: Realschulbuchhandlung, 1816, 96 pp.

Christoph Hufeland (not to be confused with Friedrich Hufeland) was a German physician, professor of pathology and therapeutics at Berlin University, and prolific writer on scientific subjects. At first, he considered Mesmer a charlatan and his theory unscientific. He then chaired a commission appointed by Berlin University in 1812 to examine the doctrine of animal magnetism. He personally approved a trip by a member of the commission, Karl Christian Wolfart, to visit Mesmer to inquire first hand about his thinking on animal magnetism. This visit led to quite a different outcome from that anticipated by Hufeland (*see* entry numbers 250 and 256). The commission eventually gave a positive judgment on the reality of animal magnetism and its power to heal. Hufeland concurred in this judgment, although he remained quite critical of certain aspects of the theory of animal magnetism. In *Auszug und Anzeig* . . . Hufeland expresses both his acceptance and his criticism. He also questions ideas held by Stieglitz (*see* entry number 252) concerning the physical agent involved in animal magnetism, and goes on to propose his own theory of the interaction between magnetizer and patient. [H]

261. ***Magnetiser's Magazine and Annals of Animal Magnetism.***
One volume only, 1816.

Published in London, the magazine was edited by F. Corbaux (1769?–1843). The periodical was of short duration and consisted in 224 pages of translation of Joseph Deleuze's *Histoire critique du magnétisme animal.* [H]

262. ***Mitchill, Samuel Latham.***
"A double consciousness, or a duality of person in the same individual." *Medical Repository* 3:185–186.

The first published account of the Mary Reynolds case, one of the earliest and best known instances of multiple personality. [H]

263. ***Montegre, Antoine François Jenin de.***
Note sur le magnétisme animal et sur les dangers que font courir les magnétiseurs à leurs patients. Paris: Faine, 1816, 8 pp.

[H]

264. ***Parrot, J. F.***
Coup d'oeil sur le magnétisme. Saint Petersburg: n.p., 1816, 65 pp.

An attempt by a physicist to reconcile the theory of animal magnetic fluid with contemporary physics. [H]

265. ***Suremain de Missery, Antoine.***
Examen de l'ouvrage qui a pour titre: "Le mystère des magnétiseurs et des somnambules dévoilé aux droites âmes et virtueuses par un homme du monde." Paris: J. G. Dentu, 1816, iv + 56 pp.

Criticism of a book by the Abbé Fustier on animal magnetism (*Le mystère des magnétiseurs* . . . , entry number 255). [H]

266. ***Vernet, Jules.***
La magnétismomanie, Comédie folie en un acte, melée de couplets. Paris: Fages, 1816, 26 pp.

[H]

267. ***Weber, Joseph.***
Der tierische Magnetismus, oder das Geheimnis des menschlichen Lebens aus dynamisch-psychischen Kräften verstandlich gemacht. Landshut: Weber, 1816.

[H]

268. ***Wolfart, Karl Christian.***
Der Magnetismus gegen die Stieglitz-Hufelandische Schrift über den thierischen Magnetismus in seinem wahren Werth behauptet. Berlin: Nikola, 1816, viii + 9–162 + (5) pp.

Wolfart, one of Mesmer's strongest supporters from Berlin University, here presents his view of the controversy taking place between Johann Stieglitz (*see* entry number 252) and Christoph Hufeland (*see* entry number 260). This work was written after Mesmer's death, and Wolfart now felt free to speak on the issues involved. He uses the occasion of the controversy to present his own views on the nature of animal magnetism and its theoretical implications. [H]

1817

269. ***Archiv für den thierischen Magnetismus.***
Vols. 1–12: 1817–1824.

This journal was continued as: *Sphinx: Neues Archiv für den thierischen Magnetismus und das Nachtleben überhaupt.* One vol. only, 1825–1826. Published in Leipzig, the editor was C. A. von Eschenmayer. [H]

270. ***Baader, Franz von.***
Über die Extase oder das Verzucktseyn der magnetischen Schlafredner. Leipzig: Reclam (parts 1 & 2), 1817; Nürmberg: Monath & Kussler (part 3), 1818.

[H]

271. ***Bapst, F. G. and Azais, Pierre H.***
Explication et emploi du magnétisme. Paris: n.p., 1817, 63 pp.

[H]

272. ***Bibliothèque du magnétisme animal, par MM. les membres de la Société du magnétisme.***

See entry number 248. [H]

273. ***Coll, ———.***
Traitement magnétique suivi d'une guérison remarquable opérée par M. Coll, archiprêtre du canton de Dangé, près Chatellerault, département de la Vienne. N.p.: n.p., 1817, 94 pp.

[H]

274. ***Deleuze, Joseph Philippe François.***
Réponse aux objections contre le magnétisme. Paris: J. G. Dentu, 1817, 51 pp.

Deleuze answers criticisms leveled against the practitioners of animal magnetism, particularly those criticisms concerning the morality of what happens between magnetizer and magnetized. [H]

275. ***Hervier, Charles.***
Théorie du mesmérisme, par un ancien ami du Mesmer, où l'on explique aux dames ses principes naturels, pour le salut de leurs familles; et aux sages de tous les pays, ses causes et ses effets, comme un bienfait de la nature qu'ils sont invités à répandre avec les précautions convenable, et d'après lesquelles plusieurs rois de l'Europe en ont encouragé l'usage dans leurs états. Paris: Agasse, 1817, 148 pp.

Hervier, an early disciple of Mesmer, had been cured of a serious illness by animal magnetism. He strongly supported Mesmer in the early years in Paris (see *Lettre sur la découverte du magnétisme animal* . . . , entry number 71). This much later treatise gives his overall view of the theory of animal magnetism. [H]

276. ***Klinger, Johann August.***
De magnetismo animali. Würzburg: F. E Nitribitt, 1817, 69 pp.

[H]

277. ***Oppert, C.***
Observations relatives à la lettre de M. Friedlander, sur l'état actuel du magnétisme en Allemagne. Paris: J. G. Dentu, 1817, (2) + 19 pp.

[H]

278. *Pfaff, C. H.*
Über und gegen den thierischen Magnetismus und die jetzt vorherrschende Tendenz auf dem Gebiete desselben. Hamburg: Perthes & Besser, 1817, xxii + (2) + 184 pp.

Pfaff, answering his own question about why anyone should write yet another book on animal magnetism, says he is writing this one because of the current lack of critical questioning about animal magnetism and its phenomena. He had first been introduced to animal magnetism by Gmelin in 1789 and he had been astounded by what he witnessed. He hoped that animal magnetism contained the core of a new and comprehensive theory of natural science. He states, however, that his hopes were dashed by subsequent experiences with somnambulists who turned out to be carrying out "juggling acts" for the benefit of their magnetizers and friends. This work is the result of Pfaff's critical attitude towards the subject. Pfaff was aware of the problem of suggestion and the tendency of the somnambulist to fulfill the subtle expectations of the magnetizer. He skeptically views the experiences of Gmelin, Fischer, F. Hufeland, and others. [H]

279. *Roullier, Auguste.*
Exposition physiologique des phénomènes du magnétisme animal et du somnambulisme: contenant des observations pratiques sur les avantages et l'emploi de l'un et de l'autre dans le traitement des maladies aigues et chroniques. Paris: J. G. Dentu, 1817, (4) + xiv + 234 + (2) pp.

[H]

280. *Weber, Joseph.*
Über Naturerklärung überhaupt und über die Erklärung der thierisch-magnetischen Erscheinungen aus dynamisch-psychischen Kraften inbesondere. Ein ergänzender Beitrag zum Archiv den thierischen Magnetismus. Landshut: Weber, 1817, 96 pp.

[H]

281. [*Wendel-Wurtz, abbé*].
Superstitions et prestiges des philosophes, ou les demonolatres du siècle des lumières par l'Auteur des Precurseurs de l'Ante-Christ. Lyon: Rusand, 1817, viii + 230 + (2) pp.

[H]

1818

282. [*Bergasse, Nicolas*].

Dialogue entre un magnétiseur qui cherche les moyens de propager le magnétisme et un incrédule qui croit l'avoir trouvé. Paris: n.p., 1818, 14 pp.

A work in favor of animal magnetism attributed to Bergasse. [H]

283. *Blatter für hohere Wahrheit: aus altern und neuern Handschrift und seltenen Buchern; mit besonderer Rücksicht auf Magnetismus.*

Vols. 1–8; 1818–1827. Neue Folge: Vols. 9–11, 1830–1832.

[H & P]

284. *Brandis, J. D.*

Über psychische Heilmittel und Magnetismus. Copenhagen: Gyldendal, 1818, (2) + ii + 172 + (1) pp.

Brandis, a Copenhagen physician, wrote the first part of this work as a series of journal articles. By way of introduction, he describes his first exposure to animal magnetism in 1785 when the famous Johann Kasper Lavater (1741–1801), a correspondent of Puységur, came to Copenhagen. Brandis had been studying the British "stroking doctor," Greatrakes, to find what healing hints might be contained in his technique. However, Lavater taught Brandis's friend, Lichtenberg, to magnetize, and Brandis found his attention diverted from Greatrakes to animal magnetism. In this work Brandis describes his own experiences in treating illnesses through "psychic" and "magnetic" means. He discusses the healing power of the will, the place of sleep and somnambulism in healing and the nature of magnetic healing. [H]

285. [*Chardel, Casimir Marie Marcellin Pierre Célestin*].

Mémoire sur le magnétisme animal, présenté à l'Académie de Berlin, en 1818. Paris: Bandoin frères, 1818, ii + 49 pp.

[H]

286. [*Dalloz, A.L.J.*]

Discours sur les principes généraux de la théorie végétative et spirituelle de la nature, faisant connaître le premier moteur de la circulation du sang, le principe du magnétisme animal et celui du sommeil magnétique, dit somnambulisme. Paris: The Author, 1818, 308 pp.

A study that is more philosophical than experimental. [H]

287. *Ebhardt, G. F.*

Theologische und philosophische raisonnements in Bezug auf den animalischen Magnetismus nebst einer Beleuchtung über Realität und Irrealität. Oder:

Blicke auf Gott, Natur und den Menschen. Leipzig: Kollmann, 1818, 119 pp. [H]

288. *Journal de la Société du magnétisme animal à Paris.*
One issue only, 1818.

This periodical appeared when the *Bibliothèque du magnétisme animal* (entry number 272) temporarily ceased publication. Since the *Bibliothèque* eventually continued, only one issue of the present journal appeared.

289. [*Sarrazin de Montferrier, Alexandre André Victor; pseudonym: A. de Lausanne*].
Eléments du magnétisme animal, ou Exposition succinte des procédés, des phénomènes et de l'emploi du magnétisme. Paris: Dentu, 1818, 56 pp.

Sarrazin de Montferrier, who wrote under the pseudonym de Lausanne, was editor of the *Annales du magnétisme animal.* [H]

290. *Virey, Julien Joseph.*
Examen impartial de la médecine magnétique, de sa doctrine, de ses procédés et de ses cures. Paris: Panckouke, 1818, 93 pp.

An important abstract from the *Dictionnaire des sciences médicales* (Vol. 29, pp. 463–558, 1818) by a member of the Faculty of Medicine of Paris. This inquiry into animal magnetism, although very critical, recognizes the reality of some of the effects attributed to that phenomenon. But rather than accepting the existence of a physical agent, magnetic fluid, that produces those effects, Virey believes that "affections" or psychological factors are sufficient explanation. [H]

291. *Zeitschrift für psychische Aerzte.*
Vols. 1–2; 1818–1819.

This journal was continued as: *Zeitschrift für psychische Aerzte, mit besonderer Berücksichtigung des Magnetismus,* Vols. 3–5: 1820–1822. Published in Leipzig, it was edited by Christian Friedrich Nasse (1778–1851). [H]

1819

292. *Deleuze, Joseph Philippe François.*
Défense du magnétisme animal contre les attaques dont il est l'objet dans le dictionnaire des sciences médicales. Paris: Berlin-Leprieur, 1819, (4) + 270 pp.

This work was written by Deleuze in response to the article on animal magnetism that appeared in the 1818 volume of the *Dictionnaire des sciences médicales* (see Virey, *Examen impartial* . . . , entry number 290). Author of the *Histoire critique du magnétisme animal* (entry number 243) and one of animal magnetism's stoutest defenders, Deleuze here attempts to answer Virey's criticisms paragraph by paragraph. Virey's article and Deleuze's response constitute one of the most intelligent and enlightening dialogues on animal magnetism to be found in the early nineteenth century. [H]

293. *Ennemoser, Joseph.*

Der Magnetismus nach der allseitiger Beziehung seines Wesens, geschichtlichen Entwickelung von allen Zeiten und bei allen Völkern wissenschaftlich dargestellt. Leipzig: F. A. Brockhaus, 1819, xxiv + 781 + (3) pp.

[H]

294. *Faria, José Custodio de, abbé.*

De la cause du sommeil lucide, ou étude de la nature de l'homme. Tome premier. Paris: Mme. Horiac, 1819, (6) + 463 + (1) pp.

This book by the Abbé Faria is one of the most important in the history of animal magnetism. Faria anticipated the views of Alexandre Bertrand by some years, contending that the true cause of the phenomena of animal magnetism was psychological. He believed that, contrary to Mesmer's teaching, there is no magnetic fluid, and that, contrary to Puységur's teaching, the power of the will of the magnetizer is not involved. In other words, Faria states that there is no external agent that produces the effects. Rather, the magnetizer makes use of *suggestion* to produce a state of "lucid sleep," as Faria calls artificial somnambulism. He says that the extraordinary powers of lucid sleep are always present in human beings, but normally unavailable. Because the sleeper does not recognize them as natural abilities, they are attributed to an external agent. Faria also states that the healing powers of "lucid sleep" are due to the very powerful effects of suggestions coming from the operator. This edition of the book is very rare. A new edition was published in 1906 (Paris: Henri Jouvet) and is important because it contains comments on Faria's doctrine by Brown-Sequard, Liébeault, Gilles de la Tourette, Pitres, Crocq, Vires, and Bernheim. Although the title suggests a volume to follow, none was ever published. [H]

295. *Lombard, A.*

Les dangers du magnétisme animal et l'importance d'en arrêter la propagation vulgaire. Paris: Dentu and Bailleul, 1819, vi + (7)–148 pp.

Lombard sees little good in the practice of animal magnetism and much danger. The book is a criticism of those who have come to accept animal magnetism as a reality and a beneficial medical technique (such as Deleuze), and it is a warning to those who may receive magnetic treatment. Not the least among the many dangers Lombard sees in animal magnetism is the threat to "virtue" of young women who submit to the ministrations of magnetizers. [H]

296. [*Sarrazin de Montferrier, Alexandre André Victor; pseudonym: A. de Lausanne*].

Des principes et des procédés du magnétisme animal, et de leurs rapports avec les lois de la physique et de la physiologie. 2 vols. Paris: J. G. Dentu, (4) + xli + (1) + 241; (4) + 314 + (1) pp.

A significant treatment of the theory and practice of animal magnetism by a man who adhered closely to Mesmer's approach and said little about somnambulism. The first volume describes in detail the techniques to be used. It contains an unusual and important section on the sensations experienced by the magnetizer when treating a patient and how those sensations may be interpreted as indications of procedures to be followed in the treatment. Montferrier claims that awareness of the utility of these subjective sensations goes all the way back to Mesmer, although this is the first detailed treatment of the subject in writing. The second volume begins with a theoretical and historical study of animal magnetism and ends with instructions about treating specific illnesses. De Lausanne, the author to whom this work is ascribed, is the pseudonym of Sarrazin de Montferrier, a mathematician and publicist who edited the first volumes of the *Annales du magnétisme animal.* [H]

1820 – 1829

1820

297. ***Archives du magnétisme animal.***

See *Annales du magnétisme animal* (entry number 248). [H]

298. ***Hénin de Cuvillers, Etienne Félix.***

Le magnétisme éclairé, ou introduction aux archives du magnétisme animal. Paris: Barrois, 1820, 252 pp.

An outspoken supporter of animal magnetism with his own peculiar views on the subject, Hénin de Cuvillers was as much criticized by the Society of Magnetism of Paris, of which he was a member, as by those who opposed animal magnetism. This was because, in his philosophical speculations and attempts to find historical antecedents to animal magnetism in the writings of the ancients, he sometimes seemed to lose touch with the everyday application of that art. Hénin de Cuvillers founded the periodical *Archives du magnétisme animal* (see entry number 248) as the vehicle for his unique views. This book is made up of selections from that journal. [H]

299. ***Nees von Esenbeck, Christian Gottfried.***

Entwickelungsgeschichte des magnetischen Schlafs und Traums. Bonn: Adolph Marcus, 1820, (8) + 159 + (2) pp.

A physician, botanist and natural philosopher, Nees von Esenbeck delivered these lectures in the summer of 1818 when he was professor of natural science and director of the botanic gardens in Erlangen. In this study of sleep, dreams and magnetic somnambulism, he reiterates Mesmer's notion of an inner sense which is capable of receiving information not available to the five senses. According to Nees von Esenbeck, this accounts for the paranormal abilities often displayed by magnetic somnambulists. [H]

300. ***Neues Archiv für den thierischen Magnetismus und das Nachtleben überhaupt.***
N.p.: n.p., 1820?

See *Archiv für den thierischen Magnetismus* (entry number 269). [H]

1821

301. ***Deleuze, Joseph Philippe François.***
Observations adressées aux médecins qui désireraient établir un traitement magnétique. Paris: Belin Le Prieur, 1821, 20 pp.

An appeal by Deleuze to the medical profession for the acceptance of animal magnetism as a healing resource. [H]

302. ***Du Potet de Sennevoy, Jules Denis.***
Exposé des expériences sur le magnétisme animal faites à l'Hotel Dieu de Paris pendant les mois d'octobre, novembre et décembre 1820. Paris: Béchet, 1821, 78 pp.

Du Potet was one of the most important and influential investigators of animal magnetism in nineteenth-century France. He had early proved himself to be an extremely effective magnetizer, and when Dr. Husson, who worked at the Hospital of the Hôtel-Dieu de Paris, was looking for someone to help him with experiments in magnetic somnambulism, Du Potet was chosen. This book is an account of those experiments. It is one of the more significant books in the field, since it marks the beginning of a series of events that led to the establishment of a new French commission to investigate animal magnetism, which eventually produced a positive report. Not only a good magnetizer, Du Potet was also an excellent lecturer and writer. His lecture demonstration in London caught John Elliotson's attention and initiated his interest in animal magnetism. He had a somewhat mystical view of animal magnetism, but his many books were for the most part very practical. Du Potet was the editor-founder of the important *Journal de magnétisme* (entry number 518). *Exposé des expériences* went through a number of editions, the fourth being included as a part of a work written by Du Potet in 1846 called *Manuel de l'étudiant magnétiseur* (entry number 534). [H]

303. ***Dupau, Jean Amédée.***
"Analyse raisonée de l'ouvrage intitulé: Le magnétisme éclairé, ou Introduction aux Archives du magnétisme animal." *Revue médical historique et philosophique* 2 (1821): 20–69.

An important review of *Le magnétisme éclairé* by Hénin de Cuvillers (*see* entry number 298). [H]

304. [*Hannapier, C. R.*].

Tératoscopie du fluide vital de la mensambulance, ou démonstration physiologique et psychologique de la possibilité d'une infinité de prodiges réputés fabuleux, ou attribués par l'ignorance des philosophes et par la superstition des ignorants à des causes fausses et imaginaires. Paris: Dentu, 1821, 392 pp.

Discusses somnambulism ("mensambulance") as a natural phenomenon that can explain many things considered to be supernatural. [H]

305. *Hénin de Cuvillers, Etienne Félix, baron d'.*

La moral chrétienne vengée, ou réflexions sur les crimes commis sous les prétextes spécieux de la gloire de Dieu et des intérêts de la religion et observations historiques et philosophiques sur les faux miracles opérés par le magnétisme animal. Paris: Barrois, 1821, 319 pp.

Another massive work by Hénin de Cuvillers on the subject of the historical antecedents of animal magnetism. In this work he describes the truth mixed with error in such practices as the "healing touch," the wearing of talismans, and the practices of magic, emphasizing the ways in which religions have misused the power inherent in their healing practices. [H]

306. *Hénin de Cuvillers, Etienne Félix, baron d'.*

Le magnétisme animal retrouvé dans l'antiquité ou dissertation historique, étymologique et mythologique sur Esculape, Hippocrate et Galien, sur Apis, Sérapis ou Osiris et sur Isis suivie de recherches sur l'alchimie. Paris: Barrois, 1821, 432 pp.

Like the author's *Le magnétisme éclairé* (entry number 298), this book is made up of selections from his *Archives du magnétisme animal.* The second edition apparently had the title: *Le magnétisme animal fantaziéxoussique retrouvé dans l'antiquité. . . .* [H]

307. [*Panin,* ———].

Abrégé de la pratique du magnétisme animal au dix-huitième et dix-neuvième siècles, ou Tableau alphabétique des principales cures opérées depuis Mesmer jusqu' à nos jours. Geneva and Paris: 1821, 225 pp.

[H]

308. *Passavant, Johann Carl.*

Untersuchungen über den Lebensmagnetismus und das Hellsehen. Frankfurt: H. L. Bronner, 1821, xii + 430 + (1) pp.

Passavant writes about the "magic power" of the human spirit and its relevance to healing and medicine. He views it as both natural and religious.

He deals with somnambulism resulting from magnetizing and describes the phenomena of sympathy (rapport) and clairvoyance. He also examines natural somnambulism and clairvoyance; clairvoyance in dreams, in sickness, near death, in contemplation; and prophetic clairvoyance. [H & P]

1822

309. *Hénin de Cuvillers, Etienne Félix, baron d'.*

Exposition critique du système et de la doctrine mystique des magnétistes. Paris: Barrois, Belin le Prieur, Treuttel et Wurtz, and Delaunay, 1822, x + (11) – 424 pp.

One of a number of books by Hénin de Cuvillers consisting of selections from his *Archives du magnétisme animal* (*see* entry number 248). This work concentrates on certain mystical aspects of the doctrine of animal magnetism considered by the author to be embodied in the writings of many religious and spiritual writers over the centuries. The last forty pages of the book give a concise summary of the views of the author on animal magnetism and its history since Mesmer. [H]

310. *Kieser, Dietrich Georg.*

System des Tellurismus oder thierischen Magnetismus. Ein Handbuch für Naturforscher und Aerzte. 2 vols. Leipzig: F. L. Herbig, (2) + xxx + (1) + 478; (2) + 602 + (3) pp.

A physician, Privy Councillor, and professor at Jena, Kieser was a strong supporter of animal magnetism and heavily involved in investigating the clairvoyant qualities of magnetic somnambulists. He called the essential agent of animal magnetism "telluric power," associating animal magnetism with a generalized magnetic force. In this lengthy and important work, he describes mineral, vegetable, and animal magnetism, attempting to provide a foundation in physics for animal magnetic phenomena. Kieser says that sleep is a general expression of magnetic life and that somnambulism is a natural product of the application of animal magnetic techniques. In describing magnetic healing, he places some emphasis on the power of belief as supplementary to the healing power of animal magnetism itself. The second volume contains a history of animal magnetism and a good bibliography. [H]

311. *Lillbopp, C.P.E.*

Die Wunder des Christentums und deren Verhältnis zum tierischen Magnetismus, mit Berücksichtigung der neuesten Wunderheilungen nach römisch-Kathol. Principien. Mainz: Muller, 1822.

[H]

312. ***Wesermann, H. M.***
Der Magnetismus und die allgemeine Weltsprache. Creveld and Cologne: Johann Heinrich Runcke (Creveld) and Johann Peter Bachem (Cologne), 1822, vii + 271 + (1) pp.

This work was intended as an introduction to the theory and practice of animal magnetism. Information and cases are drawn from a variety of sources. [H]

1823

313. ***Bertrand, Alexandre.***
Traité du somnambulisme, et des différentes modifications qu'il presente. Paris: J. G. Dentu, 1823, (4) + iv + 321 pp.

In his rather short life, the physician Bertrand wrote two important books on animal magnetism. The first of the two is one of the most thorough works on somnambulism written to that date. Here Bertrand discusses the nature of somnambulism from his own experiments and those of others. He describes the phenomena associated with artificial somnambulism, both the more ordinary phenomena and those which seem extraordinary or paranormal. Among other things, Bertrand describes somnambulists' visions of magnetic fluid emanating from the fingers of the magnetizer, a phenomenon first noted by Tardy de Montravel (*Essai* . . . , entry number 152). In his second book (*Du magnétisme animal en France* . . . , entry number 324), Bertrand will change his mind about the magnetic fluid, denying that it has any objective existence. This book also has a section on somnambulistic ecstasy. [H]

314. **[*Dalloz, A.L.J.*]**
Entretiens sur le magnétisme animal et le sommeil magnétique dit somnambulisme, dévoilant cette double doctrine et pouvant servir à en porter un jugement raisonné. Paris: Deschamps, 1823, (4) + iii + 359 + (1) pp.

[H & P]

315. ***Oegger, V.G.E.***
Traité philosophique sur la nature de l'âme et des facultés, où l'on examine le rapport qu'ont avec la morale, le magnétisme de M. Mesmer et le système de Gall. Paris: Eberhart, 1823, ii + 184 pp.

[H & P]

1824

316. *Ennemoser, Joseph.*
Historisch-psychologische Untersuchungen über den Ursprung und das Wesen der menschlichen Seele überhaupt, und über die beseelung des Kindes insbesondere. Bonn: Buschler, 1824, (1) + 129 + (1) pp.

[P]

317. *Kerner, Justinus Andreas Christian.*
Geschichte zweyer Somnambulen. Nebst einigen andern Denkwurdigkeiten aus dem Gebiete der magischen Heilkunde und der Psychologie. Karlsruhe: G. Braun, x + 452 + (1) pp.

Justinus Kerner, a physician and poet of some note, developed a strong interest in animal magnetism and particularly the clairvoyant and ecstatic phenomena associated with it. This is his first work in the area and deals with his magnetic treatment of two young women, the first treatment starting in 1816 and the second in 1822. Kerner believed the somnambulistic pronouncements of these women to be valuable sources of information about the soul and the spiritual life. [H & P]

318. [*Puységur, Armand Marie Jacques de Chastenet, marquis de.*]
Journal du traitement magnétique d'un jeune soldat, cavalier dans le régiment des lanciers de la garde royale, attaqué d'un mal à la cheville du pied, dégenéré en ulcère-fistuleux. N.p.: n.p., (1824), 76 pp.

[H]

319. *Robert, ———.*
Recherches et considérations critiques sur le magnétisme animal avec un programme relatif au somnambulisme aritificiel ou magnétique traduit du latin du docteur Metzger accompangné de notes et suivi de réflexions morales ou pensées détachées applicables au sujet. Paris: Baillière and Dentu, 1824, 396 pp.

[H]

1825

320. *Deleuze, Joseph Philippe François.*
Instruction pratique sur le magnétisme animal, suivie d'un lettre écrite à l'auteur par un médecin étranger. Paris: Dentu, 1825, (4) + 468 + (4) pp.
ENGLISH: *Practical Instruction in Animal Magnetism.* Translated by Thomas C. Hartshorn. Providence: B. Cranston & Co., 1837.

One of the most popular manuals for the practice of animal magnetism ever written. It went through at least four editions in thirty years in France and was translated into a number of foreign languages. Deleuze's instructions are clear and balanced in tone. He emphasizes that animal magnetism is intended to be a source of benefit, a healing power for those in distress. It is not, in his opinion, a good idea to try placing the individual being magnetized into a state of somnambulism simply to be able to satisfy one's curiosity about that phenomenon. The tone of the manual is reminiscent of that of Puységur, Deleuze's inspiration. The good will of the magnetizer is of paramount importance to the outcome of the treatment. Deleuze accepts the existence of a magnetic fluid and describes how objects may be employed as receptacles for that fluid and used later in the healing process. The letter referred to takes up 36 pages of the text and is from the German physician Koreff. [H]

321. ***Foissac, Pierre.***

Mémoire sur le magnétisme animal adressé à MM. les membres de l'Académie des sciences et de l'Académie de médecine. Paris: Didot le jeune, 1825, 10 pp.

In an appeal to the Academies of Science and of Medicine in Paris, Foissac calls for an investigation into the genuineness of the phenomena of animal magnetism, about which so much controversy was then raging. Foissac's proposal was debated by the Academy of Medicine and an investigatory commission was established in 1826. [H & P]

322. ***Rostan, Louis.***

Du magnétisme animal. Paris: Rignoux, 1825 (2) + 49 pp.

Rostan, physician at the Hospice de la vieillesse, wrote this article for the 1824 issue of the *Dictionnaire de médecine*. In this work Rostan is mainly concerned with the genuineness of the phenomena of magnetic somnambulism. He says that since there is so much controversy about the subject, he can only state the conclusions he has arrived at from his own experience, without claiming to have the last word on the matter. He concedes that animal magnetism does produce a "modification of the nervous system," and that this modification can bring about salutary effects on the health of the magnetic subject. But Rostan has serious doubts about the reality of the allegedly paranormal magnetic phenomena, such as clairvoyance. He points out that since the will of the magnetizer is necessarily involved in producing the magnetic state, there is a great deal of room for error in observing such phenomena, since the wish can deceive one about the fact. [H]

323. ***Sphinx: Neues Archiv für den thierischen Magnetismus und das Nachtleben überhaupt.***

N.p.: n.p., 1825.

See: *Archiv für den thierischen Magnetismus* (entry number 269). [H]

1826

324. *Bertrand, Alexandre Jacques François.*

Du magnétisme animal en France, et des jugements qu'en ont portés les sociétés savantes, avec le texte des divers rapports faits en 1784 par les commissaires de l'Académie des sciences, de la Faculté et de la Société royale de médecine, et une analyse des dernières séances de l'Académie royale de médecine et du rapport de M. Husso; suivi de considérations sur l'apparition de l'extase, dans les traitements magnétiques. Paris: J. B. Baillière, 1826, (4) + 539 pp.

Du magnétisme animal is one of the most important works on the history and theory of animal magnetism, and Bertrand's second book on the subject. Between the writing of his first book (*Traité du somnambulisme*, entry number 313) and this one, Bertrand changes his mind about the true nature of animal magnetism and magnetic somnambulism. He no longer accepts the existence of a universal magnetic fluid as the agent that produces the phenomena associated with animal magnetism, even though he maintains the genuineness of the phenomena themselves. Bertrand believes that evidence often cited in favor of the existence of the fluid (e.g., claims of somnambulists to be able to see the fluid emanating from the fingertips of the magnetizer) is largely based on preconceived ideas that affect the imaginations and expectations of both magnetizer and magnetic subject. With this acknowledgement of the importance of suggestion, Bertrand anticipates the ideas of Braid about the true nature of animal magnetic phenomena. Most of the book is devoted to the history of animal magnetism and is one of the best sources for that subject up to the time of its publication. It includes the reports of the French commissions of 1784 (including the secret report) and also a report of the Academy of Medicine written by Husson and delivered in 1825. The second part of the book takes up the subject of ecstasy and its relationship to somnambulism. This part has a very valuable discussion of the history of ecstatic phenomena over the ages. [H]

325. [*Chardel, Casimir Marie Marcellin Pierre Célestin*].

Esquisse de la nature humaine expliquée par le magnétisme animal précédée d'un aperçu du système général de l'univers, et contenant l'explication du somnambulisme magnétique et de tous les phénomènes du magnétisme animal. Paris: Dentu et Delaunay, 1826, 368 pp.

A very ambitious work intended as an analysis of human nature in the context of an understanding of the whole of nature. The last section is devoted to animal magnetism and the light it sheds on the problem. Chardel mentions an unusual case of somnambulism that continued over a period of months. [H]

326. *Deleuze, Joseph Philippe François.*

Lettre à messieurs les membres de l'académie de médecine, sur la marche qui convient de suivre pour fixer l'opinion publique relativement à la réalité du magnétisme animal, aux avantages qu'on peut en retirer, et aux dangers qu'il présente lorsqu'on en fait une application inconsidérée. Paris: Béchet jeune, 1826, 36 pp.

A letter to the French Academy of Medicine with suggestions about how to make known to the public in a balanced way the facts connected with the practice of animal magnetism. [H]

327. *Dissertation sur la médecine et le magnétisme, triomphe du somnambulisme.*

Paris: Doyen, 1826, 80 pp.

[H]

328. *Dupau, Jean Amédée.*

Lettres physiologiques et morales sur le magnétisme animal, contenant l'exposé critique des expériences les plus récentes, et une nouvelle théorie sur ses causes, ses phénomènes et ses applications à la médecine; adressées à M. le Professeur Alibert. Paris: Gabon et al., 1826, xii + (2) + 248 pp.

Dupau writes of his belief that animal magnetism has not been proven to be anything more than the work of imagination. He agrees that real effects take place, but is not convinced that they are the result of a magnetic fluid. Dupau also criticizes the statements of certain writers who have witnessed the "higher phenomena" of somnambulism, saying that their observations may have been faulty. [H]

329. *Foissac, Pierre.*

Second mémoire sur le magnétisme animal. Observations particulières sur une somnambule présentée à la commission nommée par l'Académie royale de médecine pour l'examen du magnétisme animal. Paris: n.p., 1826, 37 pp.

Foissac's first *Mémoire* (entry number 321) had led to the establishment of the Commission of the Royal Academy of Medicine to investigate animal magnetism. This second *Mémoire* supports the use of animal magnetism as a supplement to conventional medical procedures. [H]

330. *L'Hermés, Journal du magnétisme animal. Publié par une Société de Médecins.*

Vols. 1–4; 1826–1827.

An important journal on animal magnetism that included among its contributors Deleuze, Foissac and Judel. [H]

331. [*Mialle, Simon*].
Exposé par ordre alphabétique des cures opérées en France par le magnétisme animal, depuis Mesmer jusqu'à nos jours (1774–1826), ouvrage où l'on a réuni les attestations de plus de 200 médecins, tant magnétiseurs que témoins, ou guéris par le magnétisme. Suivi d'un catalogue complet des ouvrages français qui ont été publiés pour, sur ou contre le magnétisme. 2 vols. Paris: J. G. Dentu, 1826, (4) + xli + (3) + 612; (4) + 543 + (1) pp.

Mialle, a strong supporter of animal magnetism and a good scholar, describes the cures accomplished by animal magnetism from 1774 to 1826. He reports these in an order which is alphabetical by disease, starting with "abcès." Listing the person treated, the magnetizer, the procedures used, the results, and the source of his information, Mialle also has a lengthy bibliography at the end and a useful index. The work is dedicated to the memory of the Marquis de Puységur. [H]

1827

332. [*Chardel, Casimir Marie Marcellin Pierre Célestin*].
Observations de l'auteur de l'Esquisse de la nature humaine sur l'article magnétisme animal, inséré dans le 13e volume du dictionnaire de médecine par le Dr Rostan. Paris: n.p., 1827, 12 pp.

Chardel's comments on Rostan's famous essay on animal magnetism (*see* entry number 322). [H]

333. [*Crampon, ———*].
Le magnétisme animal à l'usage des gens du monde suivi de quelques lettres en opposition à ce mode de guérison. Le Havre: Chapelle, 1827, 79 pp.

The author was criticized by a local paper about certain cures he performed through animal magnetism. This is his response to that criticism. [H]

334. *Le propagateur du magnétisme animal. Journal destiné à la publication des faits et des expériences, etc., de l'histoire du magnétisme etc., de la critique des ouvrages etc. etc., par une Société de médecins.*
Vols. 1–2; 1827–1828.

Edited by Jules Du Potet and Pierre Jean Chapelain and published in Paris. [H]

335. *Lordat, Jacques.*
Réponse à la lettre de M. le Docteur Cazaintre sur un cas de transposition des sens. Montpellier: n.p., 1827, 30 pp.

A treatise confirming the reality of the transposition of the senses to different parts of the body of a magnetized subject. The author was a professor of medicine at Montpellier. [H]

1828

336. *Deleuze, Joseph Phillipe François.*
De l'état actuel du magnétisme. . . . (Paris): n.p., 1828, 24 pp.

[H]

337. *Marne, M. . . . de la.*
Étude raisonnée du magnétisme animal et preuves de l'intervention des puissances infernales dans les phénomènes du somnambulisme magnétique. Paris and Lyon: Gaume, Rusand (Paris) and Rusand (Lyon), 1828, 36 pp.

Presents a viewpoint earlier expressed by Fiard (entry number 223) and others: that animal magnetism and the phenomena of magnetic somnambulism are real, but the work of the devil. [H & P]

338. *Sue, G.A.T.*
Discours sur le magnétisme animal, lu à la séance publique de la Société royale de médecine de Marseille tenue le 11 novembre 1827. Marseilles: Achard, 1828, 24 pp.

[H]

1829

339. *Chevenix, R.*
"On Mesmerism, Improperly Denominated Animal Magnetism." *London Medical and Physical Journal,* March, June, August, October, 1829.

Chevenix had been trained in animal magnetism in Paris and in 1829 brought it to England. Although animal magnetism had been introduced in the 1790's it had not established itself as a respectable undertaking. However, when Chevenix made some demonstrations in 1829, the physician John Elliotson saw them and was impressed. This marked a new era for animal magnetism in Britain and these articles describe Chevenix's approach to the subject. [H]

340. *Du Commun, Joseph.*

Three Lectures on Animal Magnetism, as Delivered in New York, at the Hall of Science, on the 26th of July, 2d and 9th of August. New York: Berard & Mondon, 1829, 78 pp.

The earliest treatise on animal magnetism published in the United States. Du Commun, a contributor to *Annales du magnétisme animal*, practiced animal magnetism in Paris before coming to the United States and teaching French at the West Point Military Academy. These lectures are a well-constructed, popular-style introduction to animal magnetism. The first lecture covers the history of animal magnetism, the second deals with its applications and effects, while the third lecture discusses theory. Du Commun was acquainted with King Charles X, France's reigning monarch at the time of these lectures. In 1814 he presented Charles with the first three volumes of the *Annales du magnétisme animal* (entry number 248), which the king received approvingly. Charles X (previously known as the Comte d'Artois, to whom Charles D'Eslon had been personal physician) had, in 1826, appointed a commission drawn from the Royal Academy of Medicine to investigate the genuineness of the phenomena of animal magnetism. In these lectures, Du Commun states his belief that, since the king is favorable to animal magnetism, it would soon be accepted with the respect that was its due. [H]

341. *Kerner, Justinus Andreas Christian.*

Die Seherin von Prevorst; Eröffnungen über das innere Leben des Menschen und über das Hereinragen einer Geisterwelt in die unsere. 2 vols. Stuttgart and Tübingen: J. G. Cotta, 1829. ENGLISH: *Seeress of Prevorst. Being Revelations Concerning the Inner-Life of Man, and the Inter-Diffusion of a World of Spirits in the One We Inhabit.* Translated by Catherine Crowe. London: J. C. Moore, 1845.

Justinus Kerner, physician and poet, had become acquainted with animal magnetism through Eberhard Gmelin (1751–1808) and soon developed a keen interest in the apparent clairvoyant and visionary powers of somnambulists. In 1826 a young woman, Friederike Hauffe, was brought to Kerner in a state of severe depression. She was also in bad physical shape, and Kerner undertook her physical and psychological treatment. This book is an account of that undertaking. Early in his treatment of Friederike Hauffe, Kerner decided to use animal magnetism. The woman was a good subject and easily became somnambulistic. In her trance states, she had visions, premonitions, and clairvoyant experiences. Kerner believed in the genuineness of these phenomena, recording them with great care in this account. [H & P]

342. [***Villenave, Mathieu Guillaume Thérèse de***].
La vérité du magnétisme prouvée par les faits; extraits des notes et des papiers de Mme Alina D'Eldir, née dans l'Hindoustan, par un ami de la vérité; suivie d'une notice inédite sur Mesmer, qui avait été composée et mise en page pour la "Biographie Universelle." Paris: n.p., 1829, 103 pp.

[H]

1830–1839

1830

343. ***Eschenmayer, Carl Adolph von.***

Mysterien des innern Lebens; erläutert aus der Geschichte der Seherin von Prevorst, mit Berücksichtigung der bisher erschienenen Kritiken. Tübingen: Guttenberg, 1830, xviii + 176 pp.

Eschenmayer was a physician, philosophy professor, and co-editor of *Archiv für den thierischen Magnetismus* (entry number 269). Following in the vein of Jung-Stilling, Eschenmayer was interested in the occult and mystical implications of animal magnetism. Eschenmayer was a friend of Justinus Kerner (1786–1862), and this book contains comments and reflections on Kerner's famous "Seeress of Prevorst." [H & P]

344. ***Horst, Georg Conrad.***

Deuteroskoppie, oder merkwurdige psychische und physiologische Erscheinungen und Probleme aus dem Gebiete der Pneumatologie. Für Religionsphilosophen, Psychologen und denkende Aerzte. Eine nothige Beilage zur Dämono-magie, wie zur Zauber-Bibliothek. 2 vols. Frankfurt: Wilmans, 1830, 256; 264 pp.

Well known in its day, a thorough study of the phenomenon of "second sight." [P]

345. ***Macnish, Robert.***

The Philosophy of Sleep. Glasgow: W. R. M'Phun, 1830, xi + 268 pp.

Macnish's well-known treatise on sleep has sections on sleepwalking, sleep-talking and trance with comparisons to magnetic sleep. He also refers to the Mary Reynolds case of dual or alternating personality, but not by name. This omission led to confusion among some later writers who mention Mary Reynolds and "the lady of Macnish" as two separate cases of multiple personality. [H]

346. *Newnham, William.*

Essay on Superstition; Being an Inquiry into the Effects of Physical Influence on the Mind, in the Production of Dreams, Visions, Ghosts, and Other Supernatural Appearances. London: J. Hatchard and Son, 1830, xvi + 430 + (2) pp.

A little-known work of real significance for the history of both animal magnetism and of physical research. Newnham wrote the book under the patronage of the Bishop of Winchester and it was clearly an attempt to use physiological and psychological terms to explain supernatural or supernormal phenomena which embarrass the Church. Newnham does not deny the possible existence of supernormal phenomena, but he believes that since there are natural explanations for so many instances, one would have to be open to the idea that natural explanations apply to all. Newnham's chapters on sleep, dreams and somnambulism include an appreciative discussion of animal magnetism. This appeared seven years before Elliotson's invitation to Du Potet to do medical experiments with animal magnetism at London's University Hospital and started a strong wave of interest in the subject throughout England. [H & P]

347. *Le visionnaire, ou la victime imaginaire du magnétisme.* *Histoire véritable, contenant la description d'une monomanie sans exemple et dans laquelle sont consignés les lettres autographes, ainsi que les réflexions, traits de démence et récits du monomane; le tout précédé d'une esquisse sur sa vie et ses actions jusqu'à ce jour.*

Stuttgart: Charles Hoffmann, 1830, 152 pp.

[H]

1831

348. *Blatter aus Prevorst. Originalen und Lebenfruchte für* *Freunde des innern Lebens.*

Vols. 1–12; 1831–1839.

Continued as: *Magikon: Archiv für Biobachtungen aus dem Gebiete der Geisterkunde und des magnetischen und magischen Lebens. Nebst andern Zugeben für Freunde des Innern.* Vols. 1–5; 1840–1853. Published at

Stuttgart and originally edited by Justinus Kerner (1786–1862) and A.C.A. von Eschenmayer (1768–1852), this periodical published articles giving information about developments in the areas of the occult and animal magnetism. It contains stories and descriptions of cases of magnetic cures, hauntings, and mysterious phenomena in general. [H & P]

349. ***Chardel, Casimir Marie Marcellin Pierre Célestin.***
Essai de psychologie physiologique. Paris: Encyclopédie Portative, 1831, xxiv + 372 pp.

A detailed study of the nature of animal magnetism and its phenomena. In this work Chardel makes a serious attempt to incorporate animal magnetism into an elaborate philosophical and religious system. He believes in the reality of the magnetic fluid, although he recognizes the powerful influence the imagination can have over the body. Discussing questions related to apparitions and the spiritual life in general, he suggests that both magnetic cures and ecstatic spiritual communication are valid human experiences that reveal the two sides of human nature: the material and the spiritual. [H]

350. ***Husson, Henri Marie.***
Rapport sur les expériences magnétiques faites par la commission de l'Académie royale de médecine, lu dans les séances des 21 et 28 Juin, par M. Husson, rapporteur. (Paris): n.p., 1831, 96 pp.

The report of the findings of the commission set up by the Royal Academy of Medicine of Paris to look into animal magnetism. Husson was the reporter of the commission and so the report bears his name. The report was favorable to animal magnetism, describing experiences of healing through animal magnetism and instances of paranormal phenomena connected with somnambulism. The report was published in its entirety in the book by Foissac: *Rapports et discussions de l'Académie royale de médecine* (entry number 355). [H]

351. ***Kieser, Dietrich Georg.***
Über die eigenthumliche Seelenstörung der sogenannten "Seherin von Prevorst." Berlin: Vereins Buchhandlung, 1831, (3) + 52 pp.

[H & P]

1832

352. ***Brierre de Boismont, Alexandre Jacques François.***
Des hallucinations ou histoire raisonnée des apparitions, des visions, des songes, de l'extase, des rêves, du magnétisme et du somnambulisme. Paris: G. Baillière, 1832, xv + 719 pp.

A remarkably comprehensive and particularly valuable study of hallucinations. The author provides a vast panorama of instances of hallucination of all kinds over the ages, such as those associated with various types of insanity, those experienced by hysterics, the hallucinations of nightmares and dreams, hallucinations experienced in the state of ecstasy or somnambulism produced by animal magnetism, those connected with fever, etc. Brierre de Boismont believes there are two types of causes of hallucinations: the moral (or psychological) and the physical. He also describes the way hallucinations begin, how they are diagnosed and how they may be treated. [H & P]

353. *Fillassier, Alfred.*

Quelques faits et considérations pour servir à l'histoire du magnétisme animal. Thèse No. 243. Paris: Didot, 1832, 91 pp.

A medical thesis written on animal magnetism. It focusses on the phenomenon of the "transposition of the senses" first noted by Tardy de Montravel and Petetin. This refers to an apparent change of location of the sense of hearing, for example, from the ears to the stomach, that occurred while the subject was in a somnambulistic state. [H]

1833

354. *Colquhoun, John Campbell, ed.*

Report of the Experiments on Animal Magnetism, Made by a Committee of the Medical Section of the French Royal Academy of Sciences: Read at the Meetings of the 21st and 28th of June, 1831, Translated and Now for the First Time Published; with an Historical and Explanatory Introduction, and an Appendix. Edinburgh: Robert Cadell, 1833, xii + 252 pp.

An English translation of the favorable report on animal magnetism made by the French Royal Academy of Medicine (*see* Husson, *Rapport sur les expériences magnétiques* . . . , entry number 350). The translation is preceded by a long introduction by Colquhoun, who was one of animal magnetism's staunchest supporters in England. This book constitutes, in fact, the first edition of what Colquhoun would later call *Isis Revelata* (entry number 372). Colquhoun states in the title that the report is "now for the first time published" because the original French edition by Husson was very rare and never broadly distributed publicly. [H]

355. *Foissac, Pierre.*

Rapports et discussions de l'académie royale de médecine sur le magnétisme animal, recueillis par un sténographe, et publiés, avec des notes explicatives. Paris: J. B. Baillière, 1833, 561 pp.

It reports the studies and discussions carried out by the commission appointed by the academies of science and medicine of Paris to investigate animal magnetism and originally set up at the suggestion of Foissac who wrote a *mémoire* on the matter in 1825 (entry number 321). The conclusions of the commission, stated here, are very favorable to animal magnetism, pointing out its usefulness as well as its limitations. [H]

356. [*Kerner, Justinus*].
Über das Besessenseyn oder das Daseyn und den Einfluss des boesen Geisterreichs in der alten Zeit. Mit Berücksichtungen dämonischen Besitzungen der neuen Zeit. Heilbronn: Drechsler, 1833, 116 pp.

[H & P]

1834

357. *Belden, Lemuel W.*
An Account of Jane C. Rider, the Springfield Somnambulist: the Substance of which was Delivered as a Lecture Before the Springfield Lyceum, Jan. 22, 1834. Springfield, Massachusetts: G. and C. Merriam, 1834, viii + (9)–134 + (1) pp.

This appears to be one of the earliest full-length accounts of a case of spontaneous somnambulism accompanied by clairvoyance and double memory to be published in the United States. The phenomena reported are identical to those that were at the time manifesting in magnetic somnambulism, but the author was unwilling to ascribe Jane Rider's feats to that source. [H & P]

358. *Deleuze, Joseph Philippe François.*
Mémoire sur la faculté de prévision: suivi des notes et pièces justificatives recueillis par M. Mialle. Paris: Crochard, 1834, 160 pp.

Deleuze looks into the ancient history of "prevision" (foreseeing the future) and also takes note of contemporary instances. Believing that prevision is not to be explained in terms of some supernatural power, he suggests that it is the result of an instinctive collation of information available to the subject. [H & P]

359. *Du Potet de Sennevoy, Jules Denis.*
Cours de magnétisme animal. Paris: The Author, 1834, 456 pp.

The second edition of this book is called *Cours de magnétisme en 7 leçons, augm. d'un rapport sur les expériences magnétiques faites par la commission de l'Académie de médecine en 1831.* The third edition is titled

Traité complet de magnétisme animal. Each new edition was heavily revised and augmented, the third ending up with 626 pages. In its final form it has twelve lessons and it deals with the history of animal magnetism, discussing certain theoretical and practical aspects in that context. In the process Du Potet manages to mention the opinions of many of the chief French investigators of animal magnetism, and a few from outside France. In the tenth lesson, he takes up the subject of magnetic somnambulism and communication with spirits. In the eleventh he deals with hallucinations and apparitions. The final lesson is concerned with the surgical use of animal magnetism, including the less-known operations performed at Cherbourg. [H & P]

360. ***Jozwik, Albert.***

Dissertation sur le magnétisme animal, thèse soutenue à la Faculté de Paris le 13 août 1834. Paris: Didot, 1834, 12 pp.

One of a number of medical theses on animal magnetism submitted in the mid-nineteenth century. [H]

361. ***Kerner, Justinus Andreas Christian.***

Geschichten Besessener neurerer Zeit. Beobachtungen aus dem Gebiete kakodämonisch-magnetischer Erscheinungen. Nebst Reflexionen von C. A. Eschenmayer über Bessessenseyn und Zauber. Stuttgart: Wachendorf, 1834, vii + 195 pp.

An account of apparent possession of a peasant girl who passed on communications from the possessing entity who appeared to her in a monk-like costume. The entity told a sordid story of immoral sex and murder alleged to have taken place in the house now occupied by the girl and her family. Kerner relates how, at the urging of the apparition, the house was torn down, ancient bones of adults and children were found, and the possession was brought to an end. [P]

362. ***Leonard, ———.***

Magnétisme, son histoire, sa théorie, son application au traitement des maladies, mémoire envoyé à l'Académie de Berlin. Paris: Duvignau, 1834, 76 pp.

An essay written in response to a request for papers by the Academy of Berlin. The subject is animal magnetism, its history and its practical use. [H]

363. ***Maine de Biran, Marie François Pierre Gonthier de Biran dit.***

Nouvelles considérations sur le sommeil, les songes et le somnambulisme; mémoire posthume de M. Maine de Biran, lu à l'Académie des sciences morales et politiques, le 31 mai, 1834, par M. V. Cousin. N.p.: n.p., (1834), 77 pp.

A study of sleep and somnambulism written by a well-known philosopher, and published posthumously. The author discusses the state of the body during sleep, the suspension of will and the persistence of other faculties. He examines the various types of dreams and investigates somnambulism, which he considers to be a special kind of dream. [H]

364. *Simon, Claude Gabriel.*
Mémoire sur le magnétisme animal et sur son application au traitement des maladies mentales, lu au Congrès scientifique de Poitiers, le 11 septembre 1834. Paris: Guiraudet et Jouaust, 1834, 19 pp.

[H]

1835

365. *Berna, Didier Jules.*
Expériences et considérations à l'appui du magnétisme animal, thèse présentée et soutenue à la faculté de Paris. Paris: n.p., 1835, 40 pp.

In this treatise presented to the Paris Faculty of Medicine, Berna proposes to demonstrate the reality of the psychic phenomena of animal magnetism. His proposal was accepted by the Royal Academy of Medicine, which allowed Berna to use his own subjects for the experiments. [H]

366. *Les dangers du magnétisme animal.*
Paris: Leclerc et Delossy, 1835, 23 pp.

The author sees animal magnetism as dangerous from both moral and physical points of view. [H]

367. *Du Potet de Sennevoy, Jules Denis.*
Discours sur le magnétisme animal prononcé le 13 février 1835 à l'Athénée central. (Paris): Mme de Lacombe, (1835), 32 pp.

Du Potet speaks in favor of the use of animal magnetism as a medical tool and points out the flaws of conventional medicine. [H]

368. *Fechner, Gustav Theodor.*
Das Buchlein vom Leben nach dem Tode. Leipzig: Insel-Verlag, (1835?), 59 pp.

[P]

369. *Hammard, Charles Pierre Guillaume.*
Expérience sur le magnétisme animal, thèse présentée et soutenue à la Faculté de médecine de Paris. Paris: n.p., 1835, 18 pp.

[H]

370. ***Lee, Edwin.***
Animal Magnetism and Homeopathy; Being the Appendix to Observations on the Principal Medical Institutions and Practice of France, Italy, and Germany. London: Churchill, 1835, 40 pp.

[H]

371. ***Magnet.***
One vol. only; 1835.

Published in Hanover, New Hampshire. Possibly the earliest American periodical devoted to animal magnetism. [H]

1836

372. ***Colquhoun, John Campbell.***
Isis Revelata; an Inquiry into the Origin, Progress & Present State of Animal Magnetism. 2 vols. Edinburgh: Maclachlan & Stewart, 1836, lii + 395; xii + 416 pp.

A revised and much enlarged version of the *Report of the Experiments on Animal Magnetism . . . with an Historical and Explanatory Introduction and an Appendix* published by Colquhoun in 1833 (entry number 354). The introduction of that small book was expanded into this major two-volume work on the history and meaning of animal magnetism. *Isis Revelata* was the only treatise written in English which attempted to give a far-reaching exposition of the historical and philosophical context of animal magnetism. As such it furnished a strong impetus to the establishment in England of animal magnetism as a subject worthy of serious consideration. It predated Elliotson's introduction of animal magnetism as an anaesthetic into England (*see* Elliotson, entry number 474) and it prepared the way for the magnetic experiments of later British mesmerists such as Townshend, Mayo, Gregory and Haddock. [H]

373. ***Du Potet de Sennevoy, Jules Denis.***
L'Université de Montpellier et le magnétisme animal, ou une vérité nouvelle en présence de vieilles erreurs. Béziers: Carrière, 1836, 71 pp.

Description of Du Potet's encounter with the University of Montpellier and a talk he gave in Montpellier to a scientific congress. [H]

374. ***Husson, Henri Marie.***
Report on the Magnetical Experiments Made by the Commission of the Royal Academy of Medicine, of Paris, Read in the Meetings of June 21 and 28, 1831. by Mr. Husson, the Reporter, Translated from the French, and Preceded with

an Introduction, by Charles Poyen St. Sauveur. Boston: D. K. Hitchcock, 1836, lxxi + (1) + 73–172 pp.

An English translation of the favorable French report on animal magnetism produced in 1831 by Husson (*see* entry number 350). The translator is Charles Poyen, a key figure in the early popularization of animal magnetism into the United States. His long introduction is an important document in itself, and so this item is listed separately from the French entry of the Report. [H]

375. *Kerner, Justinus Andreas Christian.*
Eine Erscheinung aus dem nachtgebiete der Natur, durch eine Reihe von Zeugen gerichtlich bestätigt und den Naturforschern zum Bedenken mitgetheilt. Stuttgart and Tübingen: J. G. Cotta, 1836, xlvi + 310 pp.

[P]

376. *Kerner, Justinus Andreas Christian.*
Nachricht von dem Vorkommen des Besessenseyns eines dämonisch-magnetischen Leidens und seiner schon im Alterthum bekannten Heilung durch magisch-magnetisches Einwirken, in einem Sendschreiben an den Herrn Obermedicinalrath Dr. Schelling in Stuttgart. Stuttgart and Augsburg: J. G. Cotta, 1836, (4)+70 pp.

Kerner describes his method of treatment of apparent possession through a combination of religious, occult, and animal-magnetic techniques. [H & P]

377. *Ricard, Jean Joseph Adolph.*
Doctrine du magnétisme humain et du somnambulisme. Marseille: Vial, 1836, vi + 156 pp.

Ricard was editor of *Journal de magnétisme animal* (entry number 411) and a prolific writer in support of animal magnetism. The book contains a history of animal magnetism and a description of the techniques of Mesmer, Puységur, Deleuze, and others. [H]

378. *Schopenhauer, Arthur.*
Über den Willen in der Natur. Eine Erörterung der Bestätigungen, welche die Philosophie des Verfassers, seit ihrem Auftreten, durch die empirischen Wissenschaften erhalten hat. Frankfurt: S. Schmerber, 1836, 135 pp.

Contains a chapter called "Animalischer Magnetismus und Magie" in which Schopenhauer relates animal magnetism to his doctrine of will. [H & P]

379. *Wirth, Johann Ulrich.*
Theorie des Somnambulismus oder des thierischen Magnetismus. Ein Versuch, die Mysterien des magnetischen Lebens, den Rapport der Somnambulen

mit dem Magnetiseur, ihre Ferngesichte und Ahnungen, und ihren Verkehr mit der Geisterwelt vom Standpunkte vorurtheilsfreier Kritik aus zu erhellen und erklären für Gebildete überhaupt, und für Mediciner und Theologe insbesondere. Leipzig and Stuttgart: J. Scheible, 1836, x + 334 + (2) pp.

Wirth was a philosopher and theologian who became interested in investigating the nature of animal magnetism. He begins his treatise with an examination of the connections between magnetic phenomena and incidents described in the Bible. He then undertakes a philosophical analysis of the experience of the somnambulist. Wirth was particularly concerned with subjective elements that might affect what the somnambulist feels or perceives. In this connection, he investigated the trustworthiness of the evidence of paranormal magnetic phenomena. He states his acceptance of the reality of mental communication between magnetizer and somnambulist, attributing it to the rapport that connects them. But he also notes that he uncovered what he considered to be deliberate fraud on the part of some somnambulists, and states that magnetizers should be more careful and critical in their experiments. [H & P]

1837

380. *Bell, John.*
Animal magnetism: Past Fictions—Present Science. Philadelphia: Reprint from Select Medical Library and Eclectic Journal of Medicine, 1837, 16 pp.
[H]

381. *Durant, Charles Ferson.*
Exposition, or a New Theory of Animal Magnetism with a Key to the Mysteries: Demonstrated by Experiments with the Most Celebrated Somnambulists in America: also, Strictures on "Col. Wm. L. Stone's Letter to Doctor A. Brigham." New York: Wiley & Putnam, 1837, xi + 14–225 pp.

Durant attempts to refute the claims of the American animal magnetists. He is mainly concerned with showing that there is no reason to accept the reality of somnambulistic phenomena and clairvoyance. After conducting his own experiments, he concludes that the apparent effects of animal magnetism are simply the result of suggestibility and self-delusion. [H & P]

382. *Frère, ———.*
Examen du magnétisme animal. Paris: Gaume Frères, 1837, 172 pp.

An attempt to understand the nature of animal magnetism in the light of spiritual phenomena such as miracles, ecstasy and possession. [H & P]

383. ***Husson, Henri Marie.***

Opinion prononcée par M. Husson à l'Académie de médecine séance du 22 août 1837 sur le rapport de M. Dubois (d'Amiens) relatif au magnétisme animal. Paris: n.p., 1837, 11 pp.

Husson, who authored the report favorable to animal magnetism in 1831 (*Rapport sur les expériences magnétiques* . . . , see entry number 350), here addresses himself to a new test for the genuineness of magnetic phenomena, initiated by the physician Berna (see *Expériences et considérations* . . . , entry number 365). The findings of these new trials were negative, but Husson argues that the results have no bearing on the validity of the earlier favorable report. [H & P]

384. ***Lee, Edwin.***

Observations on the Principal Medical Institutions and Practice of France, Italy and Germany; with Notices of the Universities, and Cases from Hospital Practice. To Which is Added, An Appendix, on Animal Magnetism and Homeopathy. Philadelphia: Haswell, Barrington, and Haswell, 1837, 102 pp.

[H]

385. ***The Philosophy of Animal Magnetism Together with the System of Manipulating Adopted to Produce Ecstasy and Somnambulism — The Effects and Rationale. By a Gentleman of Philadelphia.***

Philadelphia: Merrihew and Gunn, 1837, 112 pp.

Although there is no general agreement on the matter, this book has been attributed to Edgar Allan Poe. It presents the philosophy of animal magnetism and describes how to produce magnetic somnambulism. In the view of the author, animal magnetism connects body and mind and proves the existence of the soul. [H]

386. ***Poyen Saint Sauveur, Charles.***

A Letter to Col. Wm. L. Stone, of New York, on the Facts Related in his Letter to Dr. Brigham, and a Plain Refutation of Durant's Exposition of Animal Magnetism, &c. by Charles Poyen. With Remarks on the Manner in which the Claims of Animal Magnetism should be Met and Discussed. By a Member of the Massachusetts Bench. Boston: Weeks, Jordan and Company, 1837, 72 pp.

Poyen, one of animal magnetism's principal promoters in the United States, here writes with appreciation of a treatise by William Stone (*Letter to Doctor A. Brigham, on Animal Magnetism* . . . , entry number 389). Poyen uses the occasion to take issue with a book by Charles Durant that attempts to deny the reality of animal magnetic phenomena and attributes them to suggestibility and delusion (see *Exposition, or a New Theory of Animal Magnetism* . . . , entry number 381). Poyen shows that Durant's criticisms, far from being the original insights he claims them to be, are identical to those that had been voiced as part of the controversy over animal magnetism for decades. Poyen uses material from his own classes on animal magnetism, taught at Salem, Massachusetts, to support the claim for the genuineness of the phenomena in question. [H & P]

387. *Poyen Saint Sauveur, Charles.*

Progress of Animal Magnetism in New England. Being a Collection of Experiments, Reports and Certificates, from the Most Respectable Sources. Preceded by a Dissertation on the Proofs of Animal Magnetism. Boston: Weeks, Jordan & Co., 1837, vi + (2) + 13–212 pp.

A work of considerable significance for the history of animal magnetism and spiritualism in the United States. Although Joseph Du Commun was the first to lecture on animal magnetism in America, Poyen may rightly be thought of as having done more than any other to make the phenomenon widely known there. Poyen learned the art of magnetizing in 1832 while a medical student in his native France. He developed a great enthusiasm for mesmerism, no doubt in part because of the relief it had given him from a troublesome illness. When he came to the United States a few years later, he wanted to make the virtues of animal magnetism widely known in that country. He set about the task energetically and methodically, giving lectures on animal magnetism the length of the northeastern seaboard. *Progress of Animal Magnetism* is his account of that tour. Poyen was trained in the animal-magnetic tradition of Puységur and Deleuze. Both accepted the genuineness of clairvoyant episodes in connection with somnambulism. Poyen's account shows that such clairvoyant experiences became common occurrences in the United States. Largely because of Poyen's lectures and writings, the sight of a magnetizer and his professional somnambulist partner travelling from town to town, giving medical readings and clairvoyant demonstrations, was a common one by 1848 when events around the Fox sisters gave rise to the spiritualist movement. That movement was able to spread so rapidly and become so strongly rooted in American culture largely because of the mental climate created by widespread familiarity with animal magnetism. [H]

388. *Le révélateur. Journal de magnétisme animal publié par une société de magnétiseurs.*

One vol. only; 1837.

Published in Bordeaux and edited by J.J.A. Ricard. [H]

389. *Stone, William Leete.*

Letter to Doctor A. Brigham, on Animal Magnetism: Being an Account of a Remarkable Interview Between the Author and Miss Loraina Brackett While in a State of Somnambulism. New York: George Dearborn & Co., 1837, 76 pp.

Stone describes a letter he had received telling about a number of patients in Providence, Rhode Island, who were being treated with animal magnetism under the care of several physicians. Among those patients was a blind woman, Miss Loraina Brackett, who exhibited remarkable clairvoyant powers. Although unable to see, she could apparently pick out specific colored flowers and cloths at will and know the contents of sealed letters—all while in the state of magnetic somnambulism. This letter relates the contents of an interview Stone conducted with the woman. [H]

1838

390. *Annals of Animal Magnetism.*

Vols. 1–?; 1838–?.

One of the earliest American periodicals on animal magnetism; edited by Samuel Underhill in Cleveland, Ohio. [H]

391. *Barlow, William Frederick.*

Remarks on animal magnetism. London: n.p., 1838, 16 pp.

Barlow asserts that the healing effects of animal magnetism are the result of the power of imagination. He compares cures worked through animal magnetism to those produced by the use of a magic amulet on those who believe in the power of such things. [H & P]

392. *Berna, Didier Jules.*

Magnétisme animal. Examen et réfutation du rapport fait par M.E.F. Dubois (d'Amiens) à l'Académie royale de médecine, le 8 août 1837, sur le magnétisme animal. Paris: Just. Rouvier, 1838, 116 pp.

Berna's reply to the unfavorable report of the commission that had observed his attempts to demonstrate the psychic phenomena connected with animal magnetism. [H & P]

393. *Billot, G. P.*

Recherches psychologiques sur la cause des phénomènes extraordinaires observés chez les modernes voyans improprement dits somnambules magnétiques ou Correspondance sur le magnétisme vital entre un solitaire et M. Deleuze, bibliothécaire du Muséum à Paris. 2 vols. Paris: Albanel et Martin, 1838, 342; 368 + (4) pp.

One of the first lengthy treatments of spiritist-type phenomena occurring in animal magnetic sessions. Billot describes the creation of objects from thin air during these sessions. Deleuze responds that such things have been known to happen and refers to the experiences of, among others, the surgeon Chapelain. Both Deleuze and Billot believed in communication with the spirits of deceased human beings and to a degree this anticipates the spiritualist movement that would begin some ten years later. [H & P]

394. *Charpentier, J.B.A.*

Analyse du magnétisme de l'homme; manière de l'administrer comme guérison naturelle; des effets et des phénomènes qui en résultent. Paris: Rousseau, 1838, 35 pp.

[H]

395. *Collyer, Robert Hanham.*

Lights and Shadows of American Life. Boston: n.p., 1838, 40 pp.

Collyer was an Englishman who toured the eastern United States lecturing on phrenology. In response to widespread interest, he began to include mesmerism as a lecture topic and drew large crowds. This book is the diary he kept on the tour and is interesting not only for its description of mesmerism practice, but also as a source depicting American life at the time. [H]

396. *Colquhoun, John Campbell.*

Hints on Animal Magnetism, Addressed to the Medical Profession in Great Britain. Edinburgh: Maclachlan & Stewart, 1838, 48 pp.

An open letter to the medical practitioners of England, asking that they seriously consider the evidence with regard to the efficacy of animal magnetism as a cure for illness. Colquhoun also criticizes some of the critics of his *Isis Revelata* (entry number 372), including the respected Dr. John Elliotson. [H]

397. *Defer, J.B.E.*

Expériences sur le magnétisme animal. Metz: F. Robert, 1838, 32 pp.

A treatise on the practice of animal magnetism within a group. [H]

398. *Donné, Alfred.*

Mademoiselle Pigeaire, somnambulisme et magnétisme animal. Noyon: Soulas Amoudry, 1838, 50 pp.

A compilation of three articles published in the *Journal des débats*. The author concludes that the feats supposedly performed by Mademoiselle Pigeaire are not genuine. (*See* Pigeaire, entry number 416). [H & P]

399. *Du Potet de Sennevoy, Jules Denis.*

An Introduction to the Study of Animal Magnetism. London: Saunders and Otley, 1838, (1) + xi + 388 pp.

On invitation from Dr. John Elliotson, Du Potet came to England to demonstrate animal magnetism and teach its techniques. Elliotson set him up to practice in the North London Hospital. After strong objections from some medical colleagues, Elliotson had to ask Du Potet to carry on with his work in Du Potet's own apartments in Cavendish Square. During this stay in London, Du Potet wrote this book as an introduction to animal magnetism for the English speaking world. The book begins with a French dedication, but the rest is written in English. It is a very good general treatise on animal magnetism, and is particularly useful in its sketches of the history of animal magnetism. Few books of the time give such a clear impression of the atmosphere of interest and controversy that surrounded animal magnetism in those years. The tone is personal, yet not overly subjective. One of the interesting points Du Potet makes is that in his view there are three main schools of animal magnetism: the original school of Mesmer (with emphasis on the physical fluid), the school of the Chevalier de Barberin at Lyon (emphasizing the work of the "soul" in magnetizing), and the school of the Marquis de Puységur (combining the physical treatment of Mesmer with the psychical treatment of de Barberin). The book has an appendix of reports of British practitioners in favor of animal magnetism. [H]

400. ***Reese, David Meredith.***
Humbugs of New York: Being a Remonstrance Against Popular Delusions; Whether in Science, Philosophy, or Religion. New York and Boston: J. S. Taylor (New York) and Weeks, Jordan & Co. (Boston), 1838, xii + (13)–267 + (1) pp.

[H]

401. ***Sabatier-Desarnauds, Bernard.***
Du magnétisme animal et du somnambulisme artificiel. Montpellier: Veuve Ricard, 1838, 59 pp.

[H]

402. ***Sur les faits qui semblent prouver une communication des somnambules avec let êtres spirituels et sur les consequences qu'on peut tirer de ces faits. (Extrait d'une lettre de M.***, à M. Deleuze).***
Paris: J. G. Dentu, (1838?), 63 pp.

An early discussion of communication with spirits in the magnetic state. The treatise is an extract from a letter to Deleuze and contains a response from him. [H & P]

1839

403. ***Azais, Pierre Hyacinthe.***
De la phrénologie du magnétisme et de la folie. Ouvrage dédié à la mémoire de Broussais. 2 vols. Paris: Desessart, 1839, (2) + xv + 365; (2) + 490 + (3) pp.

One of the first works to combine a knowledge of mesmerism with an interest in phrenology. [H]

404. ***Baudot, Louis Antoine.***

Quelques mots sur le magnétisme animal, suivis d'une observation de variole congénitale. Rouen: Alleaume, 1839, 16 pp.

Brief description of the effects of animal magnetism in a particular case of illness. [H]

405. ***Bird, Friedrich.***

Mesmerismus und Belletristik in ihren schadlichen Einflussen auf die Psychiatrie. Stuttgart: Hallberger, 1839, (3) + 96 pp.

[H]

406. ***Dardeps, ———.***

Aperçu de quelques expériences magnétiques faites à Nimes. Bordeaux: Ramadié, 1839, 8 pp. [H]

407. ***Dubreuil-Chambardel, ———.***

Quelques réflexions sur le magnétisme animal, thèse. Paris: n.p., 1839, 80 pp.

This physiologically oriented thesis contains only a brief mention of animal magnetism. [H]

408. ***Fischer, Friedrich.***

Der Somnambulismus. 3 vols. Basel: Schweighauser, 1839, (4) + 366; (4) + 272; (4) + 413 pp.

The first volume covers natural somnambulism in its various forms and discusses visions and hallucinations that sometimes occur in that state. The second volume discusses animal magnetism and somnabulism produced artificially. The third volume takes up clairvoyance and the phenomenon of possession. [H & P]

409. ***Frapart, ———.***

Lettres sur le magnétisme et le somnambulisme, à l'occasion de Mademoiselle Pigeaire. Paris: Dentu, 1839, vi + 7–160 pp.

An account by Dr. Frapart of experiments done with Léonide Pigeaire, daughter of Dr. Jules Pigeaire (*see* his *Puissance de l'électricité animal . . .*, entry number 416), who, when in the state of magnetic somnambulism, exhibited the apparent ability to read words and perceive objects hidden from her sight. Her father wrote a letter to the Paris Academy of Medicine (which had been investigating animal magnetism) describing his daughter's feats and suggesting that they deserved attention. The Academy's investigatory commission invited Pigeaire and his daughter to Paris to observe and test her clairvoyance. After delays due to disagreements about the conditions for the tests, the commission observed the girl's attempts to read cards with blindfolded eyes. The encounter was inconclusive and led to a heated controversy between supporters of the girl and supporters of the commissioners. Frapart was an advocate of animal magnetism. His description of the affair begins with this first edition of his letters and continues in supplementary letters through 1842. The printings of various editions at stages between 1839 and 1842 make up a complicated publication history. [H & P]

410. ***Holland, Henry.***
Medical Notes and Reflexions. London: Longman, Orme, Brown, Green & Longmans, 1839, xii + 628 pp.

Probably the first mention of Holland's theory of the brain as a double organ. [H]

411. ***Journal du magnétisme animal.***
Vols. 1–3; 1839–1842.

This journal, edited by J.J.A. Ricard, contained some notable articles, including the "Précis historique de magnétisme animal, depuis Mesmer jusqu'à présent" in Vol. 1. [H]

412. ***Lafont-Gouzi, Gabriel Grégoire.***
Traité du magnétisme animal considéré sous le rapport de l'hygiène, de la médecine légale et de la thérapeutique. Toulouse: Senac and The Author, 1839, 176 pp.

A book fanatically opposed to animal magnetism. The author sees it as a most dangerous technique, both to health and to morals. [H]

413. ***Long, H.***
Essai sur le magnétisme animal, thèse présentée et soutenue par H. Long. Montpellier: n.p., 1839, 70 pp.

One of a number of contemporary doctoral theses on animal magnetism. [H]

414. ***Meyer, Th. J.A.G.***
Natur-Analogien, oder die vornehmsten Erscheinungen des animalischen Magnetismus in ihrem Zusammenhange mit den Ergebnissen der gesammten Naturwissenschaften, mit besonderer Hinsicht auf die Standpunkte und Bedurfnisse heutiger Theologie. Hamburg and Gotha: Friedrich und Andreas Perthes, 1839, lxvi + (2) + 412 pp.

A learned attempt to explore the philosophical and theological implications of animal magnetism and somnambulism. A particularly penetrating study of rapport is included. It is an excellent treatise, written with heavy emphasis on the earlier German Writers, worthy of attention for anyone interested in the German animal-magnetic tradition. [H & P]

415. [*Mialle, Simon*].

Rapport confidentiel sur le magnétisme animal et sur la conduite récente de l'Académie royale de médecine adressé à la congrégations de l'Index, et traduit de l'Italien du R. P. Scobardi. Paris: Dentu and Germer Baillière, 1839, 164 pp.

This treatise, probably by Mialle, is written in an invented format and attributed to a fictitious "Scobardi." The author promotes the cause of animal magnetism. [H]

416. *Pigeaire, Jules.*

Puissance de l'électricité animale, ou, du magnétisme vital et de ses rapports avec la physique, la physiologie et la médecine. Paris: Dentu and Germer Baillière, 1839, (4) + 316 pp.

Pigeaire addresses himself principally to the members of the commission appointed by the Academy of Medicine to investigate animal magnetism. He states his objections to certain conditions demanded by the commission and to statements by Dubois of Amiens rejecting clairvoyant magnetic phenomena. He asserts that such paranormal phenomena are now well established. Pigeaire also writes about the curative effects of magnetic somnambulism. [H & P]

417. *Ricard, Jean Joseph Adolphe.*

Cours théorique et pratique du magnétisme animal. Toulouse: The Author, 1839, 96 pp.

[H]

418. *Werner, Heinrich.*

Die Schutzgeister, oder merkwurdige Blicke zweier Seherinnen in die Geisterwelt, nebst der wunderbaren Heilung einer zehn Jahre stumm Gewesenen durch den Lebensmagnetismus, und einer vergleichenden Uebersicht aller bis jetzt beobachteten Erscheinungen desselben. Stuttgart: J. Cotta, 1839, xxxii + 639 pp. ENGLISH: *Guardian Spirits, a Case of Vision into the Spiritual World*. Translated by A. E. Ford. New York: John Allen, 1847.

Werner describes the magnetic healing of two young women. While in the state of magnetic somnambulism they experienced Swedenborgian-type glimpses of the world of the spirits of deceased human beings. Werner also describes other healings through animal magnetism and discusses the nature of the phenomena of magnetic somnambulism. The English version contains only the cases of the two visionary girls. [H]

419. *Wilson, John.*

Trials of Animal Magnetism on the Brute Creation. London: Sherwood, Gilbert, & Piper, 1839, 48 pp.

Year 1839

This unusual little book, written by a Middlesex physician, describes experiments he performed with animal magnetism on animals. Wilson decided to try these experiments for a number of reasons. The most important reason was to see whether effects produced in animals could remove objections to the reality of the effects of animal magnetism based on possible collusion on the part of magnetizer and magnetized or deception on the part of the magnetized alone. Among the animals Wilson treated were cats, dogs, fish, a cock, macaws, a horse, pigs, and even elephants. He seems to have successfully placed all of them in a trance state using magnetic passes. [H]

1840–1849

1840

420. *Bersot, Ernest.*

Du spiritualisme et de la nature. Paris: Ladrange, 1840, lx + 362 pp.

[H]

421. *Despine, Antoine.*

De l'emploi du magnétisme animal et des eaux minérales dans le traitement des maladies nerveuses suivi d'une observation très curieuse de névropathie. Paris and Lyon: Germer Baillière, 1840, lx + 229 + (2) + (12) + (8) pp.

Despine was a physician and benefactor of the ill. This treatise discusses the practice of animal magnetism and the use of mineral waters in the treatment of nervous disorders. A certain "Estelle" was the subject used in Despine's investigations. [H]

422. *Dods, John Bovee.*

Thirty Short Sermons on Various Important Subjects, Both Doctrinal and Practical. Boston: Thomas Whittemore, 1840, xii + 348 pp.

Dods was pastor of the Universalist Society Church in Provincetown, Massachusetts, before he began his main career as a lecturer on "electrical psychology." In these early sermons, he raises questions that he will eventually answer through his exposure to the teachings of mesmerism. [H]

423. ***Du Potet de Sennevoy, Jules Denis.***

Le magnétisme animal opposé à la médecine, Mémoire pour servir à l'histoire du magnétisme en France et en Angleterre. Paris: A. Dentu, Germer Baillière, 1840, viii + 390 pp.

Du Potet travelled to many cities giving lectures on mesmerism. This book provides valuable information about some of those trips. [H]

424. ***Elliotson, John.***

Human Physiology. 5 ed. London: Longman, Orme, Brown, Green, and Longmans, 1840, 1194 pp.

John Elliotson was a notable British physician who played a crucial role in introducing animal magnetism into England. His first encounter with it was at a demonstration given in London by Du Potet on one of his lecture tours. Greatly impressed, Elliotson began to experiment on his own and came to the conclusion that animal magnetism was an art that would prove to be extremely beneficial to medical practice. From the beginning he met stiff opposition from the medical establishment, eventually having to resign his post at University College Hospital. He continued to write in support of animal magnetism and in 1843 founded England's principal mesmeric journal, the *Zoist* (entry number 490). This edition of Elliotson's famous *Human Physiology* contains a well-balanced presentation of his views on animal magnetism. [H]

425. ***Gauthier, Aubin.***

Introduction au magnétisme, examen de son existence depuis les Indiens jusqu'à l'époque actuelle, sa théorie, sa pratique, ses avantages, ses dangers et la nécessité de son concours avec la médecine. Paris: Dentu and Germer Baillière, 1840, 495 pp.

Aubin Gauthier was one of the more prolific and competent writers on animal magnetism in the nineteenth century. He had an historically oriented approach to the subject, and in this, his first book, he traces the precursors of animal magnetism in ancient medicine. He also presents a good summary of the theory and practice of animal magnetism in his own day. [H]

426. ***Guibert, Marquis de.***

Résultat des opérations magnétiques de M. le Marquis de Guibert à Fontchâteau commune de Tarascon. Tarascon: Gondart, 1840, 16 pp.

[H]

427. ***Haldat du Lys, Charles Nicolas Alexandre de.***

Recherches sur quelques phénomènes du magnétisme, le fantôme magnétique, et sur la diffraction complexe. Nancy: Grimblot, Raybois et cie., 1840, 110 pp.

[H]

428. *Kuhnholtz, Henry Marcel.*

Du magnétisme et du somnambulisme artificiel. Montpellier: Castel, 1840, 37 pp.

Kuhnholtz was on the faculty of medicine at Montpellier. He was known for his methodical, balanced approach, which is exemplified in this study of animal magnetism and artificial somnambulism. [H]

429. *Magikon: Archiv für Beobachtungen aus dem Gebiete der Geisterkunde und des magnetischen und magischen Lebens. Nebst andern Zugeben für Freunde des Innern.*

See *Blatter aus Prevorst*, entry number 348. [H & P]

430. *Roy, Emile.*

Observation de magnétisme occulte. Metz: n.p., 1840, x + 14 pp.

[H & P]

431. *Szapary, Ferencz Grof.*

Ein Wort über animalischen Magnetismus, Seelenkorper und Lebensessenz; nebst Beschreibung des ideo-somnambulen Zustandes des Fräulein Therese von B—y zu Vasarhely im Jahre 1838, und einem Anhang. Leipzig: F. A. Brockhaus, 1840, ix + (2) + 175 + (1) pp.

In this first of many books on animal magnetism by Szapary, he presents a treatment of the subject that is both practical and philosophical. He states his belief that the somnambulist can provide insight into the deeper nature of the world. He then relates the case of a nineteen-year-old woman who suffered from a spontaneous somnambulistic condition and who gave pronouncements on the nature of the soul and magnetic rapport. [H & P]

432. *Teste, Alphonse.*

Manuel pratique de magnétisme animal. Exposition méthodique des procédés employés pour produire les phénomènes magnétiques et leur application à l'étude et au traitement des maladies. Paris: J. B. Baillière, 1840, viii + 476 pp.

A practical manual on animal magnetism written by one of the most highly respected practitioners of the day. [H]

433. *Townshend, Chauncy Hare.*

Facts in Mesmerism with Reasons for a Dispassionate Inquiry into It. London: Longman, Orme, Brown, Green and Longmans, 1840, xii + 575 pp.

Townshend, a clergyman of the Church of England, was one of the most articulate British writers on animal magnetism. This book went through many editions in Britain and the United States, and it proved to be very influential in making animal magnetism a legitimate subject of interest. Townshend begins the book with a straightforward recognition of the difficulties in treating a subject which produces such unusual phenomena. He then moves on to describe those phenomena using illustrations drawn from both his experiments and those of other magnetizers. Most of the book is devoted to a description of "mesmeric sleepwalking," as he preferred to term magnetic somnambulism. Interwoven with his description are threads of semi-scientific theory about what makes such unusual phenomena possible. [H]

1841

434. ***The Animal Magnetizer: or, History, Phenomena and*** *Curative Effects of Animal Magnetism; with Instructions for Conducting the Magnetic Operation.*
Philadelphia: J. Kau jun. & Brother, 1841, 94 pp.

[H]

435. ***Bouvignier, L.J.D. de.***
Notice sur le magnétisme ou manière de se magnétiser soi-meme. Paris: n.p., 1841, 16 pp.

[H]

436. ***Burdin, Charles and Dubois, Frédéric.***
Histoire académique du magnétisme animal accompagnée de notes et de remarques critiques sur toutes les observations et expériences faites jusqu' à ce jour. Paris: J. B. Baillière, 1841, xlvii + (1) + 651 pp.

The most complete and competently written history of animal magnetism in France published to its time. After giving a history of animal-magnetic phenomena before Mesmer, the authors describe how Mesmer made his discovery and undertook to promote it to the world. They then include the four reports of 1784 (that of the Royal Academy of Sciences, the Royal Society of Medicine, the dissenting report of Jussieu, and the secret report edited by Bailly). In addition, they reprint the Extract of the correspondence of the Royal Society of Medicine, also from 1784. This is followed by a discussion of the discoveries of Puységur and the approach of Deleuze. The authors then discuss the writings of Bertrand and Georget and the events leading up to the second French commission. They reprint the favorable report of that commission made in 1826 and also the reports of two commissions set up in 1837 to investigate paranormal phenomena of animal magnetism. [H]

437. ***Charpignon, Jules or Louis Joseph Jules.***
Physiologie, médecine et métaphysique du magnétisme. Orleans and Paris: Pesty (Orleans) and Germer Baillière (Paris), 1841, (4) + 366 pp.

Charpignon was a strong supporter of animal magnetism as a genuine phenomenon. Accepting the reality of a magnetic fluid communicated from magnetizer to magnetized, he believed in many of the more extraordinary psychic phenomena attributed to magnetic somnambulists. In this book he describes experiments done to verify these phenomena and discusses the laws and conditions within which animal magnetism functions. Charpignon also deals with medical uses of animal magnetism as well as the effect of the mind on the body. [H]

438. ***Dendy, Walter Cooper.***

The Philosophy of Mystery. London: Longmans, Orme, Brown, Green and Longmans, 1841, xii + 443 pp.

A critical examination in the form of a dialogue of the question of ghosts, spectres, fantasies, mythology, and demonology. There are sections on sleep, dreams, sleepwalking, and trance. A chapter on mesmerism follows the tone of the rest of the book, seeking to bring a skeptical attitude to the extraordinary magnetic phenomena, and explaining all in terms of natural law, as opposed to supernatural intervention. A great deal of attention is devoted to detecting the presence of imagination, suggestion, and outright fraud. [H & P]

439. ***Gorgeret, P.M.E.***

Note sur le magnétisme et sur l'homéopathie, ou réponse à tout ce qui été imprimé dans les journaux de Nantes contre le magnétisme et contre l'homéopathie. Nantes: The Author, 1841, 116 pp.

[H]

440. ***Morley, Charles.***

Elements of Animal Magnetism, or Pneumatology. New York: Turner & Hughes, 1841, 23 pp.

[H]

441. ***Ordinaire, ———.***

Le magnétisme et le somnambulisme du docteur Laurent; une somnambule maconnaise. Mâcon: De Jussieu, 1841, 51 pp.

[H]

442. ***Perusson, E.***

Magnétisme animal, refus de l'Académie de médecine de constater le phénomène de la vision à travers les corps opaques. Chalon-sur-Saone: J. Duchesne, 1841, 16 pp.

[H]

443. ***Perusson, E.***

Soirées magnétiques de Monsieur Laurent à Chalon sur Saone. Chalon-sur-Saone: J. Duchesne, 1841, 18 pp.

[H]

444. ***Ricard, Jean Joseph Adolphe.***

Traité théorique et pratique du magnétisme animal, ou, Méthode facile pour apprendre à magnétiser. Paris: Germer Baillière, 1841, xii + 556 pp.

This well-constructed work begins with a useful summary of the history of animal magnetism up to 1840. The remainder of the book is taken up with lessons on the theory and application of animal magnetism. Ricard discusses the magnetic fluid, natural and magnetic somnambulism, and his own experiences with magnetic healing. He also has a chapter on "spiritualistic somnambulists" and another on magnetic ecstasy. The book is a successful attempt to give the reader an overall view of the nature and practice of animal magnetism. [H]

445. ***Sherwood, Henry Hall.***

Motive Power of Organic Life, and Magnetic Phenomena of Terrestrial and Planetary Motions, with the Application of the Ever-active and All-pervading Agency of Magnetism to the Nature, Symptoms and Treatment of Chronic Diseases. New York: H. A. Chapin & Co., 1841, 196 pp.

[H]

446. ***Tissot, Honoré.***

L'antimagnétisme animal, ou Collection de mémoires, dissertations théologiques, physico-médicales, des plus savants théologiens et médecins sur le magnétisme, la magie, les pratiques superstitieuses, etc. . . . Ouvrage utile et nécessaire spécialement aux ecclésiastiques et aux médecins. Bagnols: Alban Broche, 1841, 251 pp.

[H]

447. ***Transactions du magnétisme animal.***

One vol. only; 1841.

Published in Paris and edited by Teste. [H]

448. ***Turchetti, O.***

Cenni storico-critici sul magnetismo animale. Firenze: Tip. della Speranza, 1841, 93 pp.

[H]

1842

449. *Bourdin, Claude Etienne.*

Mémoire sur un cas d'hystérie, traité par le magnétisme animal. Paris: Closse et Gaultier-Laguionie, 1842, 24 pp.

[H]

450. *Braid, James.*

Satanic Agency and Mesmerism Reviewed. Manchester: Sims and Dinham, Galt and Anderson, 1842, 12 pp.

A work of the greatest significance in the history of hypnotism, and of utmost rarity. James Braid was the founder of modern hypnotism and was himself inventor of its name. His theories were enthusiastically taken up by such men as Azam, Broca, Richet, Charcot, Liébeault, and Bernheim. Apart from Mesmer and Puységur, no single individual had such a profound influence upon the history of hypnotism and dissociative phenomena as did Braid. Born in Fifeshire, Scotland, Braid was educated at Edinburgh. He completed his training as a surgeon and, after practicing for a while in Scotland, moved to Manchester. On November 13, 1841, Braid attended a mesmeric demonstration staged by the Frenchman, Lafontaine. On that occasion Braid thought the effects were produced by trickery, but on attending a second time he noted what he believed to be a genuine phenomenon: the mesmerized individual could not open his eyes. Intrigued by this observation, Braid set out to find the explanation. He performed some experiments with friends and relatives and arrived at what he considered to be the true cause of magnetic sleep: a rapid exhaustion of the sensory and nervous systems producing a feeling of somnolency in the mind, which then "slips out of gear." News of these experiments reached a certain Rev. H. McNeile who preached a sermon against Braid on Sunday, April 10, 1842. This occasioned the reply from Braid which is the pamphlet *Satanic Agency and Mesmerism Reviewed*. The significance of this work lies in its being Braid's first published work containing his historic theory of the cause of mesmeric phenomena and his new nomenclature for those phenomena: "hypnotic sleep," "hypnotise," and "neurohypnotism" (also used in his note "Neuro-hypnotism," *Medical Times and Gazette*: July 9, 1842). In *Satanic Agency* Braid lists three common attitudes towards mesmeric phenomena and a fourth which is his own. The first is the belief that the phenomena are due to a system of collusion and delusion. The second is that they are real but the products of imagination, sympathy, and imitation. The third attitude, that of those who accept the theory of animal magnetism, is that the phenomena are caused by the influence of a magnetic medium. Braid's own view is that they are solely attributable to a particular physiological state of the brain and spinal cord. After expounding his theory of hypnotism, Braid describes the various uses to which he has applied it: for example, extracting teeth, relief of chronic pain, removal of paralysis, and restoration of hearing and sight. This small work contains all the basic elements of the system which Braid elaborated the following year with his publication of *Neurypnology or the Rationale of Nervous Sleep* (1843) (entry number 465). It is believed that there are only two copies of *Satanic Agency and Mesmerism Reviewed* extant today. Tinterow (*Foundations of Hypnosis*, p. 317) says that A. W. Waite and W. Preyer, in searching out Braid's writings, could locate only one copy. That copy, bearing the library plate of Dr. Albert Moll, resides in Tinterow's private library. A second copy, not previously known to exist, has become known to the author of this bibliography and is privately owned. This copy contains a notation apparently written in Braid's own hand. [H]

451. *Caldwell, Charles.*

Facts in Mesmerism and Thoughts on Its Causes and Uses. Louisville, Kentucky: Prentice and Weissinger, 1842, xxx + 132 pp.

After an introduction describing his numerous involvements with controversial causes, Caldwell writes of his convictions about the genuineness of mesmerism and its great usefulness. After presenting a history of animal magnetism and its antecedents, Caldwell lists what he calls the three schools of mesmerism in existence at the time: the "genuine animal magnetists" who follow Mesmer and who hold that all effects are due to the magnetic fluid; the school who follow the "Chevalier Barbarin of Lyons," believing that the effects of mesmerism are the results of faith and volition; and the school of Puységur who accepted the existence of both physical and psychological agents. Caldwell believes the last to be the most tenable of the three positions. The final section of the book is taken up with cases of mesmeric practice, both those of the author and others. [H]

452. *Cogevina, Angelo and Orioli, Francesco*

Fatti relativi a mesmerismo e cure mesmeriche con una prefazione storico-critica. Corfu: Dala Tipografia del Governo, 1842, 349 + iii pp.

Probably the earliest book in Italian describing in some detail the application of animal magnetism for curing disease. Cogevina was a physician and Orioli was an academic of some stature. The most notable case described is that of a young woman named Elisabetta who was successfully treated by animal magnetism. The book helped ignite a strong interest in animal magnetism in Italy. [H]

453. *Douglas, James S.*

Animal Magnetism, or Mesmerism; Being a Brief Account of the Manner of Practicing Animal Magnetism; the Phenomena of that State; Its Applications in Disease, and the Precautions to be Observed in Employing It, Made so Plain that Anyone may Practice it, Experiment upon it and Test Its Effects for Himself. Hamilton, New York: J. & D. Atwood, 1842, 54 pp.

[H]

454. *Ennemoser, Joseph.*

Der Magnetismus im Verhältnisse zur Natur und Religion. Stuttgart and Tübingen: J. G. Cotta, 1842, xxii + (2) + 546 pp.

In this book Ennemoser attempts to deal with animal magnetism, not as an isolated phenomenon, but as a reality intimately connected with the essence of nature itself, and therefore with the mystical and religious awareness that springs from contact with nature. In doing this he attempts to counter superstitious beliefs that had come to be associated with magnetic phenomena, and also demonstrate to scientists that animal magnetism is a natural phenomenon. To do this Ennemoser first sketches the theory of animal magnetism and describes its principle phenomena, including magnetic sleep. He then draws connections between the phenomena described in the ancient and more recent literature of religion and occult and magnetic phenomena. In this context he attempts to prove that the phenomena of animal magnetism are simply manifestations of forces of nature and subject to natural law. The book contains a section on the "psychological explanation" of magnetic phenomena that is noteworthy. [H & P]

455. *Gauthier, Aubin.*

Histoire du somnambulisme: chez tous les peuples sous les noms divers d'extases, songes, oracles et visions; examen des doctrines théoriques et philosophiques de l'antiquité et des temps modernes, sur ses causes, ses effets, ses abus, ses avantages, et l'utilité de son concours avec la médecine. 2 vols. Paris: Félix Malteste et cie., 1842, (4) + 454 + (1); (4) + 440 pp.

One of the best and most important histories of animal magnetism and somnambulism ever written. The first part is an especially thorough examination of traces of magnetic practice among the ancients. Much of this part deals with dreams, divination, and prophecy. Gauthier's study of the use of the word "somnambulism" through the ages and in contemporary times is excellent. He has a very useful discussion of Mesmer and somnambulism and the place Puységur holds in the history of that phenomenon. He also has a valuable summary of artificial somnambulism from the time of Deleuze through the investigations of the second French commission and up to the time of his writing. Although Gauthier focuses his treatment of contemporary artificial somnambulism on the French scene, his history remains of great interest for any student of animal magnetism. [H]

456. *Journal of the Phreno-magnetic Society of Cincinnati.*

Vol. 1 (one volume only); 1842.

Published in Cincinnati. [H]

457. *The Magnet.*

Vols. 1–2; 1842–1844.

Continued as: *New York Magnet*, One vol. only; 1844. Edited by LaRoy Sunderland. [H]

458. ***Les magnétiseurs sont-ils sorciers? La France est-elle hérétique? Les memes hommes l'ont dit.***
Paris: Just. Rouvier, Leteinturier, and Huriet, 1842, 34 pp.

[H]

459. ***Mesmeric Magazine; or Journal of Animal Magnetism.***
One vol. only; 1842.

Published in Boston and edited by R. H. Collyer. [H]

460. ***M'Neile, Hugh.***
Satanic Agency and Mesmerism. A Sermon. (Liverpool): n.p., (1842), pp. 141–152.

M'Neile delivered this, the sermon that spurred Braid to write his first work on the subject: *Satanic Agency and Mesmerism Reviewed* (entry number 450), in St. Jude's Church, Liverpool. He attributes the action of animal magnetism and mesmerism to the devil, and accuses Braid of carrying out "experiments in a corner, upon (his) own servants, or upon females hired for the purpose." This sermon exists as an offprint from a source that cannot be traced. [H]

461. ***Montius, E.***
Faits curieux et intéressants produits par la puissance du magnétisme animal, ou comptes-rendus des expériences remarquable opérées en Belgique. 2 ed. Brussels: n.p., 1842, 175 pp.

Montius had a mesmeric clinic in Brussels. This work describes some of the cures he obtained there through the use of animal magnetism. He worked even in the face of prosecution by law for his practices. This is the second edition of his book; no information could be obtained about the first. [H]

462. ***Phreno-magnetic vindicator.***
Vol. 1 (one volume only??); 1842.

Published in Lexington by J. G. Forman. [H]

463. ***Tizzani, Vincenzo.***
Sul magnetismo animale, discorso istorico-critico. Letto all'Accademia di Religione Cattolica il di 21 Lūglio 1842. Rome: Salviucci, 1842, 160 pp.

[H]

464. ***Topham, William and Ward, W. Squire.***
Account of a Case of Successful Amputation of the Thigh, During the Mesmeric State, Without the Knowledge of the Patient: Read to the Royal Medical

and Chirurgical Society of London, on Tuesday, the 22nd of November, 1842. London: Baillière, 1842, 26 pp.

The original account of the operation described by Elliotson in his *Numerous Cases* (entry number 474). It was the first use of animal magnetism as a surgical anaesthesia in England. Topham was the magnetizer; Ward the surgeon. The painless operation was described in an address delivered to the Royal Medical and Chirurgical Society of London and then published in this form. The resulting controversy is related in Elliotson's book. [H]

1843

465. ***Braid, James.***

Neurypnology or The rationale of nervous sleep considered in relation with animal magnetism. Illustrated by numerous cases of its successful application in the relief and cure of disease. London: John Churchill, 1843, (2) + xxii + 265 + (1) pp.

The first full-length treatment of the subject of hypnotism by its founder. In this book Braid expands upon *Satanic Agency and Mesmerism Reviewed* (entry number 450). He further elaborates his proposed new terminology with the following vocabulary: NEURYPNOLOGY: the doctrine of nervous sleep; NEURO-HYPNOTISM: nervous sleep, a peculiar condition of the nervous system produced by artificial contrivance; HYPNOTIC: the state of nervous sleep; HYPNOTISE: to induce nervous sleep; HYPNOTISM: nervous sleep; DEHYPNOTISE: to restore from the state of nervous sleep; HYPNOTIST: one who practises neuro-hypnotism. It was Braid's intention that his terminology and theory should replace that of the animal magnetists. He meant thereby to do away with the notion of a magnetic fluid as the agent which produces mesmeric phenomena. Instead, he substituted the theory that mesmeric phenomena were of subjective origin. In *Neurypnology* Braid says that the specific cause is a rapid fatiguing of the sensory and nervous systems. In later writings Braid also gave due credit to the importance of suggestion in the hypnotic process. Braid had to face fierce opposition to his views from both mesmerists and medical colleagues. Yet he was eventually successful in having his notion of hypnotism the accepted theory to explain mesmeric phenomena—though he did not live to see this happen. Shortly before Braid's death in 1860, Dr. Azam of Bordeaux published an enthusiastic account of hypnotic experiments based upon Braid's views, with which he had become acquainted the previous year. Also in 1860, Broca reported similar successful results using Braid's methods. From that point on, Braid's fame spread rapidly. By the end of the century he was universally credited as the originator of the system which made possible the revolutionizing of medical and psychotherapeutic technique. [H]

466. *Brown, John.*
Mesmerism; Its Pretensions & Effects upon Society Considered. Boston: n.p., 1843, 8 pp.

A pamphlet containing a vigorous attack on mesmerism as practiced in the author's vicinity by two men: Small and Sharp. [H]

467. *Charpignon, Louis Joseph Jules.*
Études physiques sur le magnétisme animal soumises à l'Académie des sciences. Paris and Orleans: Germer Baillière (Paris) and The Author (Orleans), 1843, 41 pp.

Charpignon compares the results of animal magnetism to those produced by electricity and mineral magnetism. [H]

468. *Charpignon, Louis Joseph Jules.*
Lettre au docteur Frapart sur le magnétisme. Orleans: Jacob, 1843, 8 pp.

[H]

469. *Collyer, Robert H.*
Psychography, or The Embodiment of Thought; with an Analysis of Phreno-magnetism, "Neurology," and Mental Hallucination, Including Rules to Govern and Produce the Magnetic State. Philadelphia, New York, and Boston: Zieber & Co. (Philadelphia); Sun Office (New York); Redding & Co. (Boston), 1843, 44 pp.

Collyer, editor of the *Mesmeric Magazine*, wrote this book largely as a rebuttal to attacks against him and his belief in animal magnetism. The book contains testimonials to his work as a magnetizer and rules to follow in practicing magnetic healing. [H]

470. *Colquhoun, John Campbell.*
The Fallacy of Phreno-magnetism Detected and Exposed. Edinburgh: Wilson, 1843, 16 pp.

[H]

471. *Debay, A.*
Hypnologie: du sommeil et des songes au point de vue physiologique: somnambulism, magnétisme, extase, hallucination; exposé d'une théorie du fluide électro-sympathique. Paris: J. Masson, 1843, 187 pp.

Later editions used the title: *Les mystères du sommeil et du magnétisme.* [H]

472. *Dickerson, K.D.D.*
The Philosophy of Mesmerism, or Animal Magnetism. Being a Compilation of Facts Ascertained by Experience, and Drawn from the Writings of the Most Celebrated Magnetisers in Europe and America. Intended to Facilitate the

Honest Inquirer After Truth, and Promote the Happiness of Mankind, By Diffusing the Knowledge of Nature's Wisest Laws and Most Benevolent Institutions. Concord, New Hampshire: Morrill, Silsby, and Co., 1843.

Dickerson gives a history of the progress of animal magnetism in the United States from Poyen to LaRoy Sunderland, a man whom he greatly admired. It is one of the most interesting American historical compendiums because in his enthusiasm for animal magnetism the author goes out of his way to mention the chief workers and best-known somnambulists. [H]

473. *Dods, John Bovee.*

Six Lectures on the Philosophy of Mesmerism, Delivered in the Marlboro Chapel, Boston. Reported by a hearer. Boston: W. A. Hall, 1843, 68 pp.

These lectures were extremely well attended. When they were published in book form they sold out almost immediately; many more editions were then issued within a few years. In the lectures Dods expresses his belief in the power of the agent discovered by Mesmer, but rejects the term magnetism in favor of "mental electricity" or even "spiritualism." Dods' religious approach is perceivable in these lectures. He calls mesmerism a "power of God," and he connects the healing works of Jesus with this power. Dods describes six degrees of mesmerism, the sixth being clairvoyance. He also mentions surgical uses for mesmerism. [H]

474. *Elliotson, John.*

Numerous Cases of Surgical Operations without Pain in the Mesmeric State: with Remarks upon the Opposition of Many Members of the Royal Medical and Chirurgical Society and Others to the Reception of the Inestimable Blessings of Mesmerism. London: H. Baillière, 1843, 93 + (1) pp.

An important document in the history of animal magnetism. Elliotson describes the case history of an amputation of a leg above the knee while the patient was mesmerized and experienced no pain. The magnetizer was William Topham, the surgeon was W. Squire Ward. (*see* entry number 464). Elliotson also writes about the negative reactions to the operation by the medical establishment. He was indignant that such a beneficial tool could be met with what he considered to be incredible obtuseness. [H]

475. [*Grandvoinet, ———, pseudonym: Tedinngarov*].

Esquisse d'une théorie des phénomènes magnétiques. Paris: Dentu, 1843, 32 pp.

This treatise on the phenomena of animal magnetism was published in two forms in the same year: one under the pseudonym, the other under the author's real name. [H]

476. *Gregory, Samuel.*

Mesmerism, or Animal Magnetism, and Its Uses; with Particular Directions for Employing It in Removing Pains and Curing Diseases, in Producing Insen-

sibility to Pain in Surgical and Dental Operations; and in the Examination of Internal Diseases, with Cases of Operations, Examinations and Cures. Boston: Redding & Co., 1843, 16 pp.

Gregory writes about the use of mesmerism in the United States. After tracing its history from 1836, he states that it was now being used all over the country by physicians to cure illness of many kinds. Typical of writers on the subject at that time, Gregory writes about the use of medical clairvoyance, in which a mesmerized person can see or sense the internal organs of the body and the state of illness there. [H & P]

477. *Kuhnholtz, Henri Marcel.*

Analyse apologétique et critique de la brochure du docteur J. A. Tedinngarov, intitulée: Esquisse d'une théorie des phénomènes magnétiques. Paris and Montpellier: L. Castel, 1843, 15 pp.

A pamphlet on the elements essential to the action of animal magnetism. [H]

478. [*Lang, William*].

Mesmerism, its History, Phenomena, and Practice, with Reports of Cases Developed in Scotland. Edinburgh: Fraser and Co., 1843, xii + 240 pp.

Lang relates the history of animal magnetism from Mesmer on and describes the methods used by some of its principal practitioners. He also tells about the first case in which animal magnetism was used as a cure in Scotland, and other cases that followed. He has a section on phreno-magnetism (animal magnetism used in conjunction with phrenology) and a most interesting chapter on the use of animal magnetism on animals. There is also a striking description of a somnambulist who develops a second, mesmeric personality with a name different from her ordinary name. [H]

479. *Lee, Edwin.*

Report upon the Phenomena of Clairvoyance or Lucid Somnambulism (From Personal Observation). With Additional Remarks. An Appendix to the Third Edition of "Animal Magnetism." London: J. Churchill, 1843, 50 pp.

[H & P]

480. *Mesmerism and Phreno-Mesmerism. Consisting of Modes of Mesmerising.* 3 ed.

Newcastle-upon-Tyne: W. & T. Fordyce, 1843, 72 pp.

[H]

481. *Mesmerism the Gift of God: in Reply to "Satanic Agency and Mesmerism," a Sermon Said to Have Been Preached by the Rev. Hugh M'Neile: in a Letter to a Friend by a Beneficed Clergyman.*

London: n.p., 1843, 16 pp.

This letter takes issue with the same sermon to which Braid replied in 1842 (Braid: *Satanic Agency . . .* , entry number 450). [H]

482. *The Mesmerist. A Weekly Journal of Vital Magnetism.*
One vol. only, 1843.

Published in London. [H]

483. *Phreno-Magnet, and Mirror of Nature: a Record of Facts, Experiments and Discoveries in Phrenology, Magnetism &c.*
One vol. only; 1843.

Edited by Spenser Hall and published in London, the first and only volume of this journal contained twelve numbers. Hall intended to use this periodical as a way of promoting his popularly oriented system that employed both animal magnetism and phrenology. The first number contains a letter from James Braid. [H]

484. *Résimont, Charles de.*
Le magnétisme animal considéré comme moyen thérapeutique; son application au traitement de deux cas remarquables de névropathie. Paris: Germer Baillière, 1843, viii + 517 + (3) pp.

[H]

485. *A Return of Departed Spirits of the Highest Character of Distinction, as well as the Indiscriminate of All Nations, into the Bodies of the "Shakers," or "United Society of Believers in the Second Advent of the Messiah." By an Associate of Said Society.*
Philadelphia: J. R. Colon, 1843, viii + 9–52 pp.

An interesting description of the spiritism of the Shakers, an American movement which to a certain extent anticipated modern spiritualism. [P]

486. *Ricard, Jean Joseph Adolphe.*
Arrêt de la cour suprême touchant le magnétisme animal. Paris: The Author, 1843, 72 pp.

[H]

487. *Ricard, Jean Joseph Adolphe.*
Lettres d'un magnétiseur. Paris: The Author, 1843, 179 pp.

[H]

488. *Le somnambule. Journal de magnétisme.*
One vol. only; 1843–1845.

Published at Lyon and edited by Auguste Possin. [H]

489. *Sunderland, LaRoy.*

Pathetism: with Practical Instructions: Demonstrating the Falsity of the Hitherto Prevalent Assumptions in Regard to What has been Called "Mesmerism" and "Neurology," and Illustrating Those Laws which Induce Somnambulism, Second Sight, Sleep, Dreaming, Trance, and Clairvoyance, with Numerous Facts Tending to Show the Pathology of Monomania, Insanity, Witchcraft, and Various Other Mental or Nervous Phenomena. New York: P. P. Good, 1843, xvi + 247 pp.

Sunderland, a Methodist minister, attended the lectures of Charles Poyen on animal magnetism in 1837, and from that point began to incorporate that doctrine into his ideas about the effects of the mind upon the body. At the same time he developed an interest in phrenology, and combined the two into a system he labeled "phrenomagnetism." Although Sunderland later dropped phrenomagnetism, in the first edition of the present book he includes it in his overall system called "pathetism." Rejecting the idea of a magnetic fluid, Sunderland substitutes "pathetism" as the agent "by which one person, by manipulation, is enabled to produce emotion, feeling, passion, or any physical or mental effects, in the system of another" (p. 3). He sees the "sympathetic" relationship between the operator and patient, not animal magnetic fluid, as the central agent for producing these effects: "All the feelings therefore which one human being may be able to excite in the mind of another . . . are identical with this same agency. . . . If impressions be made upon the sensorium through the eye or ear, or through the nerves of sensation, the immediate agency which carries those impressions to the mind is *pathetism*" (p. 68). Sunderland had a unique view of the importance of the state of mind of those *in the environment* of the patient and operator: "Pathetism has to do with the sympathetic system, not of the operator and his patient merely, but the nervous *sympathies* and *antipathies* of every other person present" (p. 126). For Sunderland, pathetism embodies all the changes that can be brought about in the minds of men, and the physical correlates of those mental changes. In this way, the system of pathetism embodies knowledge of all the laws of life, health, and disease. [H & P]

490. *The Zoist. A Journal of Cerebral Physiology & Mesmerism, and their Applications to Human Welfare.*

Vols. 1–13; 1843–1856.

The preeminent British mesmeric journal, edited by John Elliotson and published in London. [H]

1844

491. ***L'avenir médical, Journal des intérêts de tous, avant pour but la démonstration pratique du nouvel art de guérir, l'homéopathie et le magnétisme, par la fondation d'un hopital homéopathico-magnétique pour 150 à 200 infants.***
Vols. 1–2; 1844–1845.

[H]

492. ***Barreau, Ferdinand.***
Le magnétisme humain en cour de Rome. Paris: Sagnier et Bray, 1844, 308 pp.

Written by a Catholic who tried to show that there was nothing in the doctrine of animal magnetism that was incompatible with Catholic theology. Barreau was one of a number who practiced magnetism within the Church and fought for its acceptance. [H]

493. ***Beecher, William H.***
A Letter on Animal Magnetism. Philadelphia: Brown, Bucking and Guilbert, 1844, 7 pp.

[H]

494. ***Drake, Daniel.***
Analytical Report of a Series of Experiments in Mesmeric Somniloquism, Performed by an Association of Gentlemen: with Speculations on the Production of Its Phenomena. Louisville, Kentucky: F. W. Prescott & Co., 1844, 56 pp.

A report on experiments done in Louisville, Kentucky, with a young woman who was easily magnetized. The experiments were designed to find out whether she had the ability, as had been claimed, to experience the sensations of any person put into communication with her while in the somnambulistic state. According to Drake, the experiments proved that this kind of "rapport" was a genuine fact. [H]

495. ***Gauthier, Aubin.***
Le magnétisme catholique; ou, Introduction à la vraie pratique et réfutation des opinions de la médecine sur le magnétisme; ses principes, ses procédés et ses effets. Brussels and Paris: n.p., 1844, 232 pp.

Gauthier takes exception to the prevalent view of animal magnetism among physicians. He concerns himself largely with its practice in Belgium. [H]

496. *Gauthier, Louis Philibert Auguste.*

Recherches historiques sur l'exercice de la médecine dans les temples, chez les peuples de l'antiquité, suivies de considérations sur les rapports qui peuvent exister entre les guérisons qu'on obtenait dans les anciens temples, à l'aide des songes, et le magnétisme animal, et sur l'origine des hôpitaux. Paris: Baillière, 1844, x + 264 pp.

The author compares religious healing practices of ancient times to those of his contemporary practitioners of animal magnetism. He denies that the healers of old used animal magnetism, as some had claimed. He does, however, see certain similarities in the therapeutic effects produced. [H]

497. *Johnson, Charles P.*

A Treatise on Animal Magnetism. New York: Burgess & Stringer, 1844, 96 pp.

[H]

498. [*Loubert, Jean Baptiste*].

Le magnétisme et le somnambulisme devant les corps savant, le cour de Rome et les théologiens. Paris: Germer Baillière, 1844, (4) + 702 pp.

Written from a Roman Catholic point of view, the work summarizes the history of animal magnetism, including opinions of various learned men on the subject and the conclusions of the French commissions for and against. Loubert takes the reader back to the Middle Ages and the writings of the ancients to trace magnetic-type healings before Mesmer. He then discusses the writings of some of the principal contemporary advocates of magnetism, such as J.J.A. Ricard and Aubin Gauthier, and points out the theological errors of these authors. Loubert does not, however, desire to reject magnetism and somnambulism as such, and he defends them against condemnation by the church. [H]

499. *New York Magnet.*

1844.

See *The Magnet*, entry number 457. [H]

500. *Pyne, Thomas.*

Vital Magnetism: A Remedy. London: n.p., 1844.

Information about publisher and page numbers of the first edition could not be obtained. The second edition, published by Samuel Highley, contains 94 pages. [H]

501. *Revue magnétique. Journal des faits et des cures magnétiques et somnambuliques, des théories, recherches historiques, discussions scientifiques et progrès généraux du magnétisme en France et dans les pays étrangers.*

Vols. 1–2; 1844–1846.

Edited by Aubin Gauthier and published in Paris. [H]

502. *Ricard, Jean Joseph Adolphe.*

Physiologie et hygiène du magnètiseur; régime diététique du magnétisé; Mémoires et aphorismes de Mesmer, avec des notes. Paris: Germer Baillière, 1844, xii + 216 + 228 pp.

The second part of this work was republished separately in 1846 under the title *Magnétisme animal. Mémoires et aphorismes de Mesmer suivis des procédés de D'Eslon. Nouvelle édition avec des notes* (Paris: Germer Baillière, (2) + 228 pp.). This part contains Mesmer's *Mémoire* of 1779 (entry number 10), his *Mémoire* of 1799 (entry number 211), and the *Aphorismes.* [H]

503. *Sandby, George.*

Mesmerism and Its Opponents: with a Narrative of Cases. London: 1844, x + 278 pp.

This book, written by a clergyman of the Church of England, was very influential in helping to create a favorable opinion of animal magnetism in Britain. After disposing of M'Neile's charge that mesmerism is diabolical (M'Neile: *Satanic Agency and Mesmerism: A Sermon*, entry number 460), Sandby describes his own personal experiences with animal magnetism and their influence on his opinion of the genuineness of the phenomena. Using cases from the mesmeric literature, he emphasizes both the benefits of and potential dangers of animal magnetic practice, emphasizing the need to counter superstitious attitudes in those who participate. Sandby also suggests techniques for the use of magnetizers. [H & P]

504. [*Staite, O.*].

Mesmerism, or, The New School of Arts, with Cases in Point. London: n.p., 1844, 101 pp.

A work written in verse. [H]

505. *Wigan, Arthur Ladbroke.*

A New View of Insanity. The Duality of the Mind Proved by the Structure, Functions, and Diseases of the Brain, and by the Phenomena of Mental Derangement, and Shown to be Essential to Moral Responsibility. London: Longman, Brown, Green and Longmans, 1844, xii + 459 pp.

Wigan attempts to explain "double consciousness" or what he calls "alternate consciousness" in terms of the two hemispheres of the brain. Drawing his examples from various fields, including the literature on somnambulism and what would later be called multiple personality, he suggests that there are two minds, corresponding to the two hemispheres, and that these minds can operate simultaneously and independently. This view he then contrasts with a theory of dual consciousness connected with the two hemispheres proposed by Henry Holland, who argued that although a kind of dual consciousness is involved, the mind itself is one. Wigan's book was never all that well known, although it shows up in footnotes in some later writings, such as Ribot's *Les maladies de la personnality* (1885, *see* entry number 1020). [H]

1845

506. *Amouroux, J. A.*

Le magnétisme à Chateauroux. Chateauroux: Adolph Nuret, 1845, 98 + (3) pp.

Chiefly concerned about the threat to chastity inherent in the use of animal magnetism with women, the author cites the secret report to the French king in 1784. [H]

507. *Archives de la société magnétique de Cambrai.*

Vols. 1–2; 1845–1846.

[H]

508. *Brierre de Boismont, Alexandre Jacques François.*

Des hallucinations, ou histoire raisonée des apparitions, des visions, des songes, de l'extase, du magnétisme et du somnambulisme. Paris: G. Baillière, 1845, viii + 615 pp. ENGLISH: *Hallucinations or, The Rational History of Apparitions, Visions, Dreams, Ecstasy, Magnetism, and Somnambulism.* Philadelphia: Lindsay and Blakiston, 1853.

Brierre de Boismont attempts to remove the subject of hallucination from the exclusive jurisdiction of medical pathology and place it within the domain of psychology. To do this, he examines the phenomenon historically and discusses its broad occurrence in the course of ordinary human events. After examining hallucination in connection with nervous disorder, Brierre de Boismont considers the hallucinations of dreams and nightmares and the occurrence of hallucinations in animal magnetism, somnambulism and ecstasy. There is also a section on collective hallucinations. In the latter part of the book, the causes, symptomology and treatment of hallucinations are outlined. This work enjoyed an unusual popularity and exercised a great deal of influence over subsequent thought concerning the nature of apparitions, precognition, clairvoyance, and other paranormal phenomena. [H & P]

509. ***Confessions of a Magnetiser, Being an Exposé of Animal Magnetism.***

Boston: Gleason's Pub. Hall, 1845, 50 pp.

Published anonymously, this pamphlet condemns animal magnetism as a source of depravity. The author, apparently a former magnetizer, describes how the mesmerizer is himself entranced when treating a young, beautiful woman, hardly able to resist the temptation to entice her to satisfy his base desires. The power he exercises over her thoughts and actions, according to the author, is such that she will go to any lengths to please him. [H]

510. ***Crumpe, G. S.***

Letters on Animal Magnetism. Edinburgh: W. H. Lizars, 1845, 24 pp.

[H]

511. ***Du Potet de Sennevoy, Jules Denis.***

Essai sur l'enseignement philosophique du magnétisme animal. Par le Baron Du Potet de Sennevoy. Paris: A. René et Cie, et Mensut, 1845, 356 pp.

[H & P]

512. ***Forbes, John.***

Mesmerism True—Mesmerism False: A Critical Examination of the Facts, Claims, and Pretentions of Animal Magnetism. With an Appendix Containing a Report of Two Exhibitions by Alexis. London: Churchill, 1845, 76 pp.

Forbes was editor of the *British and Foreign Medical Review* and he consistently opposed the mesmerists, whom he considered to be naive dreamers. Here, as in his journal, he attacks the views of John Elliotson. This work was also published under the title *Illustrations of Modern Mesmerism from Personal Investigation.* [H & P]

513. ***Gauthier, Aubin.***

Traité pratique du magnétisme et du somnambulisme: ou, résumé de tous les principes et procédés du magnétisme, avec la théorie et la définition du somnambulisme, le description du caractère et des facultés des somnambules, et les règles de leur direction. Paris: Germer Baillière, 1845, (2) + vi + 752 pp.

A thorough general treatise on the theory and practice of animal magnetism. Among other topics, Gauthier discusses the effect of magnetic treatments, the way the mental and physical condition of the magnetizer affects the outcome, the methods of magnetizing, the use of magnetized objects in treatment, the use of animal magnetism on animals, the phases of magnetic cures and the phenomena of magnetic somnambulism. As in most of Gauthier's work, a historical orientation for all aspects of animal magnetism is apparent. [H]

514. ***Greenhow, Thomas Michael.***

Medical Report of the Case of Miss H——— M———. London: S. Highley, 1845, 24 pp.

Account of a cure brought about through mesmerism. [H]

515. ***Grimes, James Stanley.***

Etherology; or the Philosophy of Mesmerism and Phrenology: Including a New Philosophy of Sleep and Consciousness, with a Review of the Pretensions of Neurology and Phreno-magnetism. Boston and New York: Saxton Peirce & Co., and Saxton and Miles, 1845, xvi + (17)–350 pp.

Grimes was one of the most important American writers on animal magnetism. His work was influential in Britain and Europe as well as in the United States. This book, the product of many years experience as a lecturer and mesmerizer, presents Grimes' rather unique views on mesmerism. Basing his work on the notion that all that happens in the world is to be explained in terms of matter, motion, and mind, Grimes maintains that if there are to be any effects of matter upon matter or mind upon mind, they must occur through a medium that connects them. This medium he calls the "etherium." The study of the way in which the etherium conveys impressions from one mind to another and the way in which these impressions are channeled through the physical organism, he calls "etherology." "Etheropathy" may then be said to occur whenever the etherium is forced to act in opposition to its normal mode of action. For Grimes, animal magnetism, mesmerism, pathetism, hypnotism, somnambulism, and clairvoyance are all morbid etheropathic conditions of the human constitution. From this it follows that individuals are only susceptible to being mesmerized when in a state of inner imbalance and that only through knowledge of the laws of etherology can one work with nature to benefit people needing help. Grimes combines this view of human psychology with his peculiar approach to phrenology. [H]

516. ***Haldat du Lys, Charles Nicolas Alexandre de.***

Histoire du magnétisme dont les phénomènes sont rendus sensibles par le mouvement. Nancy: Grimblot et Veuve Raybois, 1845, 51 pp. [H]

517. ***Hall, Spenser Timothy.***

Mesmeric Experiences. London: H. Baillière, 1845, viii + 103 pp.

Hall was a staunch supporter of the use of animal magnetism as a cure for disease. Believing that the opponents of magnetic cures were chiefly arrogant physicians and others with a vested interest in maintaining superiority over the common people, Hall lectured extensively on the subject throughout England. In this book, Hall writes of his experience and of the value of phrenology used in conjuction with mesmerism. [H]

518. *Journal du magnétisme, rédigé par une Société de magnétiseurs et médecins, sous la direction de M. le Baron Du Potet.*

Vols. 1–20; 1845–1860.

Continued as: *Journal du magnétisme. Sous la direction de H. Durville.* Nos. 1–4, then Vols. 22–67(?)+; 1879–1925+. This journal was edited by the Baron Du Potet and, after a brief interruption, by Hector Durville. A second series began with Vol. 16. [H]

519. *Laurent, P.*

Système des passes magnétiques, ou ensemble des procédés de magnétisation. Nantes: Forest, (1845), 15 pp.

A useful description of the magnetic "passes" used to direct animal magnetism. [H]

520. *Loisel, A.*

Observation concernant une jeune fille de dix-sept ans amputée d'une jambe à Cherbourg le 2 octobre 1845, pendant le sommeil magnétique. Cherbourg: Beaufort et Lecauf, 1845, 23 pp.

With animal magnetism as an anesthetic, Loisel carried out a number of major surgical operations at Cherbourg that were among the earliest such operations. This is an account of one of his operations. [H]

521. *Loisel, A.*

Recueil d'opérations chirurgicales pratiquées sur des sujets magnétisés. Cherbourg: Beaufort and Lecauf, 1845, 24 pp.

An account of surgical operations performed by Loisel using animal magnetism as an anesthetic. *See* entry number 520. [H]

522. *Martineau, Harriet.*

Letters on Mesmerism. London: E. Moxon, 1845, (1) + (v)–xii + 65 + (1) pp.

Martineau was one of the most unusual and influential women of nineteenth-century England. A powerful intellectual force in economics and social science, she first gained success through the publication of her *Illustrations of Political Economy* (1832–1834). In addition to writing prolifically on politics and economics, she was also a successful novelist and writer of children's stories. In 1844 Martineau was cured of a serious illness through animal magnetism, and in this collection of letters she discharges her "duty" to make the truth about mesmerism known. She had known about mesmerism for some time, and when her illness became debilitating, a medical friend brought her to the famous mesmerist Spencer Hall (*see* entry number 517), who was then lecturing at Newcastle. He mesmerized Martineau and the beneficial effects were immediate. Martineau first had her maid and then another woman mesmerize her regularly. After consistent treatment of this kind, Martineau recovered. During the mesmeric treatment, she often experienced powerful distortions of sensation which she describes in some detail in the Letters. [H]

523. ***Newnham, William.***
Human Magnetism: Its Claims to Dispassionate Inquiry: Being an Attempt to Show the Utility of Its Application for the Relief of Human Suffering. London: Churchill, 1845, vi + (1) + 432 pp.

In his *Essay on Superstition* (1830) Newnham was perhaps the first notable person in nineteenth-century England to speak highly of animal magnetism. In this work, published some fifteen years later, he takes up the subject at length, covering the principle issues of theory and practice. Of particular interest are sections on the extent to which imagination can explain the effects of animal magnetism and phreno-magnetism, a theory for which Newnham had little regard. [H]

524. ***Pellegrino, G., [pseudonym: L. Verati].***
Storia, teoria e pratica del magnetismo animale. 4 vols. Florence: n.p., 1845–1846.

A detailed study of animal magnetism, describing phenomena reported in France and Italy, and providing an account of the author's own experiences. [H]

525. ***Smith, Gibson.***
Lectures on Clairmativeness: or, Human Magnetism. New York: Searing & Prall, 1845, 40 pp.

[H]

526. ***Sunderland, LaRoy.***
"Confessions of a Magnetizer" Exposed. Boston: Redding and Co., 1845, 47 pp.

Sunderland attacks the anonymous pamphlet *Confessions of a Magnetizer, being an exposé of animal magnetism* (entry number 509) as a slander against those who competently and honorably practice animal magnetism. Sunderland claims that the pamphlet is full of falsehoods and distortions and is misleading to the public. [H]

527. ***Szapary, Ferencz Grof.***
Katechismus des Vital-Magnetismus zur leichteren Direction der Laien-Magnetiseurs. Zusammengetragen während seiner zehnjährigen magnetischen

Laufbahn nach Aussagen von Somnambulen und vieler Autoren. Leipzig: Otto Wigand, 1845, (1) + viii + 416 pp.

In this work Szapary admits that, although his first book on animal magnetism (*Ein Wort über animalischen Magnetismus*, entry number 431) appeared in 1840, it was not until 1843 that he really came to understand the true nature of the phenomenon. In that year, he had among his patients a 16-year-old girl, Auguste Kachler, whose somnambulistic pronouncements on animal magnetism took him beyond a mere "poetical" understanding to a truly scientific one. From these insights, Szapary started a new school of healing magnetism, and he credits this young woman as its true founder. In the *Katechismus*, Szapary brings together Auguste Kachler's pronouncements and those of other somnambulists and authors to construct a "catechism" of basic truths about healing magnetism. It is presented in this form for a very specific reason. Szapary states that the main difference between Mesmer's teaching and that of the new school is that according to Mesmer only a few people are capable of magnetizing others, whereas the new school teaches that everyone can do it. Because he held that anyone could magnetize, Szapary undertakes to present to the lay practitioner the basic principles of magnetic practice in a simple and readable form. [H]

528. *Szapary, Ferencz Grof.*

Die magnetische Lehre der neuen Schule in Fragen und Antworten nach den Vorlesungen . . . von ein seiner Hörer. Regensburg: Manz, 1845, 111 pp.

Four hundred forty-nine questions and answers on animal magnetic practice. This book was intended as a companion work to entry number 527. [H]

529. *Teste, Alphonse.*

Le magnétisme animal expliqué, ou Leçons analytiques sur la nature essentielle du magnétisme, sur ses effets, son histoire, ses applications, les diverses manières de la pratiquer, etc. Paris: J. B. Baillière, 1845, (3) + 479 + (1) pp.

Lectures written in an easy style and ranging over the history and theory of animal magnetism. [H]

530. *Wagner, A.*

Facts and Fallacies of Mesmerism; Demonstrated to Its Friends and Opponents. London: Stevenson, 1845, 31 pp.

[H]

1846

531. ***L'art de former les somnambules, traité pratique de somnambulisme magnétique, à l'usage des gens du monde et des médecins qui veulent apprendre à magnétiser.***
Montpellier: Pierre Grollier, 1846, 68 pp.

A practical handbook for magnetizers who work with somnambulists. It suggests what qualities those magnetizers should possess and tells how to choose and train good somnambulistic subjects. [H]

532. ***Braid, James.***
The Power of the Mind over the Body: an Experimental Inquiry into the Nature and Cause of the Phenomena Attributed by Baron Reichenbach and Others to a "New Imponderable." London, Edinburgh, and (Manchester): London: John Churchill; Edinburgh: Adam & Chas. Black; (Manchester: Grant and co.), 1846, 36 pp.

A very intelligent critique of the findings of Baron von Reichenbach concerning his newly discovered "odic" force. In Braid's mind, this was just another way of talking about the "universal magnetic fluid" of the mesmerists. Consistent with his previous writings on that subject, Braid denies that such a "new imponderable" has been proven to exist. He claims that all the phenomena adduced to establish such a proof may be accounted for in terms of the remarkable power of the human mind over the body and that it is "unphilosophical" to accept its existence unless more ordinary means of explanation fail. [H]

533. ***Carus, Carl Gustav.***
Psyche, zur Entwicklungsgeschichte der Seele. Pforzheim: Flammer and Hoffman, 1846.

Carus, a physician and philosopher, wrote books on gynecology, comparative anatomy, physiology, and psychology. Here, he describes the nature and structure of the unconscious as a repository for all of the feelings and perceptions which we once had and of which we are no longer conscious. Carus argues that communication constantly takes place between individuals on an unconscious level and that studying such communication is crucial for understanding human psychology. His ideas on the unconscious had a strong influence on Eduard von Hartmann (entry number 924). [H]

534. ***Du Potet de Sennevoy, Jules Denis.***
Manuel de l'étudiant magnétiseur, ou Nouvelle instruction pratique sur le magnétisme, fondée sur 30 années d'observation; suivi de la 4e éd. des expériences faites en 1820 à l'Hôtel-Dieu de Paris. Paris: Baillière, 1846, xii, 344 pp.

Written with the beginner in mind, the manual provides some basic theory and techniques for the neophyte in animal magnetism. It was published together with the fourth edition of Du Potet's first work: *Exposé des expériences* of 1821 (entry number 302).

535. *Elliotson, John.*

The Harveian Oration, Delivered before the Royal College of Physicians, London, June 27th, 1846. With an English Version and Notes. London: H. Baillière, (8) + 70 pp.

The Latin address recounts the shoddy treatment given to medical innovators over the centuries, and calls upon the present generation of physicians not to repeat this sorry history in their treatment of animal magnetism. The unusual procedure of publishing the address in both Latin and English was meant to make Elliotson's views more available to the public. An appendix contains a *Letter to the Royal College of Physicians, London, dated March 28, 1802* by Franz Anton Mesmer (1734–1815). Mesmer commends the doctrine of animal magnetism to the Royal College of Physicians, and states that, although some persons in England had detracted him personally, his theory nonetheless remains "undamaged in England, where the discovery has not yet been proclaimed." [H]

536. *Esdaile, James.*

Mesmerism in India, and Its Practical Application in Surgery and Medicine. London: Longman, Brown, Green, and Longmans, 1846, xxxi + (1) + 287 pp.

Esdaile, originally from Scotland, was a practicing surgeon in Calcutta, India. A physician of great skill, in 1847 he was appointed Surgeon to the Government of India. At the outset, he lists seventy-three painless surgical operations performed in the previous eight months while the patients were in a mesmeric trance, as well as eighteen cases of cures brought about by animal magnetic passes. The surgical operations included amputation of an arm, cataract operations, great toenails cut out by the root, and the removal of large tumors of up to eighty pounds. Esdaile reports that not one death occurred as a result of the operations. Esdaile then traces the history of mesmerism, gives a detailed description of how he induced the mesmeric trance, and describes the course of many cases from diagnosis through operation to recovery. Although this work, along with Elliotson's *Numerous Cases of Surgical Operations Without Pain* (entry number 474) was briefly influential, experimentation of the kind Esdaile carried out was cut short by the discovery of an effective chemical anaesthesia the very year *Mesmerism in India* was published. [H]

537. *Gauthier, Aubin.*

Réforme médicale: compérage magnétique réprimé, questions et observations d'ordre public sur la pratique du magnétisme, du mesmérisme et du somnam-

bulisme, considérée comme exercice de la médecine, etc. . . . Paris: Dondey-Dupré, 1846, ii + 44 pp.

[H]

538. *Haldat du Lys, Charles Nicolas Alexandre de.*
Deux mémoires sur le magnétisme (Recherches sur l'universalité de la force magnétique. Recherches sur l'appréciation de la force magnétique). Nancy: Grimblot et Veuve Raybois, 1846, 41 pp.

[H]

539. *Jones, Henry.*
Animal Magnetism Repudiated as Sorcery;—Not . . . Science . . . With an Appendix on Magnetic Phenomena by William H. Beecher, D.D. New York: J. S. Redfield, 1846, 24 pp.

[H & P]

540. *Leger, Theodore.*
Animal Magnetism; or Psychodunamy. New York: D. Appleton & Co., 1846, (7) + 8–402 pp.

Leger proposes a complete revision of the nomenclature of animal magnetism (apparently he had not heard of Braid and "hypnotism") and substitutes "psychodunamy" or "power of the soul." He renames all the appropriate operations, the verb being to "dunamise," etc. So he dismisses "animal electricity" (Petetin, entry number 225), "mesmerism," "pathetism" (Sunderland, entry number 489), and "etherology" (Grimes, entry number 515). In renaming the phenomenon, however, Leger did not revise the characteristics attributed to it. The book presents a very detailed and useful history of the fortunes of animal magnetism from the investigations of the favorable French commission in the 1820s to the time of its writing. It concludes with an interesting prehistory of "psychodunamy" in the medical practices of the ancients, a description of Mesmer's discovery, and an account of "psychodunamy" in England and America. [H]

541. *Loisel, A.*
Insensibilité produite au moyen du sommeil magnétique. Nouvelle opération chirurgicale faite à Cherbourg. Cherbourg: n.p., 1846.

For previous surgical operations at Cherbourg, *see* entry number 520. [H]

542. *Loisson de Guinaumont, Claude Marie Louis.*
Somnologie magnétique, ou recueil de faits et opinions somnambuliques pour servir à l'histoire du magnétisme humain. Paris: Germer Baillière, Sagnier et Bray, 1846, 324 pp.

A collection of pronouncements by a somnambulist on the nature of human magnetism and other matters. [H]

543. [*Loubert, Jean Baptiste*].

Défense théologique du magnétisme humain, ou le magnétisme est-il superstition, magie? Est-il condamné à Rome? Les magnétiseurs et les somnambules sont-ils en sûreté de conscience? Peuvent-ils être admis à la participation des sacrements? Paris: Poussielgue-Rusand, 1846, 330 pp.

Loubert defends himself against critics of the views he expressed in *Le magnétisme et le somnambulisme devant les corps savants, le corp de Rome et les théologiens* (entry number 498). Objections had been raised not only by theologians who opposed animal magnetism, but also by magnetizers who disagreed with many of his conclusions. [H & P]

544. *Magnetic and Cold Water Cure.*

Vol. 1; 1846.

One volume only (?), published in Boston and Rochester, New York. [H]

545. *Magnétisme. Insensibilité produite au moyen du sommeil magnétique. Nouvelle opération chirurgicale faite à Cherbourg.*

Cherbourg: n.p., 1846.

[H]

546. *Report of the Committee Appointed by Government to Observe and Report upon Surgical Operations by Dr. J. Esdaile, upon Patients under the Influence of Alleged Mesmeric Agency.*

Calcutta: Military Orphan Press, 1846, 29 pp.

When James Esdaile began using animal magnetism or mesmerism to anaesthetize patients for surgical operations, his apparent successes caused the government in India to determine just what was taking place. The report supported Esdaile's claims and led to the establishment of a mesmeric hospital. There he used mesmerism not only for surgery, but also to alleviate pain and even treat illnesses. [H]

547. *Roux, F.*

Coup d'oeil sur le magnétisme animal et le somnambulisme considérés sous le rapport médical et religieux. Montpellier: Boehn, 1846, 122 + (2) pp.

[H]

548. [*Tanchou, ———*].

Enquête sur l'authenticité des phénomènes électriques d'Angélique Cottin. Paris: Baillière, 1846, 54 pp.

Angélique Cottin was a peasant girl in her early teens who lived in a village near Montagne in Normandy. For about ten weeks she exhibited phenomena of an apparently "electrical" nature, beginning on January 15, 1846. Objects began to move and behave unpredictably around her, and a Dr. Tanchou decided to investigate what was happening. He discovered that her body behaved as though charged with electricity—sometimes positive and sometimes negative. He states that the electrical force emanated from the left side of her body, particularly from the area of the waist and the bend of the arm. It seemed to be strongest in the evening. [H & P]

1847

549. *Bush, George.*

Mesmer & Swedenborg; or, The Relation of the Developments of Mesmerism to the Doctrines and Disclosures of Swedenborg. New York: John Allen, 1847, x + (2) + (13)–288 pp.

A most unusual book relating the psychic and psychological phenomena of animal magnetism to the doctrines of the Swedish sage Emanuel Swedenborg. Although Mesmer published his first treatise on animal magnetism (1779) seven years after the death of Swedenborg, those who have followed the teachings of Swedenborg have always manifested an interest in mesmerism. Thus, the first known Swedish document on animal magnetism was published in Stockholm in 1788 by the "Exegetical and Philanthropical Society," a group formed to promote Swedenborg's writings (*see* entry number 186). Dr. Jung-Stilling (1740–1817), who wrote on both animal magnetism and spirit communication (*see* entry numbers 231 and 232), was well acquainted with the writings of Swedenborg. It is not surprising then, that Bush, who finds in the psychic phenomena of mesmerism a means for verifying the truth of the experiences of Swedenborg, would have attempted to unify the two streams in a cogent way. [H & P]

550. *Delaage, Henri.*

Initiation aux mystères du magnétisme. Théorie du magnétisme. Connaissance des maladies, causes et remèdes. Faits magnétiques. Vision somnambulique. Vision dans l'avenir et dans l'espace, etc. . . Rouen and Paris: A. Péron and Dentu, 1847, vi + 97 pp.

Delaage was a mystic, magnetizer, and popularizer of magnetic and spiritual subjects. A kind and modest individual, he was often consulted for spiritual advice. In his writings and life he remained a loyal Catholic and saw no contradiction between his faith and animal magnetism. This work is a general introduction to the theory and practice of animal magnetism, with advice about everything from healing to clairvoyance. [H & P]

551. ***Gentil, Joseph Adolphe.***

Magnétisme. Explication du phénomène de seconde vue et de soustraction de pensée, dont jouissent les somnambules lucides. Du magnétisme au point de vue de la thérapeutique. Marcillet, notice biographique. Paris: Albert Frères, (1847), 32 pp.

[H & P]

552. ***Lafontaine, Charles.***

L'art de magnétiser ou le magnétisme animal considéré sous le point de vue théorique, pratique et thérapeutique. Paris: Germer Baillière, 1847, vii, 364 pp.

The first book of one of the most celebrated magnetizers of the nineteenth century. Lafontaine was a stage magnetizer of great ability who toured Europe and England giving demonstrations of the power of mesmerism. It was his lecture and demonstration in England in 1841 that led James Braid to do the experiments which led to his doctrine of "hypnotism." In this book Lafontaine deals with both the theory and practice of animal magnetism. It indicates that Lafontaine was not merely a stage magnetist. He had a interest in animal magnetism as an agent for curing or alleviating illnesses, and this work is a veritable mine of information about the use of animal magnetism in Lafontaine's day. [H]

553. ***La Salzede, Charles de.***

*Lettres sur le magnétisme animal, considéré sous le point de vue physiologique et psychologique, à M. le Dr. X***.* Paris: Labe, 1847, 231 pp.

[H]

554. ***Magnétisme, insensibilité absolute produite au moyen du sommeil magnétique. Trois nouvelles opérations chirurgicales pratiquées à Cherbourg, le 4 juin 1847, en présence de plus de 60 temoins.***

Cherbourg: Beaufort et Lecauf, (1847).

[H]

555. ***Tonna, Charlotte Elizabeth (Browne).***

Mesmerism, a Letter to Miss Martineau. Philadelphia: Martien, 1847, 16 pp.

[H]

556. ***Wilson, James Victor.***

How to Magnetize, or Magnetism and Clairvoyance. A Practical Treatise on the Choice, Management and Capabilities of Subjects, with Instructions on the Method of Procedure. Revised ed. New York: S. R. Wells and Company, 1879, v + (1) + 11–104 pp.

A practical manual (which first appeared in 1847) of animal magnetism written in a very popular style. No information is available about the first edition except its date. [H & P]

1848

557. ***Almignana, Abbé.***
Magnétisme. Le Christ qualifié de magnétiseur par la synagogue et l'incrédulité modernes et le magnétisme plaidant lui-même la cause du Christ. Paris: n.p., 1848, 36 pp.

While accepting animal magnetism as a genuine phenomenon, the author, a priest, emphasizes that this does not detract from the supernatural character of the works of Christ. [H & P]

558. ***Cahagnet, Louis Alphonse.***
Guide du magnétiseur ou procédés magnétiques d'après Mesmer, de Puységur et Deleuze, mis à la portée de tout le monde suivi des bienfaits et dangers du somnambulisme. Paris: The Author, 1848, 64 pp.

A little work briefly describing how to use "passes" or special movements of the hands to magnetize, how to employ magnetism as a healing technique, and how to gain access to the spirit world through somnambulism. [H & P]

559. ***Cahagnet, Louis Alphonse.***
Magnétisme. Arcanes de la vie future devoilés, ou l'existence, la forme, les occupations de l'âme après sa séparation du corps sont prouvées par plusieurs années d'experiences au moyen de huit somnambules extatiques qui ont eu quatre-vingts perceptions de trente-six personnes de diverses conditions décédées à différentes époques, leur signalement, conversations, renseignements preuves irrécusables de leur existence au monde spirituel!. 3 vols. Paris: Baillière, (1848–1854).

The work of Cahagnet was important in providing a direct link between the tradition of animal magnetism and that of spiritualism. Cahagnet, a Swedenborgian, accepted the belief of the Swedish seer that human beings can communicate with the spirit world. It is not surprising then that when Cahagnet put individuals into a state of magnetic somnambulism, some of them began to speak of experiences of just that kind. *Arcanes de la future* is the first and most important of his works. In it he describes the visions of eight somnambulists who describe in detail their encounters with spirits of the dead. In some cases they relate information about the deceased person that seemed to be unknown to them but proved to be accurate. They also describe the nature of life beyond the grave and relay moral admonitions from those who have gone to the other side. In his work with these magnetic ecstatics, Cahagnet anticipated the spiritualist movement which was soon to sweep America and, a few years later, England, Germany and France. In France only Billot (*see* entry number 393) had carried out similar investigations with magnetic somnambulists before Cahagnet. Volume two of this work was translated into English and published under the title *The Celestial Telegraph* (London: George Peirce, 1850). [H & P]

560. *Corfe, George.*

Mesmerism Tried by the Touch-stone of Truth: Being a Reply to Dr. Ashburner's Remarks on Phrenology, Mesmerism, and Clairvoyance. London: Hatchard & son, 1848, xiv + (15)–44 pp.

[H & P]

561. *Crowe, Catherine.*

The Night Side of Nature: or, Ghosts and Ghost Seers. 2 vols. London: Newby, 1848, viii + 422; 384 pp.

A very influential work that presents a sober description of various supernormal phenomena including presentiments, trance, doppelgangers, apparitions, haunted houses, spectral lights and poltergeists. Crowe's work stirred interest in the supernormal among serious-minded individuals and helped prepare the way for what would later be called psychical research. [P]

562. *Gentil, Joseph Adolphe.*

Initiation aux mystères secrets de la théorie et de la pratique du magnétisme rendue simple et facile quant à la pratique etc., suivie d'expériences inédites faites à Monte Cristo chez Alexandre Dumas, de la biographie de J. B. Marcillet, de la visite faite au somnambule Alexis par le général Cavaignac. Paris: Robert, 1848, 100 pp.

[H]

563. *Laurent, P.*

Introduction au magnétisme animal par P. Laurent, suivie des principaux aphorismes du Dr Mesmer dictés par lui à l'assemblée de ses élèves, et dans lesquels on trouve ses principes, sa théorie et les moyens de magnétiser. Paris: Lange-Lévy, 1848, 23 pp.

[H]

564. *Newman, John B.*

Fascination, or The philosophy of Charming, Illustrating the Principles of Life in Connection with Spirit and Matter. New York: Fowler and Wells, 1848, x + (11)–176 pp.

A curious and interesting treatment of the subject of fascination, a term which the author prefers to "animal magnetism" or other alternate names. Newman traces fascination back to Adam and insists that Mesmer did not discover anything new. He points out that fascination is commonly observed in the animal kingdom and has been employed as a healing procedure among human beings from time immemorial. No earlier publication information is available. [H]

565. *Pasley, T. H.*

The Philosophy which Shows the Physiology of Mesmerism, and Explains the Phenomenon of Clairvoyance. London: Longman, Brown, Green, and Longmans, 1848, vii + 104 pp.

The author attempts to provide an elaborate explanation of the nature of animal magnetism and its effects through a "Philosophy of Mechanical Nature." His explanations are similar to those of Mesmer himself, emphasizing matter and motion as the essential elements of all magnetic action. Like Mesmer, he explains clairvoyant perception as due to the all-pervading presence of a universal fluid that the magnetized person can perceive because of the special state of the magnetized nervous system. [H & P]

566. *Ricard, Jean Joseph Adolphe.*

Esquisse de l'histoire du magnétisme humain depuis Mesmer jusqu'en 1848. Bordeaux: Institut magnétologique, 1848, 35 pp.

[H]

567. *Ricard, Jean Joseph Adolphe.*

Vade mecum du magnétiseur. Bordeaux: Institut magnétologique, 1848, 179 pp.

[H]

568. *Teste, Alphonse.*

Les confessions d'un magnétiseur: suivies d'une consultation médico-magnétique sur des cheveux de Mm. Lafarge. 2 vols. Paris: Garnier frères, 1848.

[H]

1849

569. *Angelhuber, J. F.*

Die prophetische Kraft des magnetische Schlafes, oder wunderbare Enthüllungen des Zukunft durch Somnambulen psychologische dargestellt und durch zahlreiche Beispiele bestätigt. Nebst Fingerzeigen, die zum Hochschlaf geeigneten Personen in den Zustand der clairvoyance zu versetzen. Weimar: Bernhard Friedrich Voigt, 1849, (4) + 264 pp.

Working within the magnetic tradition as elaborated by the principal German magnetists, Angelhuber describes how such things as prophecy, dowsing, physical rapport, and clairvoyance work. [H & P]

570. *Baumann, A.M.F.*

Curative Results of Medical Somnambulism, Consisting of Several Authenticated Cases, Including the Somnambule's Own Case and Cure. London: Hippolyte Baillère, 1849, 43 pp.

[H]

571. *Blakeman, Rufus.*

A Philosophical Essay on Credulity and Superstition; and also on Animal Fascination, or Charming. New York and New Haven: D. Appleton (New York) and S. Babcock (New Haven), 1849, 206 pp.

A little known treatise which discusses everything from mesmerism and quackery to ghosts and witchcraft. [H & P]

572. *Bristol Mesmeric Institute.*

Report of the First Public Meeting; the Rt. Hon. Earl Ducie, President, in the Chair. London: n.p., 1849, 12 pp.

[H]

573. *Buchanan's Journal of Man.*

Vols. 1–6; 1849–1856. Vols. 1–3; 1887–1890 (new series).

Edited by Joseph Rodes Buchanan. [P]

574. *Duncan, George W.*

A Synopsis of Mesmerism; or, Animal Magnetism, Pathetism, Electrical Psychology; or the Philosophy of Impressions. Philadelphia: n.p., 1849, 15 pp.

[H]

575. *Haddock, Joseph W.*

Somnolism and Psycheism, Otherwise Vital Magnetism, or Mesmerism: Considered Physiologically and Philosophically. With an Appendix Containing Notes of Mesmeric and Psychical Experience. London: Hodson, 1849, 73 pp.

Two lectures given by Haddock, a physician of Bolton, England, on the subject of mesmerism. In the appendix he describes some remarkable experiments done with his chief somnambulistic subject, Emma. Haddock enlarged the work for a second edition published in 1851 which enjoyed great popularity in England and the United States. The book is important both for its influence in popularizing mesmerism and also for its account of Emma's paranormal experiences. [H & P]

576. ***Lee, Edwin.***
Animal Magnetism and the Associated Phenomena, Somnambulism, Clairvoyance, etc. London: J. Churchill, 1849, vi + (3)–55 pp. [H & P]

577. ***Le magnétiseur spiritualiste, Journal rédigé par les membres de la société des magnétiseurs spiritualistes de Paris.***
Vols. 1–3; 1849–1851.

Published in Paris and edited by Louis Alphonse Cahagnet (1805–1885), this periodical was intended as a forum for short articles on animal magnetism, ecstatic somnambulism, magnetic healing and related subjects. [H & P]

578. ***Maitland, Samuel Roffey.***
Illustrations and Enquiries Relating to Mesmerism. Part I. London: William Stephenson, 1849, vi + 80 pp.

A small treatise sympathetic to mesmerism written by a British clergyman. Maitland gathers together some of the more striking accounts of mesmeric phenomena in an effort to arouse the curiosity of the reader. He treats, among other things, clairvoyance and mesmerism at a distance. Part 2 never appeared. [H & P]

579. ***Mayo, Herbert.***
Letter on the Truths Contained in Popular Superstitions, with an Account of Mesmerism. Frankfurt: J. D. Sauerlaender, 1849, iv + 5–152 pp.

Mayo was Professor of Comparative Anatomy at the Royal College of Surgeons in London. The whole first part is about various "superstitions" in which Mayo finds both truth and delusion. These include the divining rod, ghosts, and vampirism. This section is followed by a discussion of trance in its various forms. He investigates both natural trances, such as coma, and artificially induced trance, particularly that brought about by animal magnetism. Mayo accepts the reality of many of the "higher phenomena" of magnetic somnambulism and gives a detailed description of the various kinds or stages of clairvoyance. He also mentions a patient who manifested "quintuple consciousness," apparently a case of multiple personality. [H & P]

580. ***Mongruel, Louis Pierre.***
Prodiges et merveilles de l'esprit humain sous l'influence magnétique. Paris: n.p., 1849, 92 pp.

The author writes about magnetic somnambulism and the oracular states that it can produce. His wife (*see* Madame Mongruel, entry number 809) was such a somnambulist and was known in Paris as the "modern sibyl." In this work Mongruel cites certain information acquired somnambulistically that shows, he believes, that she had psychic access to secret meetings. [H & P]

581. *Olivier, Joseph.*

Traité de magnétisme, suivi des paroles d'un somnambule, et d'un recueil de traitements magnétiques. Toulouse: Jouglat, 1849, 521 pp.

[H]

582. *Pretreaux, J. D.*

Electricité naturelle, ou Mesmérisme mis en pratique à l'usage des familles. Cambrai: A. Girard, 1849, 24 pp.

[H]

583. *Reichenbach, Karl Ludwig von.*

Physikalish-physiologische Untersuchungen über die Dynamide des Magnetismus, der Electrizität, der Wärme, des Lichtes, der Krystallisation, des Chemismus in ihren Beziehungen zur Lebenskraft. 2 vols. 2 ed. Braunschweig: Friedrich Vieweg and Sohn, 1849, xiv + (4) + 218; vi + 240 pp. ENGLISH: *Psycho-physiological Researches on the Dynamides or Imponderables, Magnetism, Electricity, Heat, Light, Crystallization, and Chemical Attraction, in Their Relation to the Vital Force.* Translated by William Gregory. London: Taylor, Walton and Maberly, 1850.

Reichenbach was a German natural philosopher and industrialist, famous as the discoverer of creosote and pariffin. He liked scientific experimentation and in the 1840s questioned human perception of magnetic emanations. He came to believe that every substance in the universe gives off emanations which can be perceived by some individuals. Reichenbach called these emanations "od" and the people who can perceive them "sensitives." He believed that he had conclusively established that these individuals could see an "odyllic light" of definite color proceeding from the human fingertips, from the poles of magnets, from crystals, and from other substances. Reichenbach was certain that, in his experiments, he had removed all possibility that suggestion or ordinary perception were involved in these experiences. Sensitives, according to Reichenbach, used a faculty possessed by a great part of the population (perhaps half), and he considered the perception of "od" emanations to be a purely natural phenomenon with no supernatural implications. He tried to forcefully set himself apart from the proponents of animal magnetism, stating that his concern was simply to investigate a general physical force, and that he had no interest in the therapeutic practices of the mesmerizers. Nonetheless, it is difficult to distinguish Reichenbach's odic force from Mesmer's magnetic fluid. The similarity is reflected in general writings on human magnetism from 1850 on that often treat the two phenomena as identical. In *Physikalish-physiologische Untersuchungen*, Reichenbach's most popular work on the subject, the author presents an overview of his notion of "od" (or "odyle," as Gregory translated it) and a description of the experiments he conducted to establish its existence. There were two English translations: the one referenced above, and another translated by John Ashburner and published in

1851. Reichenbach's writings were numerous, but most of them were never translated into English. Information on the first German edition of this work is unavailable. [H]

584. *Rice, Nathan Lewis*

Phrenology Examined, and Shown to be Inconsistent with the Principles of Physiology, Mental and Moral Science, and the Doctrines of Christianity: also an Examination of the Claims of Mesmerism. New York and Cincinnati: R. Carter & Brothers (New York) and J. D. Thorpe, (Cincinatti), 1849, viii + 9–318 pp.

[H]

585. [*Rodgers, W. H.*].

Facts in Magnetism, Mesmerism, Somnambulism, Fascination, Hypnotism, Sycodonamy, Etherology, Pathetism, &c., Explained and Illustrated. Auburn, Derby: The Author, 1849, 96 pp.

[H]

586. *Scoresby, W.*

Zoistic Magnetism: Being the Substance of Two Lectures, Descriptive of Original Views and Investigations Respecting This Mysterious Agency; Delivered, by Request, at Torquay, on the 24th of April and 1st of May, 1849. London and Torquay: Longman, Brown, Green, and Longmans (London) and Cockren (Torquay), 1849, 144 pp.

Against the accusation that a supernatural, diabolical agency is involved in phenomena of animal magnetism (zoistic magnetism), Scoresby attempts to show that the phenomena are explicable in terms of the known laws of magnetism and electricity. Scoresby makes his point from experiments he himself had conducted, beginning in 1845. He distinguishes five degrees or stages of zoistic magnetism: 1) an initial effect that involves dimming of sight or warming of the forehead; 2) a closing of the eyes and inability to reopen them; 3) sleep-walking or somnambulism, with an awareness only of the magnetizer; 4) sleep-walking with the perception of sensations induced by the magnetizer, and often insensibility to pain; and 5) a sleep of insensibility, where no personal impressions affect the subject and there is total immunity to severe pain. In this fifth stage ecstatic experiences may occur. [H]

1850–1859

1850

587. ***Ashburner, John.***

Facts and observations on the mesmeric and magnetic fluids. N.P.: n.p., 1850, 16 pp.

[H]

588. ***Barth, George H.***

The Mesmerist's Manual of Phenomena and Practice; with Directions for Applying Mesmerism to the Cure of Diseases, and the Methods of Producing Mesmeric Phenomena. Intended for Domestic Use and the Instruction of Beginners. London: H. Baillière, 1850, viii + 192 pp.

Claiming no original contribution to animal magnetism, Barth sets out to make plain how mesmerism may be practiced by those who are interested. He concentrates on descriptions of the various states and stages, and depicts the expected phenomena. He also points out certain problems that can occur in the application of animal magnetism. This little manual was very popular in its day. [H]

589. ***Beesel, M.***

Darstellung und Enthüllung des Somnambulismus, mit besonderer Bezugnahme auf den Somnambulen, Stahlschmiedegesellen Carl Wilhelm Kohn. 2 vols in 1. Danzig: L. G. Homann, 1850, iv + 54 + (2); (2) + 56 + (2) pp.

[H]

590. *Benoit, Jacques Toussaint, and Biat, Chrétien.*
Communication universelle et instantanée de la pensée à quelque distance que ce soit, à l'aide d'un appareil portatif appelé Boussole Pasilalinique Sympathique. Paris: Bureau de l'Institute Polytechnique, 1850, 80 pp.

A unique treatise containing a design for communication at a distance through telepathic communication between snails. The plan called for a simple code to be transmitted from one place to another by the stimulation of a snail on the transmitting end. This code could then be perceived on the receiving end by observing the sympathetic reaction of a paired snail. The result was supposed to be a kind of "snail telegraph." [P]

591. *Braid, James.*
Observations on Trance: or, Human Hybernation. London: Churchill, 1850, 72 pp.

Here Braid deals with three cases of men who were buried alive and survived the ordeal. Drawing an analogy with the state of hibernation in animals he suggests this may be possible because the men were able to place themselves in a trance so deep that the ordinary signs of life in the body were hardly detectable. Braid also notes that the cataleptic state need not necessarily be accompanied by a loss of consciousness in human beings and suggests ways to treat pathological catalepsy. [H]

592. *Bristol Mesmeric Institute.*
Report of the Annual Meeting, May 1850; James Adam Gordon, Esq., Vice-President, in the Chair. London: n.p., 1850, 11 pp.

[H]

593. *Cahagnet, Louis Alphonse.*
Sanctuaire du spiritualisme. Étude de l'âme humaine, ses rapports avec l'univers, d'après le somnambulisme et l'extase. Paris: Germer Baillière, 1850, (6) + 585 pp.

Cahagnet's second major work, in the form of twelve "conferences", is an investigation of human nature as disclosed in the visions of ecstatics. The ecstatics discussed include those whose visions were drug induced. [H & P]

594. *Delaage, Henri.*
Le monde occulte, ou mystères du magnétisme et tableau de somnambulisme. Paris: E. Dentu, 1850, 198 pp.

[H & P]

595. *Dewey, Dellon Marcus.*
History of the Strange Sounds or Rappings, Heard in Rochester and Western New York, and Usually Called the Mysterious Noises! Which are Supposed by

Many to be Communications from the Spirit World, Together with all the Explanation than Can as yet Be Given of the Matter. Rochester: The Author, 1850, (1) + ii + (13) – 79 pp.

In 1848, the home of John D. Fox (situated in Hydesville, near Rochester, New York) was disturbed by knockings and other sounds that were believed to be produced by the spirit of a deceased man. Two of Fox's daughters, Kate (1841–1892) and Margaret (1838–1893) who seemed to be the focus of these sounds, discovered that the knocks seemed to respond to their questions. At the suggestion of their mother, they devised a simple code that would allow the knocks to spell out messages. Soon the Fox home was full of neighbors and other curious people trying to discover the source of the sounds. Eventually the daughters left home, but the knocks followed them. Kate and Margaret began to put on demonstrations in living rooms and meeting rooms for those interested in witnessing their communications with the spirit world. They began a tour that quickly brought national and then world attention to the phenomena. This was the beginning of the Spiritualist movement that over the next five years spread to countries all over the world. *History of the Strange Sounds* is probably the earliest published book on the events that took place at the Fox home. It is an important source of information about the first days and weeks in which knockings were observed. The book contains many signed testimonials from those who witnessed the early events, includes a diagram of the interior of the Fox home, and is exceedingly rare. [P]

596. ***Dods, John Bovee.***

The Philosophy of Electrical Psychology: In a Course of Twelve Lectures. New York: Fowler & Wells Co., (1850), 252 pp.

At the invitation of Senators Sam Houston and Daniel Webster, Dods delivered these lectures on his doctrine of electrical psychology to the United States Congress. He begins by indicating that for many years he had believed that electricity is the connecting link between mind and matter, which have a reciprocal action upon each other. This is the basis for his electrical psychology. By making use of electrical psychology he had been able to cure people of a great variety of diseases. Although the same "fluid" is involved in mesmerism and electrical psychology, electrical psychology is not the same as mesmerism, which is a doctrine of sympathy. Electrical psychology is a doctrine of impressions. While rapport is central to mesmerism and causes the mesmerized person to have his or her senses tied up with those of the magnetizer, in electrical psychology the subject's senses remain entirely independent of the operator. [H]

597. ***Gromier, Emile.***

Du magnétisme; qu'est-ce que le magnétisme ou étude historique et critique des principaux phénomènes qui le constituent, suivie de l'explication rationnelle qu'il convient d'en donner. Lyon and Paris: Savy, 1850, 56 pp.

A discussion of the nature of animal magnetism, drawn from an examination of some of its principal practitioners. [H]

598. ***A Key to the Science of Electrical Psychology. All its Secrets Explained, with Full and Comprehensive Instructions in the Mode of Operation and its Application to Disease, with Some Useful and Highly Interesting Experiments. Every Person an Operator. By a Professor of the Science.***
N.p.: n.p., 1850, 12 pp.

Intended to provide information about "electrical psychology," a kind of mesmeric healing technique, the pamphlet was designed to supplement the knowledge of those who had taken the practical courses in the subject. The author suggests that the subject to be operated on might place a coin on the palm of his or her hand and concentrate on it for a half hour or so, thus setting up a "galvanic current"—a practice that recalls Braid's hypnotic technique of fixing the gaze on a shiny object. [H]

599. ***Léquine, F.***
Mesmérisme à l'aide d'un bassin et d'un ventilateur inventé par F. Léquine. Paris: The Author, 1850, 4 pp.

An ill-fated attempt to revive the *baquet*, the apparatus used by Mesmer to store and distribute animal magnetic fluid. [H]

600. ***Mongruel, Louis Pierre.***
Appel de l'Union-protectrice à tous les partisans et amis du magnétisme. Protestation en faveur de la libre manifestation des croyances et de la libre application de la science de Mesmer. Rapport collectif et officiel de la commission pour servir à la défense du somnambulisme. 2 ed. Paris: M. Mongruel, 1850, 24 pp.

Information on the first edition, dated July 25, 1850, is incomplete. This second edition is dated August 1, 1850. [H]

601. ***Mongruel, Louis Pierre.***
Petit traité sur le magnétisme animal, contenant 1. un précis historique sur la matière; 2. une dissertation succincte sur les influences occultes qui dominent l'homme; 3. une courte appréciation de l'avenir du magnétisme; 4. une notice sur la sibylle moderne et sur ses facultés somnambuliques; 5. des preuves positives d'une lucidité prodigieuse et incontestable; 6. des conseils à ceux qui veulent la consulter avec fruit, etc. Paris: The Author, 1850, 15 pp.

Mongruel's wife (*see* Madame Mongruel, entry number 809) was a well-known oracular somnambulist of Paris. In this collection of essays, the author presents evidence and arguments supporting the genuineness of her utterances. [H & P]

602. *Nani, Giacomo D.*

Trattato teorico-pratico sul magnetismo animale. Turin: Ferrero E. Franco, 1850, lxiii + (1) + 305 + (2) pp.

Nani was a magnetizer in the tradition of Deleuze. [H]

603. *Schopenhauer, Arthur.*

Parerga und Paralipomena. Kleine philosophische Schriften. 2 vols. Stuttgart: Cotta, 1850.

This collection includes the essay "Versuch über das Geistersehn und was damit zusammenhängt," which is a philosophical essay on apparitions in which Schopenhauer examines the intrinsic nature of such experiences. He points out that in apparitions of spirit figures, the perceiver experiences the presence of an individual in just the same manner as he would experience the presence of a bodily figure, even though no body is present. The question for Schopenhauer boils down to whether the origin of the experience is from within the organism of the perceiver or external to it. He draws heavily on material found in the literature of animal magnetism. He sees a close relationship between animal magnetism, magic, second sight, spirit seeing, and visions of all kinds. Schopenhauer considers them all to be versions of what he calls "dreaming the real." But the fact that he relates apparitional experiences to our dreaming faculty or "dream organ," does not mean that he dismisses them as mere chimera. On the contrary, he sees good evidence that these experiences involve perceptions that correspond to realities—hence his phrase, "dreaming the real." In this way, an individual could have the dream-sensory experience of the actual presence of someone who is dead. On the other hand, Schopenhauer rejects the spiritualist view of these experiences, since he holds the notion of an immaterial substance (etherial matter) that acts on bodies to be philosophically untenable. [P]

604. *Stone, G. W.*

Electro-biology; or, the Electrical Science of Life. Liverpool: Willmer and Smith, 1850, 48 pp.

[H]

1851

605. *Atkinson, Henry George, and Martineau, Harriet.*

Letters on the Laws of Man's Nature and Development. London: J. Chapman, 1851, xii + 390 pp.

[H & P]

606. *Bennett, John Hughes.*

The Mesmeric Mania of 1851, with a Physiological Explanation of the Phenomena Produced: a Lecture. Edinburgh: Sutherland and Knox, 1851, 21 pp.

An attempt to provide a physiological explanation for animal magnetism. [H]

607. *Braid, James.*

Electro-biological Phenomena Considered Physiologically and Psychologically. Edinburgh: Sutherland and Knox, 1851, 33 pp.

In this work Braid criticizes the claims of the practitioners of "electro-biology," an approach that had its origins in the works of two Americans, J. Stanley Grimes, (1807–1903) and John Bovee Dods (1795–1872). It is Braid's contention that the claims made by this school that they produce phenomena of a hypnotic type through electrical impressions made upon the subject by the operator were not based on any kind of substantial evidence. The methods used, he argues, relied upon suggestions on the part of the operator which told the subjects what was expected of them. In other words, Braid claims that electro-biology was just an unacknowledged form of hypnotism. [H]

608. *Buckland, Thomas.*

Hand-book of Mesmerism, for the Guidance and Instruction of All Persons who Desire to Practice Mesmerism. 3 ed. London: Baillière, 1851, 72 pp.

A popular manual of animal magnetism by the Secretary to London's Mesmeric Infirmary. Following the doctrine of Deleuze and his pupils, Buckland discusses the qualities that a good magnetizer should possess. [H]

609. *Cahagnet, Louis Alphonse.*

Lumière des morts, ou études magnétiques philosophiques et spiritualistes dédiées aux libres penseurs du XIXe siècle. Paris: The Author, 1851, xiii + 322 pp.

This book deals principally with occult aspects of spiritistic magnetism. [H & P]

610. *Cahagnet, Louis Alphonse.*

Du traitement des maladies, ou Étude sur les propriétés médicinales de 150 plantes les plus usuelles par l'extatique Adèle Maginot, avec une exposition des diverses méthodes de magnétisation. Paris: G. Baillière, 1851, 212 pp.

[H & P]

611. *Capern, Thomas.*

The Mighty Curative Powers of Mesmerism: Proved in Upwards of One Hundred and Fifty Cases of Various Diseases. London: H. Baillière, 1851, xxvii + 120 pp.

A handbook of curative animal magnetism, describing treatment of conditions ranging from toothache to heart problems. [H]

612. ***Coddè, Luigi.***
Il magnetismo animale svelato ossia teoria e pratica dell'antropo-elettro-magnetismo ed una nuova meccanica delle sostanze. N.p.: Corrado, 1851.

Coddè was a magnetizer from Piedmont who developed a bio-electric theory of animal magnetism, that he discusses in this work. [H]

613. ***Colquhoun, John Campbell.***
An History of Magic, Witchcraft, and Animal Magnetism. 2 vols. London: Longman, Brown, Green & Longmans, 1851, lxiii + (1) + 326; 316 pp.

Colquhoun was an ardent partisan of animal magnetism. In this work he attempts to show that just as with many great scientific discoveries (Galileo, Harvey, etc.) in its early stages animal magnetism had been linked in the minds of its critics with magic and superstition and rejected by the scientific establishment. As these discoveries proved to be of great benefit to mankind, so will it be with animal magnetism. To substantiate this view, Colquhoun ranges through the history of the rise and development of animal-magnetic practices, harking back now and again to analogous phenomena in previous eras of history. The work is full of interesting information about magnetic practices, although somewhat uncritical in its evaluation of the genuineness of some of the facts. [H]

614. ***Concato, L.***
Sull'azione del magnetismo animale nell'umano organismo. Padua: Crescini, 1851.

[H]

615. ***Del vero spirito scientifico secondo il quale debbono essere esaminate le ragioni della frenologia e del mesmerismo. Dissertazione di W. G. già pubblicata nel Giornale Frenologico di Edinburgo. Seguita da alcuni esperimenti frenomesmerici di G. T. Hall tratti dallo stesso giornale.***
Milan: n.p., 1851.

[H]

616. ***Fechner, Gustav Theodor.***
Zend-Avesta oder über die Dinge des Himmels und des Jenseits, von Standpunkt der Naturbetrachtung. 3 vols. Leipzig: L. Voss, 1851.

Fechner was a German physicist, philosopher, and experimental psychologist. He is considered to be a founder of psychophysics, the study of variations in psychological experience that accompany controlled variation in physical events. Fechner was a spiritualist and was strongly influenced by the Swedish seer, Swedenborg, as well as the philosopher Kant (e.g., *Träume eines Geistersehers*, 1766). He also made use of the teachings of magnetic somnambulists. Fechner had a deep interest in the issue of life after death, and in this work explores the relationship between this world and the world of the "other side" of death. [P]

617. *Gregory, William.*

Letters to a Candid Inquirer, on Animal Magnetism. London: Taylor, Walton, and Maberly, 1851, xxii + 528 pp.

Gregory, a Professor of Chemistry at the University of Edinburgh and a well-respected scientist, was an enthusiastic supporter of animal magnetism. In this book he gives a clear description of the mesmeric state and the relationship between the magnetizer and magnetized. He discusses both the "lower" and "higher" phenomena of mesmerism and describes many cases illustrating both. Along with Townshend's *Facts in Mesmerism* (*see* entry number 433), this work, which in slightly abridged form went through many editions, was the best known treatment of the subject. [H]

618. *Guidi, Francesco.*

Magnetismo animale e sonnambulismo magnetico. Turin: Favale, 1851.

The first book on animal magnetism by Italy's most famous "professor of magnetology." Published in Turin, one of the more tolerant of the Italian cities with regard to animal magnetism, this work voices great enthusiasm about the beneficial effects of that art. The book attempts to cover all aspects of animal magnetism, from technique to moral considerations. Guidi would go on to write prolifically and lecture on the subject all over northern Italy. [H]

619. *Haldat du Lys, Charles Nicolas Alexandre de.*

Essai historique sur le magnétisme et l'universalité de son influence dans la nature. Nancy: Grimblot et Veuve Raybois, (1851), 23 pp.

The dating of this pamphlet is uncertain. [H]

620. *Lassaigne, Auguste.*

Mémoires d'un magnétiseur contenant la biographie de la somnambule Prudence Bernard. Paris: Dentu and Baillière, 1851, 157 pp.

A work describing ecstatic phenomena produced by magnetically induced somnambulism including a biography of Prudence Bernard, which describes remarkably successful thought transference experiments. [H & P]

621. *Lecture on Mysterious Knockings, Mesmerism, &c., with a Brief History of the Old Stone Mill, and a Prediction of Its Fall, Delivered Before the A N ti Quarian Society of*

Pappagassett . . . by Benjamin Franklin Macy D.F., D.D.F., A.S.S., Professor of Hyperflutinated Philosophy. Newport, Rhode Island: 1851, 8 pp.

A humorous parody of a lecture given by a mesmerizer. [H & P]

622. *Mongruel, Louis Pierre.*

Le magnétisme militant; origine et histoire des luttes, progrès et conquêtes de la science; le somnambulisme aux prises avec les corps savants etc. . . . résumé général des preuves historiques, philosophiques, scientifiques et juridiques, propres à démontrer l'existence du magnétisme et du somnambulisme lucide, etc. . . . Paris: The Author and E. Dentu, 1851, vi + 256 pp.

This extremely rare work deals with the issue of "lucid somnambulism" and the powers of Madame Mongruel to know hidden things and fortell the future. The book includes a lithographed portrait of the "modern sibyl," as she was called, and a reproduction of her autograph. [H & P]

623. *Rutter, John Obediah Newell.*

Magnetoid Currents, Their Forces and Directions; with a Description of the Magnetoscope; a Series of Experiments. To Which is Subjoined a Letter from William King. London: J. W. Parker & Son, 1851, 47 pp.

Rutter, an engineer, introduced the use of gas for domestic cooking and lighting. He became interested in the use of the hand-held pendulum and intrigued by the fact that its action varied from person to person. Attempting to remove the possibility of movement through muscular action, he devised an instrument (the magnetoscope) that held the pendulum stationary over a glass bowl. He obtained what he believed to be remarkable results from his experiments, observing different movements if the operator held different substances in his hand. The magnetoscope was not successful for other careful experimenters, however, and dropped into disuse. [H]

624. *Scheible, J., [ed.].*

Die sympathetisch-magnetische Heilkunde in ihrem ganzen Umfange: oder die Lehre von der Transplantation der Krankheiten, die Amulete, die Signaturen u.s.w. zum ersten Male ausführlich nach den Schriften der Paracelsisten erläutert und mit einer reichhaltigen Sammlung von Vorschriften sympathetischen Kuren ausgestattet . . . die sogenannten zauberischen Krankheiten und die magisch-magnetischen Heilungen bei den alten Hebräern. Stuttgart: n.p., 1851, 356 pp.

[H & P]

625. *Tommasi, M.*

Il magnetismo animale considerato sotto un nuovo punto di vista. Torino: Pomba, 1851, 222 pp.

[H]

626. *Wood, Alexander.*

What is Mesmerism? An Attempt to Explain its Phenomena on the Admitted Principles of Physiological and Psychical Science. Edinburgh: Sutherland & Knox, 1851, 31 pp.

[H]

1852

627. *Ballou, Aidin.*

An Exposition of Views Respecting the Principal Facts, Causes and Peculiarities Involved in Spirit Manifestations: Together with Interesting Phenomenal Statements and Communications. Boston: Bela Marsh, 1852, vi + (7) + 256 pp.

Ballou takes up the issue of the genuineness of purported communications from departed spirits. The book is a valuable compendium of cases and useful discussion of the then current issues. Ballou concludes that departed spirits do, in some cases, communicate, but, even in those genuine instances, one must exercise caution and never give up the use of one's own reason in favor of advice coming from such communications. [P]

628. *Baragnon, P. Petrus.*

Étude du magnétisme animal sous le point de vue d'une exacte pratique. Pisa: Vannucchi, 1852, 255 pp.

The second edition (1853) was augmented by a discussion of the phenomenon of the rotation of tables. [H & P]

629. *Bersot, Ernest.*

Mesmer et le magnétisme animal. Paris: L. Hachette, 1852, 192 pp.

The first edition of this book was primarily a history of mesmerism with reflections on its implications. However, with the arrival in France of the "talking tables" of American spiritualism, Bersot felt called upon to issue a new edition in 1853, supplementing the original with a lengthy section on talking table phenomena. By the fourth edition of 1876, the section on talking tables had become a lengthy treatment of the various phenomena associated with spiritualism and a thoughtful critique of their meaning. [H]

630. *Berti, A.*

Sul magnetismo animale e sul metodo per istudiarlo. Padua: Sicca, 1852.

[H]

631. *Braid, James.*

Magic, Witchcraft, Animal Magnetism, Hypnotism, and Electro-Biology; Being a Digest of the Latest Views of the Author on these Subjects. Third Edition, Greatly Enlarged, Embracing Observations on J. C. Colquhoun's "History of Magic," &c. London: John Churchill, 1852, (3) + 122 pp.

An attempt to answer certain criticisms brought to bear on Braid's doctrine of hypnotism by Colquhoun, a supporter of animal magnetism. Information is available on the third edition. [H]

632. *Dal Pozzo di Mombello, Enrico.*

Il magnetismo animale considerato secondo le leggi della natura. Siena: n.p., 1852.

The first of a number of works on animal magnetism by the author. He believed that the phenomena of animal magnetism could be explained in terms of natural scientific laws. [H]

633. *Du Potet de Sennevoy, Jules Denis.*

La magie dévoilée, ou principes de science occulte. Paris: Pommeret et Moreau, 1852, viii + 268 pp.

Du Potet, a well-known and highly successful magnetizer, here presents his views on the true nature of magic and its relationship to animal magnetism. Beginning with a long autobiographical section describing his personal evolution of thought, Du Potet discusses his belief that Mesmer's magnetic fluid was the "breath of life" and "fire of the spirit" of the ancient practitioners of magic in the Western tradition and explains how magic is undergoing a revival, largely because of the discovery of animal magnetism. Du Potet then describes the technique for carrying out magical practices and explains why they work. He also discusses the nature of the spirit world and communication with spirits. This book is one of Du Potet's most popular and remains influential among those interested in the occult. [H & P]

634. *Eagle, G. Barnard.*

Mesmerism, Clairvoyance, and Animal Magnetism Explained: Also a Treatise on Mesmerism from the Earliest Ages, Including the Life and Death of Mesmer, the Founder of the Above Science; with Instructions to Gentlemen Wishing to Introduce the New Science of Electro-biology. Including G. B. Eagle's Hand-book of magic. . . . One Hundred Beautiful Illusions are Comprehensively Described. . . . Weymouth: The Author, 36 pp.

[H & P]

635. *Ennemoser, Joseph.*

Anleitung zur mesmerischen Praxis. Stuttgart and Tübingen: J. G. Cotta, 1852, vi + 514 pp.

[H]

636. *Esdaile, James.*
The Introduction of Mesmerism, as an Anaesthetic and Curative Agent, into the Hospitals of India. Perth: Dewar, 1852, 48 pp.

An account of Esdaile's use of animal magnetism for surgery, pain relief and cure in India. [H]

637. *Esdaile, James.*
Natural and Mesmeric Clairvoyance with the Practical Application of Mesmerism in Surgery and Medicine. London: Hyppolyte Baillière, 1852, xix + (1) + 272 pp.

Esdaile, a pioneer in the use of animal magnetism as an anesthetic for surgery, discusses the "higher" and "lower" phenomena of magnetic somnambulism, with emphasis on clairvoyance, and describes how to bring about somnambulistic states. Esdaile saw animal magnetism as a physical power, one therefore capable of affecting anyone, no matter what their psychological state. Thus he believed that even sleeping people, the blind, and madmen could be magnetized. [H]

638. [*Francis, J. G.*].
Mesmerism and Spiritual Agency. London: Charles Goodall & Son, 32 pp.

[H]

639. *Gathy, August François Servais.*
Compte-rendu du banquet commémoratif de la naissance de Mesmer (118ème anniversaire) célébré le 23 Mai 1852 à Paris. Paris: Pommerat et Moreau, 1852, 15 pp.

[H]

640. *Gentil, Joseph Adolphe.*
Magnétisme. Somnambulisme. Guide des incrédules. Paris: Hubert, 1852, 192 pp.

Focuses on the more unusual phenomena of somnambulism. [H]

641. *Haldat du Lys, Charles Nicolas Alexandre de.*
Exposition de la doctrine magnétique; ou, Traité philosophique, historique, et critique du magnétisme. Nancy: Grimblot et Veuve Raybois, 1852, viii + 320 pp.

[H]

642. *Hébert de Garnay, L. M.*
Petit catechisme magnétique ou notions élémentaires du mesmérisme. Paris: n.p., 1852, 36 pp.

[H]

643. *Hudson, ———, Captain.*
An attempt to Explain Some of the Wonders and Mysteries of Mesmerism, Biology, and Clairvoyance. Chorley: George Houghton, 1852, 52 pp.

A treatise, rather flowery in style, extolling the virtues of mesmerism. [H & P]

644. *Krauss, Friedrich.*
Nothschrei eines Magnetisch-Vergifteten; Thatbestand, erklärt durch ungeschminkte Beschreibung des 36 jährigen Hergangs, belegt mit allen Beweisen und Zeugnissen. Zur Belehrung und Warnung besonders für Familienvater und Geschäftsleute. Stuttgart: 1852, 21 + vi + 22–914 + vii pp.

A most unusual book by a man who claims he was "poisoned" over a great many years by the magnetic influence of someone who was secretly in rapport with him. [H & P]

645. *Lélut, Louis François.*
Mémoire sur le sommeil, les songes et le somnambulisme. Orleans: Coignet-Darnault, 1852, 31 pp.

[H]

646. *Lovy, Jules.*
Magnétisme et somnambulisme. Paris: Vincon, 1852, 2 pp.

[H]

647. *Mackay, Charles.*
Memoires of Extraordinary Popular Delusions and the Madness of Crowds. 2 vols. London: National Illustrated Library, 1852, viii + 303; vii + 324 pp.

Mackay was a British journalist who became editor of the Glasgow *Argus* and the *Illustrated London News.* This extremely popular book is a narrative of fads and fanaticisms arising from group experience. Ranging from the alchemists to airy financial schemes, Mackay attempts to show how easily whole populations have been led astray by the enthusiastic fantasies of the few. The book contains a section on animal magnetism and the "metallic tractors" of Elisha Perkins (see entries 204–206). [H & P]

648. *Reichenbach, Karl Ludwig von.*
Odisch magnetische Briefe. Stuttgart and Tübingen: J. G. Cotta, 1852, xii + 199 + (1) pp.

A popular treatment of various aspects of Reichenbach's "Od" force (*see* Reichenbach, entry number 584). Included are subjects such as "od" and crystals, polarity, chemical phenomena, and the effects of "od" in daily life. [H]

649. ***[Sitwell, F.].***
Mesmerism Considered. Glasgow: Blackie & Son, 1852, 15 pp.

[H]

650. ***[Sitwell, F.].***
What is Mesmerism? and What its Concomitants Clairvoyance and Necromancy? 2 ed. London: Bosworth & Harrison, 1852, 32 pp.

No information available on the first edition. [H & P]

1853

651. ***Assezat, Jules, and Debuire, H.***
Magnétisme et crédulité ou solution naturelle du problème des tables tournantes. Paris: Garnier frères, 1853.

[H & P]

652. ***Bahr, Johann Karl.***
Der animalische Magnetismus und die experimentierende Naturwissenschaft. Dresden: Turk, 1853, 52 pp.

[H]

653. ***Barth, George H.***
What is Mesmerism? The Question Answered by a Mesmeric Practitioner, or, Mesmerism not Miracle: An Attempt to Show that Mesmeric Phenomena and Mesmeric Cures are not Supernatural; to Which is Appended Useful Remarks and Hints for Sufferers Who are Trying Mesmerism for a Cure. London: Baillière, 1853, ix + (3) + 150 pp.

The author is an enthusiastic supporter of animal magnetism as a means of cure, as well a means of attaining special mental states, including clairvoyance. He describes three degrees of mesmerism: 1) mesmeric sleep, resembling natural sleep, but with insensibility that can in some cases allow surgery to be performed without pain, 2) mesmeric sleep-walking, or somnambulism, with its attendant phenomena of clairvoyant vision and ecstasy, and 3) mesmeric waking state, in which the person retains full consciousness and subsequent memory, but experiences other phenomena, such as paralysis, transposition of the senses, and great suggestibility. Describing the new "table moving" fad then sweeping through England and Europe, Barth suggests that "in every second or third house has 'table moving' been practiced; it is hardly probable that any scientific phenomenon of any kind has ever been brought more generally to the test of demonstrative experiment in an equally short period of time" (p. vii). He asserts that this phenomenon is explicable in terms of the forces of mesmerism: "There can be no question that the 'table turning' will establish a general conviction of the reality of the mesmeric or animal magnetic power; — it leaves no abiding place for a doubt in the mind of any reasonable being who will trust the evidence of his senses" (p. viii). In addition to the one volume version, *What is Mesmerism?* was also published in three separate parts in 1853. [H]

654. *Beecher, Charles.*

A Review of the "Spiritual Manifestations." Read Before the Congregational Association of New York and Brooklyn. New York: G. P. Putnam, 1853, viii + 9–75 pp.

An important early atempt to explain the physical and mental phenomena of spiritualism. Arguing that these phenomena either are or are not the work of spirits, Beecher examines the not-spirit explanation and finding it wanting, argues in favor the spirit explanation. [P]

655. *Birt, William Radcliff.*

Table-Moving Popularly Explained; with an Inquiry into Reichenbach's Theory of Od Force; Also an Investigation into the Spiritual Manifestations Known as Spirit-rappings. London: n.p., 1853, 76 pp.

[H & P]

656. *Braid, James.*

Hypnotic Therapeutics, Illustrated by Cases. With an Appendix on Table-Moving and Spirit-Rapping. Reprinted from the Monthly Journal of Medical Science for July 1853. Edinburgh: Murray & Gibb, 1853, 44 pp.

Braid describes how hypnotism may be used effectively in curing diseases. He mentions the usefulness of the so-called "magnetic passes" in treatment, but denies the reality of magnetic fluid. Braid appends a section on the new phenomenon of table moving, expressing his disbelief that it is the work of spirits. [H & P]

657. *Brittan, S. B. and Richmond, B. W.*

A Discussion of the Facts and Philosophy of Ancient and Modern Spiritualism. New York: Partridge & Brittan, 1853, ix + (1) + (3)–377 pp.

An intelligent if partisan dialogue between the two authors on subjects relevant to the genuineness of spiritualism. Among other topics, the work discusses the problem of imitation while a person is in the mesmeric trance; the fluid of animal magnetism and the physical phenomena of spiritualism; Reichenbach's "Od" force as the contemporary proof of Mesmer's discovery of animal magnetism; the temperaments of mediums; and the forms of the physical phenomena of mediumship. [P]

658. ***Burq, Victor Heab Antoine.***

Métallothérapie; traitement des maladies nerveuses, paralysies, rhumatisme chronique, etc. . . du choléra, etc. . . . Paris: Germer Baillière, 1853, 48 pp.

Burq was the originator of "metallotherapy" which used metal plates, usually made of copper, to heal disease. To some degree, this approach has its roots in animal magnetism, which strongly influenced Burq, and in Perkinism, an American healing system devised by Elisha Perkins (1741–1799). This seems to be Burq's first work on the subject. [H]

659. ***Carpenter, William.***

"Electro-Biology and Mesmerism." *The Quarterly Review* 93 (1853): 501–57.

William Carpenter was a noted physiologist in Britain. In this important article, he takes up two issues: the nature of animal magnetism and the cause of certain physical phenomena associated with spiritualism, principally "table-turning" and "table-talking." With regard to the first, Carpenter reveals himself as one of the few early supporters of the ideas of James Braid in England. He states that suggestion causes the somnambulist to respond in the way the magnetizer expects. He points out that this suggestive influence might be quite subtle, so that neither magnetizer nor subject realizes it is operating. He also relates the apparent clairvoyant feats of somnambulists to subtle suggestions given the somnambulist by the magnetizer. With regard to table-turning and table-talking, he comes up with a similar explanation to that of Faraday: that the motion of the table is the result of unconscious muscular pressure exerted by the sitters. [H & P]

660. ***Der Cheiroelektromagnetismus oder die Selbstbewegung und das Tanzen der Tische (Tischrücken). Eine Anweisung in Gesellschaften das werkwürdige Phänomen einer neu entdeckten menschlischen Urkraft hervorzubringen. Nach einigen pract. Versuchen u. unter Vergleich aller bisher veröffentlichten Proben mitgeheilt.***

Berlin: Lassar, 1853, 16 pp.

[H & P]

661. ***Cohnfeld, Adalbert Salomo.***

Die Wunder-Erscheinungen des Vitalismus; Tischdrehen, Tischklopfen, Tischsprechen u. nebst ihrer rationellen Erklärung in Briefen an eine Dame. Bremen: Schunemann, 1853, 125 pp.

[H & P]

662. ***Cronaca del magnetismo animale.***

Vols. 1–2: 1853–1854.

Edited by Giuseppe Terzaghi, this is the first Italian periodical devoted to animal magnetism. [H]

663. *La danse des tables dévoilée, expériences de magnétisme animal, manière de fair tourner une bague, un chapeau, une montre, une table, et même jusqu'eux têtes des expérientateurs et celles des spectateurs.*

Paris: n.p., 1853.

[H & P]

664. *Edmonds, John W. and Dexter, George T.*

Spiritualism. With an Appendix by Nathaniel P. Tallmadge. 2 vols. New York: Partridge & Brittan, 1853 and 1855, 505; 542 pp.

A very influential work that went through numerous editions over the course of two decades. After a lengthy introduction by Edmonds and a shorter one by Dexter, the authors present a detailed description of a long series of automatic writing sessions in which Dexter was the medium for what purported to be a number of famous departed spirits, such as Emanuel Swedenborg. The subjects of their discourses were principally the world of spirits and the moral life. [P]

665. *Examen raisonné des prodiges récents d'Europe et de d'Amerique notamment des tables tournantes et répondantes, par un philosophe.*

Paris: J. Vermot, 1853, 77 pp.

The author takes a dim view of the "talking tables" fad imported from the United States. He sees this phenomenon (and animal magnetism) as the work of the devil. [H & P]

666. *Faraday, Michael.*

"Professor Farady on Table Moving." *The Athenaeum* (London), No. 1340, (July 2, 1853): 801–803.

In 1853 the fad of "table-moving" swept across the Atlantic from the United States and rapidly engulfed the whole of Britain and Europe. "Table-moving" consisted of gathering a number of people (often termed "sitters") around a table and having them attempt to get the table to rotate or rise and fall without using physical force to do so. Techniques varied greatly. For example, the sitters might spread their hands on the table, joining little finger to little finger to form a continuous circle; or they might hold their hands above the table with little or no contact with it. The practice derived from the newly formed Spiritualist movement whose central belief was that spirits of the dead can communicate with the living. Spiritualists believed that the tables were moved by spirits, and when the tables would rise and fall, causing a leg to knock against the floor, they would discern spirits messages being tapped out in code for the benefit of the "sitters." Others believed that the tables indeed did move, but that spirits were not involved; rather the physical effects were caused by some unknown force produced by the "sitters" (such as animal magnetic force, odic force, electrical force, etc.). Still others believed that no movement without physical force took place at all, and that table moving phenomena were either the result of self-deception or deliberate trickery. Since the fad had become so widespread and table moving was being attempted in many of the living rooms of the western world, a clamor arose for men of science to make a pronouncement on the reality of the phenomena. So it was that the physicist Michael Faraday, one of the most emminent scientists of the century, was persuaded to try some experiments and settle the matter "once and for all." This Faraday did, making his findings known first in a letter to the editor of the London *Times* in the June 30, 1853 issue, then in more detail in the *Athenaeum*. Bringing together a group of people as "sitters" whom he considered to be honest and who enthusiastically believed they could move the tables, Faraday concluded that the tables move, if they move, simply as a result of "quasi-involuntary" muscular exertion on the part of the sitters. While believing they were merely pressing straight down on the table, the sitters unwittingly caused their hands, which were in contact with the table, to apply force in a uniform direction of motion. Faraday's experiment was supposed to put the matter to rest, but it did not. Many protested that no general conclusions could be drawn from his very limited experiment, carried out using only one of many possible techniques. Faraday's conclusions, for example, did not pertain to those purported cases of table moving with no physical contact. Nonetheless, Faraday's reputation was such that virtually everyone who wrote about table-moving phenomena after him felt called upon to acknowledge his view and take a stand in relation to it. This early attempt to investigate the physical phenomena of spiritualism scientifically is an important milestone in the development of psychical research. The call for rational and controlled study of these experiences would grow in intensity over the next thirty years, contributing significantly to the founding in London in 1882 of the Society for Psychical Research. [H & P]

667. ***Frisz, ———.***

Les tables et les têtes qui tournent, ou la fièvre de rotation en 1853. Cent et un croquis. Paris: Librairie pittoresque, (1853?).

[H & P]

668. ***Gentil, Joseph Adolphe.***

Magnétisme, somnambulisme. Manuel élémentaire de l'aspirant magnétiseur. Paris: E. Dentu, 1853, 292 pp.

[H]

669. *Godfrey, Nathaniel Steadman.*

Table-Moving Tested and Proved to be the Result of Satanic Agency. London: Seeleys, 1853, 30 pp.

[P]

670. *Godfrey, Nathaniel Steadman.*

Table-Turning, the Devil's Modern Master-piece. Being the Result of a Course of Experiments. London and Leeds: Seeleys, 1853, 59 pp.

An argument in favor of the satanic origin of spiritualistic phenomena, based on experiments conducted by the author himself. [H & P]

671. [*Goupy,* ———].

Quaere et invenies. Paris: Ledoyen, 1853, 203 pp.

A treatise by a magnetizer largely concerned with spiritualism. There is a section on "talking tables," then just reaching France. [H & P]

672. *Guillard,* ———.

Table qui danse et table qui répond, expériences à la portée de tout le monde. Paris: Garnier, 1853, 36 pp.

[H & P]

673. *Hering, Charles E.*

Das Tischrucken. Gotha: n.p., 1853.

[H & P]

674. *Kerner, Justinus Andreas Christian.*

Die Somnambulen Tische. Zur Geschichte und Erklärung dieser Erscheinung. Stuttgart: Ebner & Seubert, 1853, (4) + xvi + 64 pp.

Kerner examines the newly imported fad of "table moving." In his discussion of the history of the phenomenon, he notes that already Athanasius Kircher (*Magnes sive de arte magnetica*, 1643) had spoken of using the magnetic power emanating from the fingers to produce the rotation of a scabbard. Kerner relates the motion of the tables to animal magnetic fluid and the physical effects it was believed to produce. He deals both with table turning and table tipping. [H & P]

675. *Magnétisme. Moyens magnétiques pour faire tourner les tables, les chapeaux.*

Paris: n.p., 1853.

[H & P]

676. *Mirville, Jules Eudes, Marquis de.*
Pneumatologie. Des esprits et de leurs manifestations fluidiques. 10 vols. Paris: H. Vrayet de Surcy, Detaroque et Wattelier, 1853–1868.

This massive work is one of the most serious attempts to investigate paranormal phenomena before the establishment of the Society for Psychical Research in 1882. Mirville's approach, which is historical, involves searching the literature for data relevant to the apparent manifestations of spirits over the ages. The first volume was written in the form of a *mémoire* addressed to the members of the Academy of Moral and Political Sciences. Mirville continued the project through thousands of pages, studying such topics as: mesmerism, table turning, various kinds of spirit possession and exorcism, the spirit rappings of spiritualism, apparitions, miraculous healings, ecstasy, and many other subjects. In the course of his work, Mirville developed his own theory concerning spirit manifestations, viewing them as genuine, but the work of evil spirits, who needed the participation of humans, since paranormal phenomena could only take place through the medium of the magnetic fluid of animal magnetism. The publication history of this work is particularly difficult to trace, since the various volumes appeared separately over a number of years, and supplements were at times added to already published parts. [H & P]

677. *Nees von Esenbeck, Christian Gottfried.*
Beobachtungen und Betrachtungen auf dem Gebiete des Lebensmagnetismus oder Vitalismus. Bremen: C. Schünemann, 1853, 136 pp.

[H]

678. *Practical Instruction in Table-moving, with Physical Demonstrations.*
London: Hippolyte Baillière, 1853, 60 pp.

After intensive experimentation, the author comes to the conclusion that table moving is a genuine phenomenon. While he cannot say how it works, he argues that it is not through "electricity, galvinism, caloric, or terrestrial magnetism." As the title indicates, the author gives detailed instructions for getting a table to move. [H]

679. *Résie, comte de.*
Lettre à M. l'abbé Croizet . . . sur le magnétisme et la danse des tables. Clermont-Ferrand: Hubler, Belle et Dubois, 1853, 15 pp.

[H & P]

680. [*Richemont, Eugène Panon Desbassayans, Comte de*].
Le mystère de la danse des tables dévoilé par ses rapports avec les manifestations spirituelles d'Amérique par un Catholique. Paris: Davarenne, 1853, 32 pp.

[P]

681. *Rogers, Edward Coit.*

Philosophy of Mysterious Agents, Human and Mundane: or the Dynamic Laws and Relations of Man, embracing the Natural Philosophy of Phenomena styled "Spiritual Manifestations." Boston: John P. Jewett, 1853, xii + 13–366 pp.

A very thorough attempt to relate the manifestations of spiritualism to natural but as yet unexplored forces. The first section is one of the most complete examinations of the history of relevant investigations to that date. Rogers covers not only investigations in the English speaking world, he also discusses the work of the French, such as Petetin and Thouvenel, and the Germans, such as Kerner and Reichenbach. The second half of the book relates these investigations to the phenomena of spiritualism. Rogers believes that they can be explained in terms of magnetic and electrical properties of the human organism as manifested through Reichenbach's "odic force." The production of the phenomena are considered to be unconscious, and so Rogers includes a section on "the action of the brain without the action of the mind." [H & P]

682. *Roubaud, Alexandre Félix.*

La danse des tables, phénomènes physiologiques démontrés par le Dr. Félix Roubaud. Paris: Librairie nouvelle, 1853, 104 + (3) pp.

[H & P]

683. *Satanic Agency and Table Turning. A Letter to the Rev. Francis Close, in Reply to His Pamphlet, "Table-turning not Diabolical."*

London: T. Bosworth, 1853, 23 pp.

[P]

684. *Schauenburg, Carl Hermann.*

Tischrücken und Tischklopfen eine Thatsache. Mit Documenten von den Herren: K. Simrock, Hoffmann von Fallersleben, O. Schade und Neusser. Dusseldorf: Arnz & Comp., 1853, 24 pp.

Schauenburg, a scholar from Bonn, agrees with Justinus Kerner that the unusual phenomena of animal magnetism are genuine, as are the newly noted phenomena of table turning and table knocking. [H & P]

685. *Schwarzschild, Heinrich.*

Magnetismus, Somnambulismus, Clairvoyance. Zwölf Vorlesungen für Aerzte und gebildete Nichtärzte. 2 vols. Cassel: Theodor Fischer, 1853, xii + 279; v + (1) + 411 + (1) pp.

The first volume of this work provides one of the better histories of animal magnetism and magnetic somnambulism produced in Germany in the nineteenth century. After a look at the history of "sympathetic" healing before Mesmer, Schwarzschild describes the discovery of animal magnetism and its rapid spread from Paris, placing special emphasis on its history in Germany. He traces the concomitant history of magnetic somnambulism and the clairvoyant experiences that accompanied that phenomenon. Schwarzschild's section on the merging of the magnetic tradition with that of the romantic-magical tradition in Germany is particularly valuable. The second volume of this work takes up the issue of the reality of the phenomena of animal magnetism, somnambulism, and clairvoyance. Schwarzschild himself believes that the phenomena are genuine, and presents a lengthy discussion of the stages of magnetic sleep, the types of clairvoyance, techniques of magnetizing, and similar subjects. [H & P]

686. *Silas, Ferdinand.*
Instruction explicative des tables tournantes, d'après les publications allemandes, américaines, et les extraits des journaux allemands, français et américains. Précédée d'une introduction sur l'action motrice du fluide magnétique, par Henri Delaage. Paris: Houssiaux, 1853, 30 + (1) pp.

[H & P]

687. [*Sitwell, F.*]
What is Mesmerism? London: Thomas Bosworth, 1853, 32 pp.

This is probably another edition of entry number 650. [H]

688. *Snow, Herman.*
Spirit-intercourse: Containing Incidents of Personal Experience while Investigating the New Phenomena of Spirit Thought and Action; with Various Spirit Communications Through Himself as Medium. Boston: Crosby, Nichols, and Company, 1853, xii + 13–184 pp.

Snow's experiences included some of the classical physical phenomena of spiritualism. At one point he examines their possible explanation by magnetism, electricity or mesmerism, but rules such explanations out because of the seemingly intelligent nature of the phenomena. [H & P]

689. *Spicer, Henry.*
Facts and Fantasies: a Sequel to Sights and Sounds; the Mystery of the Day. London: T. Bosworth, 1853, 119 pp.

This book appeared as a sequel to *Sights and Sounds* (entry number 690). [P]

690. *Spicer, Henry.*
Sights and Sounds: the Mystery of the Day: Comprising an Entire History of the American "Spirit" Manifestations. London: Thomas Bosworth, 1853, vii + 480 pp.

The best early British history of the rise and spread of spiritualism. [P]

691. *Sunderland, LaRoy.*
Book of Human Nature: Illustrating the Philosophy (New Theory) of Instinct, Nutrition, Life; with their Correlative and Abnormal Phenomena, Physiological, Mental, Spiritual. New York: Garrett & Co., 1853, xxiv + 25–432 pp.

[H & P]

692. *Sunderland, LaRoy.*
Book of Psychology. Pathetism, Historical, Philosophical, Practical; Giving the Rationale of Every Possible Form of Nervous or Mental Phenomena Known Under the Technics of Amulets, Charms, Enchantment . . . Mesmerism . . . Hallucination . . . Clairvoyance, Somnambulism, Miracles, Sympathy, etc.: Showing How These Results May Be Induced . . . and the Benevolent Uses to Which This Knowledge Should be Applied. New York: Stearns, 1853, (4) + (v–viii) + 9–120 pp.

[H & P]

693. *Szapary, Ferencz Grof.*
Magnétisme et magnéto-thérapie. Paris: n.p., 1853, 304 pp.

True to his belief that anyone can be a magnetic healer (see *Katechismus des Vital-Magnetismus*, entry number 527), Szapary presents a manual of theory and technique for those who want to practice magnetic healing. The work includes a description of qualities a magnetizer should possess and a lengthy section on somnambulism, in which Szapary discusses spasms experienced in various magnetic states. The second edition of this book (1854) was greatly augmented with a section on table-moving or, in Szapary's terms, "gyro-magnetism." Although Szapary believed animal magnetic influence could account for the movement, he was not as much interested in the movement of the table itself as he was in the "psychic community" that is formed around the experience. Those who sit around a table and concentrate on moving it themselves become, in his view, "magnetic baquets" or reservoirs of magnetid fluid. This socio-magnetic concentration of power can lead to all kinds of remarkable phenomena, such as powerful healings and even apparitions. The single person or "medium" who guides the action is simply focussing the power of the group itself. Szapary desired to develop this special "physico-psychic-magnetic" power of the "companions of the table" (the agape) and use it for the benefit of mankind. [H & P]

694. *Table Moving by Animal Magnetism demonstrated: with directions how to perform the Experiment. Also, A full and detailed account of the experiments already performed.*
London; n.p., (1853), 8 pp.

[H & P]

695. ***Table Moving, its causes and phenomena: with directions how to experiment. Profusely illustrated, on plate paper, by Nicholls.***
(London): J. Wesley, 1853, 8 pp.

[H & P]

696. ***Table Turning and Table Talking, Containing Detailed Reports of an Infinite Variety of Experiments Performed in England, France, and Germany, with Most Marvellous Results; also Minute Directions to Enable Every One to Practise Them, and the Various Explanations Given of the Phenomena by the Most Distinguished Scientific Men of Europe. Second Edition with Professor Faraday's Experiments and Explanation.***
London: Henry Vizetelly, (1853?), (2) + 200 pp.

A remarkable compendium of techniques for producing results in the parlor. The book includes a fairly impartial atempt to understand the source of the phenomena. The anonymous author's conclusion is that the movements and messages are due to the force of animal magnetism produced by the experimenters. No publication date is given in the book. [H & P]

697. ***Terzaghi, Giuseppe.***
Attestazioni di illustri scienziati ed uomini sommi in favore dell'esistenza del magnetismo animale e della sua efficacia. Milan: Guglielmini, 1853, (4) + 63 + (4) pp.

[H]

698. ***Terzaghi, Giuseppe.***
Cronica del magnetismo animale. 2 vols. Milan: Pirotta, (1853–1854).

See entry number 662. [H]

699. ***Terzaghi, Giuseppe.***
Sulla potenza motrice trasfusa dall'uomo nella materia bruta. Fenomeno comunemente noto sotto il titolo di Tavola Girante. Esperimenti ed osservazioni. Aggiunge alcune dell'illustre magnetologo prof. Lisimaco Verati ed altre del dottor Francesco Argenti, già decano della facoltà medica dell'Università di Padova e membro di varie accademie. Milan: Guglielmini, 1853.

[H]

700. ***Vollmer, Karl Gottfried Wilhelm.***
Magnetismus und Mesmerismus, oder, Physische und geistige Kräfte der Natur: der mineralische und thierische Magnetismus sowohl in seiner wirklichen Heilkraft, also in dem Missbrauch, der von Betrugern und Narren damit

getrieben worden, im Zusammenhange mit der Geisterklopferei, der Tischrückerei, dem Spiritualismus. Leipzig: Ed. Wartig, (1853) (2) + 692 pp.

[H & P]

1854

701. *Allix, Eugene.*

Guida elementare dello studente magnetizzatore. Turin: Unione tip., 1854.

[H]

702. *Almignana, abbé.*

Du somnambulisme, des tables tournantes et des médiums, considérés dans leurs rapports avec la théologie et la physique. Examen des opinions de MM. de Mirville et Gasparin. Paris: Dentu, 1854, 36 pp.

The author is described as "doctor in cannon law, magnetist-theologian, and medium." He speaks as a spiritualist medium and believes that the dead appear to the living. He holds that evidence rules out demonic interventions (Mirville) or the simple action of magnetic fluid (Gasparin) as explanations for the phenomena. [H & P]

703. *Bénézet, E.*

Des tables tournantes et du panthéisme. Paris: n.p., 1854.

[H & P]

704. *Brownson, Orestes Augustus.*

Spirit-rapper; an Autobiography. Boston: Little, Brown and Company, 1854, xi + 402 pp.

Writing from his own experience with spiritualistic seances and contact with mediums, Brownson fictionalizes the life of a spirit medium who uses spiritualism as a means to overthrow established religion. Examining various alternative explanations for paranormal phenomena, Brown dismisses the mesmeric explanation in favor of the notion that such phenomena are spirit related and that spirits are real, although the attempt to communicate with them is diabolically inspired. In the preface, Brownson cites De Mirville's *Pneumatologie* (entry number 676) as the source of his ideas on the nature of spiritualism. [H & P]

705. *Buchanan, Joseph Rodes.*

Outlines of Lectures on the Neurological System of Anthropology, as Discovered, Demonstrated and Taught in 1841 and 1842. Cincinnati: Office of Buchanan's *Journal of Man*, 1854, (2) + 16 + 384 + 16 pp.

Buchanan's "neurological anthropology" was intended to be a scientific system of philosophy that would provide a unified view of man and his nature. It was based on the conviction that in the last analysis religion and science are unified, and that neither materialism nor spiritualism, lacking an understanding of the nature of the physical side of man, could provide satisfying answers. Buchanan saw in animal magnetism a means for bringing about this unification, since it provides a way for human beings to learn about their basic physical-spiritual constitution. [H & P]

706. *Bulletin magnétique, Journal des sciences psycho-physiques rédigé par une réunion de magnétistes, de médecines, de savants, sous la direction de M. Mongruel.*
One volume only; 1854.

[H]

707. *Cahagnet, Louis Alphonse.*
Magie magnétique ou traité historique et pratique de fascination, miroirs cabalistiques, apports, suspensions, pactes, talismans, possession, envoûtements, sortilèges, etc. Paris: Baillière, 1854, 528 pp.

[H & P]

708. *Cahagnet, Louis Alphonse.*
Magnétisme. Encyclopédie magnétique spiritualiste, traitant spécialement de faits psychologiques, magie magnétique, swedenborgianisme, nécromancie, magie céleste, etc. 7 vols. Argenteuil and Paris: The author and Germer Baillière, (1854–1862), (4) + 322; (4) + 359 + (1); (4) + 356; (2) + 320; (4) + 324; (4) + 356; (4) + iii + (1) + 147 + (1) pp.

The lengthy encyclopedia is a continuation of writings begun in Cahagnet's journal *Le magnétiseur spiritualiste* (*see* entry number 577). Although Cahagnet frequently reproduces quotations relevant to the topic being discussed, nearly all the entries were written by him. As can be seen in his first book, *Magnétisme. Arcanes de la vie future devoilés* (entry number 559), he was a Swedenborgian, and his ecstatic somnambulists found confirmation of the presence and activity of spirits in this world. The seven volumes cover a broad range of subjects, such as descriptions of magnetic seances with spiritualist phenomena, magnetic healing, possession, clairvoyance, sorcerers, the Mormons, etc. It is a valuable reference work of spiritualistic phenomena and magnetic practice. [H & P]

709. *Chevreul, Michel Eugène.*
De la baguette divinatoire: du pendule dit explorateur et des tables tournantes, au point de vue de l'histoire, de la critique et de la méthode expérimentale. Paris: Mallet-Bachelier, 1854, xvi + 258 + (1) pp.

A very thorough critique of the use of the divining rod, the pendulum, and spiritualistic table-turning. Chevreul eventually rules out any supernatural or spiritistic explanation for the phenomena. He also dismisses those explanations which posit some kind of electrical or magnetic force at work in these matters. He concludes that all three types of phenomena are the result of muscular contractions produced unwittingly by the operator. [H & P]

710. *Davey, William.*

The Illustrated Practical Mesmerist, Curative and Scientific. Edinburgh: The Author, 1854, xiii + (18) + 96 pp.

Davey was an enthusiastic promotor of mesmerism in England, and in this endeavour he was a close associate of William Gregory and J. W. Jackson. This, his principal work in the area, describes how to use mesmerism to cure disease and produce psychological effects. [H]

711. *Dods, John Bovee.*

Spirit Manifestations Examined and Explained. Judge Edmonds Refuted; or, an Exposition of the Involuntary Powers and Instincts of the Human Mind. New York: De Witt & Davenport, (1854), xii + 13–252 pp.

A remarkable attempt to explain the physical and psychical phenomena of spiritualism as the result of involuntary instincts of the mind: the physical sounds being produced by an involuntary build-up of nervous electricity, and the psychical or mental abilities being the workings of the "back brain" which controls the involuntary responses of the mind. In addition to criticizing Judge Edmonds, a spiritualist, Dods also finds Reichenbach's theories of odic force useless as an explanation of the phenomena. [H & P]

712. *Gasparin, Agénor Étiene, comte de.*

Des tables tournantes, du surnaturel en général et des esprits. 2 vols. Paris: Dentu, 1854, xxiv + 565; 581 pp. ENGLISH: *Science vs. Modern Spiritualism. A Treatise on Turning Tables, the Supernatural Agent in General, and Spirits*. Translated by E. W. Robert. New York: Kiggins & Kellog, 1857.

An early attempt to explain the "supernatural" physical phenomena of spiritualism in terms of natural causes. Spiritualists held that turning tables, table levitation, knocking sounds from tables, and other seemingly inexplicable physical phenomena were the work of spirits and signs of the continuing presence in this world of those who had passed on to the next. Gasparin, while granting the reality of such curiosities as turning tables and levitation without the exercise of appropriate physical force, denies that this is the work of spirits. He places himself squarely in the tradition of orthodox animal magnetism and posits that when a group of people apply themselves to produce physical movement at a distance, the object can, in fact, be moved by the force of the "magnetic fluid" of the participants. Taking his theory further, Gasparin believed that "magnetic fluid," along with nervous excitement and hallucination, accounted for all the phenomena which spiritualists attributed to departed souls. He thereby transforms the doctrine of animal magnetism, which did so much to promote the rise of spiritualism, into a reason to deny its validity. Gasparin backs up his conjectures with a lengthy series of experiments that are described in the work. [H & P]

713. *Gentil, Joseph Adolphe.*

Magnétisme. L'âme de la terre et des tables parlantes. Paris: The Author and Dentu, 1854, 48 pp.

[H & P]

714. [*Gridley, Josiah A.*].

Astounding Facts from the Spirit World, Witnessed at the House of J. A. Gridley, Southampton, Mass., by a Circle of Friends, Embracing the Extremes of Good and Evil. The Great Doctrines of the Bible such as the Resurrection, Day of Judgment, Christ's Second Coming, Defended, and Philosophically and Beautifully Unfolded by the Spirits. . . . Southampton, Massachusetts: Josiah A. Gridley, 1854, 287 pp.

An important source for early American spiritualist phenomena.

[P]

715. *Guidi, Francesco.*

Il morbo-cholera curabile col magnetismo: memoria. Milan: Redaelli, 1854, 57 pp.

[H]

716. *Guidi, Francesco.*

Trattato teorico-pratico di magnetismo animale considerato sotto il punto di vista fisiologico e psicologico, con note illustrative e appendice. Milan: Turati, 1854, xiv + 432 pp.

Francesco Guidi, self-styled "professor of magnetology," was one of the most influential promoters of animal magnetism in mid-nineteenth century Italy. This is his general manual on the theory and practice of animal magnetism. It was placed on the Catholic Church's *Index* of forbidden books, but was nevertheless widely read. The book includes the ten lessons on animal magnetism he would give when on his lecture tour of Italy. Although Guidi began his career in animal magnetism in Turin, the book was published in Milan where he eventually settled. [H]

717. *Hewlett, Edgar.*

Personal Recollections of the Little Tew Ghost, Reviewed in Connection with the Lancashire Bogie, and the Table-Talking and Spirit-Rapping of the Present Day. London: Aylott, 1854, 24 pp.

[P]

718. ***Library of Mesmerism and Psychology in Two Volumes,*** *Comprising Philosophy of Mesmerism, On Fascination, Electrical Psychology, The Macrocosm, Science of the Soul.* 2 vols.
New York: Fowler and Wells, 1854.

A collection of separate works by different authors. It includes: Part 1 of *The Macrocosm and Microcosm* by William Fishbough; *Fascination, or the Philosophy of charming Illustrating the Principles of Life in Connection with Spirit and Matter* by John B. Newman; *The Philosophy of Electrical Psychology in a Course of Twelve Lectures* by John Bovee Dods; *Six Lectures on the Philosophy of Mesmerism* by John Bovee Dods; and *Psychology; or the Science of the Soul, Considered Physiologically and Philosophically* by Joseph Haddock. [H]

719. ***London Mesmeric Infirmary.***
Report of the Annual Meetings. 12 parts. London: 1854–1866.

The *Reports* from 1853 to 1865. Also published in one volume is the *Report* from 1865 to 1869, London, 1870. [H]

720. ***MacWalter, J. G.***
The Modern Mystery: or, Table-tapping, Its History, Philosophy, and General Attributes. London: John Farquhar Shaw, 1854, viii + 175 pp.

Relying heavily on the work of the American E. C. Rogers, MacWalter concludes that the apparent intervention of spirits in the phenomena of spiritualism is an illusion. Everything can be accounted for in terms of a natural force which is harnessed by mediums and unconsciously expressed in the phenomena. [H & P]

721. ***Il magnetofilo (Società magnetica di Torino).***
One vol. only; 1854.

The organ of the Società magnetica di Torino. [H]

722. ***Il mesmerista.***
Vols. 1–2; 1854–1855.

Published in Turin and edited by E. Allix. [H]

723. ***Michal, Victor.***
Le corps aromal, ou Réponse en un seul mot à l'Académie des sciences philosophiques à propos du concours proposé par elle sur quelques relations à l'andro magnétisme . . . Explication vraie des tables tournantes et parlantes. Paris: n.p., 1854, 23 pp.

[H & P]

724. [***Morin, Alcide***].

Comment l'esprit vient aux tables, par un homme qui n'a pas perdu l'esprit. Paris: Librairie nouvelle, 1854, 174 pp.

A personal account of the phenomenon of "talking tables." [H & P]

725. ***Morogues, Pierre Marie Sébastien Bigot, Baron de.***

Observations sur le fluide organo-électrique et sur les mouvements électrométriques des baguettes et des pendules. Bourges and Paris: Masson, 1854, 255 pp.

An attempt to study the phenomena of the divining rod and the divining pendulum scientifically. [H & P]

726. ***Noizet, François Joseph.***

Mémoire sur le somnambulisme et le magnétisme animal adressé en 1820 à l'Académie royale de Berlin. Paris: Plon Frères, 1854, xx + 428 pp.

Noizet was a general in the French army and an expert on fortifications. He was strongly influenced by the views of both Faria (entry number 294) and Bertrand (entry number 313). Nevertheless, he believed in the existence of a magnetic fluid, something Faria and eventually Bertrand denied. This book is one of the most thorough and balanced to be written by a partisan of the fluid theory. The first part discusses the faculties of the soul, the nature of sleep, and natural somnambulism. The second part deals with artificial somnambulism, which Noizet finds identical in essence to natural somnambulism. He also believes in the possibility of the transmission of thought at a distance and of clairvoyance while in the somnambulistic state. Part three takes up the issue of the magnetic fluid which Noizet terms "vital fluid" and of physical effects of "vital fluid" at a distance. [H]

727. ***Reichenbach, Karl Ludwig von.***

Der sensitive Mensch und sein Verhalten zum Ode. Eine Reihe experimentellen Untersuchungen über ihre gegenseitigen Kräfte und Eigenschaften mit Rücksicht auf die praktische Bedeutung, welche sie für Physik, Chemie, Mineralogie, Botanik, Physiologie, Heilkunde, gerichtliche Medizin, Rechtskunde, Kriegswesen, Erziehung, Psychologie, Theologie, Irrenwesen, Kunst, Gewerbe, häusliche Zustände, Menschenkentniss und das gesellschaftliche Leben im weitesten Anfange haben. 2 vols. Stuttgart and Tübingen: J. G. Cotta, 1854–1855, lv + (1) + 838; xxx + 758 pp.

[H]

728. ***Rutter, John Obediah Newell.***

Human Electricity: the Means of Its Development, Illustrated by Experiments. With Additional Notes. London: J. W. Parker & Son, 1854, vi + 182 pp.

[H]

729. ***Séré, G. Louis de.***
Application du somnambulisme magnétique au diagnostic et au traitement des maladies. Paris: Germer Baillière, 1854, 302 pp.

[H]

730. ***Szapary, Ferencz Grof.***
Das Tischrücken. (Fortsetzung.) Geistige Agapen. Plychographische Mittheilungen der Pariser Deutsch-Magnetischen Schule. Paris: The Author, 1854, xxiv + 324 + (4) pp.

A compilation of mediumistic communications written by the members of Szapary's "Deutsch-Pariser-Magnetischen Schule," and covering a variety of subjects, such as table-moving messages, automatic writing, and speaking with tongues. [H & P]

731. ***Szapary, Ferencz Grof.***
Table-Moving. Somnambulish-Magnetische Traumbedeutung. Paris: Bonaventure and Ducessois, 1854, 312 pp.

Somnambulistic writings produced by Szapary's Parisian magnetic school. [H & P]

732. ***La table parlante. Journal des faits merveilleux,*** **tables** ***tournantes, mesmérisme, somnambulisme magnétique. . .***
One vol. only; 1854–1855.

Published in Paris under the editorship of B. Du Vernet. [H & P]

733. **[*Tascher, Paul*].**
Grosjean à son évèque au sujet des tables parlantes. Paris: Ledoyen, 1854, 15 pp.

[H & P]

734. ***Townshend, Chauncy Hare.***
Mesmerism Proved True, and the Quarterly Reviewer Reviewed. London: T. Bosworth, 1854, xii + 216 pp.

[H]

735. ***L'union magnétique. Journal de la Société philanthropico-magnétique de Paris.***
Vols. 1–16; 1854–1869.

Published in Paris. [H]

736. ***Widemann, Gustav.***
Der Magnetismus des Menschen. Nurnberg: Bauer & Raspe, 1854, viii + 126 + (2) pp.

[H]

1855

737. *Beaux, Jean Jacques.*

De l'influence de la magnétisation sur le développement de la voix et du goût en musique. Paris: E. Garnot, 1855, (1) + (5)–155 pp.

[H]

738. *Braid, James.*

The Physiology of Fascination, and The Critics Criticised. Manchester: Grant & Co., 1855, 14 + 18 pp.

Two treatises combined here in one publication. The first is called "Electro-Biological Phenomena, and the Physiology of Fascination." Comparing the phenomenon of fascination, as it can be produced in animals, with that produced in human beings, Braid suggests that they have in common the state of loss of will but not of motion. The fascinated individual is not paralyzed, as is the cataleptic, but even so cannot initiate action through decision. Braid sees the state as a temporary disturbance in the nervous centers deriving from the stimulation of a strong idea. In the second part called "The Critics Criticised", Braid takes two mesmeric writers, George Sandby and Chauncy Hare Townshend, to task. Both had previously criticized Braid's notion that the effects of animal magnetism were purely psychological and did not involve a magnetic fluid. [H]

739. *Capron, Elias Wilkinson.*

Modern Spiritualism: Its Facts and Fanaticisms, Its Consistencies and Contradictions. Boston: Bela Marsh, 1855, 438 pp.

The best and most complete contemporary history of the early years of spiritualism and its spread throughout the United States. Capron covers the initial manifestations in the home of the Fox sisters, manifestations in other areas of New York State, the spread to Boston, Providence, Philadelphia and Cincinnati among other places. There is an especially interesting chapter on spiritualism in Washington, D.C. with a section on "congressional mediums." [P]

740. [*Durand, Joseph Pierre, pseudonymn: A.J.P. Philips.*]

Electro-dynamisme vital ou les relations physiologiques de l'esprit et de la matière, démontrées par des expériences entièrement nouvelles et par l'histoire raisonnée du système nerveux. Paris: J. B. Baillière, 1855, (4) + xlii + (1) + 383 pp.

Durand (de Gros) wrote this and other books under the pseudonym of A.J.P. Philips. Durand, a physiologist who had a special interest in how the body affects states of consciousness, speculates on a vital force that affects the nervous system and on the relationship between conscious and unconscious acts. His investigation of magnetic and electrical aspects of the physical organism interested him in the work of Reichenbach which in turn prepared him for his encounter with hypnotism and for the new direction for his researches (*see Cours theorique et pratique de Braidisme*, entry number 821). [H]

741. *Fabre, Pierre Honoré.*

Manuel du magnétiseur: explication physiologique des phénomènes magnétiques, utilité du somnambulisme dans l'exercice de la médecine. Saint-Quentin: Doloy et Penet ainé, 1855, 15 pp.

[H]

742. *Garcin, ———.*

Le magnétisme expliqué par lui-même, ou Nouvelle théorie des phénomènes de l'état magnétique comparés aux phénomènes de l'état ordinaire. Paris: Germer Baillière, 1855, xii + 220 pp.

A treatise developing the notion that two fluids are involved in animal magnetic phenomena rather than one. [H]

743. *Gentil, J. A.*

Orthodoxie magnétique. Catéchisme raisonné de l'aspirant magnétiseur, suivi d'un simple coup d'oeil sur le triple électro-galvanique et du pilori du magnétisme. Paris: The Author and Dentu, 1855, 63 pp.

[H]

744. *Hare, Robert.*

Experimental Investigation on the Spirit Manifestations, Demonstrating the Existence of Spirits and Their Communion with Mortals. New York: Partidge & Brittan, 1855, (9) + 5–460 pp.

The first attempt by an American scientist of note to investigate the physical phenomena of spiritualism. Hare was a graduate of Yale College and Harvard University, a medical doctor, emeritus professor of chemistry at the University of Pennsylvania, discoverer of the oxy-hydrogen blow pipe, author of over 150 articles on scientific subjects, an associate of the Smithsonian Institute, and a member of the American Association for the Advancement of Science. Although by 1888 numerous accounts of "physical" mediums apparently capable of moving physical objects at a distance by means of some unknown "spiritual" force had accumulated, Hare disbelieved these accounts and thought it necessary to put them to scientific test. He devised experiments to determine whether mediums could move objects when physical force could not be exerted upon those objects. To his surprise he discovered that they could. His experiences led to his embracing spiritualism and writing *Experimental Investigations.* The book contains five plates which depict the pieces of apparatus Hare used to carry out experiments, such as one in which a young medium was able to exert a force of eighteen pounds on a spring balance even though prevented from having any communication with the apparatus except through water. It would be twenty years before the type of painstaking experimentation carried out by Hare would come into its own with the famous researches of William Crookes in England, and when Hare submitted the results of his experiments to the American Association for the Advancement of Science, his findings were not well received. [P]

745. *Lafontaine, Charles.*

Eclaircissements sur le magnétisme. Cures magnétiques à Genève. Geneva: De Chateauvieux, 1855, vi + (3) + 10–156 pp.

An account of cures brought about through animal magnetism by this famous stage mesmerizer. The description of the cures is preceded by a history of magnetic healing among the ancients. [H]

746. *Mahan, Asa.*

Modern Mysteries Explained and Exposed. In Four Parts. I. Clairvoyant Revelations of A. J. Davis. II. Phenomena of Spiritualism Explained and Exposed. III. Evidence that the Bible is Given by Inspiration of the Spirit of God, as Compared with Evidence that These Manifestations are from the Spirits of Men. IV. Clairvoyant Revelations of Emanuel Swedenborg. Boston: J. P. Jewett, 1855, xv + 466 pp.

When Mahan undertook to investigate spiritualism, he expected to find that the phenomena were due to trick and deception. He soon had to admit, however, that this was not always the case. He concluded that some occurences were genuine and that intelligent communication was taking place but from the unconscious minds of those who were taking part in the seances rather than from disembodied spirits. The manifestations were, he believed, powered by a force of nature, the "odylic force" of Reichenbach. This force, Mahan concluded, accounted for all mesmeric and clairvoyant phenomena and for all spiritualistic manifestations. [P]

747. *Maitland, S. R.*

Superstition and Science: an Essay. London: Rivingtons, 1855, 89 pp.

A treatise skeptical of table-turning, clairvoyance, and other supernormal phenomena. [P]

748. *Mesmerism and Media, with Full Instructions How to Develop the Alleged Spiritual Rappings in Every Family.*

London: n.p., 1855, 19 pp.

[H & P]

749. *Morin, Alcide.*

Philosophie magnétique. Les révolutions du temps, synthèse prophétique du XIXe siècle. Paris: E. Dentu, 1855, vii + 336 pp.

[H]

750. ***Neilson, William.***
Mesmerism in Its Relation to Health and Disease and the Present State of Medicine. Edinburgh: Shepherd & Elliot, 1855, xi + (1) + (13)–250 + (2) pp.

A strong defense of the use of animal magnetism to cure disease. The book begins with a long section on the prejudice of the medical establishment against new ideas and the fate of mesmerism in that regard. The author tries to show that mesmerism is quite compatible with normal medical practice. He points out that those who begin their criticism of mesmeric practice with a rejection of clairvoyance are starting at the wrong end. He emphasizes the curative power of animal magnetism and the general ignorance of the facts of the matter among medical critics. [H]

751. ***Otto, Bernhard.***
Die Sprache der Verstorbenen oder das Geisterklopfen. Stimmen aus dem Jenseits und enthullte Geheimnisse des Grabes. Ein unumstösslicher Beweis für die Fortdauer der Seele nach dem Tode und deren Wiedervereinigung mit ihren Lieben. Nach gesammelten authentischen Thatsachen dargestellt. Leipzig: Gustav Ponicke, (1855), (2) + 155 pp.

An early German treatise on the phenomenon of "table talking." The author describes his experiences with the phenomenon and argues that it is an indication of the survival of the human spirit after death. [P]

752. ***Peano, C.***
Il magnetismo animale al cospetto dell'Associazione Medica degli Stati Sardi. Saluzzo: n.p., 1855.

[H]

753. ***Pezzani, André.***
Lettres à M. Lélut sur la question du sommeil, du somnambulisme et des tables tournantes, à propos de son rapport au sujet du dernier concours ouvert à l'Académie des Sciences morales et politiques (section de philosophie). Lyon: A. Vingtrinier, 1855, 79 pp.

[H & P]

754. ***Rapporto della Commissione nominata dalla sezione medica***
*della Società di Incoraggiamento di Scienze, Lettere ed Arti in Milano per l'esame delle memorie di concorso al premio proposto pel 1855 sopra un argomento di magnetismo animale.*Milan: n.p., 1855.

[H]

755. *Ricard, Jean Joseph Adolphe.*
Le magnétiseur praticien. Paris: n.p., 1855, 168 pp.

[H]

756. *Segouin, A.*
Nouvel almanach du magnétiseur. Paris: The Author, 1855, 147 pp.

[H]

757. *Séré, G. Louis de.*
Application du somnambulisme magnétique au diagnostic et au traitement des maladies, sa nature, ses différences avec le sommeil et les rêves. Paris: The Author, 1855, 302 pp.

[H]

758. [*Tascher, Paul*].
Seconde lettre de Grosjean à son évèque. Paris: Ledoyen, 1855, 23 pp.

[H]

759. *Thury, Marc.*
Les tables tournantes considérées au point de vue de la question de physique générale, qui s'y rattache. Geneva: n.p., 1855.

[H & P]

760. *Trismegiste, Johannes (pseudonym).*
Les merveilles du magnétisme et les mystères des tables tournantes et parlantes. Paris: Passard, 1855, (1) + 188 pp.

[H & P]

1856

761. *Allix, Eug.*
La verità sul magnetismo animale. Turin: n.p., 1856.

[H]

762. *Bonaventura, Isador.*
Die Mysterien Schlafes und Magnetismus oder Physik und Physiologie des magnetischen Somnambulismus. Eine auf naturwissenschaftliche Prinzipien gestutzte rationelle Erklärung der Phänomene der Schlafes und Traumes, der Ekstase und Gehergabe, der Hallucinationen und Visionen, der electrobiologischen Erscheinungen der Bewegung unbelebter Körper u.v., durch Zurück-

führung auf ihre naturlichen Ursachen. Nach Deban, Carpenter v. A., sowie nach eignen Beobachtungen. Weimar: B. F. Voigt, 1856, x + 338 pp.

[H & P]

763. *Delaage, Henri and Auguez, Paul.*
Religion, magnétisme, philosophie. Les élus de l'avenir ou le progrés réalisé par le Christianisme. Paris: E. Dentu, 1856, xi + 206 pp.

[H & P]

764. *Didier, Adolphe.*
Animal Magnetism and Somnambulism. London: T. N. Newby, 1856, iv + 319 pp.

Adolphe Didier and his brother Alexis were two of the most famous clairvoyants of their day. Adolphe was also a mesmerist, and in this book he attempts to pass on his knowledge of the theory and techniques of that "noble science" to those who already accept the reality of somnambulism and clairvoyance. [H & P]

765. [*Didier, Alexis*].
Le sommeil magnétique expliqué par le somnambule Alexis en état de lucidité. Paris: Dentu, 1856, 173 pp.

Alexis Didier (known generally to the world as "Alexis") was, in his day, the most famous mesmeric clairvoyant of France. Here Alexis describes some of his more interesting seances. [H & P]

766. *Kerner, Justinus Andreas Christian.*
Franz Anton Mesmer aus Schwaben, Entdecker des thierischen Magnetismus. Erinnerungen an denselben, nebst nachrichten von den letzten Jahren seines Lebens zu Meersburg am Bodensee. Frankfurt am Main: Literarische Anstalt, 1856, x + (2) + 212 pp.

The first really detailed biography of Mesmer. It remains to this day an important source of information about Mesmer's life and teachings, written, as it was, by a man who had access to people who personally knew the discoverer of animal magnetism. [H]

767. *Luce magnetica.*
One vol. only; 1856–?

Edited by Francesco Guidi in Turin. [H]

768. *Mozzoni, G.*
Lo psicologico. Repertorio di magnetismo ad uso di chiunque voglia e debba essere al fatto dell'origine dello scopo e dei frutti di questa scienza. Milan: n.p., 1856.

[H]

769. *Reichenbach, Karl Ludwig von.*

Odische Erwiederungen an die Herren Professoren Fortlage, Schleiden, Fechner und Hofrath Carus. Vienna: Wilhelm Braumuller, 1856, (6) + 118 pp.

[H]

770. *Reichenbach, Karl Ludwig von.*

Wer ist sensitiv, wer nicht? Oder kurze Anleitung, sensitive Menschen mit Leichtigkeit zu finden. Vienna: n.p., 1856, 70 pp.

[H]

771. *Sulla causa dei fenomeni mesmerici. 2 vols.*

Bergamo: Mazzolini, 1856, 360; 474 pp.

A thorough inquiry into the theory of mesmerism and the magnetic agency. It includes a discussion of the use of mesmerism, examining such issues as the function of the will in magnetizing. Examples such as rapport and clairvoyance are also presented. [H & P]

772. *Tiffany, Joel.*

Spiritualism Explained: Being a Series of Twelve Lectures delivered Before the New York Conference of Spiritualists. New York: Graham and Ellinwood, 1856, ix + (1) + 11–206 pp.

An attempt to present an overall philosophy of spiritualism and mediumship early in the history of the movement. It contains a chapter on marriage and free love. [P]

1857

773. *Carus, Carl Gustav.*

Über Legensmagnetismus und über die magischen Wirkungen überhaupt. Leipzig: F. A. Brockhaus, 1857, x + 306 pp.

Carus states his belief that illness does not arise from man's conscious mind but from the realm of the unconscious. Because mesmerism acts most directly on man's soul, it is the most effective possible cure for illness. Its curative powers, says Carus, apply preeminently to nervous illnesses such as hysteria and hypochrondria. He believes that mesmerism simply calls upon the unfailing healing power of the unconsciousness and allows it do it its work. [H]

774. *Davis, Andrew Jackson.*

The Magic Staff; an Autobiography of Andrew Jackson Davis. New York and Boston: J. S. Brown, B. Mars, 1857, 552 pp.

Davis began his professional career as a clairvoyant teamed up with a mesmerizer who toured the American countryside, providing diagnoses of disease and prescriptions for cure. He then began to teach in the trance state and have his pronouncements written down by a scribe. This led to the publication of a number of very popular books on various subjects from animal magnetism to life on other planets. He was considered by many to be a major precursor of the spiritualist movement in the United States, apparently predicting its coming some years before the Fox sisters heard their knockings in Hydesville, New York. This is Davis's own version of the events of his life up to 1857. [P]

775. *Grimes, James Stanley.*

The Mysteries of Human Nature Explained by a New System of Nervous Physiology: to Which is Added, a Review of the Errors of Spiritualism, and Instructions for Developing or Resisting the Influence by Which Subjects and Mediums are made. Buffalo: R. M. Wanzer, 1957, xxiv + (25) – 432 pp.

After a long explanation of human physiology and its bearing on mental functioning, Grimes attempts to explain the phenomena of spiritualism in those terms. He denies that spirits are at work in these phenomena and insists that mediums at most perform feats of clairvoyance identical to those that can be produced by any mesmeric subject. [H & P]

776. *Guldenstubbe, Louis, baron de.*

Pneumatologie positive et expérimentale: la réalité des esprits et le phénomène merveilleux de leur écriture directe démontrés. Paris: A. Franck, 1857, xxxvi + 216 p.

An attempt to trace the historical background of spiritistic thought and to compare the ancient forms of mediumship to modern spiritualism. [P]

777. *Huguet, Hilarion.*

Institut dynamo-thérapique. Du somnambulisme médical, ou esquisse de nososcopie dynamo-thérapique. Paris: The Author, 1857, 72 pp.

A description of cures worked through the somnambulist Charavet under the direction of the physician Huguet. [H]

778. *Macario, Maurice Martin Antonin.*

Du sommeil, des rêves, et du somnambulisme dans l'état de santé et de maladie, précédé d'une lettre du Dr Cerise. Lyon: Perisse frères, 1857, xliv + 307 + (1) pp.

[H]

779. *Madden, Richard Robert.*

Phantasmata or Illusions and Fanaticisms of Protean Forms Productive of Great Evils. 2 vols. London: T. C. Newby, xlii + 504; (2) + 588 pp.

Madden was an Irish surgeon who served in governmental administrative posts in regions around the world. In that capacity he had a great deal of opportunity to observe what he considered to be the "illusions" of mankind, and he wrote this book as a study of the "failings and infirmities and passions of mankind and their accompanying disorder of the imagination," leading to epidemic fanaticisms. Among these fanaticisms, he includes those of such religious figures as Joan of Arc and St. Teresa, and religious undertakings such as witch hunts and the Inquisition. Madden also discusses epidemics of a hysterical nature, such as the dancing mania and the Jansenist convulsionaries. Madden's point of view is that when imagination soars above reason it loses its bearings and the greatest evils can result. When reason is overthrown in that way, it is often only religion, and specifically Christianity, that can return it to its proper place. [H & P]

780. *Masson, Alphonsine.*

Discours prononcé par Mme Alphonsine Masson, au banquet de Mesmer de l'année 1857, rue de Richelieu. Paris: Pommeret et Moreau, 3 pp.

[H]

781. [*Rivail, Hippolyte Léon Dénizard, pseudonym: Allan Kardec*].

Le livre des esprits contenant les principes de la doctrine spirite sur la nature des esprits, leur manifestation et leurs rapports avec les hommes. . . . Écrit sous la dictée et publié par l'ordre d'esprits supérieurs par Allan Kardec. Paris: E. Dentu, 1857, (2) + 176 pp.

Rivail was the founder of modern French spiritualism and its most important early figure. An accomplished educator who had published a number of books in that field, Rivail had been pursuing a serious interest in phrenology and animal magnetism in the early 1850s when he became intrigued by the "table turning" phenomenon sweeping Europe. Encountering two young mediums who informed him that he had an important religious mission to carry out, Rivail questioned them concerning the nature of human life and spiritual development, and obtained replies through table tapping and planchette writing that were to become the basis for an elaborate philosophy of spiritualism. Some of the key elements of his philosophy were: 1) his notion that human beings are composed of three main elements —the spirit, the body, and the peri-spirit (which connects the other two and allows the spirit to act on matter); 2) the idea that the physical phenomena of mediumship (e.g., materializations, movements of objects without contact, etc.) may be explained through the action of the peri-spirit which utilizes the animal magnetic fluid of both the departed spirit and the medium; and 3) the doctrine of reincarnation that posits the repeated return of the spirit to earthly existence (this doctrine was quite foreign to spiritualism in America,

England, and Germany). *Le livre des esprits*, Rivail's first work on the subject of spiritualism which lays out his basic doctrine, was extremely influential not only in France, but in other countries as well. Topics discussed include the vital principle, the nature of spirits, reincarnation, the communication of departed spirits with this world, the occupations of the spirits in the other world, and the moral implications of these tenets. [P]

782. *Schindler, Heinrich Bruno.*

Das magische Geistesleben. Ein Beitrag zur Psychologie. Breslau: W. G. Korn, 1857, xvi + 356 pp.

Based upon ideas arising from the traditions of German romanticism and animal magnetism, and influenced by the experiments of Reichenbach, this ambitious work attempts to provide an overall understanding of human nature and the inner power of the soul. In this framework, Schindler takes up the subjects of clairvoyance, prophecy, and magic, explaining that these become available when the powers of the inner self are tapped. [H & P]

783. *The Spiritualist: Being a Short Exposition of Psychology Based upon Material Truths and of the Faith to Which It Leads.*

London: L. Booth, 1857, xvi + 111 + (1) pp.

An early British presentation of the philosophy of spiritualism, with a heavy reliance upon phenomena such as magnetism, electricity, and mesmerism to elucidate the relationship between matter and spirit. [P]

784. *Tweedie, A. C.*

Mesmerism and its Realities Further Proved by Illustrations of its Curative Powers in Disease as well as by its Development of Some Extraordinary Magnetic Phenomena in the Human Body. Edinburgh: Patox & Ritchie, 1857, 34 pp.

Description of the mesmeric treatment of a Miss Armitage and her resulting clairvoyance. [H & P]

785. *Wurm, Wilhelm.*

Darstellung der mesmerischen Heilmethode nach naturwissenschaftlichen Grundsätzen; nebst der ersten vollständigen Biographie Mesmer's und einer fasslichen Anleitung zum Magnetisieren. Munich: C. Wolf & Sohn, 1857, viii + 182 + (2) pp.

[H]

1858

786. *Auguez, Paul.*

Spiritualisme, faits curieux précédés d'une lettre à M. G. Mabru suivis de l'extrait d'un compte rendu de la fête Mesmérienne du 23 Mai 1858, et d'une relation américaine des plus extraordinaire publiés. Paris: Dentu and Germer Baillière, 1858, 88 pp.

[P]

787. *Berjot, E.*

Manuel historique élémentaire et pratique de magnétisme animal, contenant les principes généraux de l'art magnétique, l'explication des divers phénomènes qui s'y rattachent, la description des symptômes des principales maladies chroniques, leurs causes determinantes et les procédés reconnus les plus convenables à leur guérison au moyen du magnétisme par E. Berjot, membre titulaire de la Société du mesmérisme de Paris, suivi d'une dissertation sur le fluide magnétique animal par A. Bauche, membre titulaire de la même société. Paris: The Author, 1858, xi + 118 pp.

[H]

788. *Gérard, J.*

L'art de se magnétiser ou de se guérir mutuellement. Paris: Dentu, 1858.

Gérard, editor of *La revue magnétique*, spent time in the military before becoming a mesmerizer. This is his first book on the subject of animal magnetism. [H]

789. *Mabru, G.*

Les magnétiseurs jugés par eux-mêmes, nouvelle enquête sur le magnétisme animal. Ouvrage dédié aux classes lettrées, aux médecins, à la magistrature et au clergé. Paris: Mallet-Bachelier, 1858, (4) + 564 pp.

A work of formidable proportions that undertakes to discredit animal magnetism. The author believes that animal magnetism was a sham from the beginning and that it attracted many foolish or naive followers. The first half of the book is comprised largely of letters written between Mabru and Meunier. The second half attempts to give an overview of the history of animal magnetism: mesmerism and magnetic fluid; Puységur and artificial somnambulism; Dupotet and magico-magnetism; the Americans and table turning and spirit rappers; spiritualists and charlatans; magnetism and phrenology; etc. It is a useful compendium of anti-mesmeric sentiment, but does not carefully examine contradictory data. The book includes a synoptic table of the various theories professed by the different practitioners of animal magnetism. [H]

790. ***Manuale del magnetismo animale, desunto dalle più recenti*** *opere magnetiche.*
Milan: Arzione, 1858, 290 pp.

[H]

791. ***Millet, F.***
Cours de magnétisme animal en douze leçons. Paris: The Author, 1858, xx + viii + 170 pp.

[H]

792. ***Piérart, Z. J.***
Le magnétisme, le somnambulisme et le spiritualisme dans l'histoire. Affaire curieuse des possédées de Louviers. (*Explications et rapprochements avec les faits actuels, avec les phénomènes produits par M. Home.*) Paris: Au Bureau du Journal du Magnétisme, 1858, 39 + (1) pp.

[H & P]

793. ***Putnam, Allen.***
Mesmerism, Spiritualism, Witchcraft, and Miracle: A Brief Treatise Stating that Mesmerism Is a Key Which Will Unlock Many Miracles and Mysteries. Boston: B. Marsh, 1858, 74 pp.

[H & P]

794. ***Reichenbach, Karl Ludwig von.***
Die Pflanzenwelt in ihren Beziehungen zur Sensitivität und zum Ode. Eine physiologische Skizze. Vienna: Wilhelm Braumüller, 1958, viii + 122 pp.

[H]

795. ***Revue spiritualiste. Journal principalement consecré à*** *l'étude des facultés de l'âme et à la démonstration de son immortalité par l'examen raisonné des divers genres de manifestations médianimiques et de phénomènes psychiques tels que le somnambulisme, l'extase, la vue à distance, etc.*
Vols. 1–12; 1858–1869.

Continued as: *Le concile de la libre pensée. Abolition des faux dogmes et des mensonges sacerdotaux. . . .* 1870–1874. Published in Paris and edited by Z. J. Piérart. The first title had the variant: *Organ du progrès philosophique et religieux.* [P]

796. [***Rivail, Hippolyte Léon Dénizard, pseudonym: Allan Kardec***].
Instructions pratiques sur les manifestations spirites contenant l'exposé complet des conditions nécessaires pour communiquer avec les esprits et les

moyens pour développer la faculté médiatrice chez les médiums. Paris: Bureau de la Revue Spirite, 1858, (4) + 146 + (1) pp.

Generally recognized as the father of spiritualism in France, Kardec wrote this book, his second (see *Livre des Esprits*, entry number 781), as a manual of spiritistic practice. [P]

797. *Rouget, Ferdinand.*

Traité pratique de magnétisme humain, ou résume de tous les procédés du magnétisme humain, pour rétablir et développer les fonctions physiques et les facultés intellectuelles dans l'état de maladie. Paris and Toulouse: E. Dentu and Germer Baillière (Paris) and Gimet (Toulouse), 1858, 156 pp.

[H]

798. *Rovère, Jules de.*

Les magnétiseurs ont-ils tort ou raison? That is the question! Appréciation et solution en deux parties. Paris: The Author, 1858, 31 pp.

[H]

1859

799. *Wilkinson, W. M.*

Spirit Drawings: A Personal Narrative. London: Chapman & Hall, 1858, iv + 191 pp.

An account of automatic writing and automatic drawing in the family of the author that goes into useful detail about the process involved and provides important data for psychical research. The author attributes the phenomena to the work of discarnate spirits. [P]

800. *Broca, Paul.*

Sur l'anesthésie chirurgicale hypnotique. Note présentée à l'Académie des sciences le 5 décembre 1859 suivie d'une lettre adressée au rédacteur en chef du Moniteur des sciences médicales. Paris: Noblet, 1859, 16 pp.

Broca, a distinguished surgeon and anthropologist, was an authority on aphasia and a pioneer in craniology. Because of his reputation, this address was an important factor in arousing interest in hypnotism and the views of James Braid in France. It was the attention paid to hypnotism by such men as Broca that eventually induced Ambroise Liébeault (1823–1904) to experiment with the phenomenon. [H]

801. *Caroli, Giovanni M.*

Del magnetismo animale ossia mesmerismo. In ordine alla ragione e alla rivelazione. Naples: Uffizio della Biblioteca cattolica, 1859, 641 pp.

[H]

802. *Confessions of a Truth Seeker. A Narrative of Personal Investigations into the Facts and Philosophy of Spirit-intercourse.*

London: William Horsell, 1859, xxxii + 33–222 pp.

The phenomena and philosophy of spiritualism described by a spiritualist. [P]

803. *Dixon, Jacob.*

Hygienic Clairvoyance. London: W. Horsell, (1859), viii + (5)–74 pp.

[H & P]

804. *Emmons, Samuel Bullfinch.*

The Spirit Land. Philadelphia: G. G. Evans, 1859, (1) + 288 pp.

Criticizes purported spirit manifestations from the Christian point of view. Emmons explains the phenomena in terms of the effects of the imagination on the nervous system and calls for the rooting out of popular superstitions about the matter. While admitting that objects may indeed move without any physical force being applied to them, Emmons argues that it is sufficient to explain them in terms of "vital electricity" from embodied, not disembodied, spirits. [H & P]

805. *Fichte, Immanuel Hermann.*

Zur seelen lehre. Ein philosophische Confession. Leipzig: F. A. Brockhaus, 1859, xxviii + 286 pp.

Studying the phenomenon of somnambulism, Fichte drew conclusions about the nature of human consciousness. The "dark" consciousness of somnambulism is characterized by fantasies that are ordinarily not recognized by our normal consciousness but that do show themselves in such phenomena as instincts, dreams, and clairvoyance. Fichte believed that the soul has an inner body of its own which, after death, continues to exist. [H & P]

806. *Gragnon, Célestin.*

Du traitement et de la guérison de quelques maladies chroniques au moyen de somnambulisme magnétique, et à propos de MM. Calixte Renaud, de Bordeaux et Alexis de Paris. Bordeaux: G. Gounouilhou, 1859, 112 pp.

[H]

807. ***Le magnétiseur, journal du magnétisme animal.***
Vols. 1–12, 1859–1872.

Published in Geneva and edited by Charles Lafontaine. [H]

808. ***Mitchell, John Kearsley.***
"On Animal Magnetism, or Vital Induction." In *Five Essays*. Edited by S. Weir Mitchell. Philadelphia: Lippincott, 1859, pp. 143–273.

[H]

809. ***Mongruel, Madame.***
Les voix de l'avenir dans le présent et dans le passé, ou les oracles et les somnambules comparés. Paris: Dentu and The Author, 1859, 104 pp.

[H & P]

810. **[*Rivail, Hippolyte Léon Dénizard, pseudonym: Allan Kardec*].**
Qu'est-ce-que le spiritisme; introduction à la connaissance du monde invisible ou des esprits, contenant les principes fondamentaux de la doctrine spirite et la réponse aux quelques objections préjudicielles. Paris: Ledoyen, 1859.

After dealing with the objections of sceptics, Rivail presents a brief outline of spiritualistic doctrine. [P]

811. ***Vasseur-Lombard, ———.***
Principes universels du magnétisme humain appliqués au soulagement et à la guérison de tous les êtres malades. Paris: Leydoyen, 1859, 36 pp.

[H]

1860–1869

1860

812. *Azam, Eugène.*

"Note sur le sommeil nerveux ou hypnotisme." *Archives générales de médecine* (Jan. 1860): 5–24.

Azam, a physician and psychological investigator, having become acquainted with Braid's work, experimented with hypnotism and suggestion in the treatment of emotional disturbance. This article, in which Azam communicates the results of his first experiments, helped lead to the serious investigation of Braid's approach in France. Azam hoped, as did others at the time, that by using the term "hypnotism" rather than "animal magnetism" a fresh attitude could be brought to the investigation of trance phenomena, one free of the negative associations some had with mesmerism. Eventually Azam become a friend of Braid, who, three days before his death in 1860, sent Azam his last manuscript with an inscription of esteem. This manuscript, never published in English, eventually ended up in the possession of Wilhelm Preyer (1841–1897) who published a German version in *Die Entdeckung des Hypnotismus* (1881). Azam's experiments with hypnotism in 1860 eventually became known to Ambroise Liébeault (1823–1904), contributing to his initial interest in hypnotism. [H]

813. [*Bell, Robert*].

"Stranger Than Fiction," *The Cornhill Magazine* 2 (1860): 211–24.

An account of an evening with the medium D. D. Home in which Bell, a journalist, relates spectacular mediumistic feats carried out in circumstances which were believed at the time to be very unfavorable to fraud.

This article appeared with an introductory footnote by the editor William Thackeray: "As Editor of this Magazine, I can vouch for the good faith and honourable character of our correspondent, a friend of twenty-five years' standing; but as the writer of the above astounding narrative owns that he 'would refuse to believe such things upon the evidence of other people's eyes,' his readers are therefore free to give or withhold their belief." Although Thackeray knew the account he was publishing would be controversial, it turned out to be very influential in giving spiritualistic phenomena a certain credibility in England. [P]

814. *Caroli, Giovanni M.*

Filosofia dello spirito—ovvero del magnetismo animale. Naples: n.p., 1860.

[H & P]

815. *Charpignon, Louis Joseph Jules.*

Rapport du magnétisme avec la jurisprudence et la médecine légale. Paris: Germer Baillière and Duraud, 1860, 61 pp.

Among the earliest works on issues of jurisprudence in the practice of animal magnetism, this pamphlet deals largely with the responsibilities of the magnetizer and possible abuses that may occur. [H]

816. *Comet, Charles Jean Baptiste.*

La vérité aux médecins et aux gens du monde sur le diagnostic et la thérapeutique des maladies, éclairés par le somnambulisme naturel lucide. De la catalepsie, observations de facultés surnaturelles de clairvoyance, d'intuition, de prévision et d'extase. Du magnétisme animal et de ses effets. Instruction pratique sur son application au traitement des maladies. Paris: Plon, 1860, 388 pp.

[H & P]

817. *Csanady, Stephan.*

Medicinische Philosophie und Mesmerismus. Leipzig: Franz Wagner, 1860, xii + 451 pp.

[H]

818. *Demarquay, Jean Nicolas and Giraud-Teulon, M. A.*

Recherches sur l'hypnotisme ou sommeil nerveux, comprenant une série d'expériences instituées à la maison municipale de santé. Paris: J. B. Baillière, 1860, 56 pp.

A discussion of hypnotism as a surgical anesthetic. Taking note of the optimism of Velpeau, Broca, and Azam concerning the use of hypnotism to render a patient insensible, the authors undertook to confirm its efficacy for themselves. Through eighteen trials they found only one patient who became insensible to a significant degree, and that patient they considered to be hysterical. From this the authors conclude that magnetic sleep, magnetic somnambulism and hypnotism are the same phenomenon manifested in varying degrees, and that their efficacy is dependent on a condition of hysteria or near-hysteria in the patient. This conclusion is strikingly close to that which Charcot and the Salpêtrière School will reach many years later. Despite this conclusion, however, the authors still consider hypnotism to be a useful tool for treating nervous disorders, however, and the main part of the treatise is spent discussing that application. They also state that looking at a brilliant object is not necessary for the induction of hypnotism. All that is needed is a fixity of gaze. [H]

819. *Dunand, Tony.*

Magnétisme, somnambulisme, hypnotisme, considérations nouvelles sur le système nerveux, ses fonctions et ses maladies. Paris: Ledoyen, 1860, 23 pp.

Dunand, a Parisian physician, here discusses somnambulism and takes note of the "hypnotic" theory of Braid, then new to the French. [H]

820. *Dupuy, Antonin, ed.*

L'hypnotisme, compte-rendu des conférences du docteur A.J.P. Philips. Paris: Cercle de la Presse scientifique, 1860, 30 pp.

An account of the lectures of Durand de Gros (pseudonym: A.J.P. Philips) on braidism. [H]

821. [*Durand, Joseph Pierre, pseudonym: A.J.P. Philips*].

Cours théorique et pratique de braidisme ou hypnotisme nerveux: considéré dans ses rapports avec la psychologie, la physiologie et la pathologie et dans ses applications à la médecine, à la chirurgie, à la physiologie expérimentale, à la médicine légale et à l'éducation, par le docteur J. P. Philips. Paris: J. B. Baillière and Germer Baillière, 1860, xii + 180 pp.

Durand (de Gros) was one of the men who brought the teachings of Braid to the attention of the French public. This book presents three conferences Durand gave on the subject of "braidism," as hypnotism was then called in France. Because of his personal interests, Durand gave his lectures a particularly physiological slant. He concludes with a plea to physicians for further investigations of braidism. [H]

822. *Figuier, Guillaume Louis.*

Histoire du merveilleux dans les temps modernes. 4 vols. Paris: L. Hachette et Cie., (1860–1861), x + 419; (4) + 428; (4) + 407+ (1); (5) + 395 pp.

When "turning tables" arrived in France from America around 1853, Figuier became convinced that it was necessary to know the historical background of such phenomena in order to judge them with objectivity. The result was this work, divided into the following sections: Volume 1: the history of the epidemic possessions of Loudun and the Jansenist convulsionaries; Volume 2: the history of the divining rod and the Protestant prophets;

Volume 3: the history of animal magnetism; Volume 4: the history of table turning and spiritism. Volume 3 is one of the best early histories of animal magnetism. Volumes 2 and 4 give a good historical background for phenomena which began to be studied by psychical researches some ten years later. [H & P]

823. *Gigot-Suard, Jacques Léon.*

Les mystères du magnétisme animal et de la magie dévoilés ou la vérité sur le mesmérisme, le somnambulisme dit magnétique et plusieurs phénomènes attribués à l'intervention des esprits démontrés par l'hypnotisme. Paris: Labé, 1860, 114 pp.

The author intends to prove the validity of animal magnetism by placing it in the context of the western magical tradition. At the same time he brings in "hypnotism" (a new word in France) and the findings of Braid. [H & P]

824. *Gougenot des Mousseaux, Henri Roger.*

La magie au dix-neuvième siècle, ses agents, ses verités, ses mensonges. Precedée d'une lettre adressée à l'auteur par le P. Ventura de Raulica, Ancien Général de l'ordre des Théatins, Consulteur de la Sacrée Congrégation des Rites, Examinateur des Evêques et du clergé romain. Paris: Henri Plon and E. Dentu, 1860, (4) + xx + 439 pp.

Attempts to explain feats of clairvoyance and magic in terms of the fluid of animal magnetism. Gougenot des Mousseaux sees animal magnetism as an ancient magical tradition clothed in modern dress. [H & P]

825. *Guidi, Francesco.*

Il magnetismo animale considerato secondo le leggi della natura e principalmente diretto alla cura delle malattie. Con note ed un appendice sull ipnotismo. Milan: Francesco Sanvito, 1860, 574 pp.

An enthusiastic exploration of animal magnetism as providing a new way to view the world. Guidi's *Trattato teorico-pratico di magnetismo animale* (entry number 717) had been put on the *Index* of forbidden books by the Roman Catholic Church; here he emphasizes his view of animal magnetism as a natural phenomenon, in no way connected with superstition. [H & P]

826. *Guidi, Francesco.*

Prolusione sul magnetismo animale letta al Circolo Popolare di Brera in Milano la sera del 5 maggio 1860. Milan: Sanvito, 1860, 32 pp.

[H]

827. *Lovy, Jules.*

Souvenirs des banquets de Mesmer, toastes et chansons. Paris: Bureau du journal "l'Union magnétique," 1860, 72 pp.

[H]

828. ***Il magnetologo.***
One vol. only?; 1860(?).

Edited by Francesco Guidi. [H]

829. ***Mesnet, Ernest.***
Études sur le somnambulisme envisagé au point de vue pathologique. Paris: Rignoux, 1860, 30 pp.

The first work on somnambulism by this physician of the Hôtel Dieu. Mesnet began his work with hysterics in the early 1850s. In this treatise he examines somnambulism as a pathological condition related to hysteria. [H]

830. ***Morin, Alcide.***
Magie du XIXe siècle. Ténèbres. Treize nuits suivies d'un demi-jour sur l'hypnotisme. Paris: E. Dentu, 1860, xix + 280 pp.

[H & P]

831. ***Morin, André Saturnin.***
Du magnétisme et des sciences occultes. Paris: Germer Baillière, 1860, ix + 532 pp.

An important source for understanding how practitioners of animal magnetism saw themselves in relationship to the occult and supernormal phenomena. In their experiments, mesmerists encountered many phenomena that had traditionally been explained in occult or spiritist terms, and a number of mesmerists considered themselves the true inheritors of that tradition. While they believed the phenomena that occurred in the occult tradition to be genuine, because they themselves encountered similar phenomena, they strove to explain them in more "naturalistic" terms. In this book Morin expresses this viewpoint and uses the findings of the magnetists to shed light on the true nature of occult phenomena. The first half of the book is an exposition of animal magnetism and the many different forms it has taken in practice. It includes an excellent study of somnambulism and somnambulistic clairvoyance. The second part takes up the "occult sciences" and discusses such subjects as table turning, spiritualistic mediums, hallucinations, and the school of magnetic "magic" founded by Du Potet. [H & P]

832. ***Morin, André Saturnin.***
Magnétisme. M. Lafontaine et les sourds muets. Paris: G. Baillière, 1860, 24 pp.

[H]

833. ***Owen, Robert Dale.***
Footfalls on the Boundary of Another World. With Narrative Illustrations. Philadelphia: J. B. Lippincott & Co., 1860, 528 pp.

Owen was introduced to spiritualism by a fellow diplomat while serving as American minister to the Italian government at Naples. He was intrigued but skeptical. Characteristically, he undertook an historical survey of the spirit communication and, after working through "formidable piles of chaff to reach a few gleanings of solid grain," he decided he should take the matter seriously. The resultant work, which examines everything from the psychological and physiological underpinnings of psychic phenomena to the most spectacular and convincing instances of intervention by spirits from the other side of death, was one of the most popular ever written on the subject of spiritualism and did much to provide that movement with an aura of respectability. [P]

834. *Plumer, William S.*

"Mary Reynolds: A Case of Double Consciousness." *Harpers New Monthly Magazine* 20(1860): 807–12.

A detailed account of the most famous early case of multiple personality, but not completely accurate. [H]

835. *Samson, George Whitefield.*

Spiritualism Tested; or, the Facts of Its History Classified and Their Cause in Nature Verified from Ancient and Modern Testimonies. Boston: Gould and Lincoln, 1860, viii + 9–185 pp.

Samson, who taught at Columbian College in Washington, D.C., believes that the phenomena of spiritualism are genuine but not due to the intervention of spirits. He contends that all such phenomena can be explained in terms of "nervous energy." Much of the book is devoted to a study of the manifestation of spiritualistic phenomena throughout history. [P]

836. *Zuccoli, Ant.*

Le scienze occulte. Magnetismo, elettrobiologia, spiritualismo e negromanzia ossia la duplice scienza d'una levatrice. . . . Milan: Pirola, 1860, 30 pp.

[H & P]

1861

837. *Coleman, Benjamin.*

Spiritualism in America. London: F. Pitman, 1861, iv + 87 pp.

[P]

838. *Didier, Adolphe.*

Cures Effected by Animal Magnetism. London: Hippolyte Baillière and The Author, 1861, 68 pp.

This work went through many editions and changes of title, including: *Curative Mesmerism* and *Magnetism and its Healing Power.* [H]

839. *Fechner, Gustav Theodor.*

Über die Seelenfrage. Ein Gang durch die sichtbare Welt, um die unsichtbare zu finden. Leipzig: C. F. Amelang, 1861, vi + (1) + 228 + (1) pp.

Fechner, a physicist and philosopher, sought to understand the relationship between the physical and spiritual reality. His investigations led him to map out a novel field of study, labelled "psychophysics" and involving the measurement of physical concommitants of psychological acts, which became the foundation for modern experimental psychology. This work deals with the question of the soul, its nature, and its sphere of activity. Fechner believes that all things (including crystals and plants) have souls, and that the world itself, the highest being in our experience, has a soul. He attempts to explore what we know of this world to gain access to knowledge of the metaphysical world. [P]

840. *Frotté, Em.*

Une plaie professionnelle, ou la médecine exploitée par le somnambulisme. Reims: Dubois, 1861, 53 pp.

[H]

841. *Guidi, Francesco.*

Introduzione allo studio del magnetismo animale e del magnetico sonnambulismo. Naples: Nobile, 1861.

[H]

842. *Hébert, Hippolyte Joseph.*

Recherches sur l'hypnotisme et ses causes, suivies d'un discours prononcé dans l'assemblée des chirurgiens-dentistes du mois de décembre 1860. Saint-Germain-en-Laye: The Author, 1861, 31 pp.

[H]

843. *Maury, Louis Ferdinand Alfred.*

Le sommeil et les rêves. Paris: Didier, 1861, vii + 426 pp.

Maury, an engineer, scholar, and professor in the Collège de France, presented his first thoughts on sleep and dreams in periodical articles that appeared between 1848 and 1857. It is Maury's intention to give an overall theory of the nature of dreams drawn, as he says, not from speculations on theories but from methodical personal observation. This he does in the first part of his book. In the second he makes connections between his observations on dreams and what he considers to be analogous phenomena: hypnotism, somnambulism, and certain pathological conditions. His chapter on "hypnagogic hallucinations" is considered a classic. [H]

844. *Perty, Maximillian.*
Die mystischen Erscheinungen der menschlichen Natur. Leipzig and Heidelberg: C. F. Winter, 1861, xvii + 770 pp.

Perty, a professor of zoology at Bern and one of Germany's most important early psychical researchers, expounded a neoplatonic-type philosophy of the paranormal. In this book, Perty puts forward a theory of mediumship based upon an occult world view that posits a world spirit and secondary nature spirits. The book includes a large section on animal magnetism. [P]

845. [*Rivail, Hippolyte Léon Dénizard, pseudonym: Allan Kardec*].
Spiritisme expérimental. Le livre des médiums ou guide des médiums et des évocateurs, contenant l'enseignement spécial des esprits sur la théorie de tous les genres de manifestations, les moyens de communiquer avec le monde invisible, le développement de la médiumnité, les difficultés et les écueils que l'on peut rencontrer dans la pratique du spiritisme. Pour faire suite au Livre des Esprits. Paris: Didier, Ledoyen (1861), viii + 510 pp.

Rivail's second most important work after *Le livre des esprits* (entry number 782) and one of the most complete treatises ever written on mediumship. Rivail describes the nature of mediumship, the forces at work through mediums, how to develop as a medium, the qualities mediums need to have, the dangers of mediumship, and fraudulent mediums. [P]

1862

846. [*Canelle, A.*].
Du phréno-mesmérisme. Paris: E. Voitelain, 1862, 7 pp.

[H]

847. *Flammarion, Camille.*
Les habitants de l'autre monde. Études d'outre-tombe. 2 parts. Paris: Ledoyen, (1862–1863).

While a young man, Flammarion, one of France's best known astronomers, became involved with Allan Kardec's form of spiritualism. Although he never became a spiritualist, Flammarion developed a deep interest in psychical subjects and became one of the premier psychical researchers of France. *Les habitants* deals with such subjects as communications with spirits, problems of identifying spirits, spirit raps, and mediumistic writing.

Flammarion takes his material from seances with the medium Mademoiselle Huet. The first part was published in 1862; the second in 1863. [P]

848. *Lelut, Louis François.*

Physiologie de la pensée. Recherche critique des rapports du corps avec l'esprit, 2 vols. Paris: Didier, 1862.

Lelut investigates the action of electro-magnetic fluid in the nervous system and discusses various psychological phenomena such as dreams and somnambulism. This collection of studies includes the author's *Mémoire sur le somnambulisme* (entry number 646). [H]

849. *Le magnétisme. Journal des sciences magnétique, hypnotique et occultes . . . .*

One vol. only; 1862.

Published in Paris. [H]

850. *Reichenbach, Karl Ludwig von.*

Odische Begebenheiten zu Berlin in den Jahren 1861 und 1862. Berlin: E. H. Schroeder, 1862, vii + (1) + 92 pp.

[H]

851. [*Vollmer, Carl Gottfried Wilhelm, pseudonym: W.F.A. Zimmermann*].

Magnetismus und Mesmerismus oder physische und geistige Kräfte der Natur. Berlin: Thiele, 1862, 692 pp.

An exposition of animal magnetic phenomena in a fictional format. [H & P]

1863

852. *Cruikshank, George.*

Discovery concerning Ghosts: With a Rap at the "Spirit-rappers." To which is Added a Few Parting Raps at the "Rappers," Questions, Suggestions, and Advice to the Davenport Brothers. London: F. Arnold, 1863, 48 pp.

[P]

853. [*De Morgan, Sophia Elizabeth*].

From Matter to Spirit. The Result of Ten Years' Experience in Spirit Manifestations. Intended as a Guide to Enquirers. By C. D. With a Preface by A. B. London: Longman, Green, Longman, Roberts & Green, 1863, xlv + (3) + 388 pp.

Written by the wife of the renowned mathematician Augustus De Morgan, this work describes ten years of personal experiences which include rapping and table movement, automatic writing, mesmerism, and mediumistic phenomena of various kinds. Although the experiences were worth noting, the author uses them to draw rather broad philosophical conclusions. [P]

854. *Denton, William and Denton, Elizabeth M. F.*

Soul of Things: or Pyschometric Researches and Discoveries. 3 vols. Boston: Walker, Wise and Company (Vol. 1); William Denton (Vols. 2 and 3), (1863, 1873, 1874), vii + (2) + 11–370; (1) + 450 + (1); 362 + (1) pp.

Upon reading the works of Joseph Rodes Buchanan on psychometry, William Denton decided to see whether this special sensitivity could be applied specifically to geological and paleontological studies. Beginning with his sister as the sensitive or psychic, and eventually using a number of other individuals Denton recorded the impressions of geography, geology, plant and animal life, and human life they received when in contact with substances and artifacts dating from various epochs. He obtained what he believed to be remarkably accurate descriptions and recorded them in detail. It was his contention that the sciences could in this way gain valuable information otherwise completely unavailable to them. The first volume was written by Denton and his wife. The second and third were written by Denton alone. [P]

855. *Du Potet de Sennevoy, Jules Denis.*

Thérapeutique magnétique, règles de l'application dit magnétisme à l'expérimentation pure et au traitement des maladies. Spiritualisme, son principe et ses phénomènes. Paris: The Author and Dentu, 1863, 539 pp.

Du Potet's last and best major work on animal magnetism. Basically it is a summary of the experiences and ideas of all his earlier works brought together in a skillful and interesting manner. Du Potet writes both about the healing use of animal magnetism and about the wonders of magnetic somnambulism, describing ecstatic and paranormal experiences. [H & P]

856. *Gougenot des Mousseaux, Henri Roger.*

Les médiateurs et les moyens de la magie, les hallucinations et les savants; la fantôme humain et le principe vital. Paris: Henri Plon, 1863, xv + 447 pp.

[P]

857. [*Guppy, Samuel*].

Mary Jane; or Spiritualism Chemically Explained with Spirit Drawings: also Essays by, and Ideas (Perhaps Erroneous) of "*A Child at School.*" London: John King, 1863, xi + 379 pp.

[P]

858. *Home, Daniel Dunglas.*

Incidents in My Life. London: Longman, Green, Longman, Roberts and Green, 1863, xvi + 288 pp.

One of the most spectacular mediums of the nineteenth century, D. D. Home was born in Scotland and grew up in the United States. At the age of thirteen, he began to have psychic experiences and became preoccupied with the issue of life beyond death. He soon progressed beyond clairvoyant and precognitive experiences to physical mediumship, apparently producing remarkable physical effects in the presence of numerous witnesses. In 1855 Home travelled to England where he spent the rest of his life, with numerous excursions to Europe. He exercised his now famous mediumship in the living rooms of the well-to-do and the parlors of the powerful throughout England and the continent, including Russia. Home was a highly cultured man, and considered a charming and entertaining (although somewhat vain) conversationalist. Among the feats routinely performed by Home were levitations and elongations of his body, materializations of objects, such as flowers, fire-immunity, movement of heavy objects without contact, and production of mysterious lights. William Crookes studied Home in Crookes's own laboratory to ascertain whether Home's phenomena were genuine. Crookes' conclusion was that an as yet unknown psychic force was at work that produced measurable effects (see Crookes, entry numbers 939 and 959). Home was accused by contemporaries of using tricks to produce his effects, and to this day there are magicians who claim they can reproduce some of them. But in all the years Home practiced his mediumship, and in the variety of settings in which he performed his feats, Home was never detected in fraud. *Incidents in My Life* is the first of two autobiographical works, the second being *Incidents in My Life. Second series*, 1872 (entry number 948). They are a valuable, although self-serving, source of information about the life of this unusual man. The first book covers Home's life from his birth in 1833 to the death of his first wife, Alexandrina, in 1862. The contents range from a description of his daily life and travels and the many friendships he established to an account of his views concerning some of the controversial statements that were made about his mediumship. [P]

859. ***Howitt, William.***

The History of the Supernatural in All Ages and Nations, and in All Churches, Christian and Pagan: Demonstrating a Universal Faith. 2 vols. London: Longmans & Co., 1863, xix + 489; xvii + 473 pp.

In this lengthy work, Howitt attempts to show that modern spiritualism had its forerunners in every epoch of history. It is an historical study with a strong spiritualistic bias. [P]

860. **[*Krause, Ernst Ludwig, pseudonym: Carus Stern*].**

Die naturgeschichte der Gespenster. Physikalisch-physiologisch-psychologische Studien. Weimar: Voigt, 1863, x + (2) + 439 + (1) pp.

[P]

861. *Pailloux, Xavier.*
Le magnétisme, le spiritisme, et l'eglise. Paris: J. Lecoffre, 1863, xv + (1) + 460 pp.

[H & P]

862. *Perty, Maximillian.*
Die Realität magischer Kräfte und Wirkungen des Menschen gegen die Widersacher vertheidigt. Ein Supplement zu des Verfassers "Mystischen Erscheinungen der menschlichen Natur." Leipzig: Winter, 1863, (2) + 83 pp.

[P]

863. *Roux, F.*
Magnétisme, moyen d'affermir le magnétisme dans la voie scientifique. Paris: Voitelain, 1863, 32 pp.

[H]

864. *Spicer, Henry.*
Strange Things Among Us. London: Chapman & Hall, 1863, xi + 233 pp.

A popular historical discussion of psychical occurrences. [P]

1864

865. *Boret,——de.*
Lettres sur le magnétisme par le docteur de Boret, publiées pour la première fois par l'Union magnétique. Paris: Jusset and The Author, 1864, 68 pp.

[H]

866. *Brittan, S. B.*
Man and His Relations: Illustrating the Influence of the Mind on the Body; the Relations of the Faculties to the Organs, and to the Elements, Objects and Phenomena of the External World. New York: W. A. Townsend, 1864, xiv + 9–578 pp.

A discussion of the moral aspects of the effects of the mind upon the body written by a spiritualist, including sections on fascination, animal magnetism, hallucinations, somnambulism, and clairvoyance. [H & P]

867. [*Cazotte, Jacques S.*]
Témoignage spiritualiste d'outre tombe sur le magnétisme humain, fruit d'un long pèlerinage par J.S.C. publié et annoté par l'abbé Loubert. Paris: Gosselin, 1864, lvi + 134 pp.

Speculations about the nature of magnetic somnambulism. [H & P]

868. *Charpignon, Louis Joseph Jules.*
Études sur la médecine animique et vitaliste. Paris: Germer Baillière, 1864, vi + 191 pp.

A study of healing through animal magnetism and through the influence of the mind on the body. Charpignon describes recent cures in the context of early historical precedents for these approaches to healing. [H]

869. *Duroy de Bruignac, Charles Joseph Albert.*
Satan et la magie de nos jours, réflexions sur la magnétisme, le spiritisme et la magie. Paris: C. Blériot, 1864, xi + 218 pp.

The author sees the work of Satan in animal magnetism. [H & P]

870. *Fauvelle le Gallois, Auguste, ed.*
Le magnétiseur universel, recueil des progrès spiritualistes ou études sur les manifestations du spiritualisme moderne. Paris: The Author, (1864–1866).

It is difficult to decide whether this work is best considered a series of pamphlets published over a period of time or a periodical. The editor was something of a visionary and not universally respected by magnetizers. The work reflects his personal views. [H & P]

871. *Gazzetta magnetico-scientifico-spiritistica.*
One volume only, 1864–1865.

Continued as: *Salute* (*Società magnetica d'Italia residente in Bologna*). Vols 2–10; 1865–1874. [H & P]

872. *Gérard, J.*
Le magnétisme appliqué à la médecine. Paris: The Author and Dentu, 1864, 71 pp.

Gérard describes the utility of mesmerism in medical practice and criticizes those who object to it. [H]

873. *Lallart, C.*
Essai sur l'hypnotisme, nouvelle découverte précédé d'explications sur le magnétisme et le somnambulisme. Soissons: Lallart, 1864, 24 pp.

[H]

874. *Manuale del magnetismo animale. Sua storia — sue teorie — modo di magnetizzare — catechismo magnetologico — fenomeni magnetici — applicazione del magnetismo alla medicina — inconvenienti del magnetismo — conclusione — Appendice: Le tavole semoventi ovvero i miracoli del secolo XIX.*
Naples: n.p., 1864.

[H]

875. *Nichols, Thomas Low.*

A Biography of the Brothers Davenport. With Some Account of the Physical and Psychical Phenomena Which Have Occurred in Their Presence, in America and Europe. London: Saunders, Otley, and Co., 1864, viii + 360 pp.

The story of two brothers, Ira and William Davenport, who became world-famous mediums. It is written from an extremely sympathetic point of view by a man who had known the brothers only a few months when he undertook the book. Although the brothers were the subject of great controversy wherever they went, constantly being accused of fraud, Nichols found nothing of that and believed the brothers to be entirely honest in the practice of their mediumship, despite the spectacular nature of the phenomena they produced. [P]

876. *Powell, James Henry.*

Spiritualism; Its Facts and Phases. Illustrated with Personal Experiences. London: F. Pitman, 1864, (5) + viii + (2) + 168 pp.

A lengthy description of personal experiences with the phenomena of spiritualism, followed by an examination of some possible explanations. The author dismisses involuntary muscular action, mechanical contrivance, and vital magnetism as the source and defends the spiritualistic interpretation of the experiences. [P]

877. [*Rivail, Hippolyte Léon Dénizard, pseudonym: Allan Kardec*].

L'evangile selon le spiritisme contenant l'explication des maximes morales du Christ, leur concordance avec le spiritisme et leur application aux diverses positions de la vie. Paris: Dentu, 1864.

[P]

878. *Roux, F.*

Somnambulisme magnétique. Paris: Voitelain, 1864, 38 pp.

[H]

1865

879. *Babinet, Jacques.*

Les sciences occultes au 19e siècle: les tables tournantes et les manifestations prétendues surnaturelles, considérées au point de vue des principes qui servent de guide dans les sciences d'observation.—Des tables tournantes au point de vue de la mécanique et de la physiologie. Paris: Mallet-Bachelier, 1865.

[H & P]

880. ***Bauche, Alexandre.***

Causeries mesmériennes: enseignement élémentaire (histoire, théorie, et pratique) de magnétisme animal. Paris: E. Voitelain and The Author, 1865, 212 pp.

An introductory textbook of the history and nature of animal magnetism. [H]

881. ***Blanc, Hippolyte.***

Le merveilleux dans le Jansénisme, le magnétisme, le Méthodisme et le Baptisme américain; l'épidémie de Morzine, le spiritisme. Recherches nouvelles. Paris: Henri Plon, 1865, x + (2) + 445 pp.

A discussion of various phenomena associated with occult practice, along with a description of the convulsionary epidemics. [H & P]

882. ***Guyomar de la Roche Derrien,———.***

Étude de la vie intérieure ou spirituelle chez l'homme, recherches physiologiques et philosophiques sur le magnétisme, le somnambulisme et le spiritisme; théorie nouvelle de la pensée, de l'extase, de la lucidité somnambulique et médianimique, role du coeur et du cerveau. Paris: Adrien Delahaye, 1865, 40 pp.

[H & P]

883. ***Nichols, Thomas Low, ed.***

Supermundane Facts in the Life of Rev. Jesse Babcock Ferguson, A.M., LL. D., Including Twenty Years' Observation of Preternatural Phenomena. London: F. Pitman, 1865, xvi + 17–264 pp.

Includes a detailed description of mental and physical phenomena that occurred through Ferguson, a discussion of spiritualism among the Shakers, a description of how the spirits operate, and thoughts on the philosophy of spiritualism. While heavily biased in favor of spiritualism, the work provides interesting material for the psychical researcher. [P]

884. ***Pélin, Gabriel.***

Le spiritisme, la démonologie et la folie; explication de tous les faits magnétiques. Paris: Dentu, 1865, 108 pp.

[H & P]

885. ***Poincaré, Émile Léon.***

Étude physiologique sur le magnétisme animal. Nancy: Raybois, 1865, 46 pp.

Poincaré, father of the famous mathematician Henri Poincaré, was a professor in the School of Medicine at Nancy. In this, his discourse given on reception into the Académie Stanislaus, Poincaré denies the action of any kind of fluid in animal magnetism, whether the magnetic fluid of Mesmer or the nervous fluid of Rostan. He claims that all effects emanating from the animal magnetic procedure are psychogenic. While admitting that somnambulism is a reality, he rejects clairvoyance, the transposition of senses and medical diagnostic abilities as genuine powers of somnambulists. He emphasizes the importance of suggestion in the production of all the phenomena of somnambulism, declaring that somnambulists do not think or will from their own initiative, but accept ideas from others and believe them to be their own. [H]

886. ***Psyche. Deutsche Zeitschrift für Odwissenschaft und Geisteskunde.***
Vols. 1–2; 1865–1866.

Published in Grossenhain and edited by Karl Andreas Berthelen and Gottlieb Dammerung. [H]

887. **[*Rivail, Hippolyte Léon Dénizard, pseudonym: Allan Kardec*].**
Le ciel et l'enfer, ou la justice divine selon le spiritisme, contenant l'examen comparé des doctrines sur le passage de la vie corporelle à la vie spirituelle, les peines et les récompenses futures, les anges et les démons, les peines éternelles, etc. . . suivi de nombreux exemples sur la situation réelle de l'âme pendant et après la mort. Paris: Les éditeurs du "Livre des Esprits", 1865, 500 pp.

[P]

888. ***Robillard, J.B.P.***
Étude sur différents attributs de l'âme humaine, et sur la lucidité dans la veille et pendant le sommeil magnétique. Paris: E. Voitelain, 1865, 55 pp.

[H & P]

889. ***La salute. Giornale d'igiene popolare e di altre cognizioni utili.***
Vols. 1–15; 1865–1881.

[P]

1866

890. **[*Flammarion, Camille*].**
Des forces naturelles inconnues, à propos des phénomènes produits par les frères Davenport et par les médiums en général. Étude critique par Hermès. Paris: Didier, Fred. Henry, Dentu, 1866, 152 pp.

This work, attributed to Flammarion, is a description of seances held in France by the American mediums, the Davenport Brothers. In the context of his description, Flammarion discusses the nature of "psychic" (a term coined by him) or paranormal phenomena in general. [P]

891. ***Gérard, Jules.***

Médecine vitale. Réhabilitation du magnétiseur Mesmer, son baquet, sa doctrine, ses luttes et son triomphe. Paris: Librairie du Petit Journal, 1866, 33 pp.

A description of medical mesmerism by the editor of *La revue magnétique.* [H]

892. ***Gérard, Jules.***

Magnétisme organique. Le magnétisme à la recherche d'une position sociale; sa théorie, sa critique, sa pratique. Paris: Dentu and The Author, 1866, 237 pp.

[H]

893. ***Hue, Charles.***

Le vrai et le faux magnétisme, ses partisans, ses ennemis, thèse présentée à la Société de magnétisme de Paris, précédée d'un avant propos sur le fluide magnétique suivie d'aphorismes ou opinions de 60 docteurs, médecins, praticiens, prêtres, et du Pape, sur le magnétisme et le somnambulisme, et de notions sur l'origine du magnétisme, sur la Société de magnétisme et sur un project de dispensaire. Paris: Germer Baillière, 1866, 103 pp.

[H]

894. ***Lafontaine, Charles.***

Mémoires d'un magnétiseur: suivis de l'examen phrénologique de l'auteur, par le Docteur Castle. 2 vols. Paris: Baillière, 1866, 383; 409 pp.

Lafontaine practiced magnetism and toured giving lectures and demonstrations for decades. Because he was one of the most successful and well-known stage magnetizers of the century, his *Mémoire* contains much valuable information both about the practice of animal magnetism at that time and the personalities who were associated with it. [H]

895. ***Lee, Edwin.***

Animal Magnetism and Magnetic Lucid Somnambulism. With Observations and Illustrative Instances of Analogous Phenomena Occurring Spontaneously; and an Appendix of Corroborative and Correlative Observations and Facts. London: Longmans and Green, 1866, xvi + 334 pp.

An important collection of cases of magnetic somnambulism accompanied by supernormal phenomena, including a section on experiments which the French clairvoyant Alexis carried out in Brighton and Hastings in England in 1849. [H & P]

896. *Liébeault, Ambroise Auguste.*
Du sommeil et des états analogues considérés surtout au point de vue de l'action du moral sur le physique. Paris and Nancy: Victor Masson et fils (Paris) and Nicolas Grosjean (Nancy), 1866, 535 pp.

Liébeault is one of the most important figures in the history of hypnotism. In 1848, while engaged in his medical studies, Liébeault read a book on animal magnetism and was immediately intrigued. He successfully mesmerized a few people, but did not take the experiment any further at that time. He received his medical degree in 1850 and began a rural practice. In 1860 he learned of the experiments with Braid's hypnotism performed by Eugène Azam (1822–1899) (see Azam, entry number 812), and decided to see if it could be of use to him in his medical practice. Looking for willing subjects, he announced to his clientelle that patients who agreed to treatment by hypnotism would not be charged. He soon had many patients to experiment with, and for the next few years developed his own unique approach to hypnotic therapeutics. Liébeault looked on hypnotic sleep as identical to natural sleep, except that the hypnotized patient is placed in that condition by an operator and is subject to a *rapport* with that person. Liébeault became a master of hypnotising patients, and attempted the cure of a great variety of illnesses. His healing technique centered around suggestion. While the patient was in a state of hypnotic sleep, Liébeault would make simple curative suggestions appropriate to the disease. These suggestions would negate the symptoms of the disease and inculcate good habits of health maintenance. In 1864, Liébeault moved to Nancy and took some time to write a book on his work; the result was *Du sommeil.* Here he describes his method of hypnotic induction and the various states that result. He speculates on the effects that hypnotic suggestion has on the nervous system and how healing through hypnotism takes place. Liébeault also writes about the hallucinations (sometimes of spirits) that can accompany somnambulism and the central role of suggestion in creating them. There is a widely circulated tale that only one copy of *Du sommeil* was sold in the first edition (there was a second edition called *Le sommeil provoqué et les états analogues,* in 1889), but that story is probably apocryphal. In any case, Liébeault's book was not popular and he was virtually unknown among medical men for many years. Liébeault's work was eventually noted by Hippolyte Bernheim (1840–1919), then professor of medicine at Nancy. Bernheim was very impressed by Liébeault and introduced his methods into the Nancy medical hospital. The result was instant fame for Liébeault and his theory. Liébeault and Bernheim drew a small group together around them, including Henri Beaunis (1830–1921) and Jules Liégeois (1833–1908), and formed the Nancy School of hypnotism, with its emphasis on suggestion as the essential cause of hypnotic effects. The influence of this school rapidly spread throughout Europe and remained a strong force for decades. The Nancy School was opposed principally by the Salpêtrière School of hypnotism, centered around Jean Martin Charcot (1825–1893). [H]

897. *Le magnétiseur univérsel. Echo du monde magnétique.*
Vols. 1–7; 1866–1870.

Published in Paris. There is a title variant: *Organe de l'Institut magnétologique de Paris et New York*. . . . [H]

898. *Reichenbach, Karl Ludwig von.*
Aphorismen über Sensitivität und Od. Vienna: W. Braumüller, 1866, 92 pp.

[H]

899. *Rochas d'Aiglun, Eugène August Albert de.*
Les effluves odiques: conférences faites en 1866, par le baron de Reichenbach à l'Académie des Sciences de Vienne; précédées d'une notice historique sur les effets mechaniques de l'Od. Paris: Ernest Flammarion, (1866), lix + 102 pp.

A Colonel of the French army and Administrator of the École Polytechnique of Paris, de Rochas was one of the most prolific and important French writers in the field of animal magnetism and psychical research, with at least a dozen books to his credit. De Rochas was personally acquainted with virtually all the principle psychical researchers in France, England, and Germany in the late nineteenth and early twentieth centuries, and he often joined them in their investigations. This book, his first on psychical research, covers such subjects as the odic charge, the mechanical effects of human emanations on the human organism, the imposition of hands, and table turning. [H & P]

900. *Wallace, Alfred Russel.*
The Scientific Aspect of the Supernatural: Indicating the Desirableness of an Experimental Enquiry by Men of Science into the Alleged Powers of Clairvoyants and Mediums. London: n.p., 1866, vi + 57 pp.

Wallace, the famous naturalist who originated the theory of evolution by natural selection independently of Darwin in 1858, initially approached spiritualism as a materialist and sceptic, but was persuaded by the phenomena to consider the possibility of a spiritual dimension to human existence. Wallace had heard about table turning and table rapping off and on for a number of years, but in 1865, he had an occasion for first-hand experience. The results of these first encounters with the paranormal were compelling enough to lead him to investigate other mediumistic phenomena and he eventually became a convinced spiritualist, writing a number of books in its defense. This is his first work in the area of spiritualism and psychical research. Here he states the case for an objective, scientifically based study of the phenomena of mediumship. [P]

1867

901. *Ashburner, John.*

Notes and Studies in the Philosophy of Animal Magnetism and Spiritualism. With Observations upon Catarrh, Bronchitis, Rheumatism, Gout, Scrofula, and Cognate Diseases. London: H. Baillière, 1867, xxiv + 444 pp.

A London physician, and a friend of John Elliotson who wrote articles for Elliotson's mesmeric journal, the *Zoist.* Ashburner had become convinced of the reality of animal magnetism in the 1840s. In the 1850s he came to believe that spirits are able to manifest and communicate with the living, that the force involved in animal magnetism and that involved in spirit manifestations are identical, and that the phenomena manifesting in both experiences were compatible with Christian belief. *Notes and Studies* is an attempt to bring together the traditions of animal magnetism, spiritualism, and Christianity. [H & P]

902. *D'Amico, P.*

Guida teorico-pratica del magnetismo animale. Bologna: Fava e Garagnani, 1867, 92 pp.

D'Amico was one of Italy's most popular mesmerists. This book is a simple, practical manual. [H]

903. *Guidi, Francesco.*

I misteri del moderno spiritismo e l'antidoto contro le superstizioni del secolo XIX. Milan: Bettoni, 1867, 264 pp.

[H & P]

904. *Krauss, Friedrich.*

Nothgedrungene Fortsetzung meines Nothschrei gegen meine Vergiftung mit concentrirtem Lebensäther und gründliche Erklärung der maskirten Einwurkungsweise desselben auf Geist und Körper zum Scheinleben. Stuttgart: The Author, 1867, 380 pp.

Apparently a condensation of the author's earlier book on being "poisoned" by the influence of animal magnetism (see his *Nothschrei,* entry number 645). [H & P]

905. *Reichenbach, Karl Ludwig von.*

Die odische Lohe und einige Bewegungserscheinungen als neuentdeckte Formen des odischen Princips in der Natur. Vienna: Braumüller, 1867, vi + 151 pp.

[H]

1868

906. *Bacot, G. F.*

Des facultés magnétiques de l'homme des moyens divers par lesquels elles se manifestent; des conditions qu'exige leur emploi; de la responsabilité morale qu'entraine leur exercice; des services qu'on peut en attendre. Paris: The Author, 1868, 32 pp.

A treatise on animal magnetism written for the practitioner. [H]

907. *Baillif, Louis Ernest.*

Du sommeil magnétique dans l'hystérie. Thèse à la faculté de médecine de Strasbourg et soutenue publiquement le samedi 20 juin 1868, à 3 heures du soir, pour obtenir le grade de docteur en médecine. Strasbourg: n.p., 1868, (4) + 71 pp.

The author carried out prolonged experiments using animal magnetism on individuals suffering from hysteria. He recognized the importance of Braid's work with hypnotism, but did not think that Braid had successfully disproved certain phenomena attributed to animal magnetism. In his work with hysterics, Baillif found artificial somnambulism to be of some benefit, but not capable of producing a cure. [H]

908. *Colin, ———.*

Bienfaits du somnambulisme, ouvrage dédié à Mme Roger, aux amis de la verité, et aux personnes amies d'elles-mêmes. Paris: Kugelmann, 1868, 179 pp.

[H]

909. *Davis, Andrew Jackson.*

Memoranda of Persons, Places, and Events; Embracing Authentic Facts, Visions, Impressions, Discoveries, in Magnetism, Clairvoyance, Spiritualism: also Quotations from the Opposition. With an Appendix, Containing Zschokke's Great Story of "Hortensia," Vividly Portraying the Wide Difference Between the Ordinary State and that of Clairvoyance. New York: William White & Co., 1868, 488 pp.

This book, later published as *Events in the Life of a Seer*, is a compilation of entries from the author's journal. It contains records of events in Davis' life, letters, and comments on matters of personal concern. It is less philosophically ponderous than most of Davis' works, and it contains valuable information about events and people connected with spiritualism and psychic matters. [H & P]

910. *Despine, Prosper.*

Psychologie naturelle. Étude sur les facultés intellectuelles et morales dans leur état normal et dans leurs manifestations anomales chez les aliénés et chez les criminels. Tome premier. Paris: F. Savy, 1868, xviii + 616 pp.

An extremely significant contribution to the development of thought concerning artificial somnambulism and automatism. Despine distinguishes four kinds of automatism: 1) those acts that, after many repetitions under the direction of the "I," are performed solely by the organic "automaton" (e.g., running); 2) instinctive acts performed in connection with consciously intended acts (e.g., gestures while talking); 3) acts performed out of rapport or separately from conscious psychic activity simultaneously taking place (e.g., walking while thinking about a mathematical problem); and 4) acts performed by the organic "automation" while the psychic function of the brain or the "I" is suspended (e.g., actions carried out during artificial somnambulism). Insisting that actions may be automatic and yet "intelligent," Despine argues that one may produce intelligent acts that do not involve the psychic functions of the brain or the "I." One need not believe that "the intervention of the spirit is necessary in all intelligent acts of man" (p. 536). And, conversely, "to admit, with the animists, that the spirit acts without knowing it, without consciousness, is an absurdity" (p. 536). (In this, Despine anticipates Freud's view of unconscious intelligence.) Intelligence, Despine suggests, can manifest without personality, resulting from obedience to organic laws to which it is subject. From this view of automatism, Despine infers that in somnambulistic action there is a paralysis of the brain and so a non-participation of the "I" or the spirit. For that reason our understanding of the nature of somnambulistic amnesia must be revised. Somnambulistic acts are produced by the nervous system isolated from the brain, and therefore from the "I." The "I" does not "forget" what has been done while in a somnambulistic state. Rather the "I" is not involved in the action in the first place and so never has knowledge of the action. (In this, Despine anticipates Pierre Janet's notion of dissociation.) [H]

911. *Durand, Joseph Pierre.*

La philosophie physiologique et médicale à l'Académie de Médecine. Paris: Germer Baillière, 1868, 99 pp.

Contains an essay "Polyzoisme ou pluralité animale dans l'homme" which provides an interesting speculation about a possible physiological basis for the phenomenon of dissociation. [H]

912. *Lioy, P.*

Spiritismo e magnetismo. Lettura fatta a Vincenza. Florence: Sienza del Popolo, 1868, 63 pp.

[H & P]

913. ***Randolph, Paschal Beverley.***
After Death; or Disembodied Man. etc. Boston: The Author, 1868, viii + 9–260 pp.

Randolph was a physician and spiritualist. [P]

914. ***La revue magnétique. Journal des malades.***
Vols. 1–2; 1868–1870.

Edited by J. Gérard. [H]

915. **[*Rivail, Hippolyte Léon Dénizard, pseudonym: Allan Kardec*].**
La genèse, les miracles et les prophéties selon le spiritisme. Paris: A. Lacroix, Verboeckhoven et cié, 1868, 459 pp.

Rivail discusses subjects of some interest to psychical researchers, including magnetic fluids, dreams, possession, precognition, somnambulism, and apparitions. [H]

916. ***Sunderland, LaRoy.***
The Trance and Correlative Phenomena. Chicago: J. Walker, 1868, 407 pp.

[H]

917. ***Tissot, Claude Joseph.***
L'imagination: ses bienfaits et ses égarements surtout dans le domaine du merveilleux. Paris: Didier, 1868, viii + 607 pp.

[P]

918. ***Underhill, Samuel.***
Underhill on Mesmerism, with Criticisms on Its Opposers, and a Review of Humbugs and Humbuggers, with Practical Instructions for Experiments in the Science, Full Directions for Using it as a Remedy in Disease . . . the Philosophy of Its Curative Powers; How to Develop a Good Clairvoyant; the Philosophy of Seeing Without Eyes. The Proofs of Immortality Derived from the Unfoldings of Mesmerism. . . . Chicago: The Author, 1868, (3) + 5–273 pp.

Underhill, a physician and university professor in Ohio, had his first encounter with animal magnetism in 1830 when he obtained a copy of the *Lectures* of Du Commun (see entry number 340). In 1832, he began to experiment with magnetic cures and was amazed at the positive results that he obtained. In 1838, Underhill began a monthly periodical, *Annals of Animal Magnetism.* That was the first such publication in the United States. When this work was published, he had been practicing animal magnetism for over thirty years, a period, as he says, in which "Sunderland's 'Pathetism,' Dr. Braid's 'hypnotism,' Burr's 'Biology,' and Dr. Dodd's (sic) 'Psychology,' different names for the same thing, presenting phenomena, long familiar to the writer, rose and passed away. . . ." The book contains lectures Underhill delivered to discuss and demonstrate animal magnetism. It is full of detailed practical advice and contains many interesting anecdotes of animal magnetic practice from Underhill's own experience. [H & P]

1869

919. ***Bonavia, Emmanuel.***
Contributions to Christology. London: Trubner & Co., 1869, viii + 170 pp.

An examination of the miracles of Jesus from the perspective of animal magnetism. [H & P]

920. ***Chévillard, Alphonse.***
Études expérimentales sur le fluide nerveux et solution rationelle du problème spirite. Paris: V. Mason & Fils, 1869, 38 pp.

[H & P]

921. ***Dal Pozzo di Mombello, Enrico.***
Trattato pratico del magnetismo animale. Foligno: n.p., 1869.

The author develops an elaborate theory of the somnambulistic state in terms of inhibitory vibrations. Included in this theory is the proposition that in this state the individual may be able to send thoughts to others by means of vibrations. [H & P]

922. ***Dureau, Alexis.***
Histoire de la médecine et des sciences occultes. Notes bibliographiques pour servir à l'histoire du magnétisme animal. Analyse de tous les livres, brochures, articles de journaux publiés sur le magnétisme animal, en France et à l'étranger, à partir de 1766 jusqu'au 31 décembre 1868. Première partie: livres imprimés en France. Paris: The Author and Joubert, 1869, 206 pp.

The best bibliography of animal magnetism prior to Caillet's *Manuel bibliographique* (*see* entry number 1673). Dureau arranged his material in chronological order from 1775 to 1868, providing annotations (often quoted by Caillet) for many of his entries. Only one volume was published. [H]

923. ***Fahnestock, William Baker.***
Artificial Somnambulism, Hitherto called Mesmerism; or, Animal Magnetism; Containing a Brief Historical Survey of Mesmer's Operations, and the Examination of the Same by the French Commissioners. Phreno-somnambulism; or, the Exposition of Phreno-magnetism and Neurology. A New View

and Division of the Phrenological Organs into Functions, with Descriptions of Their Nature and Qualities, etc., in the Senses and Faculties. . . . Philadelphia: Barclay & Co., 1869, (2) + 43–329 pp.

Fahnestock, a physician from Lancaster, Pennsylvania, carried out many of his own experiments concentrating on the higher phenomena of magnetic somnambulism. He believed that through animal magnetism the deeper powers of seeing, hearing, and smell could be actualized and these senses made to operate at any distance. He was an adherent of a particular view of animal magnetism promoted by LaRoy Sunderland (1804–1885) called "pathetism." [H]

924. *Hartmann, Eduard von.*

Philosophie des Unbewussten. Berlin: Carl Dunker, 1869, 678 pp. ENGLISH: *Philosophy of the Unconscious; Speculative Results According to the Inductive Method of Physical Science.* 3 vols. Translated by William C. Coupland. London: Trübner & Co., 1884.

Von Hartmann was a brilliant German philosopher and original thinker whose writings strongly influenced the development of our modern notion of the unconscious. This book is the most important of his works. It went through many editions and appeared in various structural formats, from this one volume form, to a two volume edition, and finally a three volume edition, the third volume being a collection of later essays. Von Hartmann posits an intelligent but blind dynamic reality that underlies all existence, and he calls that reality the "unconscious." He begins his work with an examination of the unconscious as it manifests itself in the realm of organic life, particularly the organic life of animals. He contends that the activity of the mind we call "willing" is not the exclusive property of man, but is also participated in by animals. Von Hartmann even attributes will to the various parts of the organism, so that there are many independent wills spread throughout the body. The acts of will constantly manifested in the human organism are ordinarily not part of our conscious awareness. This means that in human beings both instinctive action and clairvoyant intuition are rightly considered manifestations of the universal unconscious will. Throughout the rest of the work von Hartmann develops this notion of the unconscious, arriving at three principal manifestations or layers: 1) the absolute unconscious that is the substance of the universe and underlies all other forms of the unconscious, 2) the physiological unconscious present in the vital actions of living organisms, and 3) the psychological unconscious which is the basis for our conscious mental activity. [H]

925. *Maggiorani, Carlo.*

La magnete e i nervosi: centuria di osservationi. Milan: F. Vallardi, 1869, 128 pp.

Investigates the possibility of using magnets to induce a nervous crisis in hysterical patients. [H]

926. *Moilin, Tony.*

Traité élémentaire, théorique et pratique du magnétisme contenant toutes les indications nécessaires pour traiter soi-même, à l'aide du magnétisme animal, les maladies les plus communes. Paris: Lacroix, 1869, viii + 335 pp.

[H]

927. *Perty, Maximillian.*

Blicke in das verborgene Leben des Menschengeistes. Leipzig and Heidelberg: Winter, 1869, ix + (3) 290 pp.

[P]

928. *Samson, George Whitefield.*

Physical Media in Spiritual Manifestations. The Phenomena of Responding Tables and the Planchette, and Their Physical Cause in the Nervous Organism, Illustrated from Ancient and Modern Testimonies. Philadelphia: J. B. Lippincott & Co., 1869, viii + 9–185 pp.

Samson explains the moving tables of spiritualism in terms of a nervous energy which he relates to animal magnetism and which he describes as akin to electricity. He also discusses clairvoyance and prophecy and finds instances of both among the ancients. [H & P]

929. [*Sargent, Epes*].

Planchette: or, The Despair of Science. Being a Full Account of Modern Spiritualism, Its Phenomena, and the Various Theories Regarding It. With a Survey of French Spiritism. Boston: Roberts Brothers, 1869, xii + 404 pp.

Written from a very pro-spiritualist standpoint, this work includes a history of spiritualism from the Fox sisters to the 1860s, an examination of somnambulism and mesmerism, and a discussion of arguments for and against spiritualism. [H & P]

930. [*Smith-Buck, George, pseudonym: Herr Dobler*].

Exposé of the Davenport Brothers. Belfast: D. & J. Allen, 1869, 48 pp.

Written by a conjurer known in his day as the "Wizard of the World," this is a detailed attempt to show how the Davenports performed their amazing mediumistic feats using the tricks of stage magicians. The work is very rare. [P]

1870–1879

1870

931. ***Adare, Windham Thomas Wyndham-Quin, Viscount.***
Experiences in Spiritualism with Mr D. D. Home. With Introductory Remarks by the Earl of Dunraven. London: Privately Printed, (1870), xxxii + 179 pp.

One of the rarest and most important early contributions to psychical research. Immigrating to England from America in 1855, Home soon attracted attention with his remarkable mediumistic feats and he became a popular house guest of the rich and influential. The phenomena that accompanied him were of the most sensational kind: levitation of furniture, writing produced without contact, fire immunity, full levitation of the body, and elongation of the body. Although subject to intimate scrutiny during many of these occurrences, Home was never caught in any kind of fraud or deception. Adare was Home's friend and sometime travelling companion. Here he describes some of the phenomena that he and others witnessed. Among the most extraordinary are Home's handling of red hot coals and his floating out one window of a house and into another. [P]

932. ***[Britten], Emma Hardinge.***
Modern American Spiritualism. A Twenty Year's Record of the Communion Between Earth and the World of Spirits. New York: The Author, 1870, (2) + viii + 9–565 pp.

A monumental study of the history of the first twenty years of the spiritualist movement in the United States written by a partisan. Unique in its time, the work is a goldmine of information about how the movement grew and spread throughout the country—with a small segment on spiritualism in Canada and South and Central America. Much of the information comes from contemporary newspaper clippings, letters, and verbal accounts of personal experiences. This influential work helped create a view of spiritualism as a cohesive movement. [P]

933. ***Rouget, Ferdinand.***

La photographie mentale des esprits dévoilée. Connaissance de la cause qui produit les effets naturels et magnétiques du spiritisme depuis l'antiquité jusqu'à nos jours. Toulouse: Bompard, 1870, 204 pp.

[H & P]

1871

934. ***Alexander, Patrick Proctor.***

Spiritualism: A Narrative with a Discussion. Glasgow: T. Murray & Sons, 1871, v + (1) + 92 p.

A favorable eye-witness account of sittings with the famous medium Daniel Dunglas Home. [P]

935. ***Coleman, Benjamin.***

The Rise and Progress of Spiritualism in England. London: Beveridge & Fraser, (1871), 59 pp.

[P]

936. ***Collyer, Robert Hanham.***

Mysteries of the Vital Element in Connection with Dreams, Somnambulism, Trance, Vital Photography, Faith and Will, Anaesthesia, Nervous Congestion and Creative Function. Modern Spiritualism Explained. 2 ed. London: Henry Renshaw, 1871, viii + 144 pp.

[H & P]

937. ***Cox, Edward William.***

Spiritualism Answered by Science. London: Longman, 1871, vii + 56 pp.

Cox was a barrister and one of the most important psychical researchers active in the period before the foundation of the Society for Psychical Research. In 1875 he founded the Psychological Society of Great Britain, intended to carry on intensive research into the nature of psychical phenomena. Its main activity, however, turned out to be the sponsorship of lectures and the society dissolved when Cox died. This is his first book on psychical research. It was revised and enlarged the following year. [P]

938. *Crookes, William.*

Psychic Force and Modern Spiritualism: A Reply to the "Quarterly Review" and Other Critics. London: Longmans, Green & Co., 1871, 24 pp.

The republication of an article that appeared in the *Quarterly Journal of Science* (edited by Crookes) in 1871. In the article, Crookes takes exception to the accusation of lack of critical acumen in experiments to discover a psychic force conducted in Crookes' *Researches in the Phenomena of Spiritualism* (*see* entry number 959). [P]

939. *Figuier, Louis.*

Le lendemain de la mort; ou la vie future selon la science. Paris: Hachette, 1871, xi + 449 pp.

[P]

940. [*Hayward, Aaron S., pseudonym:* A Magnetic Physician].

Vital Magnetic Cure: an Exposition of Vital Magnetism and Its Application to the Treatment of Mental and Physical Disease. By a Magnetic Physician. Boston: William White and Company, 1871, 216 pp.

[H]

941. *Hammond, William Alexander.*

The Physics and Physiology of Spiritualism. New York: 1871, Appleton, 86 pp.

Hammond was a Civil War Surgeon General. This book, a revised and expanded form of an article in the *North American Review* (April 1870), provides a rational explanation of "the real and fraudulent phenomena of what is called spiritualism." [P]

942. *London Dialectical Society.*

Report on Spiritualism of the Committee of the London Dialectical Society, Together with Evidence Oral and Written, and a Selection from the Correspondence. London: Longmans, Green, Reader, & Dyer, 1871, xi + 412 pp.

Along with Crookes' *Researches* (*see* entry number 959), this report was the only methodical British attempt to study the phenomena of spiritualism scientifically in the era before the formation of the Society for Psychical Research in 1882. The London Dialectical Society was a sophisticated debating society whose purpose it was to afford a hearing to subjects that were ostracized elsewhere. The stature of the individuals named to the committee appointed to study the matter assured that its findings would be given serious consideration by the public. The committee proceeded to gather evidence both from testimony and from direct experimentation. Letters and verbal testimony were received for and against the genuineness of the phenomena in question. The list of those making contributions included some of the great intellectual lights of the day and some of the best known spiritualist mediums. A subcommittee formed to carry out direct experimentation with the physical phenomena of spiritualism produced findings that were presented at the beginning of the report. The subcommittee established to its satisfaction that sounds were produced and movements of heavy bodies took place without any muscular action or mechanical contrivance. The report of the committee, basically favorable to the genuineness of the phenomena, was not accepted by the Dialectical Society as a whole and so was published by the committee on its own authority. [P]

943. *Owen, Robert Dale.*

The Debatable Land between This World and the Next. With Illustrative Narratives. New York: G. W. Carleton, 1871, xxii + (23)–542 pp.

Written eleven years after *Footfalls* (1860), Owen's first, very popular book about spiritualism, this work was one of the most learned and thorough treatises published in support of spiritualism. Aimed at the beliefs of the established protestant churches and their objections to the claims of spiritualism, Owen presents a detailed historical examination of phenomena occurring within the churches over the centuries that were identical to those claimed by the spiritualists but which the churches now rejected. For Owen, spiritualism is a natural continuation of the basic tradition of spiritual gifts within the Christian tradition. [P]

944. *Peebles, James Martin and Tuttle, Hudson, eds.*

The Year Book of Spiritualism for 1871: Presenting the Status of Spiritualism for the Current Year Throughout the World; Philosophical, Scientific, and Religious Essays; Review of Its Literature; History of American Associations; State and Local Societies; Progressive Lyceums; Lecturers; Mediums; and Other Matters Relating to the Momentous Subject. Boston: W. White & Co., 1871, 246 pp.

[P]

945. *Tuttle, Hudson.*

Arcana of Spiritualism: a Manual of Spiritual Science and Philosophy. Boston: Adams & Co., 1871, 455 pp.

Tuttle discusses the evidence for the truth of spiritualism and the framework of thought within which its phenomena can be understood. [P]

946. *Zerffi, George Gustavus.*

Spiritualism and Animal Magnetism. A Treatise on Dreams, Second Sight, Somnambulism, Magnetic Sleep, Spiritual Manifestations, Hallucinations, and Spectral Visions. London: R. Hardwicke, 1871, 148 pp.

[H & P]

1872

947. *Charcot, Jean Martin.*

Leçons sur les maladies du système nerveux faites à la Salpêtrière . . . recueillies et publiées par Bourneville. 3 vols. Paris: A. Delahaye, (1872–1887).

Charcot was one of the most highly respected neurologists of the nineteenth century. After successfully investigating a number of neurological conditions, he became particularly interested in hysteria. In 1878 he began to experiment with hypnotism, using as his subjects a number of hysterical women in his care at the Hospice de la Salpêtrière. In the course of this research, discussed in the *Leçons*, Charcot described three "stages" of hypnotism: lethargy, catalepsy, and somnambulism, each with its characteristic symptoms. Believing that these stages came in regular succession and were consistently present in hypnotic induction, Charcot held that the stages were organically determined and not the result of suggestion and that hypnotism was a pathological phenomenon, an artificially created neurosis very similar to hysteria. Charcot next turned his attention to traumatic paralysis. Noting that traumatic paralysis differed from organic paralysis but seemed to be identical with the hysterical paralysis he had observed in his patients, he succeeded in using hypnotic suggestion to reproduce the symptoms of paralysis resulting from trauma. From this he concluded that the shock of the trauma produced a spontaneous hypnotic state, and that in this state the individual induced the paralytic symptom through autosuggestion. Charcot's view of hypnotism, that it is an organically determined condition essentially related to hysteria, exerted a considerable influence on psychological investigators and became the core doctrine of the "Salpêtrière school" of hypnotism which, along with the rival "Nancy school" founded by Liébeault and Bernheim, defined the late nineteenth-century French hypnotic tradition. [H]

948. *Home, Daniel Dunglas.*

Incidents in My Life. Second Series. London: Tinsley Brothers, 1872.

The second of two autobiographical works by the famous medium D. D. Home (see *Incidents in my Life*, entry number 858). Here Home describes many seances held throughout England and Europe and gives his version of several controversial events including his expulsion from Rome at the request of the Catholic Church and his having been accused of obtaining funds from a Mrs. Jane Lyon through undue influence. [P]

949. *Tuke, Daniel Hack.*

Illustrations of the Influence of the Mind upon the Body in Health and Disease. Designed to Elucidate the Action of the Imagination. London: J. A. Churchill, 1872, xvi + 444 pp.

Daniel Hack Tuke was a London physician, author of a number of works on nervous disorders and a member of an illustrious family of Quakers which had for generations remained involved with the care of the mentally ill. Hearing of a man who had been cured of rheumatism by the shock of being in a railway accident, Tuke decided to devote his attention to the influence of the mind upon the body. The resultant work which contains numerous case illustrations, investigates the influence of the mind, the emotions, and the will on the nervous and muscular systems, and then takes up the influence of the mind on the body in the cure of disease. In a long discussion of the nature of imagination and its part in the process, Tuke compares the adherents of animal magnetism (mesmerism) to those who see purely psychological forces operating in magnetic healing (braidism). [H]

950. ***Watson, Samuel.***
The Clock Struck One, and Christian Spiritualist: Being a Synopsis of the Investigations of Spirit Intercourse by an Episcopal Bishop, Three Ministers, Five Doctors, and Others, at Memphis, Tenn., in 1855; Also, the Opinion of Many Eminent Divines, Living and Dead, on the Subject, and Communications Received from a Number of Persons Recently. New York: S. R. Wells, 1872, 208 pp.

[H]

1873

951. ***Cox, Edward William.***
What Am I? Popular Introduction to the Study of Psychology. 2 vols. London: Longmans, 1873–1874.

Cox was founder of the Psychological Society of Great Britain and one of the earliest psychical researchers in that country. In this important and rare work, he discusses the whole range of topics in psychical research, examining the psychological basis for supernormal phenomena, including dreams and somnambulism, and discussing the problem of proof of the survival of death. The first volume was published in 1873; the second in 1874. The later revised edition was titled *The Mechanism of Man.* [P]

952. ***Edmonds, John W.***
Spiritual Tracts, No. 1 – No. 13. New York: n.p., 1873, 337 pp.

A collection of letters and articles on the subject of spiritualism written by one of its strongest American advocates. These treatises were much read in their day and greatly influenced informed thought. [P]

953. ***Guidi, Francesco.***
Relazione di una interessantissima cura magnetica fatta in Berlino. Milan: Molinari, 1873, 16 pp.

[H]

954. ***Liébeault, Ambroise August.***
Ebauche de psychologie. Paris and Nancy: G. Masson (Paris) and N. Grosjean (Nancy), 1873, (4) + xvi + 202 pp.

Liébeault's attempt to frame a psychology based on his findings from hypnotic experiments. Here he presents the first form of the notion of "fixed ideas" that Pierre Janet (1859–1947) would develop so effectively. Liébeault states that ideas suggested during artificial somnambulism become fixed within the individual and remain unconscious when awakening. This fixed idea continues to operate within the individual while he carries on with his normal life and can affect that person without his becoming aware of it. [H]

955. ***Noizet, Gal.***
Mélanges de philosophie critique. Paris: Plon, 1873.

A work on spiritualism with a discussion of some paranormal phenomena. [P]

956. ***Poret, Bènigne Ernest.***
Les Puységur. Leurs oeuvres de littérature, l'économie politique et de science. Étude. Paris: Aubrey, 1873, 161 pp.

An important source of information on the life of Armand Marie Jacques de Chastenet (1751–1825), the Marquis de Puységur, discoverer of magnetic somnambulism. [H]

957. ***Sexton, George.***
Spirit Mediums and Conjurers. An Oration Delivered in the Cavendish Rooms, London, on Sunday Evening, June 15th, 1873. To Which is Appended Rules to be Observed at the Spirit Circle. London: J. Burns, (1873), 28 pp.

Written by a spiritualist, the book attempts to show that the debunking of spiritualistic phenomena by conjurors is unconvincing. Sexton's method is to expose the devices used by conjurors in their stage performances and demonstrate that their tricks are so obvious that they could in no way account for the physical phenomena of mediumship. [P]

958. ***Timmler, Julius Eduardus.***
Die Heilkraft des Lebensmagnetismus und dessen Beweiskraft für die Unsterblichkeit der Seele. 2 ed. Altenburg: Bonde, 1873, viii + 205 pp.

[H]

1874

959. ***Crookes, William.***
Researches in the Phenomena of Spiritualism (*Reprinted from the Quarterly Journal of Science*). London: J. Burns, 1874, 112 pp.

This small volume is a landmark publication in the history of psychical research. It contains an account of experiments to assess the existence of a "psychic force" carried out by one of the most respected scientists of the nineteenth century. Crookes was a genius of scientific invention and original thought. He first became well known when, at the age of twenty-nine, he discovered the element thallium. He went on to become the inventor of the radiometer, the spinthariscope, and the Crookes vacuum tube, which was instrumental in the discovery of X-rays. He was founder of the *Chemical News* and editor of *The Quarterly Journal of Science* between July, 1871 and June, 1874. In 1871, Crookes began his investigations into telekinetic "psychic force" by carrying out some experiments with the famous medium D. D. Home (1833–1886). Using many different types of apparatus, he eventually confirmed unequivocally the existence of a force which could move objects and apply pressure at a distance from the apparent source of that force. In the *Researches* he gives a detailed account of the construction of the apparatus used and of other physical circumstances of the experiments. Crookes' *Researches* made a powerful impression on the minds of intellectuals of his day and did much to create a climate receptive to the establishment of the Society for Psychical Research in 1882. [P]

960. ***Kramer, Phillipp Walburg.***
Der Heilmagnetismus; seine Theorie und Praxis. Landshut: Krüll, 1874, 86 pp.

[H]

961. ***Mesnet, Ernest.***
De l'automatisme, de la mémoire et du souvenir dans le somnambulisme pathologique, considérations médico-légales. Paris: Félix Malteste et Cie., 1874, 30 pp.

Mesnet examines the state of the senses and memory in those suffering from "pathological somnambulism." He also discusses moral and legal implications of pathological susceptibility to suggestion. [H]

962. ***Psychische Studien. Monatliche Zeitschrift, vorzüglich der Untersuchung der wenig gekannten Phänomene des Seelenlebens gewidmet.***
Vols. 1–52+; 1874–1925+.

Founded by Alexander Aksakov and published in Leipzig, this serial was continued as *Zeitschrift für Parapsychologie* in 1926. [P]

963. ***Putnam, Allen.***

Agassiz and Spiritualism: Involving the Investigation of Harvard College Professors in 1857. Boston: Colby & Rich, (1874), 70 pp.

[P]

964. ***Revue de psychologie expérimentale. Études sur le sommeil, le somnambulisme, l'hypnotisme et le spiritualisme.***

Vols. 1–3; 1874–1876.

Published in Paris and edited by T. Puel. [H & P]

965. ***Saint-Dominique, Countess de.***

Animal Magnetism (Mesmerism) and Artificial Somnambulism: Being a Complete and Practical Treatise on that Science and Its Application to Medical Purposes. Followed by Observations on the Affinity Existing Between Magnetism and Spiritualism Ancient and Modern. London: Tinsley Brothers, 1874, x + 234 pp.

[H & P]

966. ***Savile, Bourchier Wrey.***

Apparitions: a Narrative of Facts. London: Longmans, 1874, vi + 280 pp.

A collection of narratives that the author holds to be sufficient evidence for believing in the appearance of departed spirits at or near the moment of death. [P]

967. ***Sierke, Eugen.***

Schwärmer und Schwindler zu Ende des 18. Jahrhunderts. Leipzig: S. Hirzel, 1874, 462 pp.

A collection of historical studies on persons involved with unusual psychological phenomena whom the author considered to be of questionable reputation. The list of "fanatics and frauds" includes Franz Anton Mesmer and the exorcist Gassner. [H]

968. ***Vay, Adelina (Wurnbrand-Stuppach).***

Studien über die Geisterwelt. 2 ed. Leipzig: Oswald Mutze, 1874, xii + 407 pp.

A lengthy study of spiritualistic phenomena including automatic writing, automatic drawing, and spiritistic healing. [P]

969. ***Wallace, Alfred Russell.***

"A Defence of Modern Spiritualism." *Fortnightly Review* 15(1874): 630–657, 785–807.

Wallace attempts to provide spiritualism with a respectability in England that he felt it deserved but did not have. He provides a short historical sketch of spiritualism from the days of the Fox sisters and then discusses the nature of the evidence in favor of the genuineness of spiritualistic phenomena and the possibility of communication with the spirits of the dead. He also discusses the moral teachings of spiritualism, which he considers to be of a lofty nature and concludes with a plea for a serious and open-minded consideration of the subject. This article, written by a man of considerable scientific reputation and published in a periodical of stature, had a strong influence on the contemporary attitude toward spiritualism and helped prepare the way for the establishment of psychical research as a legitimate undertaking. [P]

970. *Watson, Samuel.*

The Clock Struck Three, Being a Review of Clock Struck One, and reply to It. Part II. Showing the Harmony Between Christianity, Science, and Spiritualism. Chicago: Religio-Philosophical Publishing House, 1874, 352 pp.

The author's answer to criticisms made of his earlier book *The Clock Struck One* (*see* entry number 952). [P]

1875

971. *Coste, Albert.*

Les phénomènes psychiques occultes. État actuel de la question. Montpellier: C. Coulet, etc., 1875, 198 pp.

[P]

972. *Davies, Charles Maurice.*

Mystic London: or, Phases of Occult Life in the Metropolis. London: Tinsley Bros., 1875, vii + 406 pp.

A collection of experiences, including many encounters with spiritualism and its phenomena.

973. *Fairfield, Francis Gerry.*

Ten Years with Spiritual Mediums: an Inquiry Concerning the Etiology of Certain Phenomena Called Spiritual. New York: D. Appleton & Co., 1875, 182 pp.

Fairfield attempts to describe the neuro-physiological symptoms that accompany mediumship and see to what extent the phenomena might be explained in terms of the function of the nervous system. Taking most of his data from his own personal observations and from those of medical men, he concludes that mediumistic trance and mesmeric trance are identical and that both the mental and physical phenomena of mediumship are "morbid nervous phenomena". By this, Fairfield does not mean to deny the genuineness of such phenomena but only to indicate that they should be attributed neither to the intervention of spirits, nor to a "psychic force" such as that posited by William Crookes. Rather they are an extension of certain abnormal neurological events which result in psychic and physical action at a distance. [P]

974. *Grimes, James Stanley.*

The Mysteries of the Head and the Heart Explained: Including an Improved System of Phrenology; a New Theory of the Emotions, and an Explanation of the Mysteries of Mesmerism, Trance, Mind-reading, and the Spirit Delusion. Chicago: Keen, 1875, 359 pp.

In a marked departure from his earlier work, Grimes emphasizes imagination and suggestibility in a purely psychological explanation for extraordinary phenomena in entranced subjects. [H]

975. *Harrison, William H.*

Spirit People: a Scientifically Accurate Description of Manifestations Recently Produced by Spirits, and Simultaneously Witnessed by the Author and Other Observers in London. London: Harrison, 1875, vii + 8–46 pp.

[P]

976. *Radau, Rudolphe.*

Le magnétisme. Paris: Hachette, 1875, 328 pp.

[H]

977. *Richet, Charles Robert.*

"Du somnambulisme provoqué." *Journal de l'anatomie et de la physiologie normales et pathologiques de l'homme et des animaux* 11:348–378.

Richet was a physician and physiologist of great ability who eventually become a Professor of Physiology at the University of Paris, a member of the Académie de Médecine and the Académie des Sciences and winner of the Nobel Prize in Physiology and Medicine. Even in 1875, Richet's reputation was such that this article exerted a major influence in legitimatizing the investigation of hypnotism and artificial somnambulism for the French scientific establishment. Richet's comments on automatism, hallucination, and "doubling of personality" in the somnambulistic state constitute a significant bridge between the work of Braid and the French "braidists" of the 1860s and that of Charcot, Janet, and others in the 1880s and 1890s. [H]

978. *Sargent, Epes.*

The Proof Palpable of Immortality; Being an Account of the Materialisation Phenomena of Modern Spiritualism: with Remarks on the Relations of the Facts to Theology, Morals, and Religion. Boston: Colby & Rich, 1875, 238 pp.

[P]

979. *Wallace, Alfred Russel.*
On Miracles and Modern Spiritualism. Three Essays. London: J. Burns, 1875, viii + 236 pp.

[P]

1876

980. *Azam, Eugène.*
"Amnésia périodique ou dédoublement de la vie." *Annales médico-psychologiques* 16 (1876): 5–35.

[H]

981. *Berry, Catherine.*
Experiences in Spiritualism: A Record of Extraordinary Phenomena Witnessed Through the Most Powerful Mediums, with Some Historical Fragments Relating to Semiramide, Given by the Spirit of an Egyptian Who Lived Contemporary with Her. London: Burns, 1876, 220 pp.

Berry was a well-known British medium. In this book, she describes sittings she had with the mediums Mrs. Samuel Guppy, Mrs. Thomas Everitt and others. The sittings involved spirit drawings, prophecies, healings, materializations, and spirit photography. [P]

982. *Bourneville, Désiré and Regnard, P.*
Iconographie photographique de la Salpêtrière (service de M. Charcot). I. Hystéro-épilepsie: description des attaques; les possédées de Loudun; du crucifiement. II. Épilepsie partielle et hystéro-épilepsie . . . III. Hystéro-épilepsie: Zones hystérogènes; sommeil; attaque de sommeil; hypnotisme; somnambulisme; magnétisme, catalepsie; procédés de magnétisme. 3 vols. Paris: Delahay, 1876.

[H]

983. [*Britten, Emma Hardinge*].
Art Magic: or, Mundane, Sub-mundane and Super-mundane Spiritism. A Treatise in Three Parts, and Twenty-three Sections: Descriptive of Art Magic, Spiritism, the Different Orders of Spirits in the Universe Known to be Related to, or in Communication with Man; Together with Directions for Invoking, Controlling, and Discharging spirits and the Uses and Abuses, Dangers and Possibilities of Magical Art . . . Edited by Emma Hardinge. New York: The Author, 1876, (2) + 467 pp.

A mixture of esoteric philosophy, history of spiritualism, and discussion of psychical phenomena. [P]

984. [*Britten, Emma Hardinge*].

Ghost Land: or Researches into the Mysteries of Occultism. Illustrated in a Series of Autobiographical Sketches. Translated and Edited by Emma Hardinge Britten. Boston: The Editor, 1876, 485 pp.

A rather bizarre account of experiences with spirits, apparently written by Emma Hardinge Britten. [P]

985. *Carpenter, William B.*

Principles of Mental Physiology, with Their Applications to the Training and Discipline of the Mind, and the Study of Its Morbid Conditions. 4 ed. London: Henry S. King & Co., 1876, lxiii + (1) + 737 pp.

The most important of the many editions for the subject of this bibliography is the fourth in which Carpenter introduces the concept of "unconscious cerebration" to account for the seemingly intelligent acts and communications of individuals that occur without their being aware of producing them. By "unconscious cerebration" Carpenter means automatic mental activity. Just as automatic reflex activity such as muscular contractions produced through spinal stimulation takes place outside our awareness in the lower nervous system, so also are there reflex actions on the cerebral level. These result in intellectual products that have no conscious awareness attached to them. [H]

986. *Collyer, Robert Hanham.*

Automatic Writing. The Slade Prosecution. Vindication of the Truth. London: H. Vickers, 1876, 23 pp.

[P]

987. *Eduard, Guillaume.*

Une page nouvelle de magnétisme. Sorcier malgré lui. Paris: The Author, 1876, 179 pp.

Describes techniques used by the author to perform magnetic cures. [H & P]

988. *Fechner, Gustav Theodor.*

Erinnerungen an die letzten Tage der Odlehre und ihres Urhebers. Leipzig: Breitkopf and Hartel, 1876, (4) + 55 pp.

Fechner, Professor of Philosophy at Leipzig University and founder of "psychophysics," describes the final attempts of the Baron Karl von Reichenbach (*see* entry number 583) to obtain scientific recognition for his doctrine of "odic force." [H]

989. *Grellety, J. Lucien.*

Du merveilleux, des miracles et des pèlerinages, au point de vue médical. Paris: J. B. Baillière, 1876, 87 pp.

[H & P]

990. ***Hammond, William Alexander.***
Spiritualism and Allied Causes and Conditions of Nervous Derangement. New York: G. P. Putnam's Sons, 1876, xii + 366 pp.

Hammond, a neurologist and surgeon-general of the United States Army, believed that spiritualism was dangerous. Here he discusses the various spiritualistic phenomena and concomitant delusions. [P]

991. ***Maskelyne, John Nevil.***
Modern Spiritualism: a Short Account of Its Rise and Progress, with Some Exposures of So-called Spirit Media. London: Frederick Warne, (1876), viii + 182 pp.

Maskelyne, a conjuror, demonstrates that the physical phenomena of spiritualism are in fact simply well-performed conjuring tricks. In this work, he takes a number of prominent mediums to task, including the famous Daniel Dunglas Home (1833–1886). [P]

992. ***Mitchell, G. W.***
X + Y = Z; or The Sleeping Preacher of North Alabama. Containing an Account of Most Wonderful Mysterious Mental Phenomena, Fully Authenticated by Living Witnesses. New York: W. C. Smith, 1876, 2 ed. xxi + 13–202 pp.

A detailed description of an account of a case of sleep preaching similar to that of Rachel Baker. Information on the first edition is not available. [P]

1877

993. ***Carpenter, William Benjamin.***
Mesmerism and Spiritualism, &c. Historically and Scientifically Considered. Being Two Lectures Delivered at the London Institution. London: Longmans and Green, 1877, xiv + 158 pp.

Elaborating on ideas presented in 1853 in *The Quarterly Review* (*see* entry number 659), Carpenter sides with Braid in viewing both the phenomena of mesmerism and those described by Reichenbach as psychological in nature. All spiritualistic phenomena, mental and physical, are either the result of unconscious suggestion or a product of out-and-out deception. [H & P]

994. ***Colley, Thomas.***
Later Phases of Materialisation, with Reflections to Which They Give Rise. London: J. Burns, 1877, 21 pp.

Offprint from *Human Nature*, December 1877. [P]

995. *Home, Daniel Dunglas.*

Lights and Shadows of Spiritualism. London: Virtue, 1877, xi + 412 pp.

D. D. Home, one of the most spectacular spiritualistic mediums of the nineteenth century, claims in *Lights and Shadows* that spiritualism is as old as the world. He devotes the first two parts of the book to tracing the evidence of spiritualism from ancient Assyria and Egypt through the early Christian church up to his own century. Discussing "modern spiritualism," which had begun with the "knockings" of the Fox sisters in 1848, he concentrates largely on the denunciation of fraudulent mediumship. Finding fraud to be a common phenomenon of the day. Home examines the methods of trickery used by deceitful mediums and expresses his consternation at the gullibility of much of the spiritualist flock. At the same time, however, Home is also careful to acknowledge that there are many mediums of upright character and genuine power, and he commends them for their contribution to the promotion of the "higher aspects of spiritualism." [P]

996. *Joly, Henri.*

L'imagination, étude psychologique. Paris: Hachette, 1877, (2) + ii + 264 pp.

A critical study of the imagination at work in somnambulism, belief in spirits, and other experiences. [H & P]

997. *Moses, William Stainton.*

The Slade Case; Its Facts and Its Lessons. A Record and a Warning. London: J. Burns, 1877, 40 pp.

William Stainton Moses, an ordained minister of the Chruch of England, was one of the most brilliant and highly respected spiritualists of nineteenth-century Britain. Initially opposed to spiritualism, Moses was led, through the interest of some respected friends, to investigate it more closely and soon observed things which he considered to be powerful evidence that discarnate spirits exist and communicate with the living. Less than a year after his first encounter with spiritualistic phenomena, Moses himself began to have psychic experiences and proved to be a powerful medium, regularly producing remarkable physical phenomena witnessed by many in his own home. Among others, these phenomena included repeated levitation of his own body, the appearance of lights, apports, the production of musical sounds, the mysterious appearance of writing on paper, and the movement of heavy physical objects without contact. Moses also produced a great volume of automatic writing, purportedly coming from spirits. This writing was of a cogency and moral quality that impressed many who were not inclined towards spiritualism. *The Slade Case* is a pamphlet that concerns itself with a controversial American medium, Henry Slade, who produced striking physical phenomena, but often under questionable circumstances. [P]

998. ***Parson, Frederick T.***

Vital Magnetism: Its power over Disease. A Statement of the Facts Developed by Men who Have Employed This Agent under Various Names, as Animal Magnetism, Mesmerism, Hypnotism, etc., from the Earliest Times down to the Present. New York: Adams, Victor & Co., 1877, (3) + (v)–vi + (ix)–x + 11–235 pp.

[H]

999. ***Perty, Maximillian.***

Der jetzige Spiritualismus und verwandte Erfahrungen der Vergangenheit und Gegenwart. Ein Supplement zu des Verfassers "mystischen Erscheinungen der menschlichen Natur." Leipzig and Heidelberg: C. F. Winter, 1877, xvi + 366 pp.

[P]

1000. ***Vay, Adelina (Wurnbrand-Stuppach).***

Visionen im Wasserglase. Budapest: Verein Spiriter Vorscher, 1877, 110 pp.

[P]

1001. **[*Weldon, Georgina*].**

Death-Blow to Spiritualism—Is it? Dr Slade, Messrs Maskelyne and Cooke, and Mr W. Morton. London: n.p., 1877, 204 pp.

In 1882, a much shorter version of this work was published in London by the Music and Art Association. [P]

1002. ***Wonderful Works of God. A Narrative of the Wonderful***

Facts in theCase of Ansel Bourne, of West Shelby, Orleans Co., N. Y., Who, in the Midst of Opposition to the Christian Religion, was Suddenly Struck Blind, Dumb and Deaf; and After Eighteen Days was Suddenly and Completely Restored, in the Presence of Hundreds of Persons, in the Christian Chapel, at Westerly on the 15th of November, 1857. Written under His Direction. Fall River, Massachusetts: Robertson, 1877, 40 pp.

During the 1880s, Ansel Bourne experienced one of psychology's most celebrated "fugue" episodes when he suddenly left his home, took on a new identity in another city, and set up business as a shopkeeper, with total amnesia for his former life. Eventually his memory returned just as suddenly and he had lost all knowledge of his "fugue" episode. Bourne's case was extensively discussed by William James, Richard Hodgson, and others in the late 1880s. This account of an earlier episode in his life indicates a tendency to dissociation. [H & P]

1878

1003. ***Charcot, Jean Marie.***
De la métalloscopie et la métallothérapie. Paris: G. Chamerot, 1878, 20 pp.

In his work with hysterics, Charcot revived the metallotherapy of Burq. This treatise presents the belief, somewhat reminiscent of the earlier mesmerist view, that certain substances can have a direct physical effect upon hysterical conditions. [H]

1004. ***Clarke, Edward Hammond.***
Visions: A Study of False Sight (Pseudopia). Boston: Houghton, Osgood and Company, 1878, (1) + xxii + 5–315 pp.

A valuable study of visions, concentrating on those produced by disease or chemical agents. [P]

1005. ***Fichte, Immanuel Hermann.***
Der neuere Spiritualismus, sein Werth und seine Tauschungen. Eine anthropologische Studie. Leipzig: Brockhaus, 1878, iii + 105 pp.

[P]

1006. ***Hellenbach von Paczolay, Lazar.***
Mr. Slades Aufenthalt in Wien. Vienna: 1878, 44 pp.

[P]

1007. ***Mind and Matter.***
Vols. 1–5; 1878–1883.

Published in Philadelphia. [H]

1008. ***Moses, William Stainton.***
Psychography: a Treatise on One of the Objective Forms of Psychic or Spiritual Phenomena. London: W. H. Harrison, 1878, (1) + 5–152 pp.

[P]

1009. ***Preyer, Wilhelm Thierry.***
Die Kataplexie und der thierische Hypnotismus. Jena: G. Fisher, 1878, iv + 100 pp.

[H]

1010. ***Psychological Review.***
Vols. 1–6; 1878–1883.

Originally published in London, this periodical was eventually published and distributed in various cities around the world. The primary editor was Edward W. Allen. [H]

1011. ***The Psycho-Physiological Sciences, and Their Assailants.*** *Being a Response by A.R.W, . . . J. R. Buchanan, . . . D. Lyman, . . . E. Sargent . . . ; to the Attacks of . . . W. B. Carpenter, . . . and Others.*
London: Colby and Rich, 1878, 216 pp.

[P]

1012. ***La revue magnétique. Organe du Cercle élecro-magnétique de Paris.***
Vols. 1–2; 1878–1879.

Absorbed by *Journal du magnétisme* (1879). [H]

1013. **[*Rivail, Hippolyte Léon Dénizard, pseudonym: Allan Kardec*]**
Les fluides, chapitres extraits de la Genèse. Paris: Librairie des sciences psychologiques, 1878.

[P]

1014. ***Stevens, E. Winchester.***
The Watseka Wonder: a Startling and Instructive Psychological Study, and Well Authenticated Instance of Angelic Visitation. A Narrative of the Leading Phenomena Occurring in the Case of Mary Lurancy Vennum. With Comments by Joseph Rodes Buchanan, D. P. Kayner, S. B. Brittan, and Hudson Tuttle. Chicago: Religio-Philosophical Publishing House, 1878, 26 pp.

The description of a famous case of apparent possession of a young girl by the spirit of another girl who had died a few years previously. The possessed girl, Mary Lurancy Vennum was able to identify people and objects known to the dead girl, Mary Roff, in a way that astonished the Roff family. The apparent possession ended after some months, with the Vennum girl returning to her normal state. [P]

1015. ***Zöllner, Johann Carl Friedrich.***
Wissenschaftlichen Abhandlungen. Vols. 1–3. Leipzig: L. Staackmann, (1878–1881).

Zöllner, professor of astrophysics at the University of Leipzig, was noted for his construction of an astrophotometer and expansion of the electrodynamical theory of W. Weber. Zöllner had a special interest in the phenomena of delusion, especially optical illusions. This, along with his speculations about a fourth dimension, led him to the investigation of mediumistic phenomena. Most famous and controversial of these investigations was one he conducted with the American medium, Henry Slade. These volumes are a collection of Zöllner's writings on occult issues. [H]

1879

1016. ***Burq, Victor.***

Pamphlet: I. Des origines de la métallothérapie, part qui doit être faite au magnétisme animal dans sa découverte. Le Burquisme et le Perkinisme. Paris: A. Delahaye et Lecrosnier, 1879.

First in a series of five pamphlets on metalotherapy and the only one of interest for the history of animal magnetism. It describes the connection of metallotherapy to animal magnetism and the origins of Burquism, with its roots in Perkinism. [H]

1017. ***La chaine magnétique. Organe des Société Magnétiques de France et de l'Etranger, echo des salons et cabinets de magnétisme et de somnambulisme . . . .***

Vols. 1–18; 1879–1897.

[H]

1018. ***Espinouse, A.***

Du zoomagnétisme, son existence, son utilité en médecine rendues indiscutables par les faits. Paris: Librairie du progrès, 1879, 181 pp.

After presenting the history of animal magnetism from 1784 to 1840, the author discusses its medical usefulness as demonstrated in his own work and that of others. An examination of hypnotism is also included. [H]

1019. ***Harrison, William H.***

Spirits Before Our Eyes. London: W. H. Harrison, 1879, 220 pp.

A spiritualist's study of apparitions of the dead, death-bed experiences, and hallucinations. [P]

1020. ***Hellenbach von Paczolay, Lazar.***

Die Vorurteile der Menschheit. 3 vols. Vienna: n.p., (1879–1880).

[P]

1021. ***James, John.***

Mesmerism, with Hints for Beginners. London: W. H. Harrison, 1879, vi + 102 pp.

An introductory treatise on animal magnetism which contains a section on the clairvoyance of Alexis Didier important for describing two previously unpublished experiments, both carried out with the famous French clairvoyant in James's presence. [H]

1022. *Maggiorani, Carlo.*
Influenza del magnetismo sul cervelletto: discorso accademico. Rome: Romana, 1879, 27 pp.

[H]

1023. [*Moses, William Stainton, pseudonym: M. A. (Oxon.)*].
Spirit-Identity. London: Harrison, 1879, xii + 143 pp.

Moses examines the difficulties of investigating the phenomena of spiritualism and charts his own reluctant journey towards their acceptance. He especially concerns himself with the sometimes questionable content of the messages purporting to come from spirits, and how that bears on one's judgment concerning their genuineness. [P]

1024. *Wundt, Wilhelm.*
Der Spiritismus. Eine sogenannte wissenschaftliche Frage. Offener Brief an Herrn Prof. Hermann Ulrici in Halle. Leipzig: W. Engelmann, 1879, 31 pp.

Wundt, typically considered the founder of experimental psychology, rejects the phenomena of spiritualism out of hand because he considers them to be contrary to natural law. [P]

1880–1889

1880

1025. ***Colsenet, Edmond.***
La vie inconsciente de l'esprit. Paris: Germer Baillière, 1880, vi + 279 pp.

Written at a time when the modern psychotherapeutic notion of the unconscious was being formed, the work contains material from the magnetic and hypnotic traditions, from Carpenter, and from von Hartmann, and offers philosophical observations on the nature of the unconscious. [H]

1026. ***Despine, Prosper.***
Étude scientifique sur le somnambulisme, sur les phénomènes qu'il présente et sur son action thérapeutique dans certaines maladies nerveuses du rôle important qu'il joue dans l'épilepsie, dans l'hystérie et dans les névroses dites extraordinaires. Paris: F. Savy, 1880, xii + 13–425 pp.

An important work on automatism and somnambulism. The author discusses the nature of automatism and its relationship to consciousness. He also investigates the physical and psychological states associated with somnambulism and sees a relationship between nervous disorders and states of spontaneous somnambulism. [H]

1027. ***Dunand, Tony.***
Une révolution en philosophie, résultant de l'observation des phénomènes du magnétisme animal. Étude physiologique et psychologique de l'homme. Paris: Berche et Tralin, 1880, (2) + 405 pp.

Dunand discusses the phenomena of animal magnetism from the physiological aspects of magnetic fluid to the feats of magicians. [H & P]

1028. ***Fellner, F. von.***

Animalischer Magnetismus und moderner Rationalismus: eine kultur-historische Betrachtung. Leipzig: Mutze, 1880, 76 pp.

[H]

1029. ***Heidenhain, Rudolf Peter Heinrich.***

Der sogenannte thierische Magnetismus. Physologische Beobachtungen. Leipzig: Breitkopf and Härtel, 1880, (2) + 40 pp. ENGLISH: *Hypnotism or Animal Magnetism. Physiological Observations.* Translated by L. C. Wooldridge. London: Kegan Paul, 1880.

A lecture delivered at the general meeting of the Silesian Society for Home Culture held at Breslau on January 19, 1880. Together with Braid, Heidenhain denies the existence of a magnetic fluid or any agent communicated from operator to subject. He concludes that hypnotism is the result of the inhibition of the activity of the brain. [H]

1030. ***Maggiorani, Carlo.***

Influenza del magnetismo sulla vita animale. Naples: Detken, 1880, 360 pp.

[H]

1031. **[*Moses, William Stainton, pseudonym: M. A. Oxon*].**

Higher Aspects of Spiritualism. London: E. W. Allen & Co., 1880, 124 pp.

[P]

1032. ***Richet, Charles Robert.***

"Du somnambulisme provoqué." *Revue Philosophique*, 10 (1880):337–374, 462–484.

Written five years after his first important article on artificial somnambulism (*see* entry number 978), this treatise is Richet's attempt once and for all to establish the reality and scientific importance of artificial somnambulism for still-skeptical members of the medical profession. Richet identifies three degrees of artificial somnambulism: torpor, excitation, and stupor. He states that somnambulism, properly speaking, begins with the second phase. In that phase the individual is subject to hallucinations, suggestions, and automatisms of various kinds. In the third phase an anesthesia may be produced that is complete enough to allow surgical operations. Richet also discusses the various techniques that may be employed to produce artificial somnambulism, including the fixation of the eyes, the use of a shiny object, a sudden shock of the senses, and the passes of animal magnetism. He uses this discussion of technique as a starting point for examining the various explanations of the intrinsic nature and cause of artificial somnambulism, but he does not come down in favor of any one theory. [H]

1033. [***Santini de Riols, Emmanuel Napoléon, pseudonym: J. de Riols***].

Magnétisme et somnambulisme. Méthode nouvelle facile et pratique expliquant les principes réels du magnétisme, les moyens infaillables pour arriver promptement à bien magnétiser suivis de documents historiques et de nombreuses anecdotes. Paris: Le Bailly, 1880, 32 pp. [H]

1034. ***Zöllner, Johann Carl Friedrich.***

Transcendental Physics: An Account of Experimental Investigations. From the Scientific Treatises of Johann Carl Friedrich Zöllner. Translated from the German, with a Preface and Appendices, by Charles Carleton Massey. London: n.p., 1880, xlviii + 266 pp.

The English translation of the section of Zöllner's *Wissenschaftlichen Abhandlungen* (*see* entry number 1015) that describes his experiments with the famous American medium Henry Slade. The experiments included knots produced on a cord with its ends sealed together, magnetic experiments, disappearance and reappearance of objects, slate writing and clairvoyance. Zöllner was convinced through these experiments that Slade had not been able to use any deception in the tests and yet had produced the most remarkable feats. [P]

1881

1035. ***Bäeumler, Christian Gottfried Heinrich.***

Der sogenannte animalische Magnetismus oder Hypnotismus: unter Zugrundelegung eines fuer die Akademische Gesellschaft zu Freiburg i. B. gehaltenen populären Vorträges. Leipzig: F.C.W. Vogel, 1881, 74 + (2) pp.

An examination of the psychological rather than sympathetic effects of animal magnetism employed as a cure, containing an early German appreciation of Braid. [H]

1036. ***Beard, George Miller.***

Trance and Trancoidal States in the Lower Animals. New York: W. L. Hyde & Co., 1881, 17 pp.

[H]

1037. ***Charcot, Jean Marie.***

Contribution à l'étude de l'hypnotisme chez les hystériques. Paris: Progrès medicale, 1881, (2) + 122 pp.

[H]

1038. *Dippel, Joseph.*
Der Neuere Spiritismus. Wurzburg: Woerl, 1881, 127 pp.

[P]

1039. *L'encéphale. Journal des maladies mentales et nerveuses.*
Vols. 1–9; 1881–1889.

Continued as: *Revue d'hypnologie théorique et pratique dans ses rapports avec la psychologie, les maladies mentales et nerveuses*. One vol. only; 1890. Continued as: *Annales de psychiatrie et d'hypnologie* . . . Vols. 1–5 (new series); 1891–1895. Continued as: *Revue de psychiatrie, de neurologie et d'hypnologie. Recueil des travaux publiés en France et à l'etranger*. Vols. 1–2 (third series); 1896–1897; Vols. 1–18 (fourth series); 1897–1914. Title variants of the *Revue de psychiatrie* . . . : *Revue de psychiatrie. Médecine mentale, neurologie, psychologie*; and *Revue de psychiatrie et de psychologie expérimentale*. Published in Paris. Continued as: *Revue de psychothérapie et de psychologie appliquée*. Vols. 31–34+; 1922–1925+. Title variant of the *Revue de l'hypnotisme* . . . : *Revue de l'hypnotisme et de la psychologie physiologique*. [H]

1040. *Houghton, Georgiana.*
Evenings at Home in Spiritual Séance, Prefaced and Welded Together by a Species of Autobiography. 2 vols. London: Trubner & Co. (Vol. 1) and E. W. Allen (Vol. 2), (1881 & 1882), vii + 352; 362 pp.

[P]

1041. *Ladame, Paul Louis.*
La névrose hypnotique, ou Le magnétisme dévoilé. Étude de physiologie pathologique sur le système nerveux. Paris: n.p., 1881.

[H]

1042. *Lombroso, Cesare.*
Sulla trasmissione del pensiero. Turin: n.p., 1881.

[P]

1043. *Méric, Joseph Élie.*
L'autre vie. 2 vols. Paris: Palmé, 1881.

[P]

1044. *Méric, Joseph Élie.*
Le merveilleux et la science, étude sur l'hypnotisme. Paris: Letouzey et ané, (1881), 468 pp.

Méric tries to unify the findings of the various schools of hypnotism and occult experience. [H]

1045. ***Motet, Auguste.***

Accès de somnambulisme spontané et provoqué. Prévention d'outrage public à la pudeur; condamnation; irresponsabilité; appel, infirmation et acquittement. Paris: J. B. Baillière, 1881, 16 pp.

[H]

1046. ***Perty, Maximillian.***

Die sichtbare und die unsichtbare Welt. Dieseits und Jenseits. Leipzig and Heidelberg: Winter, 1881, 320 pp.

[P]

1047. ***Preyer, Wilhelm Thierry.***

Die Entdeckung des Hypnotismus. Berlin: Paetel, 1881, (8) + 96 pp.

A tribute to the Manchester physician James Braid (1795–1860). Preyer considers Braid to have made an invaluable contribution to human psychology through his recognition of the true nature of "nervous sleep" and describes both Braid's hypnotic technique and the phenomena he produced. He also discusses at some length Braid's peculiar combination of hypnotism and phrenology, called "phreno-hypnotism." The last part of this treatise is *Über den Hypnotismus*, the German translation of a treatise on hypnotism written by James Braid and never published in English. It is dated 1860, the year of Braid's death, and its place of writing is given as Rylaw House, Oxford Street, Manchester. [H]

1048. ***Richer, Paul Marie Louis Pierre.***

Études cliniques sur l'hystéroépilepsie ou grande hystérie. Paris: A. Delahaye & E. Lecrosnier, 1881, xvi + 734 + (1) pp.

A monumental study of hysteria and hypnotism by a disciple and coworker of Jean Marie Charcot (1825–1893). In a greatly augmented second edition (Paris: A. Delahaye and E. Lecrosnier, 1885), Richer comments on the three stages of hypnotism as perceived by the Salpêtrière school, and goes into great detail concerning the physiological concomitants of those states. He delineates the various methods that may be used to induce the hypnotic state, and he also discusses hypnotic suggestion, describing the automatisms, compulsions, delusions, and hallucinations that it may produce. Richer's section on hysteria in history, concentrating on cases of possession and ecstatics, is classical, as is his section dealing with hysteria in art. [H]

1049. ***Seppilli, Giuseppe.***

Gli studi recenti sul cosi detto magnetismo animale. Turin: S. Calderini, 1881, 69 pp.

[H]

1050. ***Sully, James.***
Illusions: A Psychological Study. London: Kegan Paul, Trench, & Co., 1881, xii + 372 + (2) pp.

Sully attempts to treat physiological and psychological sources of error, such as illusions of introspection, of memory, of belief, of insight, and of dreams. Widely read by psychologists, this volume influenced those dealing with errors of perception of all kinds for decades to come. [P]

1882

1051. ***Barety, A.***
Des propriétés physiques d'une force particulière du corps humain (force neurique rayonnante) connue vulgairement sous le nom de magnétisme animal. . . . Extrait de la Gazette médicale de Paris, année 1881. Paris: Octave Doin, 1882, 40 pp.

Baréty posits the existence of a "neuric force" produced in the nervous system and radiating from the body in three areas: the eyes, the ends of the fingers, and the lungs. According to Baréty, the dynamic form of this force circulates through the body. Some substances block the force as it radiates from the body and others transmit it. The force may radiate over distances from a few centimeters to many meters, and, depending on the power of the radiation and the sensitivity of the receptor, it may be sensed by individuals in the vicinity. Baréty equated his "neuric force" with Mesmer's animal magnetism. [H]

1052. ***Beard, George Miller.***
The Study of Trance, Muscle-reading and Allied Nervous Phenomena in Europe and America, with a Letter on the Moral Character of Trance Subjects, and a Defence of Dr. Charcot. New York: n.p., 1882, 40 pp.

Beard, an American neurologist, was best known as the originator of the concept "neurasthenia," a state of physical and mental exhaustion leading to the inability to perform physical and mental work. Here he discusses the information that may be conveyed to trance subjects through muscle tension and the possibility that unconscious suggestion may be at work with emotionally disturbed patients. The pamphlet concludes with a note on Charcot's work, in which Beard defends Charcot's character and integrity as an investigator but states his belief that Charcot was insufficiently aware of this problem. [P]

1053. ***Bué, Hector Joseph.***

La vie et la santé, ou La médecine est-elle une science? Paris: A. Ghio, 1882, 150 pp.

A discussion of the medical uses of "human magnetism." [H]

1054. ***Burq, Victor.***

Des origines de la métallothérapie. Part qui doit être faite au magnétisme animal dans sa découverte. Le burquisme et le perkinisme. Paris: Delahaye et Lecrosnier, 1882, 16 pp.

[H]

1055. ***Cadwell, J. W.***

Full and Comprehensive Instructions How to Mesmerize, Ancient and Modern Miracles by Mesmerism, also Is Spiritualism True? 2 ed. Boston: The Author, 1882, 128 pp.

No information available on the first edition. [H]

1056. ***Cavailhon, Edouard.***

La fascination magnétique, précédée d'une préface par Donato et de son protrait photographié. Paris: E. Dentu, 1882, lxx + 334 pp.

Discusses the predecessors and work of the famous stage hypnotist "Donato," the pseudonym for Edouard d'Hont, and describes some of Donato's seances and writings. Cavailhon also compares the doctrine of Donato to that of Braid and includes an interesting section on "love and magnetism." [H]

1057. [***Chapman, ———***].

Confessions of a Medium. London: Griffith & Farran, 1882, xvi + 282 pp.

[P]

1058. ***Charcot, Jean Martin.***

"Sur les divers états nerveux déterminés par l'hypnotisation chez les hystériques." *Comptes-rendus hebdomadiares des séances de l'Académie des Sciences* 94(1882):403–05.

Charcot's summary statement of his views on hypnotism. He mentions three stages of hypnotism: catalepsy, lethargy, and somnambulism, describing the physiological phenomena connected with each state. This paper, delivered by a man of such high scientific reputation, did more than any other single event to get hypnotism accepted as a genuine phenomenon among the academics of the day. For more on Charcot's theory of hypnotism, see *Leçons sur les maladies du système nerveux* (entry number 948). [H]

1059. *Houghton, Georgiana.*

Chronicles of the Photographs of Spiritual Beings and Phenomena Invisible to the Material Eye, Interblended with Personal Narrative. London: Allen, 1882, x + 273 pp.

[P]

1060. *Society for Psychical Research. Proceedings.*

Vols. 1–35+; 1882–1925+.

The *Proceedings* is the most important periodical on psychical research published in any language. [H & P]

1061. *Wirth, Friedrich Moritz.*

Herrn Professor Zöllners Experimente mit dem amerikanischen Medium Herrn Slade und seine Hypothese intelligenter vierdimensionaler Wesen. Mit einer Antwort an die Herren Professoren Hernn. W. Vogel in Berlin und J. B. Meyer in Bonn. 3 ed. Leipzig: O. Mutze, 1882 xvi + 122 pp.

No information is available on the first edition. [P]

1883

1062. *Assier, Adolphe d'.*

Revenants et Fantômes. Essais sur l'humanité posthume et le spiritisme. Paris: Baillière, 1883, 308 pp. ENGLISH: *Posthumous Humanity: A Study of Phantoms*. Translated by Henry S. Olcott. London: G. Redway, 1887.

This study of ghosts and phantoms draws upon both historical accounts and the experiences of the author. Although accepting the genuineness of some apparitions, d'Assier has his own personal theory of explanation which differs from that of the spiritualists. [P]

1063. *Britten, Emma Hardinge.*

Nineteenth Century Miracles; *or*, *Spirits and Their Work in Every Country of the Earth. A Complete Historical Compendium of the Great Movement Known as "Modern Spiritualism."* Manchester, England: W. Britten, 1883, viii + 556 pp.

Whereas the author's *Modern American Spiritualism* (1870, *see* entry number 932) concentrated on the history of the first twenty years of spiritualism in the United States, this book attempts to provide a history of that movement in sixteen other countries and areas of the world up to 1883. Patently "pro" spiritualist, Britten is nonetheless remarkable in the quality of her research. Providing an array of facts connected to the history of spiritualism around the world that was unique at the time, including large sections on Germany, France, Great Britain and Australia, she also includes a lengthy updating of the history of spiritualism in America. Overall, Britten's work is particularly important for its discussion of the relation of spiritualism to other phenomena. Thus, for example, in treating spiritualism in Germany, Britten examines ties between spiritualism and the discovery and rise of mesmerism and, in the section on "spiritualism in the polynesian and West India Islands," she draws connections with the ancient beliefs of the natives of those regions. [P]

1064. *Cahagnet, Louis Alphonse.*

Thérapeutique du magnétisme et du somnambulisme appropriée aux maladies les plus communes; aidée par l'emploi des plantes les plus usuelles en médecine. Renseignements sur la compositions et sur l'application des remèdes conseillés. Planches anatomiques avec explication philosophique. Paris: Librairie Scientifico-psychologique, 1883, iv + 439 pp.

A manual describing how to best employ the somnambulistic state in the healing process and suggesting the use of other remedies as part of the healing plan. [H]

1065. *Farmer, John Stephen.*

How to Investigate Spiritualism. London: Psychological Press Association, 1883, 26 pp.

[P]

1066. *Fischer, Engelbert Lorenz.*

Der sogenannte Lebensmagnetismus oder Hypnotismus. Mainz: Franz Kirchheim, 1883, viii + 119 pp.

[H]

1067. *Guthrie, Malcolm and Birchall, James.*

"Record of Experiments in Thought-transference." *Society for Psychical Research Proceedings* 1(1883): 263–83.

First of a series of three reports on carefully executed experiments in thought-transference or telepathy conducted at Liverpool, England. The conditions are described in detail and a verbatim record of the proceedings is given. [P]

1068. *Liébeault, Ambroise August.*

Étude sur le zoomagnétisme. Paris and Nancy: Masson, 1883, 29 pp.

Originally intended as an article for the *Journal du magnétisme*, this pamphlet presents Liébeault's findings concerning the curative effects of a physical agent in animal magnetism. Impressed by the success of an acquaintance in magnetically healing infants whose intellectual development was insufficient to permit the effects of suggestion, the pioneer of suggestion asked himself whether there might not after all also be a physical agent involved in producing the effects of animal magnetism. Attempting to test the possibility that a physical agent might account for the magnetizer's success with younger children, Liébeault carried out his own experiments with infants. Much to his surprise he discovered that the action of magnetic passes produced a noticeable beneficial effect upon the infants. He ruled out the explanation that heat or pressure produced the effects, and concluded that, in addition to the effects of suggestion, there exists a "nervous action" transmitted from one individual to another by vibrations. [H]

1069. [*Moses, William Stainton, pseudonym: M. A. Oxon*].

Spirit Teachings. London: The Psychological Press Association, 1883, xii + 291 pp.

Moses' best-known book on spiritualism containing a description of the process and content of his automatic writings purported to come from spirits. The messages were of a high moral quality and impressive cogency. The communications call into question many Christian beliefs and Moses did not find that easy to accept. The book describes his own doubting questions to the spirits and the answers he received through automatic writing. [P]

1070. *Myers, Frederic William Henry.*

Essays Classical. London: Macmillan & Co., 1883, vii + 223 pp.

A collection of essays that originally appeared in various periodicals. The first is a long essay on the Greek oracles and has some relevance to psychical research. For a description of Myers's larger contribution to that field, see his *Human Personality* (1903, entry number 1525). [P]

1071. *Richet, Charles Robert.*

"La personalité et la mémoire dans le somnambulisme." *Revue philosophique* 15(1883): 225–42.

Richet states that personality is a phenomenon of memory, that personality is built on an experienced succession of memories. Personality must, he says, be distinguished from the "I." An individual may have more than one personality, but only one "I." Somnambulists may exhibit many personalities, but they are all attributed to one "I." Richet insists that the "I" endures permanently, while personalities may be abolished with the onset of amnesia. Richet also discusses what he calls "unconscious memory" (*mémoire inconsciente*), which is a memory that functions in the individual, although it is completely outside of conscious awareness. Examples of this phenomenon taken from experiments with somnambulists show, in Richet's opinion, that our acts can be determined by causes entirely unknown to us. [H]

1072. ***[Santini de Riols, Emmanuel Napoléon, pseudonym: J. de Riols].***
Spiritisme et tables tournantes. Nouvelle méthode facile et complète, expliquant les principes réels du spiritisme, les moyens infaillibles pour arriver promptement à évoquer les esprits et se mettre en rapport avec eux, suivie de la démonstration théorique et pratique du pendule-explorateur et de la baquette divinatoire. Paris: Le Bailly, 1883, 32 pp.

[P]

1073. ***Truesdell, John W.***
The Bottom Facts Concerning the Science of Spiritualism: Derived from Careful Investigations Covering a Period of Twenty-five Years. New York: G. W. Carleton, 1883, xv + (1) + 17–331 pp.

Describing interviews and seances with a variety of mediums, including the famous Henry Slade, Truesdell indicates how those who wish to investigate mediums should carry out their study. While attempting to be critical, the author begins from the assumption that spiritualistic mediumship is genuine and involves communication with the dead. [P]

1074. ***Watts, Anna Mary (Howitt).***
The Pioneers of the Spiritual Reformation. Life and Works of Dr Justinus Kerner (Adapted from the German). William Howitt and his Work for Spiritualism. Biographical Sketches. London: Psychological Press Association, 1883, xii + 325 pp.

Includes a translation of some of the writings of Justinus Kerner. [H & P]

1075. ***Yung, Emile.***
Le sommeil normal et le sommeil pathologique, magnétisme animal, hypnotisme, névrose hystérique. Paris: O. Doin, 1883, 191 pp.

[H]

1884

1076. ***Beaunis, Henri Étienne.***
Recherches expérimentales sur les conditions de l'activité cérébrale et sur la physiologie des nerfs; études physiologiques et psychologiques sur le somnambulisme provoqué. 2 vols. Paris: J. B. Baillière, (1884–1886).

The second volume of this work is an important study of artificial somnambulism. Beaunis has some particularly useful things to say about memory in somnambulism. Among other things, he discusses a "doubling of memory and consciousness." [H]

1077. *Bellanger, Augustin René.*

Le magnétisme. Vérités et chimères de cette science occulte. Un drame dans le somnambulisme, épisode historique. Paris: Guilhermet, 1884, xiv + 342 pp.

The author describes the nature of artificial somnambulism; he sees animal magnetism as an occult science. [H & P]

1078. *Belot, Camille.*

Les secrets du magnétisme. Paris: E. Dentu, 1884, 172 pp.

A popular instruction manual giving a brief description of the practice of animal magnetism. [H]

1079. *Bérillon, Edgar.*

Hypnotisme expérimental; la dualité cérébrale et l'indépendance fonctionelle des deux hémisphères cérébraux. Paris: A. Delahaye and E. Lecrosnier, 1884, viii + 192 pp.

Bérillon, a member of the "Pitié" School of hypnotism founded by Dumontpallier, was for many years the editor of the *Revue de l'hypnotisme* and was an unusually effective practitioner of medical hypnotism. In this work he discusses mental pathology, hysteria, and dreams, and puts forward an unusual thesis concerning the independent function of the two hemispheres of the brain. He believes that each hemisphere constitutes a complete and separate organ. This, he states, makes possible hemicerebral hypnotism and doubling of the personality. [H]

1080. *Bernheim, Hippolyte.*

De la suggestion dans l'état hypnotique et dans l'état de veille. Paris: Octave Doin, 1884, 110 pp.

Bernheim, Professor of Clinical Medicine at the University of Nancy, with a scientific reputation for research on typhoid fever and heart disease, began in 1882 to develop an interest in hysteria and hypnotism. He heard about the work of the physician Ambroise Liébeault (1823–1904), whose original book on hypnotism was virtually unknown in the medical world (see *Du sommeil et des états analogues*, entry number 896), but whose success in using hypnotic techniques to treat patients was difficult to ignore. Bernheim visited Liébeault to observe his approach and form his own opinion about the results. He was greatly impressed by what he saw and immediately tried Liébeault's technique himself. He soon came to the conclusion that hypnotism was not an appendage of hysteria, as the then prevailing theory of Jean Martin Charcot (1825–1893) had indicated, but a psychological state in its own right, one intimately connected with suggestion. Two years later, having continued to work closely with Liébeault, Bernheim published this book, the fundamental statement of his theory of hypnotism.

Bernheim and Liébeault combined their researches and formed what came to be known as the Nancy School of hypnotism, and *De la suggestion* may rightly be considered the foundation work of that school. Bernheim begins the book with a description of his method of hypnotic induction. He then lists Liébeault's six degrees of hypnotic sleep and, from his own experience, reclassifies those stages, noting nine in all. Bernheim then enters into a detailed discussion of the nature of suggestion and presents a definition of hypnotism: the induction of a peculiar psychical condition which increases the susceptibility to suggestion. Next Bernheim discusses the phenomena associated with hypnotic sleep, including automatic movement, automatic obedience, suggested hallucinations during hypnotic sleep, and post-hypnotic hallucinations. After some words about somnambulism and double personality, Bernheim notes the physiological effects produced by hypnotism. This is followed by a discussion of suggestion in the waking state and an answer to the criticisms leveled against his views by Charcot and the men at the Salpêtrière. Bernheim then provides a brief history of animal magnetism and hypnotism up to his own time. Next there is an important chapter on automatisms and suggestion and a concluding discussion of the moral and legal aspects of hypnotic suggestion. With *De la suggestion* the stage is set for the controversy between Bernheim and Liébeault and their followers at Nancy, and Charcot and his followers at the Salpêtrière, with the Nancy School insisting that all the phenomena of hypnotism arise from suggestion, and the Salpêtrière School claiming that the phenomena of hypnotism are based on physiological processes that create an artificial neurosis comparable to an hysterical attack. This controversy was pivotal to the discussion of the theory of hypnotism for many years. [H]

1081. ***Bottey, Fernand.***
Le "magnétisme animal"; étude critique et expérimental sur l'hypnotisme, ou sommeil nerveux provoqué chez les sujets sains, léthargie, catalepsie, somnambulisme, suggestions, etc. Paris: E. Plon, Nourrit, 1884, (6) + iii + (1) + 282 pp.

[H]

1082. ***Du Prel, Karl Ludwig August Friedrich Maximillian Alfred.***
Die Philosophie der Mystik. Leipzig: Ernst Gunther, 1884, xii + 548 pp. ENGLISH: *The Philosophy of Mysticism.* Translated by C. C. Massey. London: George Redway, 1889.

On somnambulism and its implications for understanding the nature of human consciousness. Although the work is philosophical in orientation, Du Prel bases his analysis on a broad foundation of experimentation with somnambulistic states. Du Prel speaks about the division of the ego which is observed not only in somnambulism, but also in more ordinary experiences such as dreams. This division or splitting of the ego leads to the conclusion that human beings are very complex, with an unconscious component that determines much of what happens in conscious life. In his discussion of the unconscious, the influence of Kant, Schopenhauer, and Eduard von Hartmann can be discerned. But Du Prel's examination of somnambulism and the nature of consciousness takes him a step beyond their speculations. [H]

1083. *Edard, Guillaume.*

La vie par la magnétisme et l'électricité. Paris: The Author, 1884, xvi + 599 pp.

A general work on animal magnetism and its history by a man who describes himself as a magnetizer-electrician. [H]

1084. *Gurney, Edmund.*

"The Problem of Hypnotism." *Society for Psychical Research Proceedings* 2(1884): 265–92.

Edmund Gurney was one of the founders of the Society for Psychical Research in England and served as its first honorary secretary. He was considered by his coworkers to be an investigator and writer of extraordinary brilliance. Between 1874 and 1878 he observed a great number of spiritualist seances. After the foundation of the Society, he continued these investigations, but also turned his attention to the phenomenon of hypnotism. Gurney died an untimely death at age 41, apparently through an accidental overdose of a painkiller. This article is a significant contribution to knowledge of hypnotism. In the article Gurney cautions against a too rapid formulation of theories about the nature of hypnotism. He says that a more useful approach would be to outline certain well-defined problems and see what can be clarified in those specific areas. He then goes on to discuss such important issues as the influence of the operator on the subject, the nature of automatism, and the problem of post-hypnotic memory. [P]

1085. *Gurney, Edmund.*

"The Stages of Hypnotism." *Society for Psychical Research Proceedings* 2(1884): 61–72.

The first of a number of important articles by Gurney on hypnotism published in the *Proceedings.* Gurney describes his experiments with the various "depths" of hypnotism and their relationship to such things as suggestibility and memory. [H]

1086. *Guthrie, Malcolm.*

"An account of some experiments in thought-transference." *Society for Psychical Research Proceedings* 2(1884): 24–43.

A second report on experiments done in thought transference at Liverpool, England, under the supervision of Malcolm Guthrie and James Birchall. The experiments involved the telepathic transmission of simple drawings. The report includes reproductions of the original drawings that were attempted to be conveyed by the transmitter and the resulting drawings executed by the receiver. The results are striking. [P]

1087. ***Magnin, Paul de.***

Étude clinique et expérimentale sur l'hypnotisme; de qq. effets des excitations périphériques chez les hystéro-épileptiques à l'état de veille et d'hypnotisme. Paris: A. Delahaye et E. Lecrosnier, 1884, 97 pp.

A study of hypnosis that presents the position of the Pitié School of hypnotism, a group quite close in teaching to that of the Salpêtrière School. Although it admits of a number of intermediary states between the three great phases of the latter school, it is in agreement that hypnotism is a hysterical condition and presents certain consistent physical symptoms. [P]

1088. ***Maricourt, R. de.***

Souvenires d'un magnétiseur. Paris: Plon, 1884, 315 pp.

[H]

1089. ***Moutinho, An.***

Magnetismo animale. Principios de magnetologia o methodo facil de aprender a magnetisar segundo os systemas de Mesmer, Puysegur, Deleuze, de Lausanne, Rostan, Brivasac, Ricard, Du Potet, Gauthier e Lafontaine. Lisbon: n.p., 1884.

[H]

1090. ***Myers, Frederick William Henry.***

"On a Telepathic Explanation of Some So-called Spiritualistic Phenomena." *Society for Psychical Research Proceedings* 2(1884): 217–37.

F.W.H. Myers was one of the most important psychical researchers of the late nineteenth century. He was a founder of the Society for Psychical Research, a poet, a classical scholar, and considered by some to be one of the most significant psychologists of the time. Driven by the need to determine what can be discovered empirically about the nature of human personality and its possible survival of death, Myers devoted his considerable energies to the investigation of all phenomena that might clarify those issues. He investigated somnambulists, mediums, and all types of psychic and psychological phenomena. Myers originated a number of terms that came to be used commonly by psychical researchers, such as: "telepathy," "supernormal," and "veridical." In this article, Myers examines the phenomenon of automatic writing to try to discover whether, as the spiritualists claimed, the writer conveys messages from the spirits of the dead. He concludes that unless there is content completely unknown to the writer, there is no reason to assume that anything more than the unconscious self of the writer is involved in the process. If, he says, there is evidence of material coming through automatic writing that could not have been obtained through normal means, it is still not established that spirits are at work. Telepathy may be a sufficient explanation. This article is a good example of the kind of critical thinking and good writing so characteristic of Myers. [P]

1091. *Pennell, Henry Cholmondeley, ed.*

"Bringing it to Book:" Facts of Slate-writing through Mr W. Eglinton. Edited by H. Cholmondeley-Pennell. Being Letters Written by the Hon. Roden Noel, Charles Carleton Massey, George Wyld, the Hon. Percy Wyndham, and the Editor. London: Psychological Press Association, (1884), 23 pp.

[P]

1092. *Perronnet, Claude.*

Le magnétisme animal. Lons-le-Saunier: n.p., 1884, 63 pp.

[H]

1093. *Richet, Charles Robert.*

L'homme et l'intelligence. Fragments de physiologie et de psychologie. Paris: Félix Alcan, 1884, vii + 5–570 pp.

Articles by Richet previously published in either the *Revue philosophique* or the *Revue des deux mondes* and covering issues from the "poisons of intelligence" (alcohol, opium, etc.) to artificial somnambulism. Richet's sections on possession experiences are especially well done. [H]

1094. *Richet, Charles Robert.*

"La suggestion mentale et le calcul des probabilités." *Revue philosophique*, 18(1884): 609–74.

Richet here describes the first use of statistical analysis to evaluate the results of experiments in mental suggestion, a kind of telepathy. [P]

1095. *Rieger, Conrad.*

Der Hypnotismus. Psychiatrische Beiträge zur Kentniss der sogenannten hypnotischen Zustände. Nebst einem physiognomischen Beitrag von Dr. Hans Virchow. Jena: Gustav Fischer, 1884, (4) + 151 pp.

Oriented to a physiological account of hypnotism the author is hesitant to accept hypnotic phenomena as involving a special state. [H]

1096. *Simony, Oskar.*

Über spiritistische Manifestationen vom naturwissenschaftlichen Standpunkt. Vienna: A. Hartleben, 1884, 48 pp.

[P]

1097. *Society for Psychical Research. Journal.*

Vols. 1–+; 1884–1925+.

Along with the *Proceedings* of the Society for Psychical Research (*see* entry number 1060), this journal provides a gold mine of information about serious investigation of paranormal phenomena. [H & P]

1098. ***Tuke, Daniel Hack.***
Sleep-walking and Hypnotism. London: J. & A. Churchill, 1884, vi + (2) + 119 + (1) pp.

Tuke discusses both spontaneous and artificially induced somnambulism, attempting to provide an accurate picture of the state of the senses during natural sleepwalking, free from what he considers to be misconceptions that have sprung up around the phenomenon. When describing the hypnotic state, he discusses "double consciousness" and relates it to the two sides of the brain. Tuke's lavish use of illustrative cases contributes much to the value of the book. [H]

1099. ***Voisin, A.***
Étude sur l'hypnotisme et sur les suggestions chez les aliénés. Paris: G. Rougier, 1884, 16 pp.

[H]

1100. ***Welton, Thomas.***
Mental Magic. A Rationale of Thought Reading, and Its Attendant Phenomena, and Their Application to the Discovery of New Medicines, Obscure Diseases, Correct Delineations of Character, Lost Persons and Property, Mines and Springs of Water, and All Hidden and Secret Things: To Which is Added the History and Mystery of the Magic Mirror. London: Redway, 1884, 178 pp.

[P]

1885

1101. ***Beaunis, Henri Etienne.***
"L'expérimentation en psychologie par le somnambulisme provoqué." *Revue philosophique* 20(1885): 2–36, 113–134.

Beaunis attempts to show the usefulness of artificial somnambulism or hypnotism as a tool for experimental research in psychology. He points out how the researcher can use hypnotism to produce a kind of "moral vivisection" of experimental subjects and make certain psychological phenomena happen at will under the eyes of the experimenter. [H]

1102. ***Bourru, Henri and Burot, P.***
"De la multiplicité des états de conscience chez un hystéro-épileptique." *Revue Philosophique* 20(1885): 411–16.

Describes certain peculiarities of consciousness experienced by a young man called "V" who had begun to undergo "attacks" of various kinds after being bitten by a snake. "V" would experience changes of states of consciousness accompanied by different kinds of anaesthesia and muscular contraction. The changes in consciousness also involved marked alterations of personality, each personality being associated with a certain age, the youngest being fourteen. Six distinct states were identified. This is an interesting early case of a type of multiple personality precipitated by a physical trauma. [H]

1103. ***Buchanan, Joseph Rodes.***

Manual of Psychometry: The Dawn of a New Civilization. Boston: The Author, 1885, (3) + 212 + 194 + 94 pp.

The word "psychometry", coined in 1842, literally means "measuring the soul." Buchanan, however, altered that definition to "measuring by the soul" and claimed that every field of knowledge could benefit from the psychic capacity innate in some individuals. These gifted people, one of whom was Buchanan's wife, could use their gifts to provide important data in the fields of medicine, physiology, history, biography, paleontology, geology, astronomy, and theology, among others. The book details the history and philosophy of this faculty and its practical uses. [P]

1104. ***Colas, Albert.***

L'hypnotisme et la volonté. Paris: A. Ghio, 1885, 36 pp.

[H]

1105. ***Dallmer, Oskar.***

Das Problem des Gedankenlesens. Munich: n.p., 1885.

[P]

1106. ***Deher, Eugen.***

Der Hypnotismus, seine Stellung zum Aberglauben und zur Wissenschaft. Halle: Carl Marhold, (1885?), (2) + 33 pp.

Attempts to do a philosophical-scientific analysis of the relationship between psychic phenomena and hypnotism. [H]

1107. ***Delanne, Gabriel.***

Spiritisme devant la science. Paris: Dentu, 1885, 472 pp.

[P]

1108. ***Edard, Guillaume.***

Vitalisme curatif par les appareils électro-magnétiques. Paris: Passy, 1885, lxxxii + 166 pp.

By a member of the Société mesmérienne of Paris describing the use of some unusual electrical contraptions to guard against illnesses of various kinds. [H]

1109. ***Gurney, Edmund.***
"Hallucinations." *Society for Psychical Research Proceedings* 3(1885): 151–89.

An excellent treatment of a difficult subject. Gurney ranges over the main issues that must be raised in any attempt to understand the various types of hallucination, concentrating on the problem of the origin of the hallucinatory experience. [P]

1110. ***Gurney, Edmund and Myers, Frederic William Henry.***
"Some Higher Aspects of Mesmerism." *Society for Psychical Research Proceedings* 3(1885): 401–23.

Gurney and Myers take up three aspects of mesmerism: 1) the mesmeric treatment of disease, 2) silent "willing" and "willing" at a distance, and 3) clairvoyance. [H & P]

1111. ***Guthrie, Malcolm.***
"Further Report on experiments in thought-transference at Liverpool." *Society for Psychical Research Proceedings* 3(1885): 424–52.

The last of three reports on experiments done at Liverpool, England on thought-transference. This set of experiments focused on the telepathic conveyance of thoughts, images, pain, tastes, smells, and diagrams. As in previous trials, striking results were obtained. [P]

1112. ***Hartmann, Eduard von.***
Moderne Probleme. Leipzig: Wm. Friedrich, 1885.

A treatise on somnambulism by the author of the famous *Philosophy of the Unconscious* (*see* entry number 924). Von Hartmann sees the special knowledge available to the somnambulist as similar to the knowledge available to animals rather than as the higher knowledge that some had claimed it to be. [H & P]

1113. ***Hartmann, Eduard von.***
Der Spiritismus. Berlin: Friedrich, 1885, 118 pp. ENGLISH: *Spiritism*. Translated by C. C. Massey. London: The Psychological Press, 1885.

Von Hartmann discusses the phenomena of spiritualism, including materializations and transfigurations. Provisionally accepting the phenomena as genuine, von Hartmann takes issue with the "spirit hypothesis" which posits that discarnate human spirits are the source of the experiences. He attributes everything to the minds of the medium and sitters, suggesting that some of the phenomena are due to the action of a "nerve force" that can produce mechanical effects. The apparent intelligence operating in these instances is due, says Hartmann, to the unconscious operation of the medium's unrealized somnambulistic consciousness, working outside the awareness of the normal consciousness. [P]

1114. *Hellenbach von Paczolay, Lazar.*

Geburt und Tod als Wechsel der Anschauungsform, oder die Doppel-Natur des Menschen. Vienna: W. Braumüller, 1885, vi + 325 pp.

Hellenbach was a Hungarian born philosopher whose work was significant for psychical research. Here, he suggests that perception at a distance, automatic writing, and trance speaking indicate a double nature in human beings and that through birth and death there is no change beyond an alteration in the method of perception. [P]

1115. *Hovey, William Alfred.*

Mind-Reading and Beyond. Boston: Lee and Shepard, 1885, (2) + 201 pp.

The book is drawn from the material published by the Society for Psychical Research on thought reading. [P]

1116. *Ireland, William W.*

The Blot upon the Brain: Studies in History and Psychology. Edinburgh: Bell & Bradfute, 1885, viii + 374 pp.

A study of subconscious phenomena producing hallucinations of sight and hearing. The author discusses experiences of this kind reported byMuhammed, Luther, Swedenborg and St. Joan. [H & P]

1117. *Mind in Nature: a Popular Journal of Psychical, Medical and Scientific Information.*

Vols. 1–2; 1885–1887.

Published in Chicago and edited by J. E. Woodhead. [H]

1118. *Myers, Frederick William Henry.*

"Automatic Writing." *Society for Psychical Research Proceedings* 3(1885): 1–63.

The second part of a previous article titled "On a Telepathic Explanation of certain so-called Spiritualistic Phenomena" (*see* entry number 1091). Here Myers continues his argument that unless automatic writing contains information completely unknown to the writer, there is no reason to think it is anything but the production of that person's unconscious self. While Myers believes that in some cases there is "supernormal" content in automatic writing, he does not assume that it originates from spirits. [H & P]

1119. *Prince, Morton.*

The Nature of Mind and Human Automatism. Philadelphia: Lippincott, 1885, x + 173 pp.

The first published work of one of the most important figures in the study of the unconscious. The first part of the book, "The Nature of Mind," contains a key concept, the notion of a dynamism of ideas, that ideas once formed possess a drive of their own that tends to expression. In the second part of the book, "Human Automatism," Prince elaborates this concept into a theory of automatism. [H]

1120. *Ribot, Théodule Armand.*

Les maladies de la personalité. Paris: Alcan, 1885, 174 pp. ENGLISH: *The Diseases of Personality.* Chicago: Open Court, 1910.

In this important book, Ribot investigates organic and psychological disorders of the mind. Discussing the identity and unity of the mind, as well as conditions that involve a breakdown of that unity, he describes various examples of duality of personality and examines possession, mysticism, and hypnotism. [H]

1121. *Sergi, Giuseppe.*

L'origine dei fenomeni psichici e loro significazione biologica. Milan: Fratelli Dumodard, 1885, xxiv + 452 pp.

[P]

1122. *Underhill, Leah.*

The Missing Link in Modern Spiritualism. Revised and Arranged by a Literary Friend. New York: Knox & Co., 1885.

Leah Underhill was the married name of the oldest of the famous Fox sisters, initiators of the spiritualistic movement in the United States. Although she was not, as were her two sisters, mediumistic, she played a key role in her younger sisters' promotion. This book, although not written by Leah, was produced under her direction and from materials she provided. It is a personal narrative of the birth of spiritualism and the events involving herself and her sisters. [P]

1123. *Wallace, Mrs. Chandos Leigh (Hunt).*

Private Instructions in the Science and Art of Organic Magnetism. London: The Authoress, 1885, xvi + 211 pp.

[H]

1886

1124. *Barth, Henri.*

Du sommeil non naturel: ses diverses formes. Thèse présentée au concours pour l'agregation. Paris: Asselin et Houzeau, 1886, (4) + 186 pp.

A study of natural somnambulism, pathological attacks of sleep, and artificial somnambulism. Comparing certain similar phases of these various kinds of "non-natural" sleep, Barth discusses the therapeutic usefulness of hypnotism or "artificial nervous sleep." [H]

1125. *Beaunis, Henri Etienne.*
Le somnambulisme provoqué: études physiologiques et psychologiques. Paris: Baillière, 1886, 250 pp.

Beaunis worked at Nancy with Liébeault, Bernheim and Liégeois. Here, he describes procedures used to hypnotize his subjects and presents the results of a series of hypnotic experiments, with a long section on suggestions and their effects on the hypnotized subject. [H]

1126. *Berjon, A.*
La grande hystérie chez l'homme. Phénomènes d'inhibition et de dynamogénie, changements de la personnalité, action des médicaments à distance. D'après les travaux de MM. Bourru et Burot. Paris: J. B. Baillière, 1886, 80 pp.

[H]

1127. *Bernheim, Hippolyte.*
De la suggestion et de ses applications à la thérapeutique. Paris: Octave Doin, 1886, iii + 428 pp. ENGLISH: *Suggestive Therapeutics*. Translated from the second revised edition by Christian Herter. New York: Putnam's, 1897.

The first part of this work is a republication of Bernheim's *De la suggestion dans l'état hypnotique et dans l'état de veille*, 1884 (*see* entry number 1081). In this part Bernheim views the nature of hypnotism in sharp contrast to that of Jean Martin Charcot (1825–1893) which was dominant at the time. Charcot believed hypnotism to be a pathological condition based on physiological processes, an artificial neurosis found only in hysterics. Bernheim, on the other hand, sees hypnotism as a purely psychological condition, one essentially related to suggestion. His definition of hypnotism is: the induction of a peculiar psychical condition which increases the susceptibility to suggestion. This basic difference of opinion constitutes the core of the dispute that would continue for many years between the followers of Charcot (the Salpêtrière School of Hypnotism) and the followers of Bernheim (the Nancy School of Hypnotism). In the second part of the book, Bernheim discusses suggestion as a therapeutic agent. Beginning with a look at the traditional role of the imagination in healing through the ages, Bernheim places modern suggestive therapeutics in that historical stream. This introduction is followed by a long section on cases of specific illnesses Bernheim treated through hypnotic suggestion or through waking suggestion (Bernheim insists that the hypnotic state is not essential for effective treatment through suggestion). The book concludes with a chapter on the dangers to be guarded against in the use of hypnotism and suggestion. This work became the basic text used by the adherents of the Nancy School and holds a unique place in the history of hypnotism. [H]

1128. ***Binet, Alfred.***

La psychologie du raissonnement. Recherches expérimentales par l'hypnotisme. Paris: Germer Baillière, Félix Alcan éditeur, 1886, 171 pp.

Based on laboratory experiments in hypnotism carried out by Binet and Charles Féré (1852–1907). The experiments were made at the Salpêtrière using "hysterico-epileptic" women as subjects. Taking hypnotically induced hallucinations as a starting point, Binet examines the relationship between perception and reasoning, concluding that reasoning is an organization of images, and as such is physiologically based. [H]

1129. ***Brackett, Edward A.***

Materialized Apparitions: If not Beings from Another Life, What are They? Boston: Colby and Rich, 1886, 182 pp.

First hand accounts of materializations of the human form in seances with a number of mediums. In his discussion of possible explanations of the phenomenon, the author makes it clear that he believes them to be genuine and the work of spirits of the departed. [P]

1130. ***Chazarain, Louis Théodore and Décle, Charles.***

Découverte de la polarité humaine, ou démonstration expérimentale des lois suivant lesquelles l'application des aimants, de l'électricité, et les actions manuelles ou analogues du corps humain déterminent l'état hypnotique et l'ordre de succession de ses trois phases. . . . Paris: O. Doin, 1886, 29 pp.

[H]

1131. ***Cullere, A.***

Magnétisme et hypnotisme: Exposé des phénomènes observés pendant le sommeil nerveux provoqué au point de vue clinique, psychologique, therapeutique et médico-légal avec un résumé historique du magnétisme animal. Paris: Librairie J. B. Baillière et Fils, 1886, viii + 381 + (7) pp.

A study written for the layman. It discusses research findings in hypnotism by men such as Charcot and Bernheim and places them in the context of the history of animal magnetism. [H]

1132. ***David, Pierre.***

Magnétisme animal, suggestion hypnotique et post-hypnotique. Son emploi comme agent thérapeutique. Narbonne: F. Pons, 1886, iv + 40 + 121 pp.

[H]

1133. ***Delboeuf, Joseph Remi Léopold.***

"La mémoire chez les hypnotisés." *Revue Philosophique* 21(1886): 441–72.

Starting with the fact that upon wakening somnambulists typically do not remember what happened while in the somnambulistic state, Delboeuf attempts to discover just what this means. He concludes that the "hypnotic dream" is the same as the "natural dream." That is, somnambulists will recall what happened when somnambulistic in the same way that they might recall a regular dream. This recall of somnambulistic experience is subject to the same vagaries of memory as regular dreams. [H]

1134. *Delboeuf, Joseph Remi Léopold.*

Une visite à la Salpêtrière. Brussels: n.p., 1886.

[H]

1135. *Dujardin, Edouard.*

Les phantises. Paris: Léon Vanier, 1886, 176 pp.

[P]

1136. *Du Prel, Carl Ludwig August Friedrich Maximillian Alfred.*

Justinus Kerner und die Seherin von Prevorst. Leipzig: n.p., 1886.

[H & P]

1137. *Durville, Hector.*

Lois physiques du magnétisme. Polarité humaine. Traité expérimental et thérapeutique de magnétisme. 2 ed. Paris: Librairie du Magnétisme, 1886, viii + 181 pp.

Probably the second edition of entry 1138. [H]

1138. *Durville, Hector.*

Traité expérimental et thérapeutique de magnétisme. Paris: n.p., 1886, viii + 181 pp.

Hector Durville's first book on animal magnetism. It presented a framework for his teachings that would be revised and filled out for decades to come. [H]

1139. *Farmer, John Stephen.*

'Twixt Two Worlds: a Narrative of the Life and Work of William Eglinton. London: Psychological Press, 1886, (5) + 196 pp.

Although Farmer had complete faith in the genuineness of the mediumship of the book's subject, William Eglinton, others were less convinced and Eglinton's mediumship was the subject of some controversy. Although this well-produced, large format book is very easy reading and provides a good many interesting details of Eglinton's life, the critical questions that need to be answered with regard to the reliability of these phenomena are not raised. [P]

1140. *Gibier, Paul.*

Le spiritisme (fakirisme occidental): étude historique, critique et expérimentale. Paris: Octave Doin, 1886, xxxii + (33)–398 pp.

One of the better nineteenth century French histories of spiritualism. Gibier argues that the phenomena reported by spiritualists have received sufficient verification to be accepted as genuine and indicates four possible explanations: 1) they are produced by a special fluid coming from medium and sitters, 2) they are produced by the devil, 3) they are produced by "elementals" (non-human spirits), 4) they are produced by the spirits of the dead. Gibier claims that he has not decided in favor of any one of these theories, but is still investigating their respective merits. [H]

1141. *Gley, E.*

"A propos d'une observation de sommeil provoqué à distance." *Revue philosophique* 21(1886): 425–28.

A description of producing artificial somnambulism at a distance, that is, with the operator and subject at locales distant from each other and with the subject ignorant of the time of attempted induction. This experiment was one of a number carried out at the time to determine the genuineness of the phenomenon. [*See*, for example, entry numbers 1146, 1147, 1154.] [H & P]

1142. *Gurney, Edmund.*

"Peculiarities of Certain Post-hypnotic States." *Society for Psychical Research Proceedings* 4(1886): 268–323.

Gurney investigates the manner in which post-hypnotic suggestions are carried out and the degree of intelligent involvement that is indicated. [H]

1143. *Gurney, Edmund.*

"Stages of Hypnotic Memory." *Society of Psychical Research Proceedings* 4(1886): 515–31.

Gurney describes experiments that indicate that hypnotic subjects develop multiple separate memory tracts corresponding to various depths of hypnosis. [H]

1144. *Gurney, Edmund, Myers, Frederic W. H., and Podmore, Frank.*

Phantasms of the Living. 2 vols. London: Trubner and Co., 1886, lxxxiii + (1) + 573; + xxvii + (1) + 733 + (3) pp.

With Myers's *Human Personality* (*see* entry number 1525), this work stands as the most important ever written in the field of psychical research. It was published with the sanction of the Society for Psychical Research and was the result of a long and laborious collection of first hand evidence of psychic occurrences. Myers was responsible for the lengthy introduction and the forty page "Note on a Suggested Mode of Psychical Interaction" in the second volume; Podmore collected and sifted through a large part of the evidence used in the book; and Gurney wrote all of the text apart from the sections done by Myers. As Myers says in the Introduction, the work embraces all transmissions of thought and feeling from one person to another by means other than through the recognized channels of sense, and this includes apparitions. Myers had already coined the word "telepathy" to denote these transmissions. The authors included apparitions of the living in this study, but excluded apparitions of the dead. The area of investigation is in this way limited to mind-to-mind transmissions between living persons. In Myers's introductory remarks, he undertakes to show how a study of these phenomena must be approached if it is to be scientific. He points out that in the literature of all ages there have been references to experiences which are, at least at first glance, "supernormal" (another term coined by Myers). The question is how to determine whether or not such events may be explained in terms of known factors and laws. His answer is to tackle the problem by examining contemporary data for which first hand witnesses are available. That is precisely what the Society for Psychical Research did. *Phantasms of the Living* is the result of their findings. Approximately two thousand cases were considered to be well enough attested to be used in the study. Along with these data, the investigators examined experiments with telepathy carried out under controlled conditions. The results were summarized in three points: 1) experiments show that telepathy is a fact of nature; 2) testimony proves that phantasms (impressions, voices, or figures) of people who are undergoing some crisis (particularly death) are perceived by individuals close to them with a frequency which mere chance cannot explain; and 3) these phantasms, whatever else they may be, are instances of supersensory action of one mind on another. The bulk of the two volumes is devoted to presenting and commenting on the data of telepathy and phantasms collected by the Society. Gurney begins with some notes of caution concerning the implications of the findings. He then moves on to chapters on experiments in thought transference, spontaneous telepathy, transference of mental pictures and of motor effects, dreams, hypnagogic images, hallucinations, chance coincidence, cases with single and multiple percipients, and collective cases. There is a large supplement which includes additional cases from most categories covered previously. Never before or since *Phantasms of the Living* had such a massive, well thought out, and well written piece of research been produced in the field of psychical research. [P]

1145. *Héricourt, J.*

"Un cas de somnambulisme à distance." *Revue Philosophique* 21(1886): 200–203.

One of a number of important articles written in 1886 on the subject of the induction of somnambulism at a distance. Héricourt describes one such case that occurred in 1878. The somnambulist was a twenty-four year old woman, Madame D. [H & P]

1146. ***Janet, Pierre.***

"Les actes inconscients et le dédoublement de la personnalité pendant le somnambulisme provoqué." *Revue philosophique* 22(1886): 577–92.

The first of three important articles on unconscious acts occurring during somnambulism (see "L'anesthésie systématisée," (entry number 1176) and "Les actes inconscients," (entry number 1211) and the first of many writings that would establish Janet as the preeminent explorer of the "subconscious" (a term that Janet coined to designate the realm of mental activity outside conscious awareness) and a psychologist of great stature. Janet stands at the meeting point between the animal magnetic tradition, with its discovery and exploration of artificial somnambulism, and the great modern psychological systems. His early works strongly affected the thinking of Richet, Ochorowitz, and F.W.H. Myers, and were important sources for Freud, Jung, and Adler. In this article which explores actions deriving from a source outside the awareness of the individual who performs them, Janet discusses suggestions made in the hypnotic state and the products of automatic writing to see what light they throw on subconscious processes. He also notes the apparent creation of a secondary personality by means of suggestion and calls this and related phenomena the "doubling of the personality." Janet's master work, *L'automatisme psychologique* (1889), brought to fruition ideas put forward in this and the two articles that followed. [H]

1147. ***Janet, Pierre.***

"Deuxième note sur le sommeil provoqué à distance et la suggestion mentale pendant l'état somnambulique." *Revue philosophique* 22(1886): 212–23.

Janet's second talk on somnambulism brought about at a distance (see "Note sur quelques phénomènes de somnambulisme," entry number 1148) delivered at a meeting of the Société de Psychologie Physiologique. Here he describes his experiments again with Madame B. to reconfirm his earlier finding but with a new element: the issuing of mental commands to carry out some action. This second set of experiments, performed with additional precautions, was successful. [H & P]

1148. ***Janet, Pierre.***

"Note sur quelques phénomènes de somnambulisme." *Revue philosophique* 21(1886): 190–98.

This article reproduces a talk given by Janet at a meeting of the Société de Psychologie Physiologique under the presidency of Charcot in late 1885. Janet was at that time already engaged in the investigations with somnambulism that would culminate in three important articles on the subject published in the Revue philosophique from 1886 to 1888 (q.v.). In this talk he tells of experiments done with a certain Madame B who was very susceptible to being put in the somnambulistic state and a good subject for experiments. Janet talks about a peculiarity that began to occur as he and others worked with her. They discovered what seemed to be an unusual responsiveness to the thoughts of the person who put her into the trance state, so much so that it seemed she could be made somnambulistic by the operator's merely thinking about it. To test this matter some experiments were tried in which the operator was in another location at some distance from the house of the somnambulist and he attempted to entrance her at set times unknown to her. The experiments proved successful to the degree that Janet thought the phenomenon worthy of further study. Janet offered no theory as to how this was possible. [H]

1149. ***Lombroso, Cesare.***
Studi sull 'ipnotismo, con ricerche oftalmoscopiche del Reymond e dei prof. Bianchi e Sommer sulla polarizzazione psichica. 2 ed. Turin: Fratelli Bocca, 1886, 72 pp.

No information available on the first edition. [H]

1150. ***Le magnétisme. Revue générale des sciences physio-psychologiques.***
One vol. only; 1886. Published in Paris and edited by "Donato" [Alfred Edouard d'Hont (1840–1900)].

[H]

1151. ***Morselli, Enrico.***
Il magnetismo animale, la fascinazione e gli stati ipnotici. Turin: Roux E. Favale, 1886, viii + 427 pp.

Morselli was the director of the Psychiatric Clinic at the University of Genoa and a highly respected researcher. This study of animal magnetism and hypnotism is a balanced and cautious examination of all aspects of the phenomena, including the paranormal. [H & P]

1152. ***Myers, Frederic William Henry.***
"Multiplex Personality." *Society for Psychical Research Proceedings* 4(1886): 496–514.

Myers begins to develop ideas about the complex nature of human personality that he will bring to fruition with his notion of the "subliminal consciousness" in later articles. [H]

1153. ***Myers, Frederick William Henry.***
"On Telepathic Hypnotism, and its Relation to Other Forms of Hypnotic Suggestion." *Society for Psychical Research. Proceedings* 4(1886): 127–88.

This article contains Myers' personal observations of experiments done by Pierre Janet and a Doctor Gilbert at Havre in 1886 to determine whether it is possible to induce a state of hypnotic trance in a subject separated from the operator by a considerable distance. He points out that mesmerists had for a long time claimed to be able to use animal magnetism at great distances, since they believed that the magnetic influence that proceeds from their fingers and eyes could operate through any physical obstacle. Myers also states that many "suggestionists" (those who follow the teachings of Liébeault and Bernheim) also claim to be able to exercise their suggestive power at great distances. Myers believes, however, that both schools of thought have been insufficiently critical in their observations and in their use of terms to bring the discussion to a definite conclusion. From the experiments done at Havre and the evidence of other observers, Myers concludes that telepathic hypnotism at a distance may be possible. In this process of discussing the issue, Myers provides a useful sketch of the possible modes of induction of hypnotic phenomena. [H & P]

1154. ***Perronnet, Claude.***
Force psychique et suggestion mentale, leur demonstration, leur explication, leurs applications possible, à la thérapeutique et à la médecine légale. Paris: J. Lechevalier, 1886, 72 pp.

[H & P]

1155. ***Preyer, Wilhelm.***
Die Erklärung des Gedankenlesens. Leipzig: T. Greiben, 1886, (8) + 70 pp.

[H & P]

1156. ***Revue de l'hypnotisme expérimental et thérapeutique.***
Vols. 1–24; 1886–1910.

Continued as: *Revue de psychothérapie et psychologie appliquée.* Vols. 25–29; 1910–1915. Continued as: *La psychologie appliquée.* Vols. 1–2; 1920–1921. [H]

1157. ***Richet, Charles.***
"Un fait de somnambulisme à distance." *Revue philosophique* 21(1886), 199–200.

Describes a case of apparent induction of somnambulism at a distance carried out by Richet. [H & P]

1158. ***Zanardelli, D.***
La verità sull 'ipnotismo. Rivelazioni. Rome: Reggiani e soci, 1886, 74 pp.

One of Italy's more famous mesmerists, Zanardelli toured Italy, France, and Spain with his wife, giving demonstrations of the marvels of hypnotism. This book describes those appearances. [H]

1887

1159. *Azam, Etienne Eugène.*

Hypnotisme, double conscience, et altérations de la personnalité. Paris: Librairie J.-B. Ballière et fils, 1887, 284 pp.

A work significant for establishing a firm connection between multiple personality and the somnambulistic state. Dr. Azam was a professor of surgery at the Bordeau Medical School and the first in France to reproduce the hypnotic experiments of Braid. Here, he describes the results of his study of one of the most remarkable cases of multiple personality yet recorded: that of Felida X. He first met Felida when she was a young girl in 1858, and he was still observing her in 1887 when the book was published. Felida had two distinct personalities, each with its own characteristics. One was called her "primary personality," the other her "secondary condition"—which Azam considered to be a state of total somnambulism. In her secondary condition Felida exhibited personality traits very different from those of her primary personality. Although her primary personality could not remember anything of what happened while she was in her secondary condition, her secondary self had full memory of both states. This "dédoublement de la personnalité," as Azam termed it, continued throughout Felida's lifetime. Over the years the secondary personality gradually gained the upper hand, so that eventually her primary personality showed itself only rarely. Throughout the book, Azam points out the similarities between Felida's secondary condition and the state of artificial somnambulism as described by Braid. The connection between dual personality and somnambulism had been made before. Azam, in this work, for the first time provides a solid scientific basis for the view that multiple personality was something other than madness or possession—that it was, in fact, a special case of somnambulistic consciousness. [H]

1160. *Barety, A.*

Le magnétisme animal: étudié sous le nom de force neurique: rayonnante et circulante dans ses propriétés physiques, physiologiques et thérapeutiques. Paris: Octave Doin, 1887, (iv) + (xvi) + 662 + (2) pp.

A massive elaboration of the small treatise that Barety wrote in 1882 (*see* entry number 1051). Based on observations made principally on hysterical individuals, he claims to have discovered a "neuric force" circulating through and radiating from the body. Claiming that "neuric force" had until then been incorrectly dubbed "animal magnetism," he relabels it and extends his terminology to magnetic sleep or artificial somnambulism, calling it, in turn, "neuric sleep." [H]

1161. ***Belfiore, Giulio.***
L'ipnotismo e gli stati affini. Napoli: Luigi Pierro, 1887, x + 447 pp.

[H]

1162. ***Binet, Alfred and Féré, Charles Samson.***
Le magnétisme animal. Paris: Ancienne Librairie Germer Baillière et Cie; Félix Alcan, éditeur, 1887, (8) + 283 + (1) pp. ENGLISH: *Animal Magnetism.* New York: D. Appleton, 1890.

A well researched and widely read treatment of animal magnetism and hypnotism by two respected psychological investigators. The book includes a brief but informative history of those phenomena up to the time of its publication. It also contains a treatment of the stages of hypnotism, the nature of hypnotic suggestion, and the therapeutic application of hypnotism. The theoretical understanding of hypnotism presented is strongly influenced by the Salpêtrière School. [H]

1163. ***Björnström, Fredrik Johan.***
Hypnotismen, dess utveckling och nuvarande standpunkt. Popular framstallning. Stockholm: Gerber, 1887, 201 pp. ENGLISH: *Hypnotism: Its History and Present Development.* Translated by Baron Nils Posse. New York: Humboldt Publishing Co., 1887.

A short but competent treatment of the history, methods of induction, stages, and effects of hypnotism. It includes a lengthy section on the uses of hypnotic suggestion. Björnström drew his cases from his own work and from the literature of hypnotism. [H]

1164. ***Bourru, H. and Burot, P.***
La suggestion mentale et l'action à distance des substances toxiques et médicamenteuses. Paris: Baillière, 1887, 310 pp.

One of the most thorough studies of the effects of various substances on the body when placed in the vicinity of the subject. The authors discuss the emanations given off by plants, metals, and various chemicals and attempt to determine what effects they have on the physical organism. [H]

1165. ***Delboeuf, Joseph Remi Léopold.***
De l'origine des effets curatifs de l'hypnotisme; étude de psychologie expérimentale. Paris: Germer Baillière, 1887.

[H]

1166. ***Dessoir, Max.***
"Experiments in Muscle-reading and Thought-transference." *Society for Psychical Research Proceedings* 4(1887): 111–26.

Dessoir discusses ways in which muscle-reading might account for apparently telepathic communication. He also describes a series of experiments in which apparent telepathic communication took place without the benefit of any such normal means of knowledge. [H]

1167. ***Durville, Hector and Constantin, Comte de.***
Discours prononcé à l'inauguration de la Société magnétique de France, le 7 octobre 1887. Paris: n.p., (1887).

[H]

1168. ***Figuier, Louis.***
Les mystères de la science. 2 vols. Paris: Librairie illustrée, (1887).

[H & P]

1169. ***Fontan, Jules and Segard, Charles.***
Eléments de médecine suggestive, hypnotisme et suggestion: faits cliniques. Paris: Octave Doin, 1887, 306 pp.

[H]

1170. ***Fonvielle, W. de.***
Les endormeurs. La verité sur les hypnotisants, les suggestionistes, les magnétiseurs, les Donatistes, les Braidistes, etc. Paris: Henry du Parc, (1887), 308 pp.

A critical evaluation of the nature of hypnotism and somnambulism. The author sees imagination as the central element. [H]

1171. ***Gessmann, Gustav Wilhelm.***
Magnetismus und Hypnotismus: eine Darstellung dieses Gebietes mit besonderer Berücksichtigung der Beziehungen zwischen dem mineralischen Magnetismus und dem sogenannten thierischen Magnetismus oder Hypnotismus. Vienna, Pest, and Leipzig: A. Hartleben, 1887, xiii + (3) + 216 pp.

[H]

1172. ***Gilles de la Tourette, George Albert Edouard Brutus.***
L'hypnotisme et les états analogues au point de vue médico-légal, les états hypnotiques et les états analogues, les suggestions criminelles, cabinets de somnambules et sociétés de magnétisme et de spiritisme, l'hypnotisme devant la loi. Paris: E. Plon, Nourrit et cie., 1887, (4) + xv + (1) + 534 pp.

Considered the best and most thorough work on the moral and legal aspects of the practice of animal magnetism and hypnotism. In order to establish the context, the author begins with a description of the history of animal magnetism and hypnotism and a discussion of the nature of the hypnotic state. He is an adherent of the school of Salpêtrière and he uses the hypnotic stages as described by Charcot as his framework. Gilles de la

Tourette next considers the nature of hypnotic suggestion and susceptibility to suggestions while in the hypnotic state. He discusses the possibility of resistance to suggestion and the importance of amnesia for this issue. To throw light on moral and legal responsibility of the hypnotic subject, Gilles de la Tourette then examines what he calls states "analogous" to hypnotism, principally natural somnambulism, hysteria, and dual personality. In the next section of the book, the author investigates the benefits and dangers of hypnotism, emphasizing the responsibility of the hypnotizer for making the situation safe. The final section takes up important legal aspects of hypnotic practice: hypnotism and the commission of crimes, the possible abuse of the public through hypnotic demonstrations, the legal regulation of hypnotism, and the sexual violation of hypnotic subjects. [H]

1173. ***Gurney, Edmund.***

"Further Problems of Hypnotism." *Mind* 47(1887): 212–332.

[H]

1174. ***Gurney, Edmund.***

Tertium Quid: Chapters on Various Disputed Questions. 2 vols. London: Kegan Paul, Trench & Co., 1887.

[P]

1175. ***Hodgson, Richard.***

"The Possibilities of Mal-observation and Lapse of Memory from a Practical Point of View." *Society for Psychical Research Proceedings* 4(1887): 381–495.

Hodgson was born in Australia and received his higher education at Cambridge University. A charter member of the Society for Psychical Research and one of its chief investigators he was particularly interested in the detection of trickery and was instrumental in exposing an incident of fraud perpetrated by Madame Blavatsky. Until his investigation of Mrs. Leonora Piper (1859–1950), a mental medium from Boston (see Hodgson, 1892 and 1898), Hodgson found little that he considered to be genuinely paranormal. This very adept article discusses possible sources of error in observation of the phenomena of spiritualism. [H]

1176. ***Janet, Pierre.***

"L'anesthésie systématisée et la dissociation des phénomènes psychologiques." *Revue philosophique* 23(1887): 449–72.

The second of three important articles written by Janet for the *Revue philosophique* on subjects relating to "doubling of the personality" and dissociated acts. In this article he concentrates on certain physiological effects of suggestion made to susceptible individuals. These effects include blindness, numbness, and paralysis. He also discusses experiments done with what Bernheim called "negative hallucinations" in which certain objects or individuals are made to disappear from the perception of the subject through suggestion. All these conditions are included by Janet under the term "anaesthesias." They are, says Janet, examples of experiences being "dissociated" from the conscious awareness of the individual but remaining within the purview of "another consciousness" different from the individual's primary consciousness. Janet's use of the word "dissociation" with this particular meaning seems to be the first such use in the literature of modern psychology. In this article Janet also begins to form the outlines of a psychotherapy based upon the bringing of problematic dissociated material into the primary consciousness. [H]

1177. ***Johnson, Franklin.***

The New Psychic Studies in their Relations to Christian Thought. New York and London: Funk and Wagnalls, 1887, 91 pp.

Discusses the implications for Christian thought of such things as mesmerism, clairvoyance, apparitions, and hauntings. [H & P]

1178. ***Luys, Jules Bernard.***

Les émotions chez les sujets en état d'hypnotisme. Études de psychologie expérimentale faites à l'aide de substances médicamenteuses ou toxiques impressionnant à distance les réseaux nerveux periphériques. Paris: Baillière, 1887, 106 + (6) pp.

Luys, a highly respected French neurologist who practiced at the Salpêtrière and the Charité, here discusses the transmission of states of emotion from one hypnotized person to another and the effects on the nervous system of certain substances and apparatuses placed at a distance from a subject. Materials used included laudanum and glass tinted with various colors. Luys adhered to a description of the successive phases of hypnotism that was a modified form of the Salpêtrière School. The principal subject for his experiments was a somnambulist he calls "Esther." One of the interesting outcomes of his experimentation was his development of a unique version of the notion of "doubling of the personality." In Luys' view, in the doubled personality opposite emotions resided in the left and right halves of the body. [H]

1179. ***Magini, Giuseppe.***

Le maraveglie dell' ipnotismo. Sommario dei principali fenomeni del somnambulismo provocato, e metodi di sperimentazione. Turin: E. Loescher, 1887, 49 pp.

[H]

1180. ***Massey, Charles Carleton.***

Zöllner. An Open Letter to Professor George S. Fullerton of the University of Pennsylvania, Member and Secretary of the Seybert Commission for Investigating Modern Spiritualism. (London?): n.p., 1887, 40 pp.

The Seybert Commission was appointed by the University of Pennsylvania to study the validity of spiritualism. The preliminary report was published in 1887 and its findings were negative (see Seybert Commission, 1191). The *Report* contained a statement by Fullerton attributed to Massey, deriving from a conversation the two had in 1886. The issue was the mental competency of Professor Zöllner (1834–1882), whose experiments with the medium Henry Slade were criticized in the *Report*. Massey asserts the inaccuracy of remarks attributed to himself, and takes issue with Fullerton's conclusion that Zöllner was insane at the time of the experiments. Massey takes pains to cite evidence of those who knew Zöllner at the time in question that indicates he was of sound mind. [H & P]

1181. *Mesnet, Ernest.*

Étude médico-légale sur le somnambulisme spontané et le somnambulisme provoqué. Paris: G. Masson, 1887, 39 pp.

[H]

1182. *Moréty, G.*

Le magnétisme triomphant: exposé historique et critique de la question. Paris: Auguste Ghio, 1887, (4) + vi + (1) + 131 pp.

Contains a valuable section on the hypnotist "Donato" (real name: Edouard d'Honte) who was enthralling audiences all over Europe at the time. The author also discusses the notion of "fascination" which was central to Donato's presentation. [H]

1183. *Moutin, Lucien.*

Le nouvel hypnotisme. Paris: Perrin et Cie, 1887, (6) + 220 pp.

Moutin believed that hypnotism and animal magnetism are two quite distinct phenomena operating in quite different ways. He writes appreciatively of Braid's work and the importance of his discoveries concerning the psychology of "nervous sleep." But he emphasizes that although in hypnotism the production of a state of sleep is essential for healing, in animal magnetism healing does not require a change of consciousness. Moutin also reviews effects he believes to be produced by animal magnetism that could not be accounted for by hypnotism, such as magnetic effects at a distance without the knowledge of the subject. [H & P]

1184. *Ochorowicz, Julian.*

De la suggestion mentale. Paris: Octave Doin, 1887 (6) + vi + (1) + 558 + (2) pp. ENGLISH: *Mental Suggestion*. Translated by J. Fitzgerald. New York: Humboldt, 1891.

The most thorough and comprehensive book on mental suggestion to appear in the nineteenth century. A mental suggestion is one communicated by one individual to another without the benefit of the senses; there is, therefore, an apparently direct communication of the suggestion from one mind to another. The phenomenon was first noted in the tradition of animal magnetism in 1784 in Puységur's *Memoire* (*see* entry number 105) where Puységur speaks of being able to communicate his wishes mentally to his patient, the peasant Victor Race. From that time on, literally hundreds of magnetizers noted the same phenomenon, and it became one of the most controversial aspects of animal magnetism discussed in the literature of the subject. In this study, Ochorowicz attempts to deal definitively with the issue not only by recounting the experiences of others, but principally through his own painstaking experiments. His conclusion is that mental suggestion is a reality and must be taken into account in any adequate theory of human psychology. Included in the book is a detailed description of the experiments conducted by Janet assisted by Ochorowicz, F.W.H. Myers, and others at Havre concerning suggestion at a distance. These experiments and related experiences were described in articles appearing in the *Revue philosophique* in 1886 written by Janet, Gley, Héricourt, Richet and Beaunis. Ochorowicz also includes a long section on possible explanations of the phenomena of mental suggestion. [H & P]

1185. ***Regnard, Paul.***
Les maladies épidémiques de l'esprit. Sorcellerie, magnétisme, morphinisme, délire des grandeurs. Paris: E. Plon, 1887, iii–xii + 429 + (2) pp.

[H & P]

1186. ***Revue des sciences hypnotiques.***
Vols. 1–2; 1887–1888.

Published in Paris, this periodical describes itself as a monthly that treats magnetism, braidism, hypnotism, fascination, hypnosis, ecstasy, suggestion, somnambulism (both natural and artificial), lethargy and catalepsy, legal medicine, physiological psychology, the study of psychic substances, nervous phenomena, mental illnesses, metalloscopy and metallotherapy, and moral and therapeutic applications. [H]

1187. ***Rochas d'Aiglun, Eugène August Albert de.***
Les forces non définies: recherches historiques et expérimentales. Paris: G. Masson, 1887, (4) + 392 + (22) pp.

De Rochas attempts to provide an overall view of the magnetic forces at work in the organisms of living things. In this context he deals with the power of hypnotism over the mind of the hypnotic subject and includes a section on the phases of hypnotism, focussing on the "state of credulity" of the hypnotized person and the power of suggestion. [H]

1188. ***Rossi-Pagnoni, Francesco and Moroni, L.***
Alcuni saggi di medianita ipnotica. Pesaro: Stab. Annesio Nobili, 1887, 16 pp.

[H & P]

1189. ***Sallis, Johann G.***

Der tierische Magnetismus und seine Genese. Ein Beitrag zur Aufklärung und eine Mahnung an die Sanitätsbehorden. Leipzig: Gunther, 1887, 53 pp.

[H]

1190. ***Ségard, Ch. and Fontan, J.***

Eléments de médecine suggestive. Hypnotisme et suggestion—faits cliniques. Paris: Doin, 1887, xv + 306 pp.

[H]

1191. ***Seybert Commission.***

Preliminary Report of the Commission Appointed by the University of Pennsylvania to Investigate Modern Spiritualism in Accordance with the Request of the Late Henry Seybert. Philadelphia: J. B. Lippincott, 1887, 160 pp.

Henry Seybert, a spiritualist, left a bequest of $60,000 to the University of Pennsylvania to be used for the maintenance of a chair to be known as the "Adam Seybert Chair of Moral and Intellectual Philosophy," with the condition that the incumbent, either alone or with a commission of the university faculty, make a study of all religious systems, particularly modern spiritualism. A commission was appointed and began its work in 1884, resulting in this *Report.* It studied a number of mediums first hand over a period of three years. The representative of the testator on the commission repeatedly protested that certain members had as their only object the discrediting of spiritualism. He requested that those members be removed, but they were not. The findings of the commission were totally negative. An appendix contains a controversial discussion of the sanity of Prof. Johann Zöllner, a German professor of physics and astronomy who investigated the American medium Henry Slade. Spiritualists claimed that the intentions of Mr. Seybert were never fairly carried out. [P]

1192. ***Theobald, Morell.***

Spirit Workers in the Home Circle. An Autobiographic Narrative of Psychic Phenomena in Family Daily Life Extending over a Period of Twenty Years. London: Fisher Unwin, 1887, xvi + 310 pp.

[P]

1193. ***Vallombroso, ———.***

L'ipnotismo e magnetismo svelato e spiegato sulle teorie di Donato Guidi e Mesmer. Milan: n.p., 1887.

[H]

1194. ***Younger, D.***

Instructions in Mesmerism &c. The Magnetic and Botanic Family Physician, and Domestic Practice of Natural Medicine, with Illustrations Showing Var-

ious Phases of Mesmeric Treatment, Including Full and Concise Instructions in Mesmerism, Curative Magnetism, and Massage. London: E. W. Allen, 1887, 151 pp.

[H]

1888

1195. *Binet, Alfred.*

Études de psychologie expérimentale. Le fétichisme dans l'amour. La vie psychique des micro-organismes. L'intensité des images mentales. Le problème hypnotique. Note sur l'écriture hystérique. Paris: Octave Doin, 1888, 307 pp.

[H]

1196. *Bourru, Henri and Burot, P.*

Variations de la personnalité. Paris: J. B. Baillière, 1888, 314 + (2) pp.

A significant attempt to discover the nature and origin of states of multiple personality, including a much expanded description of a case of multiple personality first reported in the *Revue philosophique* (1102). To this is added a discussion of further experiences with multiple consciousness in patients whom the authors have treated with the aid of hypnotism. [H]

1197. *Carpenter, William Benjamin.*

Nature and Man, Essays Scientific and Philosophical. With an Introductory Memoire by J. Estlin Carpenter. London: Kegan Paul, Trench & Co., 1888, vi + 483 pp.

Carpenter discusses a variety of topics, including the automatic execution of voluntary movements, the influence of suggestion in modifying and directing muscular movement and independently of volition, and the doctrine and limits of human automatism. [H]

1198. *Coste de Lagrave, ———.*

Hypnotisme, états intermédiaires entre le sommeil et la veille. Paris: J. B. Baillière, 1888, 160 pp.

A study of various degrees of consciousness and the condition of the will in hypnotism. [H]

1199. *Cumberland, Stuart.*

A Thought-Reader's Thoughts: Being the Impressions and Confession of Stuart Cumberland. London: Sampson Low, Marston, Searle, & Rivington, 1888, xi + (1) + 326 pp.

An exposé of the life of the author as a fraudulent medium, providing a detailed account of the tricks and deceptions that he and others used to convince the gullible that they had psychic powers. His debunking of all spiritualistic phenomena as false came to be known as "cumberland-ism." [P]

1200. ***Davenport, Reuben Briggs.***
The Death-Blow to Spiritualism: Being the True Story of the Fox Sisters, as Revealed by the Authority of Margaret Fox Kane and Catherine Fox Jencken. New York: Dillingham, 1888, 247 pp.

Forty years after the Hydesville "knockings" in the home of the Fox sisters led to the modern spiritualist movement, the two sisters involved met with the public to say that they had made it all up. Margaret made the explanation, stating that she and Kate had made the "spirit" sounds by cracking their joints. Kate was present, but made no comment. A short time after the "exposé," Kate stated in a letter that the sisters (both subject to alcoholism) were out of money, and the public confession had netted them a thousand dollar fee. She goes on to say that she (Kate) could perhaps now bring in an income by proving Margaret's exposé to be false. The whole affair had been initiated when Margaret had agreed to work with Reuben Davenport on the confession. Shortly after the public meeting, his book, embodying the exposé, was published. [P]

1201. ***Delboeuf, Joseph Remi Leopold.***
L'hypnotisme et la liberté des représentations publiques. Liège: Ch. Aug. Desoer, 1888, 111 pp.

Delboeuf was a determined advocate of the freedom to practice both hypnotism and animal magnetism in society and addresses that issue in this pamphlet. [H]

1202. ***Dessoir, Max.***
Bibliographie des modernen Hypnotismus. Berlin: Carl Duncker, 1888, 94 pp.

A useful bibliography listing 801 titles of books and articles on hypnotism, animal magnetism and closely related areas. The titles are subdivided into these categories: General, Medicine, Magnetism and Hypnotism, Physiology, Psychology and Pedagogy, Action at a Distance, Modern Mesmerists, and Various Topics. There is an index of authors and interesting statistics providing information about the various schools of hypnotism, the various countries of origin, the various languages of the writings, and the various years of publication. [H]

1203. ***Dufay, ———.***
"Deux cas de somnambulisme provoqué à distance." *Revue Scientifique* 25(1888): 240–43.

In this communication, originally sent to the Société de Psychologie Physiologique, Dufay describes two cases of artificial somnambulism induced when the subject was at some distance from the operator. In one of the two cases, the distance was 112 kilometers. [H & P]

1204. ***Gurney, Edmund.***
"Hypnotism and Telepathy." *Society for Psychical Research Proceedings* 5(1888): 216–59.

A significant study of the relationship between hypnotic trance and the transmission of thought or suggestion at a distance. [H & P]

1205. ***Gurney, Edmund.***
"Recent Experiments in Hypnotism." *Society for Psychical Research Proceedings* 5(1888): 3–17.

A description of experiments conducted to discover the degree of intelligence involved in automatism and to study the production of local anesthesia and rigidity through hypnotism. [H]

1206. ***Home, Madame Daniel Dunglas.***
D. D. Home: His Life and Mission. London: Trübner & Co., 1888, viii + 428 pp.

A biography of the famous medium written by his second wife. She covers the whole of his life, from his birth in Scotland and his childhood in America to his death from consumption in 1886. It is a valuable source of detailed information about Home, containing many letters, as well as details of conversations and seances. [P]

1207. ***Hont, Alfred Edouard d', pseudonym: Donato.***
La lumière sur le magnétisme, ses défenses et ses ennemis. Neuchatel: n.p., 1888.

[H]

1208. ***Hubbell, Walter.***
The Great Amherst Mystery. A True Narrative of the Supernatural. Chicago: Brentano's, 1888, 168 pp.

A description of a classical poltergeist case that occurred at Amherst, Nova Scotia in the Teed family between 1878–1879. The phenomena centered around a Mrs. Cox, sister of Mrs. Teed. Hubbell concluded that there was no fraud involved and that ghosts were producing the disturbances. [P]

1209. ***L'initiation, Revue philosophique indépendente des hautes études, hypnotisme, force psychique, théosophie, kabbale gnose, franc maçonnerie, sciences occultes.***
Vols. 1–96; 1888–1912.

Continued as: *Mysteria. Revue mensuelle illustrée d'études initiatiques.* Vols. 1–2; 1913–1914. [H & P]

1210. ***James, Constantin.***

L'hypnotisme expliqué dans sa nature et dans ses actes. Mes entretiens avec S. M. l'Empereur Don Pedro sur le Darwinisme. Paris: Librairie de la Société Bibliographique, 1888, 92 pp.

Touches a wide array of subjects including hypnotism as taught by the School of Salpêtrière. [H]

1211. ***Janet, Pierre.***

"Les actes inconscients et la mémoire pendant le somnambulisme." *Revue philosophique* 25(1888): 238–79.

In the last of three important articles written by Janet for the *Revue philosophique* from 1886 to 1888, he discusses four kinds of acts that can be unconsciously performed by an individual: 1) those inculcated by posthypnotic suggestion, 2) those inculcated through anaesthesia, 3) those inculcated through "distraction," and 4) those produced spontaneously. He then continues to investigate the phenomenon of "doubling of the personality" and the kinds of amnesia involved. This becomes a particularly intriguing question when the investigator discovers, as did Janet, more than one "secondary" or somnambulistic personality in an individual. In describing the individual's actions in this state, Janet speaks of the person acting unconsciously or "subconsciously" and of "subconscious acts." This seems to be Janet's first use of the term "subconscious," which was to become pivotal to his subsequent writings and so influential to those who would carry out similar investigations. [H]

1212. ***Krafft-Ebing, Richard von.***

Eine experimentelle Studie auf dem Gebiete des Hypnotismus. Stuttgart: Ferdinand Enke, 1888, 80 pp.

[H]

1213. ***Luys, Jules Bernard.***

"Sur l'état de fascination déterminé chez l'homme à l'aide de surfaces brillantes en rotation (action somnifère des miroirs à alouettes)." *Compte rendus et mémoires des sciences de la Société de biologie* 107(1888): 449.

[H]

1214. ***Mitchell, Silas Weir.***

"Mary Reynolds: a case of double consciousness." *Transactions of the College of Physicians of Philadelphia* 10(1888): 366–89.

The most accurate account of the most famous early case of multiple personality. Mary Reynolds is the "lady" mentioned by Robert Macnish in his *Philosophy of Sleep* of 1830 (*see* entry number 345). [H]

1215. ***Revue spirite. Journal d'études psychologiques.***
Vols. 1–57; 1858–1914. Vol. 58; 1915. Vols. 60–69+; 1917–1925+.

Published in Paris and edited by Hippolyte Rivail (Allan Kardec). [P]

1216. ***Richet, Charles.***
"Expériences sur le sommeil à distance." *Revue Philosophique* 25(1888): 434–52. [H & P]

1217. ***Richet, Charles.***
"Relation de diverses experiences sur la transmission mentale, la lucidité, et autres phénomènes non explicables par les données scientifiques actuelles." *Society for Psychical Research Proceedings*, 5(1888): 18–168.

A lengthy description of experiments in thought transference and clairvoyance carried out by Richet. It is an important study, giving details of experimental conditions and results. [P]

1218. ***Richmond, Almon Benson.***
What I saw at Cassadaga Lake: A Review of the Seybert Commissioners' Report. Boston: Colby & Rich, 1888, 244 pp.

[P]

1219. ***Schrenck-Notzing, Albert Phillibert Franz von.***
Ein Beitrag zur therapeutischen Verwerthung des Hypnotismus. Leipzig: F.C.W. Vogel, 1888, (4) + 94 + (2) pp.

A small but well-researched treatise on the therapeutic use of hypnotism from the time of Braid. Schrenck-Notzing discusses its history in France, Germany, Italy, Belgium, Holland, England and other countries. He also provides a useful bibliography of relevant literature from each country. [H]

1220. ***Wetterstrand, Otto Georg.***
Om Hypnotismens anvandande i den praktiska medicinen. Stockholm: J. Seligmann, (1888), iv + 90 pp. ENGLISH: *Hypnotism and Its Application to Practical Medicine.* New York: The Knickerbocker Press, 1897.

[H]

1221. ***Wollny, F.***
Sammlung von Actenstücken, als da sind: Eingaben und Adressen in Sachen der gemeingefährlichen Einwirkungen durch Magnetisation auf telepathischen Wege, an verschiedene Behörden, Vereine &c. gerichtet. Leipzig: n.p., 1888, 32 pp.

[H & P]

1889

1222. ***American Society for Psychical Research (Old) Proceedings.***
Published as one volume; 1885–1889.

The only volume of the *Proceedings* of the first or "old" American Society for Psychical Research, considered a branch of the original Society for Psychical Research in Britain. After this brief initial existence, it was simply absorbed into the parent society and only became an independently functioning body publishing its own *Proceedings* in 1907 (see American Society for Psychical Research *Proceedings*). This volume contains articles by Richard Hodgson, Josiah Royce, William James, and others. [H & P]

1223. ***Babinski, J.***
"Grand et petit hypnotisme." *Archives de neurologie* 17(1889): 92–108, 253–69.

[H]

1224. ***Belfiore, Giulio.***
Grande isteria ed ipnotismo. Studio medico legale su Paolo Conte imputato di truffa in danno del dr. Fusco. Naples: n.p., 1889.

Study of a celebrated legal case involving the "hysteric" Paolo Conte. [H]

1225. ***Bérillon, Edgar, ed.***
Congrès international de l'hypnotisme expérimental et thérapeutique. Comptes rendus. Paris: O. Doin, 1889, 368 pp.

[H]

1226. ***Coste, Marie Leon.***
L'inconscient, étude sur l'hypnotisme. Paris: J. B. Baillière, 1889, 159 pp.

[H]

1227. ***La curiosité. Journal de l'occultisme scientifique.***
Vols. 1–36; 1889–1924.

Edited by Ernest Bosc and later continued as: *La curiosité. Revue des sciences psychiques.* [P]

1228. ***Delboeuf, Joseph Remi Léopold.***
Le magnétisme animal: à propos d'une visite à l'école de Nancy. Paris: Germer Baillière, 1889, 128 pp.

Delboeuf was a successful practitioner of medical hypnotism and an acute observer of the psychological factors involved in hypnotic phenomena. An independent thinker who wanted to observe matters for himself, he visited both the Salpêtrière and Nancy to see how they conducted their hypnotic experiments. In this work he begins with a description of his earlier visit to the Salpêtrière, expressing strong criticism of the methods used there. He was keenly aware of the suggestibility of the Salpêtrière subjects and felt their responses in states of hypnotism were unknowingly conditioned by the observers. His view of the work of Liébeault and Bernheim is more sympathetic, although he believes they underrate the intelligence and awareness of their subjects. The book also has a section on the views of Liégeois on moral and legal aspects of hypnotism. [H]

1229. *Dessoir, Max.*
Das Doppel-Ich. Berlin: n.p., 1889, 42 pp.

Discusses the double (or multiple) streams of consciousness revealed by experiments in dissociation to be present in all people. Evidence for this phenomenon is taken from the data of hypnotism, hysteria, and the automatisms (such as automatic writing) associated with normal individuals. This significant work was expanded in a number of editions. [H]

1230. *Drayton, H. S.*
Human Magnetism. Its Nature, Physiology and Psychology. Its Uses, as a Remedial Agent, in Moral and Intellectual Improvement, etc. New York: Fowler & Wells Company, 1889, iv + 5–168 pp.

Drayton, a member of the New York Medical Society, had a particular interest in nervous conditions. In this book he writes about the practical application of both animal magnetism and hypnotism. In the latter area he was strongly influenced by the school of Charcot. [H]

1231. *Du Prel, Karl Ludwig August Friedrich Maximillian Alfred.*
Das hypnotische Verbrechen und seine Entdeckung. Munich: Verlag der Academischen Monatschefte, 1889, 105 pp.

[H]

1232. *Forel, August.*
Der Hypnotismus: seine Bedeutung und seine Handhabung. In kurzgefasster Darstellung. Stuttgart: Ferdinand Enke, 1889, 88 pp.

Forel, Professor of Psychiatry at the University of Zurich and Director of the Burghölzli mental hospital, was well known for his work on the anatomy of the brain. He became interested in hypnotic phenomena and in 1887 visited Bernheim to study his technique. Forel became very adept at hypnotism and used it in his psychiatric work and as a treatment for physical illnesses. The main part of this work appeared as an article in the *Zeitschrift für die Gesammte Strafrechtswissenschaft* and is concerned largely with the forensic aspects of hypnotism. Following the approach of the Nancy School of hypnotism, Forel emphasizes the place of suggestion in the induction of hypnotism and the implications of suggestion for personality responsibility on the part of the hypnotized. The book became very well known and was often revised and augmented, soon reaching editions of several hundred pages. [H]

1233. *Gibier, Paul.*

Physiologie transcendantale. Analyses des choses. Essai sur la science future, son influence certaine sur les religions, les philosophies, les sciences et les arts. Paris, Philadelphia and Madrid: Dentu, J. B. Lippincott, and Fuentes y Capdeville, (1889), 270 pp.

Gibier was a French biologist of some note who had strong spiritualist beliefs. In this book he attempts to present a comprehensive theory of the nature of the universe, in which there is matter and spirit connected by an "animic" or ethereal energy. [P]

1234. *Gilles de La Tourette, George Albert.*

Documents satiriques sur Mesmer. (Paris): n.p., (1889).

A very unusual and rare publication containing thirteen reproductions of eighteenth-century engravings satirically depicting Mesmer and animal magnetism. There is an accompanying explanation of the engravings which apparently came from the collection of Charcot. [H]

1235. *Janet, Pierre.*

L'automatisme psycholoqique: essai de psychologie expérimentale sur les formes inférieures de l'activité humaine. Paris: Félix Alcan, 1889, 496 pp.

A major contribution to the scientific study of human automatism. It is a fine piece of research that influenced virtually all later research on the subject. At the same time, this book can be considered the summation and culmination of the best of the research on animal magnetism, somnambulism, and hypnotism that had preceeded it. Although his approach was innovative, Janet was aware of his indebtedness to that tradition. Written as a thesis for a doctorate in philosophy at the Sorbonne, this was Janet's first full length book, the continuation of a line of investigation begun in his three important articles published in 1886, 1887, and 1888 in the *Revue philosophique* (*see* entry numbers 1149, 1177, 1212). Here he examines those human acts which, while bearing the earmarks of intelligence, yet bypass the will and escape conscious awareness. Janet calls these acts "psychological automatisms." To clarify the subject, he divides the book into two parts: the first dealing with total automatisms, and the second with partial automatisms. The first part treats of catalepsy and somnambulism, and contains an important chapter on suggestion. The second part includes discussions of distraction and subconscious acts, post-hypnotic suggestion, hysterical anesthesias, and "psychological disaggregation." As examples of the latter, Janet examines the topics of madness, fixed ideas, hallucinations, possessions, and the phenomena of spiritualism. *L'automatisme psychologique,* with its broad historical scope and original approach to research, provides an important link between the descriptive psychology of the magnetizers who studied somnambulistic automatisms and the scientific psychology of the late nineteenth century. [H]

1236. *Langsdorff, Georg von.*

Zur Einführung in das Studium des Magnetismus, Hypnotismus, Spiritualismus nebst Kritik von drei Broschuren und eines Buches des magnetischen Heilers Dr. Jul. Ed. Timmler in Altenburg. Berlin: Karl Siegismund, 1889, 28 pp.

A pamphlet on somnambulism and its relationship to spiritualistic mediumship and other extraordinary phenomena. [H & P]

1237. *Liégeois, Jules.*

De la suggestion et du somnambulisme dans leurs rapports avec la jurisprudence et la médecine légale. Paris: O. Doin, 1889, viii + 758 pp.

Liégeois, an adherent of the Nancy School of hypnotism, was more interested in the forensic than the physiological aspects of hypnotism. Following the techniques of Bernheim, Liégeois experimented with suggestion and its implications for personal responsibility on the part of the hypnotized subject. He discovered that fifteen to twenty percent of his subjects could be placed in a somnambulistic state, in which they were extremely susceptible to hypnotic and posthypnotic suggestion. While his subjects were somnambulistic, Liégeois was able to create all sorts of hallucinations and induce various degrees of amnesia. He also experimented with suggestion in the waking state, obtaining striking results. Liégeois carried out other experiments with hypnotic suggestion, discovering that he was able to get subjects to carry out suggestions at a distance in space (by using the telephone) and at a distance in time (finding that suggestions could be executed long after they were given). Liégeois's experimentation with criminal suggestions were imaginative and bold, at times crossing beyond the boundaries of human decency. He succeeded in inducing hypnotized individuals to commit what they believed were criminal acts and discovered that, when awakened, they could experience complete amnesia for those actions. He concluded that it is possible for one person, through hypnotic suggestion, to induce another to commit criminal acts which are contrary to the conscience of the hypnotized person. In these cases, the person who made the suggestions should be considered the responsible person and the hypnotized subject should be exonerated. To obtain realism in his experiments, Liégeois seemed ready to go to great lengths, sometimes showing little concern for the dignity or distress of his subjects. Nevertheless, Liégeois's work was extremely helpful for forensic psychologists and retains its value today. [H]

1238. ***Lombroso, Cesare and Ottolenghi, S.***

Nuoovi studi sull 'ipnotismo e sulla credulità. Turin: Bocca, 1889, 52 pp.

[H & P]

1239. ***Marrin, Paul.***

L'hypnotisme théorique et pratique, comprenant les procédés d'hypnotisation. Paris: Ernest Kolb, (1889) x + 336 pp.

Marrin accepts the three phases of hypnotism taught by Charcot, but divides the third phase into three: hyperesthesia, exaltation, and intelligence. His book contains interesting comments on Faria's importance to the history of hypnotism and includes criticisms of the work of Luys. [H]

1240. ***Moll, Albert.***

Der Hypnotismus. Berlin: H. Kornfeld, 1889, vii + (1) + 279 + (1) pp. ENGLISH: *Hypnotism.* Translated from the second German edition. London: Walter Scott Ltd., 1890.

Moll was a physician and psychiatrist in Berlin whose experimentation and research in hypnotism earned worldwide respect. He worked in cooperation with a number of other investigators in the field, and in the introduction to this book, he expresses his thanks to August Forel and Max Dessoir. Moll begins this work with a brief description of the history of hypnotism in France and Germany, and moves on to an examination of the methods used for hypnotic induction. He then discusses the characteristics of the state of hypnotism, describing both physiological and psychological symptoms. In this section Moll has a discussion of suggestion in which he credits the mesmerists as the discoverers of post-hypnotic suggestion. Moll next takes up the theory of hypnotism and the possibility of simulating the hypnotic state. After examining the medical uses of hypnotism and legal aspects, Moll inquires about the status of animal magnetism, pointing out errors in experiments that have been cited as confirmation of the existence of a magnetic fluid. This work became a very popular general treatise on hypnotism, going through many revised editions over the next twenty-five years. [H & P]

1241. ***Morand, J. S.***

Le magnétisme aimal. Étude historique et critique. Paris: Garnier frères, 1889, viii + 498 pp.

[H]

1242. ***Morselli, E. and Tanzi, E.***

Contributo sperimentale alla fisio-psicologia dell'ipnotismo. Ricerche sul polso e sul respiro negli stati suggestivi dell'ipnosi. Milan: Dumolard, 1889, 27 pp.

A careful experimental study of the physiological phenomena of hypnotism. [H]

1243. [*Moses, William Stainton, pseudonym: M. A. (Oxon.)*]

Second Sight: Problems Connected with Prophetic Vision and Records Illustrative of the Gift, Especially Derived from an Old Work not now Available for General Use. London: London Spiritualist Alliance, etc., (1889), 35 pp.

[P]

1244. *Myers, Frederick William Henry and Barrett, William Fletcher.*

"D. D. Home, His Life and Mission." *Society for Psychical Research Journal* 4(1889): 101–36.

In the process of reviewing Madame Home's biography (*see* Madame D. D. Home, 1888) of her husband, the famous medium Daniel Dunglas Home (1833–1886), Myers and Barrett use the occasion to publish eyewitness testimony of some of his more remarkable mediumistic feats. They discuss all possible natural explanations for the things he did, including concealed props, sleight of hand, and hypnotic suggestion. They conclude that Home's phenomena, while not miraculous, must be considered supernormal (not explainable through known physical laws). [H]

1245. *Pélin, Gabriel.*

Homo duplex: note physiologique sur l'organisme humain presentée aux Facultés de Médecine et à l'Académie des Sciences. Paris: L. Sauvaitre, 1889.

Pélin rejects the supernatural view and explains spiritualism and madness by a sort of double life in man. [H & P]

1246. *Raue, Charles Godlove.*

Psychology as a Natural Science Applied to the Solution of Occult Psychic Phenomena. Philadelphia: Porter & Coates, 1889, 541 pp.

An attempt to understand the mental and physical phenomena of mediumship in terms of a comprehensive human psychology. [P]

1247. *Renterghem, Albert Willem van and Eeden, van.*

Clinique de psychothérapie suggestive fondée à Amsterdam. Brussels: A. Manceaux, 1889, iv + 7–92 pp.

A lengthy version of this work was published in Paris in 1894. [H]

1248. *Richmond, Almon Benson.*

What I saw at Cassadaga Lake: 1888. Addendum to a Review in 1887 of the Seybert Commissioners' Report. Boston: 1889, Colby & Rich, 163 pp.

In answer to the scepticism of the Seybert Commissioners Report (1887). Richmond published this report of spiritualistic phenomena that he considered genuine. It is a collection of descriptions of events at the spiritualist town of Lily Dale (Cassadaga Lake in New York State), newspaper articles, editorials, and letters. (*See* Richmond: *What I saw at Cassadaga Lake*, entry number 1218.) [P]

1249. *Simonin, Amédée H.*

Solution du problème de la suggestion hypnotique. La Salpêtrière et l'hypnotisme. La suggestion criminelle. La loi doit intervenir. Paris: E. Dentu, 1889, 133 pp.

[H]

1250. *Touroude, A.*

L'hypnotisme, ses phénomènes et ses dangers. Étude. Paris: Bloud et Barral, (1889), 277 pp.

[H]

1251. *Tuckey, Charles Lloyd.*

Psycho-Therapeutics; or Treatment by Sleep and Suggestion. London: Baillière, Tindall and Cox, 1889, xii + 80 pp.

This treatise on hypnotism as a therapeutic agent, which went through many greatly expanded editions, emphasizes the influence of the mind on the body and the beneficial effects of hypnotic suggestion on many kinds of diseases. [H]

1890 – 1899

1890

1252. ***Aide, Hamilton.***

"Was I Hypnotized?" *The Nineteenth Century* 27 (1890):576–81.

Basing his account on contemporary notes, Aide writes of a séance with the famous medium D. D. Home that he attended many years earlier. Stating that he is not a spiritualist and does not believe that spirits produced the astounding physical events that he observed in Home's presence in full lighting, he poses the question of whether he and the other witnesses were hypnotized in some unknown way and in that way deceived about what had taken place. [P]

1253. ***Aksakov, Alexander Nikolaevich.***

Animismus und Spiritismus. Versuch einer kritischen prufung der mediumistischen Phanomene mit besonderer Berucksichtigung der Hypothesen der Hallucination und des Unbewussten. Als Entgegnung auf Dr Ed. v. Hartmann's Werk: "Der Spiritismus." Translated from the French by Gregor Constantin Wittig. 2 vols. Leipzig: Oswald Mutze, 1890, xlv + (2) + 768 pp.

Aksakov, Imperial Counsellor to the Czar of Russia, was a Swedenborgian, a spiritualist, and an important psychical researcher. Because of the censorship of spiritualistic literature in Russia, Aksakov centered his work in Germany. There he founded the *Psychische Studien* (later called the *Zeitschrift für Parapsychologie*), for decades Germany's chief periodical of psychical research. Aksakov carried out numerous investigations of mediums and translated many works on psychical research into Russian. *Ani-*

mismus und Spiritismus, an important work in the field of psychical research, takes up the issue of whether or not the dead communicate with the living through mediums, and looks into the genuineness of other psychic phenomena. Taking Eduard von Hartmann's *Der Spiritismus* (1885, entry number 1113), as his point of departure, Aksakov examines the history of the "anti-spiritists" and especially von Hartmann's contention that all spiritualistic phenomena can be explained in terms of purely human powers, natural and paranormal. Conceding that natural and paranormal human powers are an important element in the phenomena of spiritualism, Aksakov nonetheless insists that some phenomena can only be explained in terms of the action of spirits. In this he includes both the extra-corporeal action of the spirit of a living person (animism) and outside intervention from disembodied spirits (spiritism). In marshalling his evidence, he brings to bear both his own and others' investigations. *Animismus und Spiritismus* was originally written in Russian and then translated into French in 1890. The German version was translated from the French. Neither the Russian or French first editions appear to be extant. The earliest available French version is dated 1895 (*Animisme et spiritisme*, Paris: G. Leymarie). [P]

1254. ***Bentivegni, Adolf von.***
Die Hypnose und ihre civilrechtliche Bedeutung. Leipzig: E. Gunther, 1890, 66 pp.

[H]

1255. ***Binet, Alfred.***
On Double Consciousness. Experimental Psychological Studies. Chicago: Open Court, 1890, 93 pp.

A significant contribution to the experimental study of the subconscious. Binet was the director of the laboratory of physiological psychology at the Sorbonne and a psychologist of international renown. Although best known for his work on intelligence testing, his contributions to the study of artificial somnambulism and psychological automatisms are significant. This collection of essays, originally written in English and contributed to *The Open Court* during 1889–1890, reports studies of subconscious states and secondary consciousness carried out with both hysterical and healthy individuals. Binet believed that his work proved conclusively that doubling of consciousness takes place in both hysterical and healthy subjects. [P]

1256. ***Bonjean, Albert.***
L'hypnotisme, ses rapports avec le droit et la thérapeutique, la suggestion mentale. Paris: Germer Baillière and Félix Alcan, 1890, ix + 376 + (4) pp.

[H]

1257. ***Congrès International de 1889. Le magnétisme humain appliqué au soulagement et à la guérison des malades, Rapport Général, d'après le compte rendu des séances du Congrès.***
Paris: Georges Carré, 1890.

The volume, with a preface by Camille Flammarion, is a collection of discourses, mostly by practicing French magnetizers. The emphasis is on the healing power of magnetism and its physical effects. [H]

1258. ***Congrès international de psychologie physiologique. Comptes rendus.***
Paris: n.p., 1890, 159 pp.

Proceedings of the first of the international congresses of psychology. This congress devoted considerable attention to research on hypnotic phenomena. [H]

1259. ***Delboeuf, Joseph Remi Leopold.***
Magnétiseurs et médecins. Paris: Ancienne Librairie Germer Baillière, 1890, 115 + (1) pp.

Concerned to protect the freedom of action of those who support and practice hypnotism and animal magnetism, Delboeuf defends the celebrated "Donato" (whose real name was Alfred Edouard d'Hont) against the condemnation of Ladame. [H]

1260. ***Dessoir, Max.***
Erster Nachtrag zur Bibliographie des modernen Hypnotismus. Berlin: Carl Duncker, 1890, 44 pp.

Organized in terms of the categories employed in Dessoir's *Bibliographie des mordernen Hypnotismus* (1888, *see* entry number 1202), this supplement contains 382 additional titles published between 1881 and 1890. [H]

1261. ***Du Prel, Karl Ludwig August Friederich Maximillian Alfred.***
Studien aus dem Gebiete der Geheimwissenschaften. 2 vols. Leipzig: W. Friedrich, 1890, viii + 252; vi + 247 pp.

[P]

1262. ***Durville, Hector.***
Le magnétisme humain considéré comme agent physique. Paris: Librairie du magnétisme, 1890, 36 pp.

[H]

1263. ***Flournoy, Theodore.***
Métapsychique et psychologie. Geneva: H. Georg, 1890, 133 pp.

Flournoy was professor of psychology at the University of Geneva, author of a number of medical and psychological works, founder of the *Archives de Psychologie*, and a psychical researcher of some note. This is his first book in the area of psychical research. [P]

1264. ***Foveau de Courmelles, François Victor.***
L'hypnotisme. Paris: Hachette, 1890, (4) + vi + 326 + (2) pp. ENGLISH: *Hypnotism.* Translated by Laura Ensor. London: George Routledge and Sons, 1891.

A popularly written description of animal magnetism and hypnotism. Although the author, a physician, drew his information from interviews with magnetizers, medical practitioners of hypnotism, and academics, he claims to have been able to test many of the assertions concerning hypnotism for himself. The first part of the book contains a history that is important for its description of the school of hypnosis of the Hospital de la Charité. [H]

1265. ***Foveau de Courmelles, François Victor.***
Le magnétisme devant la loi. Paris: Georges Carré, 1890, 40 pp.

[H]

1266. ***Gardy, Louis.***
Cherchons. Réponse aux conférences de M. le Professeur Emile Yung sur le spiritisme. Geneva: R. Burkhardt, 1890.

[P]

1267. ***Giornale del magnetismo ed ipnotismo.***
Vol. 1; 1890.

Continued as: *Magnetismo ed ipnotismo.* Vols. 2–3; 1891–1892. In turn, this was continued as: *l'Ipnotismo.* Vols. 4–5; 1893–1894. This journal was the organ of the Società medico-psicologica Italiana and published in Florence. [H]

1268. ***Home, Madame Daniel Dunglas.***
The Gift of D. D. Home. London: Kegan Paul, Trench, Trubner and Co., 1890, viii + 388 pp.

Two years after her first book (*D. D. Home: his life and Mission*, 1888, entry number 1206) on her famous medium husband, Daniel Dunglas Home (1833–1886), Madame Home published this new collection of evidence of his powers, the first published records of mediumistic sessions Home held in America very early in his career. [P]

1269. [*Huguеny, Charles*].

Recherche des bases qui la constituent et des lois qui gouvernent l'univers physique et moral et l'homme en particulier, par un étudiant de 90 ans. Nancy: Berger-Levrault, 1890, xxx + 128 pp.

Includes, among other things, the reproduction of important documents related to the beginning of the Société harmonique des amis réunis de Strasbourg in 1785, a society that was very influential in promoting animal magnetism in its early years. [H]

1270. *James, William.*

"The Hidden Self." *Scribner's Magazine* 7 (1890):361–73.

William James was Professor of Psychology and Philosophy at Harvard University and a scholar, thinker, and writer of international renown. He founded the first laboratory of experimental psychology in the United States (1876) and his monumental *Principles of Psychology* (1890, entry number 1271) was a major force in shaping modern American psychology. As a philosopher, he developed the doctrine of pragmatism and wrote the extremely influential *Varieties of Religious Experience* (1902). James was also an important psychical researcher, helping to found the American Society for Psychical Research in 1884 and later serving a term (1894–1895) as president of the Society for Psychical Research in London. His interest in psychical research evolved directly from his psychological studies and his desire to pursue every possible avenue for understanding the nature of the human psyche and its hidden dynamics. Although sceptical when he began his first significant project in psychical research, an investigation in 1885 of the famous American medium, Leonora Piper (1859–1950), James eventually concluded that she possessed some power as yet unexplained. In this important article, James discusses the findings of Janet, Binet, and others concerning the existence of the subconscious. Emphasizing the importance of discoveries of a second stream of consciousness which can be tapped in certain conditions such as the state of hypnotism, James sees in this secondary or, as he calls it, "submerged consciousness" a key to understanding such diverse phenomena as possession and multiple personality. He calls the coexistence of a secondary self or selves with the primary self a "splitting of the mind," and urges further investigation to determine more about the relationship between hypnotic trance and these subconscious states. [H]

1271. *James, William.*

The Principles of Psychology. 2 vols. New York: Henry Holt, 1890, xii + 689; vi + 688 + (14) pp.

The *Principles* is a milestone in the field of psychology and an important resource for those who study hypnotism, consciousness, and the psychological aspects of psychical research. The chapter on "Consciousness of Self," with its discussion of the normal plurality of selves, delusions, alternating selves, mediumship, and possession, is a classical piece of psychological writing, developing a keen examination of experience into a striking psychological synthesis. [H]

1272. *Lehmann, Alfred Georg Ludwig.*

Die Hypnose und die damit verwandten normalen Zustände. Vorlesungen gehalten auf der Universität Kopenhagen im herbste 1889. Leipzig: O. R. Reisland, 1890, viii + 194 pp.

[H]

1273. *Luys, Jules Bernard.*

Leçons cliniques sur les principaux phénomènes de l'hypnotisme dans leurs rapports avec la pathologie mentale. Paris: Georges Carré, 1890, xv + (1) + 287 + (1) pp., with 12 plates.

An important study of the production of psychotic symptoms in a non-psychotic subject through hypnotic suggestion. Luys was able to reproduce the principal symptoms of the psychoses, including hallucinations, delirium and irresistible compulsions merely by suggesting them to an entranced individual. [H]

1274. *Myers, Frederick W. H., Lodge, Oliver, Leaf, Walter, and James, William.*

"A Record of Observation of Certain Phenomena of Trance." *Society for Psychical Research. Proceedings* 6 (1890):436–659.

A detailed examination of the mediumship of one of the most remarkable mental mediums of the last century: Mrs. Leonora Piper (1859–1950), a Boston housewife. Observed in séance over decades and even followed for a lengthy period by detectives hired by the Society for Psychical Research, Mrs. Piper was the most thoroughly investigated medium in the annals of psychical research. Although she produced astounding information about hundreds of deceased people with whom she had had no acquaintance, at times even providing information presumably unknown to any living person at the time of the séance, no hint of trickery or deceit was ever discovered. This article is the first lengthy study of her mediumship and contains details of information given by the medium while in trance, verification of that material, and a careful account of the time, place, and circumstance of the séances. [P]

1275. *Preyer, Wilhelm Thierry.*

Der Hypnotismus: Vorlesungen gehalten an der K. Friedrich-Wilhelms-Universität zu Berlin. Vienna: Urban & Schwartzenberg, 1890, vi + 217 + (1) pp.

A very good presentation of the history and description of the nature of hypnosis. The history, starting with Mesmer, contains an appreciative look at Braid. More importantly, Preyer also includes his German translation of a treatise by Braid on the difference between nervous sleep (hypnotism) and normal sleep that he obtained in 1881 from Braid's estate. The original has never been found and no English version was ever published. [H]

1276. ***Psiche. Ipnotismo, magnetismo, spiritismo.***
One vol. only; 1890.

Published in Rome and edited by Sara Carlotta Di Capua and then Efisio Ungher, this periodical continued as a spiritualist organ under the title *Sfinge.* [H & P]

1277. ***Revue d'hypnologie théorique et pratique dans ses rapports avec la psychologie, les maladies mentales et nerveuses.***
1890.

See *L'encéphale* (1881, entry number 1039). [H]

1278. ***Rose, W.***
Mesmerism or Animal Magnetism, Hypnotism and Thought Reading. Lowestoft: Mickleburgh & Co., (1890) 16 pp.

[H & P]

1279. ***Schroeder, H. R. Paul.***
Die Heilmethode des Lebensmagnetismus. Theorie und Praxis besprochen und mit eine Nachweise über den wesentlichen Unterschied zwischen Hypnotismus und Heilmagnetismus versehen. Leipzig: n.p., 1890, 104 pp.

[H]

1280. ***Smith, R. Percy and Myers, A. T.***
"On the Treatment of Insanity by Hypnotism." *Journal of Mental Science* 36 (1890):191–212.

[H]

1281. ***Stefanoni, Luigi.***
Magnetismo e ipnotismo svelati. Storia critica. Rome: Voghera Carlo, 1890, vii + 475 pp.

A critical study of the phenomena of animal magnetism as described by practitioners such as Guidi, D'Amico, Pilati, and Zanardelli. Citing evidence he considers indicative of collusion and trickery on the part of the practitioners, the author concludes that the phenomena are not genuine. [H & P]

1891

1282. *Annales de psychiatrie et d'hypnologie.*
1891.

See *L'encéphale* (1881, entry number 1039). [H]

1283. *Annales des sciences psychiques. Recueil d'observations et d'expériences paraissant tous les deux mois (et mensuellement consacré aux recherches expérimentales et critiques sur les phénomènes de télépathie, lucidité, prémonition, médiumnité, etc.)*
Vols. 1–29; 1891–1919.

Continued as: *Revue métapsychique*. 1919–1925+. The most important French journal on psychical research, founded by Charles Richet and X. Darieux. From 1905, the *Annales* was edited by Cesar Baudi de Vesme, after absorbing his *Revue des études psychiques*. An English translation edition of the *Annales* called *Annals of Psychic Science* was published from 1905–1910. In 1920 the *Annales* was replaced by the *Revue métapsychique*, which took over for the *Institut métapsychique international. Bulletin*, and functioned as the official publication of the Institut Métapsychique International. [P]

1284. *Bernheim, Hippolyte.*
Hypnotisme, suggestion, psychothérapie; études nouvelles. Paris: Octave Doin, 1891, ii + 518 pp.

In this work, Bernheim places an even greater emphasis on the importance of suggestion than he did in his earlier writings. After an historical review of the use of suggestion in healing among the Egyptians and the Hebrews, Bernheim notes that successful treatment by suggestion can be carried out in the waking state and need not involve hypnotism at all. He follows this with important observations on hallucinations and amnesia, and discusses the problem of the moral responsibility of a subject in the hypnotic state. The final section, the largest of the book, presents observations on the treatment by suggestion of a variety of illnesses ranging from organically based conditions to hysteria. [H]

1285. *Bettoli, Parmenio.*
Rivelazioni ed insegnamento del givoco col simulare i fenomeni magnetici e ipnotici. Milan: Verri, 1891, 70 pp.

A description of tricks used by individuals who claimed to be able to produce mediumistic phenomena. [H & P]

1286. ***Delboeuf, Joseph Remi Leopold.***
L'affaire des magnétiseurs de Braine-le-Chateau. Examen critique du la rapport des médecins experts. Liége: Emile Pierre & Frère, 1891, 8 pp.

[H]

1287. ***Denis, Léon.***
Après la mort. Exposé de la philosophie des esprits, ses bases scientifiques et expérimentales, ses consequences morales. Paris: Librairie des sciences psychologiques, 1891, 432 pp.

[P]

1288. **[*Donavan, ———*].**
Revelations of a Spirit Medium; or Spiritualistic Mysteries Exposed. A Detailed Explanation of the Methods Used by Fraudulent Mediums. St. Paul, Minnesota: Farrington & Co., 1891, vi + 7–324 pp.

A detailed account of methods used by bogus spiritualist mediums to deceive their audiences. Written by a fake medium, himself well acquainted with those practices. Although there has been some dispute concerning the name of the author, Donavan seems the most likely choice. The book became very rare, apparently because many copies were destroyed by those trying to protect their trade secrets. [P]

1289. ***Ferret, J.***
La cause de l'hypnotisme. 2 ed. Paris: Téqui, 1891, 366 pp.

Combines descriptions of hypnotic séances with speculations about the nature of will and its effects in the spiritual realm. No information is available on the first edition. [H & P]

1290. ***Fronda, R. and Grimaldi, A.***
Trasmissione del pensiero e suggestione mentale. Studio sperimentale e critico . . . e seguito da alcune indagini fatte sullo stesso soggetto a richiesta. Naples: n.p., 1891, 83 pp.

[P]

1291. **[*Gilles de la Tourette, George Albert*].**
Nouveaux documents satiriques sur Mesmer. (Paris): n.p., (1891).

Containing reproductions of three eighteenth-century engravings satirically portraying Mesmer and animal magnetism, this work is undoubtedly the sequel to the earlier (1889, entry number 1234) work by Gilles de la Tourette which contains thirteen such reproductions. Like the earlier work, it is extremely rare, apparently not to be found in any public library or collection. [H]

1292. ***Grossmann, Jonas [ed.].***
Die Bedeutung der hypnotischen Suggestion als Heilmittel. Gutachten und Heilberichte der hervorragendsten wissenschaftlichen Vertreter des Hypnotismus der Gegenwart. Berlin: Bong & Co., 1891, xviii + 160 pp.

[H]

1293. ***Hartmann, Eduard von.***
Die Geisterhypothese des Spiritismus und seine Phantome. Leipzig: Wilhelm Friedrich, 1891, 126 pp.

An answer to the criticism of Aksakov concerning von Hartmann's original book on spiritualism. In that book (*Der Spiritismus*, 1885, entry number 1113) von Hartmann had argued against the position that spiritualistic manifestations, such as messages from the dead and clairvoyant knowledge, proved the existence of discarnate spirits. In his *Animismus und Spiritismus* (1890, entry number 1253) Aksakov had taken issue with some of von Hartmann's points and suggested that von Hartmann had not taken certain data into account. Here von Hartmann presents further arguments in support of his original position. [P]

1294. ***Kiesewetter, Carl.***
Geschichte des neureren Occultismus. Geheimwissenschaftliche Systeme von Agrippa von Nettesheym bis zu Carl du Prel. Leipzig: Wilhelm Friedrich, (1891), xiv + 801 pp.

Investigates "modern occultism," the German equivalent of psychical research. An important reference work for the German tradition in the study and use of animal magnetism. [H & P]

1295. ***Kingsbury, George Chadwick.***
The Practice of Hypnotic Suggestion; Being an Elementary Handbook for the Use of the Medical Profession. Bristol: J. Wright, 1891, 206 pp.

[H]

1296. ***Laurent, Emile.***
Les suggestions criminelles. Viols. Faux et captations. Viols moraux, les suggestions en amour. Gabrielle Fenayrou et Gabrielle Bompard. Lyon and Paris: Storck and Société d'éditions scientifiques, 1891, 56 pp.

[H]

1297. ***Liébeault, Ambroise August.***
Thérapeutique suggestive: son mécanisme. Propriétés diverses du sommeil provoqué et des états analogues. Paris: Octave Doin, 1891, vii + 308 pp.

Liébeault continues the discussion of suggestion undertaken in previous works, relating the effectiveness of suggestion in treatment of disease to the fluctuation of "nervous force" it produces in the organism. Liébeault also discusses legal aspects of hypnotism, "zoomagnetism," and "lucidity" or clairvoyance in the hypnotic state (a subject that he had not previously examined in detail). In his chapter on "zoomagnetism," Liébeault contradicts the conclusions he had reached in his *Étude sur le zoomagnétisme* (1883, entry number 1068). In that work he had acknowledged the existence of a physical, vibratory agent which, along with suggestion, could be a source of healing. He had become convinced of the reality of this agent through experiments done with infants, who were benefited through the application of animal magnetic techniques. Here Liébeault admits that further experimentation had revealed to him that, astonishing as it may seem, even infants less than two years old have sufficient awareness to respond to suggestion. Liébeault concludes the book with his "confession of a hypnotizer" previously published (1886, entry number 1156) in the *Revue de l'hypnotisme*. [H]

1298. ***Luys, Jules Bernard and Encausse, Gérard.***
Du transfert à distance à l'aide d'une couronne de fer aimanté, d'états névropathiques variés, d'un sujet à l'état de veille sur un sujet à l'état hypnotique. Clermont (Oise): Daix frères, 1891, 4 pp.

[H]

1299. ***Magnetismo ed ipnotismo.***
1891.

See *Giornale del magnetismo ed ipnotismo* (1890), entry number 1267. [H]

1300. ***Marryat, Florence.***
There is No Death. London: Farran, 1891, 365 pp.

Marryat was an English medium and author who knew many of the most famous mediums of the day. This book describes sittings with, among others, William Eglinton, Mrs. Guppy Volckman, and William Fletcher. [P]

1301. ***Myers, Frederic William Henry.***
"Science and a Future Life." *The Nineteenth Century* 28 (1891):628–47.

On the relationship between science and psychical research. To the question of whether science can have anything to say about the issue of human survival after death, Myers answers that a scientific attitude and scientific methods can be brought to bear upon the study of supernormal phenomena that seem at first glance to indicate that people continue to exist after death. Myers believes that findings concerning telepathy are relevant to this issue because, if human beings can communicate at a distance without the benefit of the bodily senses, then perhaps communication with those who are at the greater distance involved with death is also possible. [P]

1302. *Pitres, Albert.*

Leçons cliniques sur l'hystérie et hypnotisme: faites à l'Hôpital Saint-André de Bordeaux. 2 vols. Paris: Octave Doin, 1891, (12) + 531; (4) + 551 pp.

[H]

1303. *Regnier, Louis Raoul.*

Hypnotisme et croyances anciennes. Paris: Lecrosnier et Babe, 1891, xxiii + 223 pp.

The author, an adherent of the Salpêtrière school of hypnotism, provides a long introduction focusing on the nature and effects of hypnotism, with many photographic plates showing individuals in various hypnotic attitudes. The main section of the book is concerned with the history of hypnotism, with emphasis on the traces of hypnotic practices to be found in ancient religious and occult lore, and is followed by a lengthy bibliography. [H]

1304. *Reichel, Willy.*

Der Heilmagnetismus. Berlin: K. Siegismund, (1891), 47 pp.

[H]

1305. *Reichenbach, Karl Ludwig von.*

Ein Schwerer sensitiv-somnambuler Krankheitsfall geheilt ausschliesslich mittelst einfacher Anwendung der Gesetze des Odes. Für Physiker, Physiologen, Psychologen, Psychiatriker und insbesondere für Ärzte. Hrgb. von A. Freiherr von Schrenck-Notzing. Leipzig: Ambr. Abel, 1891, vii + (1) + 160 pp.

[H]

1306. *Rochas d'Aiglun, Eugène Auguste Albert de.*

Le fluide des magnétiseurs. Précis des expériences du Baron de Reichenbach sur les propriété physiques et physiologiques, classées et annotées par le Lt-Colonel A. de Rochas. Paris: Georges Carré, 1891, 180 pp.

[H]

1307. *Stead, William Thomas.*

Real Ghost Stories: a Record of Authentic Apparitions. Being the Christmas Number of the Review of Reviews. Collated and Edited by W. T. Stead. London: Carlyle, 1891, 110 pp.

Stead was a famous British journalist who was also a champion of spiritualism. He was the editor-founder of the *Review of Reviews* and of *Borderland*, a quarterly devoted entirely to psychical matters. This book, along with *More Ghost Stories* published the following year, constitutes what is probably the first systematic collection of cases of apparitions, hauntings, astral projection, premonitions, and clairvoyance in the English speaking world. [P]

1308. *Tarchanoff, Jean de or Tarkhanov, Ivan Romanovich.*

Hypnotisme, suggestion et lecture des pensées. Tr. du Russe par Ernest Jaubert. Paris: G. Masson, 1891, viii + 163 + (1) pp.

The author taught at the Imperial Academy of Medicine in St. Petersburg. After examining hypnotism and suggestion, Tarchanoff takes up the issue of thought reading and concludes that the whole phenomenon is based on self deception. Basing his conclusions on work done by an American investigator named Bird, Tarchanoff argues that the supposed thought reader in fact gains his information from the minute unconscious muscular movement of the person who is supposed to be communicating telepathically. [H & P]

1309. *Weatherly, Lionel Alexander.*

The Supernatural? With Chapters on Oriental Magic, Spiritualism, and Theosophy, by J. N. Maskelyne. Bristol: J. W. Arrowsmith, (1891), xv + 273 pp.

Concerns itself with sense deceptions, how they are produced and how they may account for such things as visions of ghosts and the apparently marvellous phenomena of mediumship. Both Weatherly and Maskelyne believe that these things may be explained in terms of hallucinations or trickery. Maskelyne was a magician by trade and claimed to be able to reproduce any mediumistic feat through the art of conjuring. [P]

1892

1310. [*Bathurst, James*].

Atomic-Consciousness. An Explanation of Ghosts, Spiritualism, Witchcraft, Occult Phenomena, and All Supernormal Manifestations. Whimple, Exeter: Harris & Haddon, 1892, 284 pp.

A highly original, if eccentric, attempt to account for everything from ghosts to mesmerism in terms of a philosophy of a universal mental force that pervades the universe and connects all minds. [H & P]

1311. *Binet, Alfred.*

Les altérations de la personnalité. Paris: Baillière, 1892, viii + 323 + (1) pp. ENGLISH: *Alterations of Personality.* Translated by Helen Green Baldwin. With Notes and a Preface by J. Mark Baldwin. New York: D. Appleton and Co., 1896.

Binet's second important work on dissociated subconscious states (see *On Double Consciousness*, 1890, entry number 1255). Binet provides a summary of previous work in this area and discusses his own ongoing experiments. He divides the material into sections on "successive personalities" and "coexistent personalities." The former contains a description of induced and natural somnambulism, with an account of Azam's famous case of multiple personality, Félida X. The latter section includes important material on distraction, unconscious action, automatic writing, and the plurality of consciousness in healthy subjects. The final section of the book presents descriptions of Binet's experiments with suggestion and induced hallucination, with a small discussion of division of personality in spiritism. [H & P]

1312. *Brofferio, Angelo.*
Per lo spiritismo. Milan: Domenico Briola, 1892, 16 pp.

[P]

1313. *Delboeuf, Joseph Remi Leopold.*
L'hypnotisme devant les chambres legislatives belges. Paris, Brussels, and Liége: Félix Alcan (Paris), P. Weissenbruch (Brussels), Ch.-Gus. Desoer (Liége), 1892, 80 + (1) pp.

[H]

1314. *Ermacora, Giovanni Battista.*
Fenomeni rimarchevoli di medianità osservati senza medi di professione. Turin: Baglione, 1892, 36 pp.

[P]

1315. *Garrett, Julia E.*
Mediums Unmasked. An Exposé of Modern Spiritualism. By an Ex-Medium. Los Angeles: H. M. Lee & Bro., 1892, 56 pp.

[P]

1316. *Green, J. H.*
"Hypnotic Suggestion and Its Relation to the Traumatic Neurosis." *The Railway Age and Northwestern Railroader* 23 (1892):106–13.

The author believes that victims of railway accidents often end up in a state of psychological shock, a kind of trance which makes the victim subject to hypnotic suggestion. In the care of these victims, therefore, it is necessary to guard against suggestions that might lead them to consider their injuries to be worse than they are. [H]

1317. *Hodgson, Richard.*
"A Record of Observations of Certain Phenomena of Trance." *Society for Psychical Research. Proceedings* 8 (1892): 1–167.

The continuation of an article written by F.W.H. Myers et al. in 1890 (*see* entry number 1274) concerning the investigation of the remarkable mental medium from Boston, Mrs. Leonora Piper (1859–1950). In 1887 Hodgson was sent by the Society for Psychical Research to Boston to act as secretary for the newly formed American Society for Psychical Research. In May of that year he met Mrs. Piper and began what would be a fifteen-year-investigation of her mediumship. In this careful preliminary report of his findings, Hodgson does not reach definite conclusions about the genuineness of the material obtained in the seances. However, the article is replete with statements of individuals who sat with the medium and who had no doubt that she had obtained information from or about deceased persons that she could not have come to through ordinary means. Hodgson wrote a second report on Mrs. Piper in an article in the *Proceedings* in 1898 (*see* entry number 1439). [P]

1318. *Janet, Pierre.*

État mental des hystériques. Les stigmates mentaux, and *État mental des hystériques. Les accidents mentaux*. 2 vols. Paris: Rueff et Cie, (1892) and 1894, 233 + (3); 304 pp. ENGLISH: *The Mental State of Hystericals. A Study of Mental Stigmata and Mental Accidents*. 2 vols. in 1. Translated by Caroline Rollin Corson. New York and London: G. P. Putnam's Sons, 1901.

A continuation of work on subconscious states in hysteria begun by Janet in his articles in the *Revue philosophique* (1886, 1887, 1888 entries 1148, 1176, 1211) and *L'automatisme psychologique* (1889 entry number 1235). As in this earlier work, Janet attempts, insofar as possible, to remain on the level of description and to avoid metaphysical speculation concerning the nature of hysteria. Presenting his theory of hysteria in descriptive terms, he argues that the central element is a contraction of the patient's field of consciousness. This contraction defends the patient against the emergence into consciousness of unacceptable, disturbing subconscious sensations. Although he makes no attempt to spell out the mechanism, Janet sees the source of this contraction as cortical. The first volume of the work, which contains a preface by Jean-Martin Charcot (1825–1893) indicating that the studies had been carried out at The Salpêtrière and first described in lectures delivered there by Janet in 1892, deals with "mental stigmata." Mental stigmata are characteristics of hysteria, essential to the condition, that are experienced by the patient as a generalized discomfort difficult to describe. The second volume considers "mental accidents," or symptoms of hysteria that are superadded to the condition and experienced as transient and painful to the patient. Among the "mental stigmata" Janet includes anesthesias, amnesias, abulias, motor disturbances, and modifications of character. "Mental accidents" include subconscious acts, fixed ideas, attacks, somnambulisms, and deliriums. Although Janet cites the work of other researchers copiously, he concentrates on the cases he himself has treated and he draws his descriptions from his own observations. [H & P]

1319. ***Janet, Pierre.***
"Le spiritisme contemporain." *Revue philosophique* 33 (1892):413–42.

[P]

1320. ***Joire, Paul.***
Précis théorique et pratique de neuro-hypnologie; études sur l'hypnotisme et les différents phénomènes nerveux, physiologiques et pathologiques qui s'y rattachent: physiologie, pathologie, thérapeutique, médecin légale. Paris: A. Maloine, 1892, 327 pp.

[H]

1321. ***Laurent, Louis Henri Charles.***
Des états seconds: variations pathologiques du champ de la conscience. Paris: Octave Doin, 1892, 179 pp.

[H]

1322. **[*Leroux, Auguste, pseudonym: Rouxel*].**
Rapports du magnétisme et du spiritisme. Paris: Librairie des Sciences psychologiques, 1892, xiii + 354 pp.

[H & P]

1323. ***Moll, Albert.***
Der Rapport in der Hypnose. Untersuchungen über den thierischen Magnetismus. Leipzig: Ambr. Abel, 1892, 242 pp.

This work also appeared as *Schriften der Gesellschaft für psychologische Forschung* (Vol. 3/4: 277–514). Based largely on original experiments conducted by Moll, it focuses on the nature of rapport in hypnosis. Moll points out that although the term "rapport" in its most general sense refers to a special connection between hypnotist and subject, it has been given a bewildering variety of nuances of meaning over the previous hundred years, and he hopes to bring some clarity to the confusion. He distinguishes "isolated rapport," in which the subject relates only to the hypnotist, from other kinds of rapport, in which the subject may be aware of other persons. Moll also examines those instances in which the hypnotized individual has rapport with no one, a circumstance arising in autohypnosis. In the process of discussing rapport, Moll examines an issue that he considers essentially related: whether there is such a thing as animal magnetism and whether an animal magnetic fluid exists. This question reduces itself to the issue of whether, as the Nancy School teaches, the hypnotic state is brought about through purely psychological means, or whether there is some physical or semi-physical agent involved. Moll states that he has not been able to observe the phenomena so often cited by the magnetists as proof of their theory. However, he does not intend this finding necessarily to imply the disproof of the reality of animal magnetism. [H]

1324. ***Myers, Frederick William Henry.***
"The Subliminal Consciousness." *Society for Psychical Research. Proceedings* 7 (1892):298–355.

The first of four very important articles by Myers on the subliminal consciousness published in the *Proceedings* of the Society for Psychical Research. Here Myers develops his notion of a hidden part of the human psyche that functions "below the threshold" of consciousness and has powerful effects on all facets of human life. The part of us that we are aware of, the supraliminal consciousness or part "above the threshold" of consciousness, is what we ordinarily consider to be our whole self. But in Myers' view, the subliminal consciousness is a much greater and richer part of our being. Myers' concept of the subliminal self goes beyond the notion of the "subconscious" developed by Janet and the "unconscious" of Freud. He includes in it not only subconscious mental activity, instinctual drives, and repressed "fixed ideas," but also those intuitive flashes that characterize genius and even knowledge gained through some kind of supernormal faculty. [H & P]

1325. ***Myers, Frederick William Henry.***
"The Subliminal Consciousness." *Society for Psychical Research. Proceedings* 8 (1892):436–535.

The second of Myers' series of four articles on the subliminal consciousness (*see* entry number 1324), rich in case material illustrative of the action of a hidden, subliminal mental faculty. [H & P]

1326. ***Nizet, Henri.***
L'hypnotisme: étude critique. Brussels: C. Rozez, (1892) 304 pp.

[H]

1327. ***Reichel, Willy.***
Der Magnetismus und seine Phänomene. Berlin: Siegismund, 1892, 84 pp.

[H]

1328. ***Rochas d'Aiglun, Eugène Auguste Albert de.***
Les états profonds de l'hypnose. Paris: Chamuel, 1892, 117 pp.

This important little book discusses the stages of hypnotism derived from the author's own extensive experimentation. De Rochas has been underestimated as an investigator of hypnotic and psychic phenomena, and this, like most of his works, is highly original and worthy of note. Here, De Rochas sets out the stages or phases of hypnotism as he sees them: 1) a state of credulity, 2) catalepsy, 3) somnambulism, 4) rapport, 5) a state of sympathy to contact, 6) a state of lucidity, and 7) a state of sympathy at a distance. Each phase, he says, is followed by an intervening state of lethargy. Referring to the latter four phases as the "profound states of hypnosis," he examines them in detail. In addition to a discussion of these phases, the book contains a chapter on the "exteriorization of sensibility" and one on ecstasy. [H]

1329. *Schrenck-Notzing, Albert Phillibert Franz.*

Die Suggestions-Therapie bei Krankhaften Erscheinungen des Geschlects-sinnes, mit besonderer berücksichtigung der conträren Sexualempfindung. Stuttgart: Ferdinand Enke, 1892, xvii + 314 pp. ENGLISH: *Therapeutic Suggestion in Psychopathia Sexualis.* Translated by Charles Gilbert Chaddock. London (?): F. A. Davis, 1895.

[H]

1330. *Sinnett, Alfred Percy.*

The Rationale of Mesmerism. Boston and New York: Houghton, Mifflin and Company, 1892, (5) + 232 + (6) pp.

A general treatment of animal magnetism by a theosophist which contains a good bibliographical chapter entitled "the real literature of mesmerism." As was so often the case, however, Sinnett gets Mesmer's name wrong, calling him "Frederick Anthony Mesmer." [H & P]

1331. *Stead, William Thomas.*

More Ghost Stories: A Sequel to Real Ghost Stories. (London) New York: Publishing office of the Review of Reviews, 1892, 104 pp.

See *Real Ghost Stories* (1891, entry number 1307). This was the New Year's number of the *Review of Reviews.* [P]

1332. *Wundt, Wilhelm Maximillian.*

Hypnotismus und Suggestion. Leipzig: Wilhelm Engelmann, 1892, 110 pp.

[H]

1333. *Zeitschrift für Hypnotismus, Suggestionstherapie, Suggestionslehre und verwandte psychologische Forschungen.*

Vols. 1–3; 1892–1894.

Continued as: *Zeitschrift für Hypnotismus, Psychotherapie, sowie andere psycho-physiologische und psychopathologische Forschungen.* Vols. 4–10; 1896–1902. Continued as: *Journal für Psychologie und Neurologie. Zugleich Zeitschrift für Hypnotismus.* [H]

1893

1334. *Azam, Etienne Eugène.*

Hypnotisme et double conscience origine de leur étude divers travaux sur des sujets analogues. Paris: Félix Alcan, 1893, viii + 375 pp.

In this volume, Azam brings together material on hypnotism and "double consciousness" or multiple personality from periodical articles, proceedings of conferences, and his own earlier book on the same subject, *Hypnotisme, double conscience et altérations de la personalité*, published in 1887 (*see* entry number 1159). [H]

1335. *Baraduc, Hippolyte Ferdinand.*

La force vitale. Notre corps vital fluidique, sa formule biométrique. Paris: Georges Carré, 1893, viii + 224 pp.

The first of many books by Baraduc on what he called a "biometric vital force" that emanates from living beings. He constructed an instrument called the "biometer" supposed to detect the presence of a nervous force and other vibrations acting beyond the human body. [H & P]

1336. *Bruni, Em.*

Il magnetismo smascherato e svelato. Florence: Salani, 1893, 16 pp.

[H]

1337. *Bué, Alphonse.*

Le magnétisme curatif. Manuel technique. Paris: Chamuel, 1893, xxii + 196 pp.

A manual for the practitioner of "curative magnetism." The author describes the best techniques for applying this healing art and discusses phenomena associated with artificial somnambulism. [H]

1338. *Coates, James.*

How to Mesmerise. London and Glasgow: H. Nesbit & Co., 1893, viii + 120 pp.

Coates, a prolific writer of "how to" books, describes how to apply animal magnetism to subjects, both human and animal. He also deals with the "higher phenomena" of somnambulism such as clairvoyance and prevision. His advice about mesmerizing animals is both interesting and amusing. Information on the first printing is unavailable. [H & P]

1339. *Coates, James.*

How to Thought-read: a Manual of Instruction in the Strange and Mystic in Daily Life, Psychic Phenomena, Including Hypnotic, Mesmeric, and Psychic States, Mind and Muscle Reading, Thought Transference, Psychometry, Clair-

voyance, and Phenomenal Spiritualism. London and Glasgow: H. Nisbet & Co., (1893), 128 pp.

[H & P]

1340. *Cullerre, Alexandre.*
La thérapeutique suggestive et ses applications aux maladies nerveuses et mentales, à la chirurgie, à l'obstétrique et à la pédagogie. Paris: Librairie J.-B. Baillière et Fils, 1893, ix + (1) + 11–318 pp.

[H]

1341. *Delanne, Gabriel.*
Le phénomène spirit: témoignages des savants. Étude historique, exposition méthodique de tous les phénomènes, discussion des hypothèses. Conseils aux médiums, la théorie philosophique. Paris: Chamuel, 1893, vii + 296 pp.

Delanne, an eloquent promoter of the cause of spiritualism, made significant contributions to psychical research. In this book he discusses such phenomena as human levitation, transmission of thought, apparitions and materializations. [P]

1342. *Ermacora, Giovanni Battista.*
Attività subconsciente e spiritismo. Rome: Balbi, 1893, 14 pp.

[P]

1343. *Fajnkind, Stephanie.*
Du somnambulisme dit naturel (noctambulisme), ses rapports avec l'hystérie et l'attaque hystérique à forme somnambulique. Paris: n.p., 1893, 151 pp.

[H]

1344. *Feytaud, Urbain.*
Le spiritisme devant la conscience. Paris: Chamuel, 1893, 208 pp.

[P]

1345. *Hart, Ernest.*
Hypnotism, Mesmerism and the New Witchcraft. London: Smith, Elder & Co., 1893, vii + 182 pp.

A collection of articles, written by Hart for *Nineteenth Century* and the *British Medical Journal*, which contains a strong denunciation of the findings of psychical research and a rejection of the "higher phenomena" of magnetic somnambulism. Hart argues that all apparent cases of supernormal events are simply the result of intended or accidental suggestions given to persons who are either hysterical or in a hypnotic state. [H & P]

1346. *Hudson, Thomson Jay.*

The Law of Psychic Phenomena: A Working Hypothesis for the Systematic Study of Hypnotism, Spiritism, Mental Therapeutics, etc. Chicago: A. C. McClurg, 1893, xvii + (1) + 19–409 pp.

Hudson was an American thinker who developed a comprehensive explanation for a variety of extraordinary phenomena based upon a theory that human beings possess two minds: an objective mind which functions in everyday life, and a subjective mind which functions on a subconscious level and which has the memory of all of the experiences of the individual. This book is his first and his masterwork. Hudson used his two-mind theory to oppose the spiritualistic view of the world, insisting that apparent communication with the dead is self delusion. He contended that all of the information supposedly coming from spirits could be discovered in the subjective minds of the persons involved in the seances. This book, widely read, was very influential among Hudson's contemporaries. [H & P]

1347. *Kiesewetter, Carl.*

Die Entwicklungsgeschichte des Spiritismus von der Urzeit bis zur Gegenwart. Leipzig: Max Spohr, 1893, 50 pp.

[P]

1348. *Kiesewetter, Carl.*

Franz Anton Mesmer's Leben und Lehre. Nebst einer Vorgeschichte des Mesmerismus, Hypnotismus und Somnambulismus. Leipzig: Max Spohr, 1893, 180 pp.

Kiesewetter was a noted historian of the occult. This book is a useful biography of Mesmer and a history of his teaching. Kiesewetter's chief contribution, however, is an interesting prehistory of animal magnetism which describes practices and phenomena centuries before Mesmer. [H]

1349. *Lazare, Bernard.*

La télépathie et le néo-spiritualisme. Paris: Librairie de l'art independant, 1893, 36 pp.

[P]

1350. *Myers, Frederic William Henry.*

Science and a Future Life: With Other Essays. London: Macmillan, 1893, 243 pp.

A collection of essays originally appearing in the *Nineteenth Century* and the *Fortnightly Review*. The first, entitled "Science and a Future Life," is the only one concerned with psychical research. It addresses the question: What does science have to say about man's survival of death? [P]

1351. ***Myers, Frederick William Henry.***
"The Subliminal Consciousness." *Society for Psychical Research. Proceedings* 9 (1893):2–128.

The third of four important articles by Myers on the subliminal consciousness. Here he examines the evidence of subliminal mental activity in hysteria. As is usual with Myers, the article is replete with case material. [H & P]

1352. ***Rochas d'Aiglun, Eugène Auguste Albert de.***
Les états superficiels de l'hypnose. Paris: Chamuel, 1893, 149 pp.

In *Les états profonds de l'hypnose* (1892, entry number 1328), de Rochas listed eight stages or phases of hypnotism. This volume focuses on the first three: 1) credulity, 2) catalepsy, and 3) somnambulism, which de Rochas calls the "superficial states of hypnosis." It also includes a long section on the nature of hypnotic suggestion. [H]

1353. ***Savage, Minot Judson.***
Psychics: Facts and Theories. Boston: Arena Publishing Co., 1893, x + (5)–158 pp.

Savage was a minister of the Unitarian church and an early member of the American Society for Psychical Research. This is his first book on psychical research. [P]

1354. ***Sextus, Carl.***
Hypnotism, Its Facts, Theories and Related Phenomena: with Explanatory Anecdotes, Descriptions and Reminiscences. Chicago: C. Sextus, 1893, viii + 9–304 pp.

A noteworthy publication on hypnotic practice in America. While presenting well-known material on the nature of artificial somnambulism and hypnotic states, Sextus also includes chapters on unusual subjects such as one on magnetization of snakes in the United States and the introduction of hypnotism into Chicago that make for interesting reading. [H]

1355. ***Souriau, Paul.***
La suggestion dans l'art. Paris: Ancienne librairie Germer Baillière, Félix Alcan, éditeur, 1893, (4) + 348 pp.

This unusual contribution investigates the place of hypnotism, fascination, and suggestion in art. After discussing how the experience of art can produce hypnotic effects, Souriau notes the crucial role of suggestion in visual arts, musical arts, writing, and acting. He also examines the way a sympathetically attuned individual identifies with elements in the artistic presentation, and relates this to a doubling of the personality and a kind of possession experience. This little known book provides a unique perspective on the human experience of art. [H]

1356. ***Vincent, Ralph Harry.***
The Elements of Hypnotism: the Induction of Hypnosis, Its Phenomena, Its Dangers and Value. London: Kegan Paul, Trench, Trubner & Co., 1893, xiv + 269 pp.

[H]

1894

1357. ***Benedikt, Moriz.***
Hypnotismus und Suggestion. Eine klinisch-psychologische Studie. Vienna: M. Breitenstein, 1894, ii + 91 pp.

[H]

1358. ***Cocke, James Richard.***
Hypnotism, How It is Done; Its Uses and Dangers. Boston: Arena Publishing Co., 1894, v + (1) + 373 pp.

Cocke covers practically every aspect of the phenomenon from methods of hypnotic induction to theories on the nature of hypnotism. Well researched and eclectic in orientation, this is one of the best general treatises on hypnotism to appear in the United States to its date. [H]

1359. ***Crocq, Jean.***
L'hypnotisme et le crime. Brussels: H. Lamertin, 1894, iv + vii + 5–300 pp.

[H]

1360. ***La curiosité. Revue des sciences psychiques.***
Vols. 1–4; 1894–1898.

Edited by Ernest Bosc and published in Nice. [P]

1361. ***Dailey, Abram Hoagland.***
Molly Fancher, the Brooklyn Enigma. An Authentic Statement of Facts in the Life of Mary J. Fancher, the Psychological Marvel of the Nineteenth Century. Unimpeachable Testimony of Many Witnesses. Brooklyn: Eagle, 1894, (1) + xiii + (1) + 262 pp.

A case history of a multiple personality accompanied by clairvoyant experiences. [H & P]

1362. ***Du Prel, Karl Ludwig August Friedrich Maximillian Alfred.***
Die Entdeckung der Seele durch die Geheimwissenschaften. 2 vols. Leipzig: E. Gunther, (1894 and 1895).

[P]

1363. ***Durand de Gros, Joseph Pierre.***
Le merveilleux scientifique. Paris: Félix Alcan, 1894, 344 pp.

A summary of the views of Durand de Gros on mesmerism, braidism (as the French sometimes termed hypnotism), and spiritism. [H & P]

1364. ***Fugairon, Louis Sophrone.***
Essai sur les phénomènes électriques des êtres vivants comprenant l'explication scientifique des phénomenènes dits spirites. Paris: Chamuel, 1894, 202 pp.

[H & P]

1365. ***Glendinning, Andrew [ed.].***
The Veil Lifted. Modern Developments of Spirit Photography. A Paper by J. Traill Taylor Describing Experiments in Psychic Photography, Letter by the Rev. H. R. Haweis, Addresses by James Robertson, Glasgow, and Miscellanea by the Editor, Andrew Glendinning. London: Whittaker & Co., 1894, xii + 164 pp.

A collection of writings on an unusual phenomenon; the production of forms on film which apparently cannot be accounted for in normal terms. These forms were thought to be produced by spirits and often were the likenesses of deceased persons. Taylor, the editor of the *British Journal of Photography*, describes his investigation of the photographic medium David Duguid and his inability to account for certain "extras" that were produced on the films used in the experiment. Other papers in this collection were equally affirmative in their views on the matter. [P]

1366. ***Lang, Andrew.***
Cock Lane and Common Sense. London: Longmans, Green & Co., 1894, xvi + 357 pp.

Andrew Lang was a literary genius and author of over fifty books in anthropology, mythology, folk lore and other fields. In 1906, he joined the Society for Psychical Research, contributing many articles to its *Journal* and *Proceedings* and eventually serving as its president. *Cock Lane and Common Sense* is a collection of essays that had previously appeared in various journals. The subject is spiritualism, ancient and contemporary. Lang's approach is to use material that is as solidly based in true experience as possible and in the preface he discusses a number of problems regarding the truthfulness in such accounts. These include the inaccuracy of human testimony; the possibility of hallucination, and fraud. Taking these difficulties into account, Lang presents accounts of spiritualistic experiences which he nonetheless considers worthy of serious consideration. They range from ancient neo-platonic spiritistic séances to the contemporary phenomena of table turning. [P]

1367. *Lillie, Arthur.*

Modern Mystics and Modern Magic; Containing a Full Biography of the Rev. William Stainton Moses, etc. London: S. Sonnenschein & Co., 1894, 172 pp.

Sketches of the lives and ideas of some important modern mystics: Swedenborg, Boehme, Madame Guyon, the "Illuminati", Madame Blavatsky, and William Stainton Moses. The latter sketch is by far the most valuable and is in fact one of the most complete descriptions of the life and experiences of Stainton Moses available. As such it is an important contribution to the history of psychical research. [P]

1368. *Mesnet, Ernest.*

Outrages à la pudeur. Violences sur les organes sexuels de la femme dans le somnambulisme provoqué et la fascination. Étude médico-légale. Paris: Rueff, 1894, xxiv + 267 pp.

On the moral and legal misuse of hypnotism, with emphasis on sexual violation. [H]

1369. *Metzger, Daniel.*

Essai de spiritisme scientifique. Paris: Librarie des Sciences Psychologiques, (1894), xii + 455 pp.

A detailed study of spiritualism and its philosophy. The author discusses the dangers of spiritualism and possibilities of hallucination and illusion in spiritualistic manifestations. Phenomena discussed include telepathy, apparitions, automatic writing, spirit photography, and materializations. [P]

1370. *Parish, Edmund.*

Über die Trugwahrnehmung (Hallucination und Illusion) mit besonderer Berücksichtigung der internationalen Enquete über Wachhallucinationen bei Gesunden. Leipzig: A. Abel, 1894, 246 pp. ENGLISH: *Hallucinations and Illusions.* London: Walter Scott, 1897.

An important study of hallucinations based upon the author's own experiments and a study of the literature of Germany, France, England, and America. He concludes that all hallucinations arise while the subject is in a state of dissociation of consciousness in which the higher nerve elements that carry on the work of association are exhausted. [P]

1371. *Podmore, Frank.*

Apparitions and Thought-Transference: An Examination of the Evidence for Telepathy. London: Walter Scott, 1894, xiv + 401 pp.

Although a civil servant by occupation, Podmore devoted a great deal of time to the first hand investigation of psychical phenomena and became one of the most important figures in the early history of the Society for Psychical Research. Although initially a spiritualist, the exposure of a number of well-known mediums as frauds led him to become an extreme sceptic. He developed such severe criteria for judging the genuineness of mediumistic experiences that some believed they were impossible to meet. This scepticism undoubtedly increased the credibility of the Society for Psychical Research as a seriously critical investigative body. Although Podmore had contributed to *Phantasms of the Living* (1886, *see* entry number 1144), *Apparitions and Thought-Transference* is his first solely authored book. Podmore provides a solid description of the history of the question from the time of the early mesmerists and points out reasons for caution in accepting the conclusions of those earlier investigators. He then arrives at a position that he was to maintain throughout all of his works: that alleged apparitions and communications from the dead are explicable in terms of the operation of telepathy. [P]

1372. *Renterghem, Albert Willem van and Eeden, F. van.*

Psycho-thérapie: communications statistiques, observations cliniques nouvelles. Compte rendu des résultats obtenus dans la clinique de psycho-thérapie suggestive d'Amsterdam, pendant la deuxième période. Paris: Sociétés d'éditions scientifiques, 1894, vi + 293 pp. [H]

1373. *Schrenck-Notzing, Albert Philibert Franz.*

Der Hypnotismus im Münchener Krankenhause (Links der Isar). Eine kritische Studie über die Gefahren der Suggestivbehandlung. Leipzig: A. Meiner, 1894, 39 pp. [H]

1374. *Society for Psychical Research.*

"Report on the Census of Hallucinations." *Society for Psychical Research. Proceedings* 10 (1894):25–422.

The most comprehensive survey of paranormal experiences ever published, this statistical study was begun in 1889 by a committee under the chairmanship of Henry Sidgwick, and a group of 410 collectors of material. A questionnaire was used to which 17,000 replies were received, 2,272 of which report some kind of paranormal hallucination. After a long period of sifting and confirming the affirmative replies, 1600 of these were considered reliable enough to be used in the study, which analyzes hallucinations according to type and circumstance and contains a lengthy discussion of the results. The discussion was illustrated with numerous details of individual experience. [P]

1375. Stoll, Otto.

Suggestion und Hypnotismus in der Völkerpsychologie. Leipzig: K. F. Koehler, 1894, xii + 523 pp.

A very important study in the social psychology of suggestion. Stoll investigates the place of hypnotic and waking suggestion in the development of the religious and occult practices and beliefs of all ages. In this context, he discusses the phenomena of hetero-suggestion, auto-suggestion, and group-suggestion. The work is extremely well researched, a scholarly accomplishment of the first order. [H]

1376. ***Young, J. F. and Robertson, R.***

The Divining Rod: Its History — With Full Instructions for Finding Subterranean Springs, and Other Useful Information. Also an Essay Entitled: Are the Claims and Pretensions of the Divining Rod Valid and True? by E. Vaughan Jenkins. (London): J. Baker, 1894, 138 pp. [P]

1895

1377. ***Acevedo, M. Otero.***

Los Espiritus. 2 vols. Madrid: Revista psicologica, "La Irradiacion", 1895, 371; 256 pp.

A history of spiritism with accounts of experiments of psychical researchers from England, Germany, France, Italy, and Spain. [P]

1378. ***Binet, Alfred.***

"Contribution à l'étude de la soi-disant télépathie." *Annales des sciences psychiques* 5 (1895): 193–99.

[P]

1379. Coste, Albert.

Les phénomènes psychiques occultes, état actuel de la question. Montpellier and Paris: C. Coulet (Montpellier) and Masson (Paris), 1895, 226 pp.

[P]

1380. ***Durand de Gros, Joseph Pierre.***

Suggestions hypnotiques criminelles. Paris: Félix Alcan, 1895, 16 pp.

[H]

1381. ***Durville, Hector.***

Bibliographie du magnétisme et des sciences occultes. Paris: Librairie du magnétisme, 1895, 36 pp.

[H]

1382. ***Durville, Hector.***

L'enseignement du magnétisme. Paris: Librairie du magnétisme, 1895, 27 pp.

[H]

1383. ***Durville, Hector.***

Traité expérimental de magnétisme avec portrait de l'auteur et figures dans le texte. Cours professé à l'école pratique de magnétisme et de massage. Physique magnétique. 2 vols. Paris: Librairie du magnétisme, (1895 and 1896), 324; 360 pp.

With the second two volumes (see Durville: *Traité expérimental de magnétisme*, entry number 1435), this work is one of the most detailed and complete treatises on the practice of animal magnetism and magnetic healing ever written. [H]

1384. ***Erny, Alfred.***

Le psychisme expérimental, étude sur les phénomènes psychiques. Paris: E. Flammarion, 1895, (3) + iii + 232 pp.

The author attempts to take critical but fair notice of the phenomena of spiritualism and to determine their value for psychical research. [P]

1385. ***Hudson, Thomson Jay.***

A Scientific Demonstration of the Future Life. Chicago: A. C. McClurg and Co., 1895, 326 pp.

Whereas in *The Law of Psychic Phenomena* (1893, entry number 1346) Hudson attempted to show that all of the phenomena of spiritualism could be explained in terms of "this-wordly" factors, in the present work he intends to demonstrate that human beings are nevertheless destined to a life beyond death. His demonstration is not, however, based on proofs cited by spiritualists, such as communications from the dead, which Hudson considers to be superstition. Rather it is founded on an analysis of the physical, intellectual, and psychical structure of human beings. Hudson contends that any other conclusion is logically and scientifically untenable. [P]

1386. ***Janet, Pierre.***

"J. M. Charcot, son oeuvre psychologique." *Revue Philosophique* 39 (1895):569–604.

Janet criticizes Charcot's methodology in his investigation of hypnotism and hysteria. He finds fault with Charcot's attempt to provide a description of mental conditions based on the same model as that used to categorize organic disease. Janet also points out that Charcot was misled in his conclusions because of a lack of awareness of what was happening to his patients on the ward. Charcot did not realize that his hysterics were often "magnetized" by various individuals and that his "three stages" of hypnotism were in fact behaviors learned in those magnetic sessions, not organically determined responses. [H]

1387. [*Leroux, Auguste, pseudonym: Rouxel*].
Histoire et philosophie du magnétisme chez les anciens et chez les modernes. 2 vols. Paris: Librairie du magnétisme, 1895 and 1896, 359; 324 pp.

A competent history of animal magnetism, tracing both ancient precursors and the development of animal magnetism from the time of Mesmer. [H]

1388. *Myers, Frederick William Henry.*
"The Subliminal Self." *Society for Psychical Research. Proceedings* 11 (1895):334–593.

The last of four important articles by Myers on the subliminal consciousness. Here he amasses a great many instances of apparent precognition and retrocognition to show that the subliminal consciousness seems to have access to information of a supernormal kind. (For a description of Myers's notion of the subliminal consciousness, see entry number 1324.) [H & P]

1389. *Preyer, Wilhelm Thierry.*
Ein merkwürdiger Fall von Fascination. Stuttgart: Ferdinand Enke, 1895, vi + (2) + 55 pp.

Description of a case of fascination using the eyes and waking suggestion. The subject was Freulein Ellida Hill. [H]

1390. *Rivista di studi psichici. Periodico mensile dedicato alle ricerche sperimentali e critiche sui fenomeni di telepatia, chiaroveggenza, premonizione, medianità, etc.*
Vols. 1–6; 1895–1901.

Published in Padua and Turin and edited by G. B. Ermacora and Giorgio Finzi, followed by Cesare di Vesme. [P]

1391. *Rochas d'Aiglun, Eugène Auguste Albert de.*
L'extériorisation de la sensibilité. Étude expérimentale et historique. Paris: Bibliothèque Chacornac, 1895, viii + 250 pp.

"Exteriorization of sensibility" refers to the extension beyond the physical body of the ability to sense or perceive in a manner analogous to sensation. It was first noted and named as such by Paul Joire in his *Précis théorique et pratique de neuro-hypnology* (1892, entry number 1320). In the present book, De Rochas discusses the history and nature of the phenomenon, his own experiments with it and its relation to his stages of hypnotism (*see* entry number 1328). Asserting that it has been well established that specially sensitive individuals perceive emantions proceeding from the organisms of living beings and pointing out that some investigators believe that a hypnotic subject in the state of rapport with the hypnotizer may be able to experience the sensations of the hypnotizer, de Rochas is led to the belief that the power of sensation may be extended beyond the skin of the body. This phenomenon, he notes, is not new, and he traces the history of analogous phenomena back hundreds of years, citing among other sources, literature that treats of the powder of sympathy and sympathetic medicine in general. In this regard, he devotes a full chapter to the magnetic medicine of William Maxwell. [H & P]

1392. *Savino, E.*

Il magnetismo, l'ipnotismo e lo spiritismo, ovvero Satana e la moderna magia. Benevento: n.p., 1895.

Reports many supernormal phenomena accomplished in the magnetic or hypnotic state and attributes them to the power of Satan. [H & P]

1393. *Schrenck-Notzing, A. von, Grashey, ———, Hirt, ———, and Preyer, Wilhelm.*

Der Prozess Czynski. Thatbestand desselben und Gutachten über Willensbeschränkung durch hypnotisch-suggestiven Einfluss abgegeben vor dem oberbayerischen Schwurgericht zu München. Stuttgart: Ferdinand Enke, 1895, iii + 102 pp.

[P]

1394. *Soloviev, Vsevolod Sergyeevich.*

A Modern Priestess of Isis. London: Longmans, Green, and Co., 1895, xix + (1) + 366 pp.

In 1884 the Society for Psychical Research undertook an investigation of "theosophical phenomena." At that time a committee was set up by the Society to determine the genuineness of the psychic happenings said to take place around Madame Blavatsky. The finding of the committee, published in its *Proceedings*, was basically that she was a charlatan and that many of the apparently supernormal phenomena were deliberately staged. In a preface, the Society's president, Henry Sidgwick, states that Mr. Soloviev's study of Madame Blavatsky, at least in certain parts, constituted a supplement to the statement of results given by the Society. The original longer version appeared in eight monthly installments in the *Russky Vyestnik* (Russian Messenger) in 1892 and was abridged and translated on behalf of The Society by Walter Leaf. [P]

1896

1395. *Aksakov, Alexander Nikolaevich.*
Un cas de dématérialisation partielle du corps d'un médium: enquête et commentaires. Paris: L'art Indépendant, 1896, 221 pp. ENGLISH: *A Case of Partial Dematerialization of the Body of a Medium.* Translated by Tracy Gould. Boston: Banner of Light Publishing, 1898.

A description and discussion of a rarely reported phenomenon: the apparent partial dematerialization of the body of a medium during a materialization séance. The medium was Madame d'Espérance (whose real name was Elizabeth Hope), and the séance was conducted in December of 1893 in Finland. The French edition was translated from German. [P]

1396. *Baraduc, Hippolyte Ferdinand.*
L'âme humaine, ses mouvements, ses lumières et l'iconographie de l'invisible fluidique. Paris: Georges Carré, 1896, 299 pp.

Baraduc claimed he was able to photograph rays emanating from the astral bodies of human beings. The book contains 70 photographic plates. [P]

1397. *Baraduc, Hippolyte Ferdinand.*
L'iconographie en anses de la force vitale cosmique et la respiration fluidique de l'âme humaine. Son atmosphère fluidique. Paris: Georges Carré, 1896, 89 pp.

[P]

1398. *Cleveland, William.*
The Religion of Modern Spiritualism, and Its Phenomena Compared with the Christian Religion and Its Miracles. Cincinnati: The Light of Truth Publishing Co., 1896, 400 pp.

The author presents spiritualism as a continuation of Christianity and a purification of an excessive and oppressive orthodoxy. [P]

1399. *Crocq, Jean.*
L'hypnotisme scientifique. Paris: Société d'éditions scientifiques, 1896, xi + 451 pp. [H]

1400. *Dulora de la Haye, ———.*
Somnambulisme et magie. Paris: Fayard, 1896, 238 pp.

A popularly written book on animal magnetism and hypnotism, with emphasis on the more spectacular phenomena. [H & P]

1401. ***Harte, Richard.***
The New Spiritualism. (London): Office of "Light," (1896), 35 pp.

[H]

1402. ***Hypnotic Magazine.***
1896–1897, Vols. 1–2.

Edited by Sidney Flower in Chicago. [H]

1403. ***Moutin, Lucien.***
Le diagnostic de la suggestibilité. Paris: Société d'éditions scientifiques, 1896, 110 pp.

Written by a practitioner in the magnetic tradition, this volume is principally concerned with determining the degree of a subject's sensibility to magnetic influence. [H]

1404. ***Pictet, Raoul.***
Étude critique du matérialisme et du spiritualisme par la physique expérimentale. Geneva: H. Stapelmohr, 1896, 596 + (3) pp.

[P]

1405. ***Revue de psychiatrie, de neurologie et d'hypnologie. Recueil des travaux publiés en France et à l'étranger.***
1896.

See: *L'encéphale* (1881), entry number 1039. [H]

1406. ***Rochas d'Aiglun, Eugène Auguste Albert de.***
L'extériorisation de la motricité. Recueil d'expériences et d'observations. Paris: Chamuel, 1896, viii + 432 pp.

"Exteriorization of motricity" refers to the extension of motor force beyond the periphery of the body. The theory behind this term is that when a telekinetic individual moves objects without contact, the motor nerves of that individual are involved. De Rochas wrote this book with the intention of establishing the fact that inert objects can be moved through a force that emanates from the physical organism of some people. The first part of the book is largely a description of séances with the famous Italian medium, Eusapia Palladino (1854–1918), whose physical mediumship was investigated by more panels of experts than any other medium of the day. She produced some physical phenomena that were never explained, but she was also detected in fraud. De Rochas provides detailed accounts of experiments and comments by observers present and discusses the occasional presence of fraud. The second part of the book deals with the turning tables of 1853, the British investigation of physical phenomena (the report of the Dialectical Society and the work of William Crookes), and the phenomena produced by a number of mediums such as Henry Slade and Donald MacNab. [P]

1407. ***Sausse, Henri.***

Biographie d'Allan Kardec. Discours prononcé à Lyon, le 31 mars 1856. Tours: E. Arrault, (1896), 32 pp.

[P]

1408. ***Schrenck-Notzing, Albert Phillibert Franz.***

Über Spaltung der Personlichkeit (Sogennantes Doppel-Ich). Vienna: Holder, 1896, 23 pp.

[H]

1409. ***Suggestive Therapeutics.***

Vols. 1–3; 1896–1897.

Continued as: *Hypnotic Magazine* Vols. 4–5; 1898. Continued as: *Journal of Medical Hypnotism* Vol. 1; 1898. Continued as: *Journal of Magnetism* Vol. 1 and Vols. 10–12; 1901. Merged into: *New Thought* Vol. 1 and Vols. 10–19; 1901–1910. [H]

1410. ***Vesme, Cesare Baudi di.***

Storia dello spiritismo. 2 vols. Turin: Frassati, (1896 and 1897) xvi + 379; 574 pp. ENGLISH: *A History of Experimental Spiritualism.* 2 vols. London: Rider & Co., 1931.

This wide ranging work takes up the history of spiritualism with the specific aim of assessing whether "experimental spiritualism," (the direct experience of the paranormal phenomena of spiritualism, including apparent communication with departed spirits) can show mankind the true meaning of life and inject a note of optimism into human existence, a task which modern materialism had not been able to accomplish. Di Vesme finds the key in the investigation of paranormal phenomena. These lead to the conclusion that life continues beyond the grave and that the events of this life must be understood in terms of that broader context. Volume one deals with primitive man and present primitive cultures. Volume two examines spiritualism in antiquity. [P]

1411. ***Zeitschrift für Hypnotismus, Psychotherapie, sowie andere psycho-physiologische und psychopathologische Forschungen.***

1896.

See: *Zeitschrift für Hypnotismus, Suggestionstherapie, Suggestionslehre und verwandte psychologische Forschungen* (1892, entry number 1333). [H]

1897

1412. ***Bernheim, Hippolyte.***
L'hypnotisme et la suggestion dans leurs rapports avec la médecine légale. Nancy: A. Crépin-Leblond, 1897, 102 pp.

A talk originally delivered at the twelfth Congrès international de médecine held at Moscow in 1897. [H]

1413. ***Betiero, T. J.***
Practical Essays on Hypnotism and Mesmerism. Chicago: (Zelah and Co.), 1897, 139 pp.

[H]

1414. ***Coates, James.***
Human Magnetism or How to Hypnotise. A Practical Handbook for Students of Mesmerism. London: George Redway, 1897, 80 pp.

A reworking of material contained in Coates's *How to Mesmerize* (1893, *see* entry number 1338). The author's "how to" manuals were very much in demand. [H]

1415. ***Delanne, Gabriel.***
L'évolution animique. Essai de psychologie physiologique suivant le spiritisme. Paris: Chamuel, 1897, 368 pp.

Delanne presents a guide for living based on a spiritistic view of human nature. The sections on artificial somnambulism, possession, and madness are of interest to the psychical researcher. [H & P]

1416. ***Encausse, Gérard.***
Du traitement externe et psychique des maladies nerveuses. Aimants et couronnes magnétiques. Mirroirs. Traitement diététique. Hypnotisme. Suggestion. Transferts. Paris: Chamuel, 1897, 208 pp.

[H]

1417. ***Gasc-Desfosses, Edouard.***
Magnétisme vital: contributions expérimentales à l'étude par le galvanomètre de l'électro-magnétisme vital; suivies d'inductions scientifiques et philosophiques. Paris: Société d'éditions scientifiques, 1897, xviii + 336 pp.

An attempt to relate the phenomena of animal magnetism to electromagnetism. [H]

1418. ***Hope, Elizabeth, [pseudonym: Mme. D'Esperance].***
Shadowland, or Light from the Other Side. London: George Redway, 1897, xxii + 414 pp.

A classical autobiographical account of a non-professional medium who played an important part in experiments in psychical research. As a child she was aware of "shadow people," individuals who, though not frightening to her, nevertheless did not have substantial bodies and seemed to be deceased individuals. As a young woman she accidentally discovered what appeared to be remarkable telekinetic powers, and from that time she began to develop as a materialization medium. She was eventually able to produce what appeared to be misty human forms in three dimensions. Investigated by many psychical researchers, she came to be highly respected for her honesty. Aksakov's book *Un cas de dématérialsation partielle du corps d'un médium* (1896, entry number 1395) is about a bizarre incident in her physical mediumship. She wrote a second volume of *Shadowland*, but the manuscript was destroyed in Germany during the First World War. [P]

1419. *Hodgson, Richard.*

"A Further Record of Observations of Certain Phenomena of Trance." *Society for Psychical Research. Proceedings* 13 (1897):284–582.

A follow-up report on the medium Leonora Piper (1859–1950) supplying information supplementing that of Hodgson's report of 1892 (*see* entry number 1317). In this article Hodgson dicusses possible alternative explanations, expressing his opinion that the "spirit hypothesis" seems most likely. [P]

1420. *Lang, Andrew.*

The Book of Dreams and Ghosts. London: Longmans, Green & Co., 1897, xviii + 301 pp.

Lang did not intend this book to be just a collection of odd stories. He meant it to be a systematic study of true accounts of unusual events gathered from a literature spanning a number of centuries and examined from an anthropological point of view. It deals with poltergeists, animal ghosts, crystal gazing (a subject of particular interest to Lang), bilocation, hauntings and other phenomena. Lang examines his material critically, but does not rule out the possibility that the accounts refer to genuine paranormal experiences. [P]

1421. *Mason, Rufus Osgood.*

Telepathy and the Subliminal Self. New York: Henry Holt and Company, 1897 (1) + vii + 343 pp.

The author, a physician, attempts to bring findings from hypnosis and double personality together with those from psychical research proper in a systematic analysis of the subliminal or subconscious self. He has researched his subject well and is successful within his own terms of reference. It is a valuable book and one that must be considered an essential source in tracing the connections between psychical research and investigations arising out of animal magnetism and hypnosis. [H & P]

1422. ***Morselli, Enrico Agostino.***
I fenomeni telepatici e le allucinazioni viridiche. Osservazioni critiche sul neo-misticismo psicologico. Florence: Landi, 1897, 58 pp.

A study of telepathy and apparitions by one of the most cautious and careful of the Italian psychical researchers. [P]

1423. ***Müller, Rudolf.***
Das hypnotische Hellseh-Experiment im Dienste der naturwissenschaftlichen Seelenforschung. 3 vols. Leipzig: A. Strauch, (1897–1900).

Volume 1: *Der Veränderungsgesetz* (1897); Volume 2: *Das normale Bewusstsein* (1898); Volume 3: *Wille; Hypnose; Zweck* (1900). [H & P]

1424. ***Ochorowicz, Julian.***
Magnetismus und Hypnotismus. Leipzig: Oswald Mutze, 1897, vi + (2) + 138 pp.

Ochorowicz presents first the history of animal magnetism and then that of hypnotism. He defines animal magnetism as the healing art practiced by using magnetic passes and hypnotism as the special state of consciousness which is usually called somnambulism. He describes the progression from Puységur through Petetin to Faria, Noizet, Bertrand, Deleuze, etc. Ochorowicz acknowledges the importance of Braid and his development of the notion of hypnotism, and investigates the most significant results of hypnotism. [H]

1425. ***Parish, Edmund.***
Zur Kritik des telepathischen beweis Materiels. Leipzig: n.p., 1897, 48 pp.

A criticism read before the Psychological Society of Munich of the findings of the "Report on the Census of Hallucinations" (*see* entry number 1374). Parish contends that there are good reasons to be skeptical of the evidence for telepathy presented in that report. [P]

1426. ***Podmore, Frank.***
Studies in Psychical Research. London: Kegan Paul, Trench, Trubner & Co., 1897, ix + 458 pp.

One of the best books on psychical research written before 1900. Podmore treats most of the major areas of investigation and, as usual, his approach is skeptical in the extreme. His historical and investigative research is excellent. In addition to interesting chapters on secondary personalities and obsession in which he attempts to show how these psychological phenomena cast light on spiritualistic experiences, subjects include the physical phenomena of spiritualism, poltergeists, telepathy, ghosts, haunted houses, and premonitions. As in his previous book, Podmore concludes that only telepathy has so far been established as a proven fact. [P]

1427. ***Randall, Frank H.***
Practical Instruction in Mesmerism. Westminster: Roxburge Press, 1897, (8) + 113 pp.

[H]

1428. ***Rochas d'Aiglun, Eugène Auguste Albert de.***
La lévitation. Paris: P. G. Leymarie, 1897, 110 pp.

Collects information regarding the paranormal levitation of objects and living beings. Among the incidents described is the bodily levitation of the famous medium, Daniel Dunglas Home (1833–1886). [P]

1429. ***Trufy, Charles.***
Causeries spirites. Paris: Chamuel, 1897, 263 pp.

[P]

1898

1430. ***Aksakov, Alexandr Nikolaevich*** **[*ed.*].**
Der Spiritualismus und die Wissenschaft. Experimentelle Untersuchungen über die psychische Kraft, von William Crookes, nebst bestätigenden Zeugnissen des Physikers C. F. Varley. Prüfungs-Sitzungen des Mr. D. D. Home mit den Gelehrten zu St. Petersburg und London. Translated from the Russian into English by Gregor Constantin and edited by Alexandr Aksakov. 2 ed. Leipzig: Oswald Mutze, 1898, xxiii + 125 pp.

[P]

1431. ***Aksakov, Alexandr Nikolaevich.***
Vorläufer des Spiritismus. Hervorragende Falle willkürlicher mediumistischer Erscheinungen aus den letzten drei Jahrhunderten. Translated from the Russian with a supplement by Feilgenhauer. Leipzig: Oswald Mutze, 1898, xvi + 356 pp.

Aksakov attempts to provide a panorama of spiritistic phenomena from 1661, through the Rochester knockings of 1848, and up to his own time. He recounts many cases of mediumistic happenings which are drawn from the better known literature and includes material, chiefly Russian in origin, never before published, some with extensive documentation. [P]

1432. ***Belfiore, Guilio.***
Magnetismo e ipnotismo. Milan: Ulrico Hoepli, 1898, (1) + (v)–viii + 377 pp.

[H]

1433. *Bérillon, Edgar.*
L'hypnotisme et l'orthopédie mentale. Paris: Rueff, 1898, 48 pp.

[H]

1434. *Binet, Alfred.*
"La suggestibilité au point de vue de la psychologie individuelle." *Année psychologique* 5 (1898):82–152.

[H]

1435. *Durville, Hector.*
Traité expérimental de magnétisme avec figures dans le teste. Cours professé à l'école pratique de magnétisme et de massage. Théories et procédés. 2 vols. Paris: Librairie du magnétisme, (1898–1904), 360; 332 pp.

With the first two volumes (*see* Durville: *Traité expérimental de magnétisme,* 1895, entry number 1383), this work is one of the most detailed and complete treatises on the practice of animal magnetism and magnetic healing ever written. [H]

1436. *Ermacora, Giovanni Battista.*
La telepatia. Padua : L. Crescini, 1898, 150 pp.

A collection of articles on thought transference written by one of the more important Italian psychical researchers. It deals largely with the methods and conditions necessary for a scientific study of telepathy and also presents a review of the evidence in favor of the reality of telepathic phenomena. [P]

1437. *Fontenay, Guillaume de.*
À propos d'Eusapia Paladino. Les séances de Montfort-L'Amaury (25–28 Juillet 1897). Compte rendu, photographies, témoignages et commentaires. Paris: Société d'éditions scientifiques, 1898, xxx + 281 pp.

This book describes séances with the famous Italian medium Eusapia Palladino (1854–1918). [P]

1438. [*Geley, Gustave, pseudonym: "Dr. Gyel"*].
Essai de revue générale et d'interprétation synthétique du spiritisme. Lyon: L. Bourgeon, 1898, 128 pp.

A somewhat philosophical investigation of the doctrine of spiritism. Geley questions the nature of the soul and its connection with the material world, the existence of a "perispirit" or astral body, and the relevance of subconscious states to the issue of mediumship. This was the author's first book on psychical research. [P]

1439. *Hodgson, Richard.*

"A Further Record of Observations of Certain Phenomena of Trance." *Society for Psychical Research. Proceedings* 13 (1898):284–582.

A continuation of articles written by F.W.H. Myers et al. in 1890 (*see* entry number 1274) and by Hodgson himself in 1892 (*see* entry number 1317) describing the investigation of the remarkable mental medium from Boston, Leonora Piper (1859–1950). Hodgson summarizes the findings of the previous articles and proceeds to give a lengthy description and analysis of his séances with the medium from 1892 to 1895. Although in his article of 1892 Hodgson did not state any definite conclusions about the genuineness of Mrs. Piper's mediumship, in this article he says that he has no doubt that the chief communicators who manifest through Mrs. Piper are indeed the deceased persons they claim to be, and that they have survived death and are able to communicate with the living. [P]

1440. *James, William.*

Human Immortality: Two Supposed Objections to the Doctrine. Boston and New York: Houghton, Mifflin & Co., 1898, 70 pp.

[P]

1441. *Janet, Pierre.*

Névroses et idées fixes. I. Études expérimentales sur les troubles de la volonté, d'attention, de la mémoire, sur les émotions, les idées obsédantes et leur traitement, par Dr Pierre Janet. II. Fragments des leçons cliniques du mardi sur les névroses, les maladies produits par les émotions, les idées obsédantes et leur traitement. Par Dr F. Raymond et le Dr Pierre Janet. 2 vols. Paris: Félix Alcan, 1898, iii + 492; x + 559 pp.

A classical study of the subconscious elements at work in the formation of neuroses. Janet formulated the concept of "subconscious fixed ideas" in his *L'automatisme psychologique* (1889, entry number 1235). This is the notion that an idea, excluded from personal consciousness because of its disturbing content, can develop unobserved in a hidden part of the personality and work subconsciously to produce all manner of neurotic symptom. In the present work he develops this insight into a comprehensive theory of the neurosis. The first volume is Janet's systematic presentation of his ideas, while the second constitutes a kind of clinical appendix to the first, containing short clinical lessons along with a description of cases treated by Janet and F. Raymond in their laboratory at the Salpêtrière. In presenting his theory, Janet discusses the nature of fixed ideas, how they manifest in symptoms, how they may be uncovered, and the treatment to be used in their management. Janet distinguishes between "hysteria" and "psychasthenia" and describes "psychasthenia" as a state involving "mental weakness" manifesting in obsessions, phobias, and other symptoms. "Hysteria," on the other hand, is characterized by a narrowing of the field of consciousness manifesting in such symptoms as anesthesias, abulias, and amnesias. Treatment of the neurosis must, Janet asserts, involve the bringing to consciousness of the fixed ideas, the dissolution of those fixed ideas, and the reeducation of the patient for healthy living. *Névroses and idées fixes* became an influential textbook for the therapy of the neurosis and helped established Janet as one of the nineteenth century's preeminent clinicians. [H]

1442. *Journal of Medical Hypnotism.*

1898.

See *Suggestive Therapeutics* (1896), entry number 1409. [H]

1443. *Lang, Andrew.*

The Making of Religion. London: Longmans, Green and Co., 1898, vii + 380 pp.

Presents anthropological material related to the rise of religion and examines it in a new light. Lang deplored the fact that so many anthropologists did not make use of recent findings in the field of psychology and psychical research to throw light on the experiences of "savages" relevant to the development of religion. He believed that the work of psychologists on trance states and secondary personalities and the information gathered by psychical researchers on mediumship, possession, and apparitions should be taken into account in any serious investigation of primitive attitudes toward the supernatural. That is his goal in *The Making of Religion.* [P]

1444. *Lehmann, Alfred Georg Ludwig.*

Aberglaube und Zauberei von den ältesten Zeiten an bis in die Gegenwart. Stuttgart: Ferdinand Enke, 1898, xii + 556 pp.

A study of superstitions and the occult over the ages. Lehmann believes that most superstitions arose as erroneous attempts to find scientific explanations for the nature of things. He includes among the "occultists" modern investigators of the paranormal such as William Crookes (1832–1919), whose investigations of "psychic force" he attempts to discredit. Among the superstitions studied, Lehmann includes modern spiritualism, which he considers to be *a priori* inadmissible. [P]

1445. *Louis, Eugène Victor Marie.*

Les origines de la doctrine du magnétisme animal: Mesmer et la Société de l'Harmonie. Paris: Société d'éditions scientifiques, 1898, 56 pp.

A well-researched history of Mesmer and of animal magnetism during Mesmer's lifetime. The treatise was written as a doctoral thesis for the Faculté de Médecine de Paris. [H]

1446. ***Magnetic Journal.***
Vols. 1–3 (?); 1898–1900 (?).

Published in Nevada, Missouri. [H]

1447. ***Majewski, Adrien.***
Médiumnité guérissante par l'application des fluides éléctrique, magnétique et humain. Paris: P. G. Leymarie, (1898), 67 pp.

A discussion of human magnetism and electricity as it applies to healing. It is based on the notion of fluidic emanations proceeding from the human body. [H & P]

1448. ***Robinson, William Elsworth.***
Spirit Slate Writing and Kindred Phenomena. New York: Munn Co., 1898, v + 148 pp.

A description of conjuring tricks used in connection with slate writing and other phenomena associated with spiritualistic mediumship. [P]

1449. ***Schofeld, Alfred Taylor.***
The Unconscious Mind. London: Hodder and Stoughton, 1898, xv + 436 pp.

[H]

1450. ***Sidis, Boris.***
The Psychology of Suggestion: A Research into the Subconscious Nature of Man and Society. New York: D. Appleton, 1898, (2) + x + 386 pp.

A student of William James (1842–1910) at Harvard, Sidis went on to become a pioneer in psychological experimentation in the laboratory. His main area of interest was suggestion and its connection with hypnotism and hypnoid states and this book is one of the most important works on suggestion written in the English language. Sidis describes his experiments with suggestion and his conclusion that within the human mind coexist two streams of consciousness that constitute the waking and the subwaking self. The waking self is the ruling self, having the power to investigate its own nature, create ideals, and struggle for them. The subwaking self is stupid, lacking in critical sense, and highly suggestible. The submerged, subwaking self has a broader awareness than that of the waking self, knowing the life of the latter, while the waking self knows nothing of the life of the subwaking self. In addition to this discussion of the self and suggestion, the book includes an important section on social suggestibility and crowd phenomena. [H]

1451. ***Verworn, Max.***
Beiträge zur Physiologie des Centralnervensystems. Erster Theil. Die sogenannte Hypnose der Thiere. Jena: G. Fischer, 1898, iv + 92 pp.

[H]

1899

1452. ***Bormann, Walter.***
Der Schotte Home, ein physiopsychischer Zeuge der Transscendenten im 19. Jahrhundert. Leipzig: Oswald Mutze, 1899, 92 pp.

[P]

1453. ***Braid, James.***
Braid on Hypnotism. Neurypnology; or The Rationale of Nervous Sleep Considered in Relation to Animal Magnetism or Mesmerism and Illustrated by Numerous Cases of Its Successful Application in the Relief and Cure of Disease. A New Edition, Edited with an Introduction Biographical and Bibliographical Embodying the Author's Later Views and Further Evidence on the Subject. By Arthur Waite. London: G. Redway, 1899, xii + 380 pp.

Republication of Braid's *Neurypnology* which had appeared only in a first edition (1843, entry number 465). It is accompanied by important comments and notes by Arthur Waite (1857–1942). Waite lists and summarizes all of the work of Braid known to him and examines the main themes that run through them. He has a very valuable biography of Braid at the beginning. [H]

1454. ***Delanne, Gabriel.***
L'âme est immortelle. Paris: Chamuel, 1899, 468 pp.

This study of the nature of the soul has as its thesis the fact that human beings have a "second body" or "subtle body" that is a vehicle of the soul and that survives death. This enveloping form is called by Delanne the "perispirit." In this book he presents experimental evidence for the existence of such a vehicle, drawing largely on the work of de Rochas and Luys. [P]

1455. ***Du Prel, Karl Ludwig August Friedrich Maximillian Alfred.***
Die Magie als Naturwissenschaft. 2 vols. Jena: H. Costenoble, 1899.

[P]

1456. ***Du Prel, Karl Ludwig August Friedrich Maximillian Alfred.***
Der Tod, das Jenseits und das Leben im Jenseits. Munich: The Author, 1899, 119 pp.

[P]

1457. ***Flammarion, Camille.***
L'inconnu et les problèmes psychiques. Paris: Ernest Flammarion, (1899), xiv + 587 pp.

Flammarion was a famous French astronomer, spiritualist, and psychical researcher strongly influenced by the first great French spiritualist, Allan Kardec (Hyppolyte Rivail) who thought that human survival of death was as much a matter of scientific proof as of belief. In this book, Flammarion attempts to establish the reality of telepathy, clairvoyance, premonitions, and other psychic phenomena. Given the genuineness of such experiences, one must, Flammarion believes, conclude that the soul can exist apart from the body and can act without its benefit. [P]

1458. ***Geley, Gustave.***
L'être subconscient. Essai de synthèse explicative des phénomènes obscurs de psychologie normale et anormale. Paris: Félix Alcan, 1899, 191 pp.

Geley attempts to explain the phenomena of mediumistic trance in terms of a vital energy emanating from the medium and those around and subconsciously directed by the medium. [H & P]

1459. ***Goodrich-Freer, Adela and Bute, John Marquess of.***
The Alleged Haunting of B——— House, Including a Journal Kept During the Tenancy of Colonel Lemesurier Taylor. London: G. Redway. 1899, 249 pp.

One of psychical research's most detailed studies of an allegedly haunted house. The authors include testimony from many first hand witnesses, a detailed description of certain experiments, including seismographic measurement, carried on within the house to determine the source of the phenomena. [P]

1460. ***Goodrich-Freer, Adela.***
Essays in Psychical Research, by Miss X. 2 ed. London: G. Redway, 1899, xv + 330 pp.

A collection of essays by the author concerning subjects related to psychical research. Miss Goodrich-Freer was a well-respected member of the Society for Psychical Research and her investigations of hauntings, crystal gazing, the divining rod and other subjects were significant contributions to the literature. No information is available on the first edition (1899). [P]

1461. *Hélot, Charles.*
Le diable dans l'hypnotisme (Soustraction hypnotique de la conscience. Hypnotisme médical. Evocation du démon. Suggestion, etc. . .). Paris: Bloud et Barral, 1899, 64 pp.

[H & P]

1462. *Jage, Gustav.*
Die Heilmethode des Lebensmagnetismus. Leipzig: Oswald Mutze, 1899.

[H]

1463. *Savage, Minot Judson.*
Life Beyond Death: Being a Review of the World's Beliefs on the Subject, a Consideration of Present Conditions of Thought and Feeling, Leading to the Question as to Whether It can be Demonstrated as a Fact: to Which is Added an Appendix Containing Some Hints as to Personal Experiences and Opinions. New York: G. P. Putnam's Sons, 1899, xv + 336 pp.

A serious study of the question of human survival of death. Beginning with a discussion of the nature of science and a rationale for the claim that psychical research can be said to be a science, Savage proceeds to his own personal belief based on the available evidence that human beings do survive death and that the dead communicate with the living. He does not hold, however, that this opinion has been scientifically demonstrated. [P]

1464. *Schroeder, H. R. Paul.*
Geschichte des Lebensmagnetismus und des Hypnotismus. Von Urfang bis auf den heutigen Tag. Leipzig: Arwed Strauch, 1899, (1) + (5) + 5–681 pp.

The best overall history of animal magnetism and its predecessors before 1900. Schroeder begins with the Egyptians and includes the chief French and German figures in the mesmeric tradition. The work contains 29 portraits. [H]

1465. *Spiritisme moderne. Katie King. Histoire de ses apparitions d'après les documents anglais avec illustrations par Un Adepte.*
Paris: Librairie des sciences psychiques, P. G. Leymarie, 1899, 150 pp.

A collection of articles from the *Revue spirite* concerning the British medium Florence Cook and the materialized apparition, Katie King, who "manifested" in the medium's séances. [P]

1466. *Webb, Arthur L.*
Somnambulism. Chicago: C. H. Kerr, 1899, 45 pp.

[H]

1900 – 1909

1900

1467. ***Binet, Alfred.***
La suggestibilité. Paris: Schleicher frères, 1900, (3) + 391 + (2) pp.

Binet attempts to draw a clear distinction between suggestion and hypnotism and to concentrate on an experimental study of the former. From experiments carried out with 46 subjects, Binet concludes that suggestibility is a normal human quality in the healthy as well as the ill and that it is possible to measure the degree of suggestibility and calculate a suggestibility index for individuals. Here, Binet also describes ingenious methods for producing automatisms, inducing fragments of the organism to act spontaneously and in an automatic way. [H]

1468. ***Cook, William Wesley.***
Practical Lessons in Hypnotism. Chicago: Thompson & Thomas, (1900).

[H]

1469. ***De Laurence, Lauron William.***
Hypnotism, a Complete System of Method, Application and Use. Chicago: Alhambra Book Company, 1900, 188 pp.

De Laurence was a popularizer of theosophical bent who was self-described as the "Moses of the Hindus, High-caste Adept and Famous Occult Magician." Although this work emphasizes the supernormal aspects of hypnotic phenomena, it is a competent and readable summary of information about hypnotism. [H]

1470. ***Flournoy, Theodore.***

Des Indes à la planète Mars; étude sur un cas de somnambulisme avec glossolalie. Geneva and Paris: Ch. Eggiman (Geneva) and Félix Alcan (Paris), 1900, xii + 420 pp.

A classic in the field of psychical research. Flournoy, Professor of Psychology at the University of Geneva and highly respected in his field, heard about a medium in Geneva who exhibited phenomena that he thought deserving of psychological investigation. Her name was Catherine Elise Muller (called "Hélène Smith" in the book to protect her identity). Flournoy was admitted to her circle in 1894 and attended her séances on and off for five years. A number of "departed spirits" manifested through the medium, including Cagliostro and Marie Antoinette. Flournoy concluded that these personalities were the subconscious productions of the medium's mind and not the actual people they claimed to be. Muller also believed she was in touch with life on Mars, and while in trance she spoke a language she claimed to be martian. Through a laborious analysis of the syntax of this language, Flournoy discovered beneath the apparently alien form a structure basically French, the native tongue of the medium. The publication of the book alienated Muller and her friends, who considered themselves betrayed by someone they had befriended. The work is a prototypical study of mediumship from a psychological point of view. Although some questions remained about sources for certain obscure historical information that came through Muller, the overall impression was that she was a sincere but self-deluded medium. [P]

1471. ***[Hall, Caxton].***

Mesmerism & Hypnotism. An Epitome of all the Best Works on the Hypnotic Phases of Psychology in the Form of Question and Answer. By an Adept. Blackpool: Ellis Family, 1900, 71 pp.

[H]

1472. ***Loewenfeld, L.***

Somnambulismus und Spiritismus. Wiesbaden: Bergmann, 1900, 57 pp.

A short treatise on the relationship between somnambulism and such spiritualistic phenomena as telepathy, clairvoyance, and the physical phenomena of mediumship. [H & P]

1473. ***Neal, E. Virgil, and Clark, Charles S. [eds.].***

Hypnotism and Hypnotic Suggestion. A Scientific Treatise on the Uses and Possibilities of Hypnotism, Suggestion and Allied Phenomena. By Thirty Authors. 5 ed. Rochester, New York: State Publishing Company, 1900, xiii + 259 pp.

An excellent collection of essays on hypnotism. Subjects range from the hypnotism of animals to various aspects of hypnotic suggestion. Authors include Max Dessoir, Thomson Jay Hudson, William Newbold, Robert Yerkes and Carl Sextus. Information on the first edition is unavailable. [H]

1474. ***Ottolenghi, Salvatore.***
La suggestione e le facoltà psichiche occulte in rapporto alla pratica legale e medico-forense. Turin: Fratelli Boca, 1900, xvi + 712 pp.

[H & P]

1475. ***Pike, Richard.***
Life's Borderland and Beyond. London: Simkin & Marshall, 1900, 312 pp.

A collection of experiences related to dying, death, and the apparition of the dead to the living. The book was used by William Barrett in his *Death Bed Visions* (1926). [P]

1476. ***Quackenbos, John Duncan.***
Hypnotism in Mental and Moral Culture. New York and London: Harper and Brothers, 1900, xi + (1) + 290 + (1) pp.

A significant study of the use of hypnotic suggestion as a means of removing anti-social and criminal impulses and creating responses based on conscience. Among the anti-social impulses treated by Quackenbos are the cigarette habit, alcoholism, kleptomania, habitual lying, and "moral anaesthesia." He also discusses the use of hypnotic suggestion in treating the insanities and in training "erratic, backward, and unmanageable children." At the end of the book, Quackenbos includes an examination of hypnotic suggestion as an aid to creative inspiration and the development of the arts. [H]

1477. ***El Sol. Revista de historia magnetismo, estudios psiquicos.***
One vol. only; 1900.

Published in Lima and edited by Carlos Paz Soldan. [H & P]

1478. ***Tournier, Anna Marie Valentin.***
Philosophie du bon sens. Le spiritisme devant la raison. Le Dieu de la République. L'infaillibilité papale. Qu'était Jésus? Souvenires inédits sur la 32me demi-brigade. Edition posthume. Tours: Anna Tournier, 1900, xxii + 775 pp.

[P]

1479. ***Villari, Luigi Antonio.***
Spiritismo e magnetismo. Note e confronti polemici. Trani: Vecchi, 1900, 66 pp.

[H & P]

1901

1480. ***Bozzano, Ernesto.***
Lo spiritismo di fronte alla scienza: note polemiche. Genoa: Unione tip., 1901, 54 pp.

[P]

1481. ***Congrès international de l'hypnotisme expérimental et thérapeutique. Procès-verbaux sommaires. Edited by Edgar Bérillon.***
Paris: Imprimerie nationale, 1901.

The second International Congress met in Paris in 1900. [H]

1482. ***Halphide, Alvan Cavala.***
The Psychic and Psychism. Chicago: The Author, 1901, xi + (1) + 19–228 pp.

Describing the twentieth century as the "psychic century," the author attempts to provide a sort of textbook of "psychism," the study of the inner life of the psychic. This includes an investigation of the character of people who produce psychic phenomena, the data of psychic occurences, and the laws and conditions governing that data. Halphide believes that all psychic phenomena can be accounted for as subconscious creations of the mind. [P]

1483. ***Henry, Victor.***
Le langage martien, étude analytique de la genèse d'une lange dans un cas de glossolalie somnambulique. Paris: J. Maisonneuve, 1901, xx + 152 pp.

[P]

1484. ***Hubbell, Gabriel G.***
Fact and Fancy in Spiritualism, Theosophy and Psychical Research. Cincinnati: Clarke, 1901, 208 pp.

The author points out frauds and deceptions often associated with mediumistic phenomena, but concludes that some occurrences cannot be explained in these terms. [P]

1485. ***Hutchinson, Horatio Gordon.***
Dreams and Their Meanings. With Many Accounts of Experiences Sent by Correspondents and Two Chapters Contributed Mainly from the Journals of the Psychical Research Society on Telepathic and Premonitory Dreams. London: Longmans, Green & Co., 1901, 320 pp.

[P]

1486. ***Hyslop, James Harvey.***
"A Further Record of Observations of Certain Trance Phenomena." *Society for Psychical Research. Proceedings* 16 (1901):1–649.

Hyslop writes about his experiences with the medium Leonora Piper (1859–1950) and of the implications of those experiences for psychical research. He includes a lengthy discussion of the "telepathy" as opposed to the "spirit" hypothesis to account for the material obtained from this remarkable medium and makes it clear that he favors the latter explanation. [P]

1487. ***James, William.***
"Frederic Myers's Service to Psychology." *Society for Psychical Research. Proceedings* 17 (1901):13–23.

An eloquent memoriam that places Myers in the history of the psychological study of the unconscious. James regarded Myers's investigation of the "subliminal consciousness" to be of such importance that the task of mapping its extent should thenceforth be called "the problem of Myers." [P]

1488. ***Jastrow, Joseph.***
Fact and Fable in Psychology. Boston and New York: Houghton, Mifflin and Company, 1901, xvii + (1) + 375 + (1) pp.

A collection of essays on subjects relating to psychical research by one of its most serious critics. It takes up such subjects as the logic of telepathy and the psychology of spiritualism, which Jastrow considered a disease of society. [P]

1489. ***Journal of Magnetism.***
See *Suggestive Therapeutics* (1896, entry number 1409).

[H]

1490. ***Lowenfeld, L.***
Der Hypnotismus.Handbuch der Lehre von der Hypnose und der Suggestion, mit besonderer Berücksichtigung ihrer Bedeutung für Medicin und Rechtspflege. Wiesbaden: J. F. Bergmann, 1901, xii + 522 pp.

A massive handbook on the use of hypnotism in medicine and psychotherapy. Lowenfeld includes a well-constructed discussion of the theory of the nature of hypnotism. [H]

1491. ***Luce e ombra. Revista mensile illustrata di scienze spiritualiste.***
Vols. 1–25 +; 1901–1925 +.

Published in Milan. In 1931 the name was changed to *Ricerca psichica . . .*

[P]

1492. *Mason, Rufus Osgood.*

Hypnotism and Suggestions in Therapeutics, Education, and Reform. New York: H. Holt & Co., 1901, vii + 344 pp.

[H]

1493. [*Metzger, Daniel*].

Autour des "Indes à la Planète Mars." Geneva and Paris: Leymarie, 1901.

[P]

1494. *New Thought.*

1901.

See *Suggestive Therapeutics* (1896, entry number 1409). [H]

1495. *Prince, Morton.*

"The Development and Geneology of the Misses Beauchamp: A Preliminary Report of the Case of Multiple Personality." *Society for Psychical Research. Proceedings* 15 (1901): 466–83.

The first account of the famous "Miss Beauchamp" case of multiple personality investigated by this important figure in American psychopathology. For the fuller account, see *The Dissociation of a Personality,* 1905, entry number 1559. [H]

1496. *Quatrième congrès international de psychologie. Compte rendu.* Edited by Pierre Janet.

Paris: Félix Alcan, 1901, iii + 814 pp.

At the Second International Congress of Psychology held at London under the auspices of the Society for Psychical Research in 1892, issues related to psychical research first became an important part of the agenda. Although psychical research continued to be prominent in this, the fourth congress, Janet, who was Secretary of the congress, removed 30 articles related to psychical research for the proceedings. This is indicative of the then current struggle between supporters of psychical research and those members of the academic establishment who opposed its legitimation. [P]

1497. [*Raupert, John Godfrey Ferdinand*].

The Dangers of Spiritualism, Being Records of Personal Experiences with Notes and Comments. By a Member of the Society for Psychical Research. London: Sands, 1901, 153 pp.

The author attests to the reality and objectivity of many paranormal phenomena, but points out the danger of prolonged experience with these manifestations. [P]

1498. ***Reid, Hiram Alvin.***
Unseen Faces Photographed. A Condensed Report of Facts and Findings During a Year of Most Critical Research into the Phenomena Called Spirit Photography. Los Angeles and Pasadena: n.p., 1901, 52 pp.

[P]

1902

1499. ***Congrès international de l'hypnotisme experimentale et thérapeutique. Comptes rendus.*** Edited by Edgar Bérillon and Paul Farez.
Paris: Revue de l'hypnotisme, 1902, 320 pp.

This second congress, held in Paris in 1900, included among its officers such prominent figures as Liébeault, Bernheim, Richet, Liégeois, Pitres, and Richer. The overall feeling of the participants is reflected in the opening remarks by the president of the congress, F. Raymond, who declares that hypnotism has finally come into its own, being now established as a legitimate branch of neurology. [H]

1500. ***Delanne, Gabriel.***
Recherches sur la médiumnité. Étude de travaux des savants, l'écriture automatique des hystériques, l'écriture mécanique des médiums, preuves absolues de nos communications avec le monde des esprits. Paris: Librairie des sciences psychiques, 1902, xii + 515 pp.

A work of great significance in psychical research. In this book, Delanne, although writing from an avowedly spiritualist point of view, shows that he is able to bring a keen investigative eye to the study of mediumship. *Recherches sur la médiumnité* is a work of impressive scholarship in which Delanne shows a familiarity with the psychological literature relating to automatisms and the influence of the subconscious or conscious acts. Beginning with a comparison between the productions of mediums and hysterical automatisms and a discussion of the light shed on mediumistic phenomena by the work of Binet, Janet, and others, Delanne examines cases of mediumship that give evidence of clairvoyance, telepathy, precognition and other paranormal phenomena and looks at the possible influence of suggestion on these phenomena. The last part of the book is concerned with proofs that the dead communicate with the living, bringing to bear evidence that Delanne believes can only be explained in terms of spirit communication. [P]

1501. ***Duff, Edward Macomb and Allen, Thomas Gilchrist.***

Psychic Research and Gospel Miracles. A Study of the Evidences of the Gospel's Superphysical Features in the Light of the Established Results of Modern Psychical Research. New York: Thomas Whittaker, 1902, xii + (2) + 5–396 pp.

The findings of psychical research to date. The authors then point out the ways they believe these findings verify the miraculous works attributed to Christ. [P]

1502. ***Ellsworth, Robert G.***

A Key to Hypnotism. A Complete and Authentic Guide to Clairvoyance, Mesmerism and Hypnotism. Philadelphia: David McKay, 1902, iii + iii + 11–140 pp.

Ellsworth draws a clear distinction between mesmerism and hypnotism, seeing mesmerism as involving a physical exchange and hypnotism as an induced psychological state. Given the distinction, he argues that one can mesmerize but not hypnotize a young child (under three). With regard to hypnotism, Ellsworth seems most strongly influenced by the findings of Charcot. With regard to mesmerism, his emphasis is on the physical and curative effects. [H & P]

1503. ***Flournoy, Théodore.***

Nouvelles observations sur un cas de somnambulisme avec glossolalie. Geneva: C. Eggimann, 1902, (101)–255 pp.

This study, first published in the *Archives de psychologie de la Suisse Romande,* provides further observations and reflections on the mediumship of "Hélène Smith" (see Flournoy's *Des Indes à la planète Mars,* 1900, entry number 1470). Flournoy's conclusions remain as he had expressed them in the earlier work, but he adds further information concerning various aspects of the case. Among other things, he quotes the findings of Professor Victor Henry on the "martian" language developed in Hélène Smith's mediumship (*see* Henry, *Le langage martien,* 1901, entry number 1483). Henry traced what he believed to be the unconscious origins of that phenomenon. Flournoy's further elaboration of material related to the "personalities" that communicated through the medium strengthens, in his view, the case for the unconscious origin of these figures. [P]

1504. ***Harte, Richard.***

Hypnotism and the Doctors. 2 vols. London: L. N. Fowler & Co., 1902, xxxii + 128; iii + (1) + 253 pp.

An excellent study of hypnotism and the response it elicited from the medical profession that was originally intended as a three volume work. Of the three volumes, however, only two appeared. Volume one deals with Mesmer, his theory and the history of his work, and with Puységur and the consequences of his discovery of artificial somnambulism. Volume two takes up the findings of the second French commission, French practitioners such as Lafontaine and Du Potet, Braid's discoveries, and certain American magnetizers. Volume three, which never appeared, was to have focused on modern hypnotism: Liébeault, Charcot, the various schools of hypnotism, and "recent mesmerism." [H]

1505. *Journal für Psychologie und Neurologie. Zugleich Zeitschrift für Hypnotismus.*

1902.

See *Zeitschrift für hypnotismus, Suggestionstherapie, Suggestionslehre und verwandte psychologische Forschungen* (1892), entry number 1333. [H]

1506. *Jung, Carl Gustav.*

Zur Psychologie und Pathologie sogenannter occulter Phänomene. Leipzig: Oswald Mutze, 1902, (2) + 121 pp. ENGLISH: "On the Psychology and Pathology of So-called Occult Phenomena." In *Psychology and the Occult.* Translated by R.F.C. Hull. Princeton, New Jersey: Princeton University Press, 1977.

Jung's dissertation for his doctorate in medicine, written under the direction of Eugen Bleuler (1857–1939). Here Jung discusses examples of double consciousness and somnambulism and treats in great detail the case a young female medium whose seances Jung had himself witnessed. He attempts to explain the phenomena observed in terms of psychological mechanisms such as "cryptomnesia," the emergence into consciousness of forgotten information without the medium recognizing that information as memory. [P]

1507. *Moll, Albert.*

Gesundbeten Medizin und Okkultismus. Berlin: H. Walther, 1902, 48 pp. ENGLISH: *Christian Science, Medicine, and Occultism.* Translated by F. J. Rebman. London: Rebman, 1902.

[H & P]

1508. *Podmore, Frank.*

Modern Spiritualism, a History and a Criticism. 2 vols. London: Methuen and Co., 1902, xviii + 307; xii + 374 pp.

Podmore was a very good historian and he brings that ability to bear in this work. After providing a brief survey of early psychic phenomena, he launches into an investigation of the era beginning with Mesmer and ending in 1900 and connects the rise of modern spiritualism to the spread of mesmerism throughout France, Germany, Britain, and the United States. From its beginnings as a movement with the events that took place in the home of the Fox sisters in New York state in 1848, he traces the spread of spiritualism over the "ground prepared by mesmerism" throughout the United States and eventually to England. Podmore then categorizes the various kinds of mediumship that developed within the movement and discusses the principal mediums of the nineteenth century from the point of view of the reliability and genuineness of their phenomena. As in his earlier writings, Podmore holds that mediumship may be adequately explained in terms of fraud, bad observation, or, in the case of some mental phenomena, unconscious telepathy. [P]

1509. [*Rebell, Hughes*].

Les anglaises chez elles. Le magnétisme du Fouet, ou les indiscrétions de Miss Darcy, traduit de l'anglais par Jean de Villiot. Paris: Charles Carrington, 1902, 283 pp.

[H]

1510. *Rochas d'Aiglun, Eugène Auguste Albert de.*

Les frontières de la science. 2 vols. Paris: Librairie des sciences psychologiques, 1902 and 1904, 126; 212 pp.

A valuable collection of information on the state of psychical research at the turn of the century. [H & P]

1511. *Sage, Michel.*

Madame Piper et la Société Anglo-Américaine pour les Recherches Psychiques. Paris: Leymarie, 1902, viii + 272 pp.

A rather uncritical treatment of the phenomena connected with Leonora Piper (1859–1950), one of the most famous mental mediums. [P]

1512. *Savage, Minot Judson.*

Can Telepathy Explain? Results of Psychical Research. New York: G. P. Putnam's Sons, 1902, xvi + (2) + 248 pp.

Savage here presents evidence for the reality of the spirit world. He attempts to show that the facts cannot be adequately explained simply by telepathy but must involve the action of discarnate human spirits. [P]

1513. *Sidis, Boris.*

Psychopathological Researches: Studies in Mental Dissociation. New York: G. E. Stechert, 1902, xxii + 329 + (20) pp.

A significant study of the nature of mental dissociation and synthesis, of material on manifestations of the subconscious in conscious life. Sidis examines issues such as elements of intelligence in dissociated subconscious systems and subconscious motor operations, and raises questions about the place of mental dissociation in psychotic states. This volume also includes essays on various aspects of dissociation by William A. White and George M. Parker. [H]

1514. ***Vaschide, Nicolas and Piéron, Henri.***
"Contribution à l'étude expérimentale des phénomènes télépathiques." *Bulletin de l'Institut général de Psychologie* 2 (1902):117–41.

[P]

1903

1515. ***Bennett, Edward T.***
The Society for Psychical Research: Its Rise and Progress and a Sketch of its Work. With Facsimile Illustrations of Three Pairs of the Thought-Transference Drawings. London: R. Brimley Johnson, 1903, 57 pp.

Bennett describes himself as "assistant secretary to the Society for Psychical Research from 1882 to 1902." Since he was with the Society from its inception, he was in a good position to give a sketch of its history. Much of this little work is devoted to evaluating progress in the investigation of telepathy, hypnotism, and communication with the dead, but Bennett also includes a section of Myers' concept of subliminal self. [P]

1516. ***Bozzano, Ernesto.***
Ipotesi spiritica e teoriche scientifiche. Geneva: Donath, 1903, 531 pp.

An excellent study of spiritualistic mediumship by the dean of Italian psychical researchers. Bozzano discusses, among other things, the mediumship of Eusapia Palladino and possible explanations for the phenomena that she produced. He also includes a philosophical discussion of alternative frameworks for understanding how spiritualistic phenomena are possible. [P]

1517. ***Bramwell, J. Milne.***
Hypnotism: Its History, Practice and Theory. London: G. Richards, 1903, xiv + 478 pp.

Bramwell was a practitioner of medical hypnosis with a strong interest in its history. In this volume, one of the most important and informative histories of hypnotism, he uses an historical analysis to evaluate the true nature of the phenomenon. Beginning with Mesmer, he traces developments through the Abbé Faria, Elliotson and Esdaile, to Liébeault and the formation of the Nancy School, to whose insights along with those of Braid, he was obviously partial. He did not believe the mesmerists proved their case for a magnetic fluid; neither did he agree with the pathologically oriented views of the school of Charcot. He does treat them all, however, and quite fairly. His description of the history of medical hypnotism is valuable and his discussion of the various explanatory theories of hypnotism is enlightening. [H]

1518. ***Bruce, H. Addington.***

The Riddle of Personality. New York: Funk and Wagnalls, 1903, 247 pp.

Bruce looks into the "nature and destiny of human personality" as illuminated by psychological and psychic experiences. Drawing mainly on the animal magnetism/hypnotism tradition on the one hand and the mediumistic tradition on the other, he comes to the conclusion that human personality not only survives but continues to develop beyond death. [H & P]

1519. ***Grasset, Joseph.***

L'hypnotisme et la suggestion. Paris: O. Doin, 1903, 534 pp.

[H]

1520. ***Grasset, Joseph.***

Leçons de clinique médicale faites à l'hôpital Saint-Eloi de Montpellier. Montpellier and Paris: Goulet & fils and Masson, 1903, vi + 755 pp.

Grasset, a student of hypnotism and hysteria, takes up issues related to psychical research, such as spiritualism and hauntings. [H & P]

1521. ***Hudson, Thomson Jay.***

The Law of Mental Medicine: the Correlation of the Facts of Psychology and Histology in Their Relation to Mental Therapeutics. Chicago: A. C. McClurg, 1903, xix + 281 pp.

A study of healing through the power of the mind. Using his notion of a double mental organization in human beings, Hudson provides a comprehensive view of healing through suggestion and the action of the subjective mind. [H]

1522. ***Janet, Pierre.***

Les obsessions et la psychasthénie. I. Études cliniques et expérimentales sur les idées obsédantes, les impulsions, les manies mentales, la folie du doute, les tics, les agitations, les phobies, les délires du contact, les angoisses, les sentiments d'incomplétude, la neurasthénie, les modifications des sentiment du réel, leur pathologie et leur traitement. Par le Dr Pierre Janet. II. Fragments de leçons cliniques du mardi sur les états neurasthéniques, les aboulies, les sentiments d'incomplétude, les agitations et les angoisses diffuses, les algies, les phobies, les délires du contact, les tics, les manies mentales, les folies du doute, les idées obsédantes, les impulsions, leur pathogenie et leur traitement. Par le Dr F. Raymond et le Dr Pierre Janet. 2 vols. Paris: Félix Alcan, 1903, xii +764; xxiv +543 pp.

Continues the formulation of a comprehensive theory of the neurosis and therapeutic treatment begun in *Névroses et idées fixes* (1898, entry number 1441). As in that earlier work, the first volume provides a systematic presentation of Janet's ideas, while the second volume is a sort of clinical appendix to the first, containing short clinical lectures along with a description of cases treated by Janet and Raymond in their laboratory at the Salpêtrière. Among new ideas developed in this work is that of "presentification," the process of making present a state of mind or a group of phenomena. It is a synthetic power, one that operates on the highest level of the psyche and is at the heart of the ability to grasp reality. This level is at the top of a five tier hierarchy of psychic functioning. Functioning at the top level requires a greater degree of psychological tension. As one descends through lower, less reality oriented levels of functioning, the tension or force required is less. With the concept of hierarchical levels of psychic functioning, Janet revises his view of mental energy, paying more attention to the quality than to the quantity of energy expended. The concept also provides him with a solid psychological framework for explaining the nature of obsessions and related psychasthenic disturbances. [H]

1523. *Lang, Andrew.*

The Valet's Tragedy and Other Studies. London: Longmans, Green and Co., 1903, x + (4) + 336 pp.

A collection of historical mysteries, some of which have supernatural or paranormal overtones. The accounts are factual, but the mysteries are not solved. [P]

1524. [*Lunt, Edward D.*].

Mysteries of the Seance and Tricks and Traps of Bogus Mediums. A Plea for Honest Mediums and Clean Work. Boston: Lunt Bros., 1903, 64 pp.

[P]

1525. *Myers, Frederic William Henry.*

Human Personality and Its Survival of Bodily Death. 2 vols. London: Longmans, Green, and Co., 1903, xlvi + 700; xx + 660 pp.

Without question the most important single work in the field of psychical research. Myers, one of the founders of the Society for Psychical Research, served as its secretary from 1888–1899 and its president for the year 1900. During the 1880s and 1890s he authored many articles on psychical research in the Society's *Proceedings* and in other periodicals such as the *Nineteenth Century* and the *Fortnightly Review*. With Edmund Gurney and Frank Podmore, he co-authored the monumental *Phantasms of the Living* (1886, entry number 1144). In his day Myers was a noted classical scholar and a poet of distinction. His academic training was at Cambridge University, the birthplace of the Society for Psychical Research. He was a fellow of the University and a lecturer in classics there for some years. Myers was a man of great intellectual versatility and accomplishment. Theodore Flournoy called Myers one of the most remarkable personalities of his time in the realm of mental science. William James said of Myers that he was a wary critic of evidence, a skilful handler of hypotheses, a learned neurologist and an omnivorous reader of biological and cosmological matter. *Human Personality* was published after Myers' death in 1901. He had nearly completed the writing by that time and he had entrusted the press work to Miss Alice Johnson. She and Richard Hodgson carried out whatever editing was still needed after Myers' death. *Human Personality* is equally a work of psychology and of psychical research. Myers draws upon the vast literature of both fields and finds a place for every phase of human experience through his unifying concept of the "subliminal self." According to Myers, the supraliminal self is that self with which we ordinarily identify. It embraces all our conscious, purposeful dealings with reality. The subliminal self, on the other hand, represents man's central, greater self. Myers sees the subliminal self as the enveloping mother-consciousness in each of us. From this matrix our supraliminal conscious self is precipitated like a crystal. The supraliminal self has as its job to deal with the vicissitudes of the world we live in. Its task is, therefore very precise, and over the aeons the supraliminal self has evolved to its present form to do that task better and better. The supraliminal self, the self we ordinarily call "I", is a very limited and specialized production of the greater subliminal self. The subliminal self is the source of all those experiences and impulses which arise unplanned and unbidden from within—ranging from primitive animal impulses to flashes of genius and clairvoyant intuition. It is the source of the worst and of the greatest in us, and its territory is as yet largely unexplored. William James said, "For half a century now, psychologists have fully admitted the existence of the subliminal mental region . . . but they have never definitely taken up the question of the extent of that region, never sought explicitly to map it out. Myers definitely attacks the problem which, after him, it will be impossible to ignore. What is the precise constitution of the subliminal–such is the problem which deserves to figure in our science hereafter as the problem of Myers." In *Human Personality* Myers puts together a remarkable compendium of the literature of psychical and psychological research up to the end of the nineteenth century, with his concept of the subliminal self providing the unifying framework for the vast array of material. The work features chapters on disintegrations of personality (including the best summary of the literature of multiple personality to that time), genius, sleep, hypnotism (containing an extensive study of the history and theories of hypnotism, along with Myers' own hypotheses about its nature), sensory automatisms (including the phenomena of clairvoyance, crystal gazing, telepathy, apparitions of the living and hallucinations), motor automatisms (including automatic writing, automatic drawing and automatic speaking), phantasms of the dead (hauntings and apparitions of departed spirits), and a chapter on trance, possession and ecstasy (with a discussion of mediumship, spirit-possession, and visionary hypnotic states). The appendices with extensive descriptions of cases equal the length of the main text. The publication of *Human Personality* marked a high point in the history of psychical research. Myers undertook to assimilate and unify the work that researchers had carried out over the previous thirty years. His book became the point of reference for every later attempt to synthesize the data of psychical research. [H & P]

1526. ***Maxwell, Joseph.***
Les phénomènes psychiques: recherches, observations, méthodes. Paris: Félix Alcan, 1903, xi + 317 pp.

Maxwell was a distinguished French lawyer who devoted a great deal of his time to the pursuit of psychical research. He investigated many mediums and came to conclusions that were quite original. He was one of the few researchers to take up the problem of mediumistic "personification," that is, the way the phenomena in séances present themselves as coming from an individual intelligence. He noted that the personification seemed to vary in its form depending on the composition of the group carrying out the sitting. In the book Maxwell also introduces a helpful system of classification for psychic phenomena, making a distinction between physical and mental phenomena. Charles Richet, who had great respect for Maxwell, took up this classification and it became the accepted one for many years. [P]

1527. ***La medianité: rivista mensile di scienze psico-fisiche e morali.***
One vol. only; 1903

Edited by Enrico Carreras and published in Rome, this periodical called itself "a monthly review of spiritualism, telepathy, telekinesis, clairvoyance, premonitions, somnambulism, animal magnetism, hypnotism, and positive philosophy." [H & P]

1528. ***Podmore, Frank and Cook, E. Wake.***
Spiritualism: Is Communication with the Other World an Established Fact? Pro – E. Wake Cook . . . Con – Frank Podmore. London: Isbister & Co., 1903, 238 pp.

[P]

1529. ***Schiller, Ferdinand Canning Scott.***
Humanism: Philosophical Essays. London: Macmillan, 1903, xxvii + 297 pp.

A philosophical study of human life, strongly influenced by the pragmatism of William James. The author's discussion of the possibility of a future life is relevant to the attempt of psychical research to determine whether people survive death, and if so, in what form. [P]

1530. ***Sollier, Paul Auguste.***
Les phénomènes d'autoscopie. Paris: Alcan, 1903, 175 pp.

[P]

1904

1531. *Baraduc, Hippolyte Ferdinand.*

Les vibrations de la vitalité humaine. Méthode biométrique appliquée aux sensitifs et aux névroses. Paris: J. B. Baillière, 1904, viii + 280 pp.

Baraduc describes the results of his experimentation with fluidic emanations from the human body and interprets them in occult terms. [H & P]

1532. *Bennett, Edward T.*

Twenty Years of Psychical Research: 1882–1901. With Facsimile Illustrations of Thought-Transference Drawings. London: R. Brimley Johnson, 1904, 66 pp.

Bennett's long association with the Society for Psychical Research is apparent in this little work, which provides a very brief history of psychical research to 1901 and then discusses progress made in telepathy, hauntings, and the physical phenomena of spiritualism. [P]

1533. *Bolton, Gambier.*

Psychic Force, an Experimental Investigation of a Little-known Power. London: n.p., 1904, 96 pp.

[P]

1534. *Champville, Gustave Fabius de.*

Pour transmettre sa pensée; notes et documents sur la télépathie ou transmission de pensée. Paris: Librairie du magnétisme, (1904), 49 pp.

[P]

1535. *Chautard, E.*

Les révélations d'un magnétiseur, trucs ingénieux employés au théatre pour obtenir les phénomènes de la transmission de pensée, du magnétisme et de l'hypnotisme. Montceau-les-Mines: (A. Lemoine) Imprimerie nouvelle, Charles & Delorme, 1904, vii + 67 pp.

[H & P]

1536. *Denis, Léon.*

Dans l'invisible. Spiritisme et médiumnité. Traité de spiritualisme expérimental. Les faits, les lois. Phénomènes spontanés. Typtologie et psychographie. Les fantômes des vivants et les esprits des morts, la médiumnité à travers les âges. Paris: Leymarie, 1904, 466 pp.

[P]

1537. ***Funk, Isaac K.***

The Widow's Mite and Other Psychic Phenomena. New York: Funk and Wagnalls, 1904, xiv + 538 pp.

Funk, who was the director and principal owner of Funk and Wagnalls Publishing Company, had a strong personal interest in psychical research. This book, written from a peculiarly American point of view and citing mainly American cases, presents the results of twenty-five years of investigation of psychical phenomena including telepathy, clairvoyance, clairaudience, apparitions, secondary personalities and possession, communication with the dead, and spirit photography. Of particular interest is a section that deals with a spiritualist family "circle" which met regularly to experiment with psychic phenomena. This circle was responsible for the finding of a "widow's mite," an ancient coin which had been put away in a safe and forgotten place. [P]

1538. ***Grasset, Joseph.***

Le spiritisme devant la science. Montpellier and Paris: Coulet et fils (Montpellier) and Masson (Paris), 1904, xxix + 392 pp.

[P]

1539. ***Hudson, Thomson Jay.***

The Evolution of the Soul and Other Essays. Chicago: A. C. McClurg, 1904, xi + (1) + 344 pp.

A posthumously published collection of essays that deals with areas touched on in Hudson's previous publications, such as science and the future life, man's psychic powers, prophecy, and hypnotism. [H & P]

1540. ***Journal de psychologie normal et pathologique.***

Vols. 1–22 +; 1904–1925 +.

Founded and edited by Pierre Janet and Georges Dumas. Vols. 13–16 (1916–1919) were not published. [H]

1541. ***Maxwell, Joseph.***

Un récent procès spirite. L'aventure du médium aux fleurs. Bordeaux: G. Gounouilhou, 1904, 40 pp.

Description of the trial of a Berlin medium, Anna Rothe, convicted of fraud. [P]

1542. ***Raupert, John Godfrey Ferdinand.***

Modern Spiritism: A Critical Examination of Its Phenomena, Character, and Teaching in the Light of Known Facts. London: Sands & Co., 1904, vi + (2) + 248 pp.

Raupert expresses his alarm at the spread of spiritualism and what he considers its undermining of religious belief. He denies the truth of its basic tenet that the spirits of the dead communicate with the living. After examining the phenomena of spiritualism, the form of mediumistic communications, and their content, Raupert concludes that they cannot be due to the action of departed spirits and that they directly contradict the fundamental truths of the Catholic faith. [P]

1543. ***Renterghem, Albert Willem van.***
Kort Begrip der psychische Geneeswi jze voodracht gehouden op uitnoodiging van het Bestur van "Ons Huis Buiten de Muiderpoort" . . . Amsterdam: F. van Rosen, 1904, 127 + (1) pp.

[H]

1544. ***Riko, A. J.***
Handbuch zur ausübung des Magnetismus, des Hypnotismus, der Suggestion, der Biologie und verwandter Fächer. Leipzig: M. Altmann, 1904 (4) + 167 pp.

A German translation of The Third Dutch edition. [H]

1545. ***Rossi, Pasquale.***
Les suggesteurs et la foule. Psychologie des meneurs: artistes, ovateurs, mystiques, guerriers, criminels, écrivains, enfants, etc. . . . Paris: A. Michalon, 1904, xii + 222 pp.

[P]

1546. ***Sage, Michel.***
Le sommeil naturel et l'hypnose. Leur nature, leurs phases. A qu'ils nous disent en faveur de l'immortalité de l'âme. Paris: Félix Alcan & Leymarie, 1904, 367 pp.

[H & P]

1547. ***Sidis, Boris and Goodhart, Simon P.***
Multiple Personality: an Experimental Investigation into the Nature of Human Individuality. New York: D. Appleton, 1904.

An important study of a case of multiple personality and the nature of dissociation. The case described is that of Rev. Thomas Carson Hanna who, in 1897, struck his head in an accident and from that time began to live a dual life. He manifested a new state of consciousness in which he remembered nothing of his previous life and had to relearn basic language and motor skills. After a period of time, this "secondary state" began to alternate with his "primary state," in which he regained the memories of his life and could exercise his normal motor skills. Through intensive therapeutic work, the two states were eventually fused into one. This case is described in the center section of the book, preceded by Sidis' discussion of neuron organization in its relationship to multiple personality and followed by Sidis' treatment of "consciousness and multiple personality." This later section contains extremely important material on conscious and subconscious elements in multiple personality and on the place of "hypnoid" and "hypnoidic" states in extreme forms of dissociation. Here Sidis makes some remarkable statements about the normality of multiplicity in human psychological experience, suggesting that the phenomena of multiple personality, "far from being mere freaks, monstrosities of consciousness, . . . are in fact shown to be necessary manifestations of the very constitution of mental life. *Multiple consciousness is not the exception, but the law. . . . One great principle must be at the foundation of psychology, and that is the synthesis of multiple consciousness in normal, and its disintegration in abnormal mental life*" (pp. 364–65). [H]

1905

1548. *Annals of Psychic Science.*

1905.

See *Annales des sciences psychiques* (1891, entry number 1283). [P]

1549. *Baylina, Ignacio Ribera.*

Patologia de las enfermadades epilépticas y mentales, con un estudio del hombre en su modo de ser fisico moral. Tratado de psico-terapia practica y racional, en el que se determinan las leyes que anteceden a los fenomenos del sonambulismo, hipnotismo, sugestion y su relativa afinidad con las enfermedades morales. Barcelona: n.p., 1905, 336 pp.

[H]

1550. *Bekhterev, Vladimir Mikhailovich.*

Die Bedeutung der Suggestion im sozialen Leben. Wiesbaden: J. F. Bergmann, 1905, ix + 142 pp.

Bekhterev, founder of the Psychoneurological Institute in St. Petersburg, was renowned for his system of "reflexology," the study of reflex action of the brain as the source of psychological phenomena. This work examines the implications of the use of suggestion in groups and in society at large. [H]

1551. *Bennett, Edward T.*

Automatic Speaking and Writing: A Study. London: Brimley Johnson and Ince, 1905, 72 + (2) pp.

A valuable discussion of the phenomena of automatic speaking and writing written by a careful researcher. Bennett examines the literature of the field and the various possible sources for automatic productions, from the writer's own subconscious mind to departed spirits. [P]

1552. ***Coates, James.***
The Practical Hypnotist. Concise Instructions in the Art and Practice of Suggestion; Applied to the Cure of Disease, the Correction of Habits, Development of Will-Power, and Self-Culture. London: Nichols & Co., 1905, iv + 5–64 pp.

Coates had written a book on mesmerism in 1893 (*see* entry number 1338). In this "how to" manual, a genre for which Coates was famous, he takes up the subject of suggestion and its practical application to cure problems related to bad habits. [H]

1553. ***Driesch, Hans Adolf Eduard.***
Der Vitalismus als Geschichte und also Lehre. Leipzig: J. A. Barth, 1905, x + 246 pp.

Driesch, elected for a term as President of the British Society for Psychical Research in 1926, was Professor of Philosophy at the University of Leipzig and one of Germany's most important psychical researchers. Trained as a biologist, he had become convinced that a mechanistic explanation of the nature of life was insufficient and that a unifying, non-material principle must give order and direction to life processes. This philosophical orientation led him to tackle basic issues concerning the methodology and the nature of scientific proof in psychical research. In this volume, Driesch examines the historical and philosophical relevance of vitalism. [P]

1554. ***Filiatre, Jean.***
Hypnotisme et magnétisme. Somnambulisme, suggestion et télépathie. Influence personnelle . . . resumant . . . tous les connaissances humaines sur les possibilités, les usages et la pratique de l'hypnotisme . . . du magnétisme, de la suggestion et de la télépathie. . . . Saint-Etienne: Genest, (1905), xxii + 405 pp.

A very detailed, practical treatise on hypnotism and animal magnetism. Filiatre describes the techniques used by Mesmer, Puységur, Deleuze, Du Potet, Lafontaine, and many others, and provides a long and useful discussion of psychic phenomena sometimes associated with hypnotism. [H & P]

1555. ***Harnack, Erich.***
Studien über Hautelektrizität und Hautmagnetismus des Menschen. Nach eigenen Versuchen und Beobachtungen. Jena: Gustav Fischer, 1905, (8) + 65 pp.

A late attempt to test the theories of Karl von Reichenbach (1788–1869). The author examines the fingertips as a source of electricity and investigates electrical and magnetic movement originating from the body. [H]

1556. ***Hyslop, James Hervey.***
Science and a Future Life. Boston: Herbert Turner & Co., 1905, xi + (1) + 372 pp.

Hyslop was one of the most important American psychical researchers. He was professor of logic and ethics at Columbia University from 1889–1902. Although his interest in psychical research began with some striking sittings with the famous American medium Leonore Piper in 1888, he did not begin writing on the subject until after his retirement from teaching. In 1906, he organized the new American Society for Psychical Research (there had been an earlier short-lived Society from 1885 to 1890) which published its first *Journal* in January, 1907. *Science and a Future Life* is Hyslop's first book on psychical research and it is one of his most important. Here he deals with the issue of the evidence for survival after death as found in mediumistic communications. Although taking up the problem in its most general form, he concentrates on the material received through Mrs. Piper, examining the various possible explanations for the data, in particular the spiritistic versus the telepathic hypothesis. [P]

1557. ***Jastrow, Joseph.***
The Subconscious. Boston and New York: Houghton, Mifflin & Co., 1905.

An important work on subconscious phenomena and the nature of the subconscious. Criticizing the notion of the "subliminal self" proposed by Myers to account for the same set of mental phenomena, Jastrow sees it as a natural function that is affected by experience much like the conscious mind. [H]

1558. ***Magnin, Emile.***
L'art et l'hypnose. Interprétation plastique d'oeuvres littéraires et musicales. Geneva: Atar, 1905, 463 pp.

[H]

1559. ***Prince, Morton.***
The Dissociation of a Personality: a Biographical Study in Abnormal Psychology. New York: Longmans, Green, 1905, x + 575 pp.

Prince was an important figure in American psychology and a pioneer in the study of mental dissociation. Having studied at Vienna, Strasbourg, Paris, and Nancy, as well as at Harvard Medical School, he developed a special interest in the investigations of Charcot and Janet related to hypnotism and hysteria, an interest that helped give direction to his own original research. A founder and long-time editor of the *Journal of Abnormal Psychology*, president of the American Psychological Association, and a member of the American Society for Psychical Research, Prince eventually taught at Harvard and became Professor of Neurology at Tufts. In 1901, in the *Proceedings* of the British Society of Psychical Research, Prince published his first essay on one of his patients, a woman given the fictitious name of Miss Beauchamp, who exhibited four different personalities. One of the personalities, "Sally," had a co-conscious existence, that is, she persisted as a separate self with a differentiated self-conciousness while each of the others was present. *The Dissociation of a Personality*, an important account of an early attempt to understand and determine what should be the outcome of therapeutic work with multiple personality, describes Prince's discovery of Miss Beauchamp's personalities, the course of her therapy, and Prince's theoretical reflections on the nature of multiple personality. [H]

1560. ***Richet, Charles Robert.***
"Faut-il étudier le spiritisme?" *Annales des sciences psychiques* 15 (1905): 1–41.

[P]

1561. ***Rivista delle riviste di studi psichici. Rassegna mensile di 300 reviste di studi psichici.***
Vols. 1–?; 1905–?

A review of psychical research published in Rome. Complete information is unavailable. [P]

1562. **[*Sidgwick, Arthur and Sidgwick, Eleanor Mildred*].**
Henry Sidgwick: a Memoir. London: Macmillan, 1905, 633 pp.

Henry Sidgwick was a founder of the Society for Psychical Research and its first president. A professor of philosophy at Cambridge and a man of high reputation, he was an important figure in establishing the credibility of the Society as a serious and critical investigator of the paranormal. This biography, written by his wife and son, describes the broad range of his interests and his particular concern for psychical research, which he considered "profoundly important to mankind." [P]

1563. ***Thomas, Northcote Whitridge.***
Crystal Gazing: Its History and Practice, with a Discussion of the Evidence for Telepathic Scrying. London: W. Alexander Moring, 1905, xlvii + 162 pp.

An excellent historical study of "scrying" in its various forms over the centuries. Although there is a brief discussion of the evidence for telepathy in scrying, the book deals mainly with crystal visions and the use of the black stone or shew stone to obtain visual images. [P]

1564. ***Thomas, Northcote Whitridge.***
Thought Transference: a Critical and Historical Review of the Evidence for Telepathy, with a Record of New Experiments 1902–1903. London: Alex Moring, 1905, viii + 214 pp.

[P]

1565. ***Vial, Louis Charles Emile.***
Les erreurs de la science. Paris: The Author, 1905, 293 pp.

Vial, a philosopher and cosmologist, deals with a variety of issues, including sleep, hypnosis, telepathy, and psychic radiations. [H & P]

1566. ***Wallace, Alfred Russel.***
My Life: a Record of Events and Opinions. 2 vols. London: Chapman & Hall, Ltd., 1905, xii + 435; viii + 459 pp.

Volume two of this very readable autobiography contains three chapters on Wallace's involvement with and his views on spiritualism. Wallace's support was one of the major factors in gaining spiritualism wide acceptance in late nineteenth-century Britain and obtaining for it a hearing from a number of the then prominent men of science. This, in turn, helped pave the way for the development of psychical research. [P]

1906

1567. ***Adamkiewicz, Albert.***
Pensée inconsciente et vision de la pensée. Essai d'une explication physiologique du processus de la pensée et de quelques phénomènes surnaturels et psychopathiques. Paris: J. Rousset, 1906, 98 pp.

[P]

1568. ***Bennett, Edward T.***
The Physical Phenomena Popularly Classed under the Head of Spiritualism, with Facsimile Illustrations of Thought Transference Drawings and Direct Writing. London: T. C. & E. C. Jack, 1906, 140 pp.

[P]

1569. ***Christison, John Sanderson.***
The Tragedy of Chicago. A Study in Hypnotism. How an Innocent Young Man, Richard Evans, was hypnotised to the Gallows. Chicago: The Author, 1906, 90 pp.

[H]

1570. *Coates, James.*

Seeing the Invisible: Practical Studies in Psychometry, Thought-transference, Telepathy, and Allied Phenomena. London: L. N. Fowler & Co., 1906, xvi + 298 pp.

Coates was a prolific writer of "how-to" books. Here he brings together a great deal of information about psychometry (sensing vibrations from objects) and telepathy. [P]

1571. *Dalgado, D. G.*

Mémoire sur la vie de l'abbé de Faria. Explication de la charmante légende du chateau d'If dans le roman "Monte-Cristo." Paris: Henri Jouve, 1906, x + 124 pp.

Dalgado taught at the Royal Academy of Sciences at Lisbon. This important book is one of the few sources of information concerning the life of the José Custodio de Faria, whose *De la cause du sommeil lucide* (1819, *see* entry number 294) guaranteed him a central position in the history of hypnotism. [H]

1572. *Grasset, Joseph.*

Le psychisme inférieur. Étude de physiopathologie clinique des centres psychiques. Paris: Chevalier et Rivière, 1906, iii + 516 pp.

[P]

1573. *Hennig, Richard.*

Der moderne Spuk-u. Geisterglaube. Eine Kritik und Erklärung der spiritistischen Phänomene. II. Teil des Werkes "Wunder und Wissenschaft." Mit einem Vorwort von Dr. Max Dessoir. Hamburg: Ernst Schultze, 1906, 367 pp.

A friendly critique of spiritualism. Hennig presents a very thorough treatment of errors such as the tendency to personify the unknown inherent in the spiritualistic evaluation of certain phenomena, a far ranging examination of trance and spirit possession, and a long section on the medium Hélène Smith. [P]

1574. *Hyslop, James Hervey.*

Borderland of Psychical Research. Boston: Herbert B. Turner & Co., 1906, viii + (2) + 425 pp.

An attempt to elucidate issues in "normal" and "abnormal" psychology through the data of psychical research. Areas covered include sense perception, memory, dissociation, hallucinations, secondary personality, hypnotism, and reincarnation (a subject with which psychical researchers did not often deal). [P]

1575. *Hyslop, James Hervey.*

Enigmas of Psychical Research. Boston: Herbert B. Turner & Co., 1906, x + (2) + 432 pp.

Hyslop here notes that in his first book, *Science and a Future Life* (1905, *see* entry number 1556), he had given only a sketchy description of phenomena bearing on telepathy, apparitions, and related areas. His intention here is not to attempt to amass facts to prove survival beyond death, but to provide evidence that these phenomena are in need of further investigation. Subjects covered include crystal gazing, clairvoyance, premonitions, and various mediumistic phenomena. [P]

1576. *Lapponi, Giuseppe.*

Ipnotismo e spiritismo. 2 ed. Rome: Desclee, Lefebvre e C., 1906, 16 pp.

Lapponi, a Catholic author, writes of many remarkable spiritistic phenomena, which he believes to be the work of the devil. No information on the first edition (1897) is available. [H & P]

1577. *Malgras, J.*

Les pionniers du spiritisme en France: documents pour la formation d'un livre d'or des sciences psychiques recueillis par J. Malgras. Paris: Librairie des sciences psychologiques, 1906, 479 pp.

[P]

1578. *Verrall, Margaret de Gaudrion Merrifield.*

"On a Series of Automatic Writings." *Society for Psychical Research. Proceedings* 20 (1906): 1–432.

A detailed discussion of automatic writing produced by the author, including a discussion of "cross-correspondences" or information in automatic writings from diverse mediums that seem to come from a common source. [P]

1907

1579. *Abbott, David Phelps.*

Behind the Scenes with Mediums. Chicago: Open Court Publishing Co., 1907, (2) + (iii)–vi + 328 pp.

A compendium of tricks and deceptions used by some mediums to produce apparent spiritualistic phenomena. Written by an astute observer, the book is an exposé of techniques that range from filching personal papers from the pockets of sitters to switching slates with purported spirit writing. [P]

1580. ***American Society for Psychical Research. Journal.***
Vols. 1–19+; 1907–1925+.

[P]

1581. ***American Society for Psychical Research. Proceedings.***
Vols. 1–19+; 1907–1925+.

[P]

1582. ***Bayley, William.***
"Some Facts in Mesmerism." *American Society for Psychical Research. Proceedings* 1 (1907): 8–22.

[H & P]

1583. ***Bonnaymé, Ernest.***
La force psychique, l'agent magnétique et less instruments servant à les mesurer. Cusset: Imprimerie Bouchet, 1907, 38 + (1) pp.

This pamphlet contains descriptions of devices purportedly capable of measuring psychic and magnetic forces. A second edition published in 1908 was expanded to 220 pages. [H & P]

1584. ***Carrington, Hereward.***
The Physical Phenomena of Spiritualism, Fraudulent and Genuine: Being a Brief Account of the Most Important Historical Phenomena; a Criticism of Their Evidential Value, and a Complete Exposition of the Methods Employed in Fraudulently Reproducing the Same. Boston: Herbert H. Turner & Co., 1907, 426 pp.

Born in England, Carrington came to the United States at the age of eighteen and immediately embarked on a long career as a researcher and writer. After a period of skepticism, he became convinced that some of the reported spiritualistic phenomena were genuine and went on to establish the American Psychical Institute which gave weekly lectures, put on monthly demonstrations, and periodically published bulletins on matters related to psychical research. The Scientific Council of the Institute included many pre-eminent psychical researchers from around the world, such as Max Dessoir, Hans Driesch, Paul Joire, Oliver Lodge, Joseph Maxwell, Eugene Osty, Charles Richet, Harry Price, and even Pierre Janet. Carrington was particularly interested in studying mediums and their psychic feats. Becoming fairly competent at conjuring, he used his knowledge to expose the tricks of fraudulent mediums, but he also believed that he came across much that was genuine. *The Physical Phenomena of Spiritualism*, Carrington's first book and one of the best books on the subject, concentrates on the deceptions employed by some mediums, but also deals with what Carrington thought to be genuine. [P]

1585. ***Dalgado, D. G.***

"Braidisme et fariisme, ou la doctrine du Dr. Braid sur l'hypnotisme comparée avec celle de l'abbé de Faria sur le sommeil lucide." *Revue de l'hypnotisme expérimental et thérapeutique* 21(1907): 116–23, 132–46.

Dalgado compares the method and theory of hypnotism of two of the most important figures in the history of the subject: the abbé José Custodi de Faria (1755–1819) and James Braid (1795–1860). Noting that Braid considered the cause of hypnotism to be "psycho-physiological," whereas Faria believed the cause to be "psychical," Dalgado makes a statement by statement comparison between Braid's *Neurypnology* (1843, *see* entry number 465) and Faria's *De la cause du sommeil lucide* (1819, *see* entry number 294). Based on this analysis, he argues that Faria had anticipated Braid's ideas on nearly every essential point and that his explanation of the cause of hypnotism and his method of induction were superior to Braid's. [H]

1586. ***Flammarion, Camille.***

Les forces naturelles inconnues. Paris: Ernest Flammarion, 1907, xi + 604 pp. ENGLISH: *Mysterious Psychic Forces.* Boston: Small, Maynard and Company, 1907.

In the introduction to this work, Flammarion refers to his book of the same title written forty years earlier when psychical research was still in its formative stages, points out that things have progressed a great distance since then, and indicates that the aim of this book is to provide a panorama of that progression. In the process Flammarion discusses the Allen Kardec spiritualist group, the experiments of De Gasparin and Thury, the British investigators, Eusapia Palladino, mediumistic fraud, and possible explanations for psychic phenomena. [P]

1587. ***Funk, Isaac K.***

The Psychic Riddle. New York: Funk and Wagnalls, 1907, viii + 243 pp.

Funk's second book on psychical research. Here he continues the discussion of psychical research as a science begun in *The Widow's Mite* (1904, *see* entry number 1537) and writes at length about the phenomenon of "direct voice." [P]

1588. ***Grasset, Joseph.***

L'occultisme hier et aujourd'hui, le merveilleux prescientifique. Montpellier: Coulet et fils, 1907, 435 pp.

[P]

1589. ***Janet, Pierre.***

The Major Symptoms of Hysteria. Fifteen Lectures Given in the Medical School of Harvard University. New York: Macmillan, 1907, x + (2) + 345 pp.

A series of excursions into the realm of hysteria drawing upon the vast experience of the author. Janet justifies his choice of hysteria as the subject for these lectures by pointing out that most of the major contributors to modern psychotherapy took the study of hysteria as their starting point. He examines the "major symptoms" of hysteria: monoideic somnambulisms, polyideic somnambulisms, dual personalities, convulsive attacks, fits of sleep, motor agitations, contractures, paralysis, anesthesias, and disturbances of vision, speech and alimentation. He concludes his lectures with a description of the chief characteristics of hysteria and a discussion of "contraction of the field of consciousness" and suggestion. [H]

1590. ***Lowenfeld, Leopold.***

Somnambulismus und Spiritismus. Wiesbaden: J. F. Bergman, 1907, 71 pp.

[H & P]

1591. ***Moutin, Lucien.***

Le magnétisme humain, l'hypnotisme, et le spiritualisme moderne considérés aux points de vue théorique et pratique. Paris: Perrin et Cie., 1907, (4) + 477 pp.

Moutin distinguishes between animal magnetism and hypnotism, claiming that animal magnetism produces phenomena that hypnotism never will. Pointing out that there is "nothing new under the sun," Moutin refers to authors before Mesmer who had discussed an animal magnetic fluid, including among Mesmer's precursors the "Professor of Physiology Lecot" [i.e., Nicolas Claude Le Cat (1700–1768)] who in 1767 (in his *Traité des sensations et des passions*), twelve years before Mesmer's first *Mémoire* (1779), wrote of an "animal fluid" with the same characteristics. Moutin claims that too much emphasis has been placed on suggestion, to the detriment of the study of a "transmissible agent" (magnetic fluid) that passes between operator and subject. He relates psychic phenomena, such as telepathy, clairvoyance, mental suggestion, and other phenomena of spiritualism, to animal magnetism and claims that they cannot be fully understood until a thorough investigation of animal magnetism has been conducted. [H]

1592. ***Pfungst, Oskar.***

Das Pferd des herrn von Osten (der kluge Hans), ein experimentellen tier-und Menschen-psychologie. Leipzig: Johann Ambrosius Barth, 1907, 193 pp. ENGLISH: *Clever Hans (The Horse of Mr. Von Osten): a Contribution to Experimental Animal and Human Psychology*. Translated by Carl L. Rahn. New York: Holt, 1911.

A study of one of the famous Elberfeld horses who were supposed to be able to communicate through a code tapped out by their hoofs. In 1891 William von Osten discovered that his horse, Kluge Hans, would count objects placed before him with his hoofs. Von Osten went on from there to develop a more subtle communication with the animal, with the horse apparently able to signal answers to questions in German and solve difficult mathematical problems. Over the years the phenomenon was studied by numerous scientists, many of whom attested to the horse's remarkable feats. Here, Pfungst comes down on the negative side of the controversy, suggesting that when Kluge Hans did give correct answers it was because von Osten unconsciously provided cues to the animal to indicate when he should start and stop in giving the tapped code. The controversy continued for many years after the appearance of this study, and remains an issue among psychical researchers to this day. [P]

1593. ***Piéron, H.***
"Grandeur et décadence des Rayons N: histoire d'une croyance." *Année psychologique* 13 (1907): 143–69.

[P]

1594. ***Prince, Morton, Münsterberg, Hugo, Ribot, Théodule Armand, Jastrow, Joseph, and Janet, Pierre.***
"A Symposium on the Subconscious." *Journal of Abnormal Psychology* 2 (1907): 22–43, 58–80.

An important collection of articles on the subconscious by some of the leading minds in the field. Of great interest are the articles by Janet and Prince. Janet, inventor of the term "subconscious," emphasizes that he has from the beginning used the notion of the subconscious as a crystallization of the results of observations, not as a metaphysical theory. Prince takes up the issue of an adequate explanation for subconscious phenomena, insisting that the purely physiological interpretation is insufficient and that it is necessary to postulate the existence of dissociated ideas. [P]

1595. ***Revue générale des sciences psychiques.***
Vols. 1–3; 1907–1911.

Published in Paris and edited by E. Bosc. [P]

1596. ***Rouby, ———.***
Bien-Boa et Ch. Richet. Lisbon: n.p., 1907, 60 pp.

[P]

1597. ***Schiller, Ferdinand Canning Scott.***
Studies in Humanism. London: Macmillan, 1907, xvii + 492 pp.

A collection of essays on subjects of interest to psychical researchers: truth, dreams, dissociation of personality, and psychical research. [P]

1598. ***Wells, David Washburn.***
Psychology Applied to Medicine. Introductory Studies. Philadelphia: F. A. Davis Company, 1907, xiv + 141 pp.

A good introductory study of hypnotism and suggestion and their psychotherapeutic applications. [H]

1599. ***Zurbonsen, Friedrich.***

Das zweite Gesicht (Die "Vorgesichten") nach Wirklichkeit und Wesen. Cologne: n.p., (1907) 108 pp.

[P]

1908

1600. ***Baraduc, Hippolyte Ferdinand.***

Mes morts, leurs manifestations, leurs influences, leurs télépathies. Paris: Leymarie, 1908, 86 pp.

[P]

1601. ***Barrett, William Fletcher.***

On the Threshold of a New World of Thought: An Examination of the Phenomena of Spiritualism. London: Kegan Paul, Trench, Trubner & Co., 1908, xv + 127 pp.

William Barrett, for over thirty years Professor of Physics at the Royal College of Science in Dublin, was one of the founders of the Society for Psychical Research, editor of the Society's *Journal* from 1884 to 1899, and a key figure in the initiation of the American Society for Psychical Research in 1885. Barrett began his study of psychical research in 1874 and after a long and thorough investigation became convinced of the reality of the phenomena of spiritualism. He held there was evidence for the existence of a spiritual world, survival of death, and communication with departed spirits. This book, Barrett's first major work on psychical research, has as its basis a lecture delivered some fourteen years earlier in which he presents evidence regarding the genuineness of psychic phenomena. This evidence is drawn from his own experiments and from investigations undertaken principally by members of the Society for Psychical Research, much of it having already appeared in the *Proceedings* of the Society. Barrett adopts the spiritualistic explanation for the phenomena as a working hypothesis, insisting that objective students of the facts must give this hypothesis serious consideration. A thoroughly revised version of this work, published in 1918 under the title *On the Threshold of the Unseen: An Examination of the Phenomena of Spiritualism and the Evidence for Survival after Death,* takes the same position but cites the findings of ten additional years of psychical research in confirmation of his spiritualistic viewpoint. [P]

1602. ***Bates, E. Katharine.***
Do the Dead Depart? New York: Dodge, 1908, 263 + (3) pp.

An investigation of spirit return based on twenty-five years investigation, with an interesting chapter on "guardian children." [P]

1603. ***Beers, Clifford Whittingham.***
A Mind that Found Itself: An Autobiography. With Introductory letters by William James. New York: Longmans, Green, and Co., 1908, ix + 363 pp.

[H]

1604. ***Bennett, Edward T.***
The Direct Phenomena of Spiritualism—Speaking, Writing, Drawing, Music and Painting: a Study. With Facsimile Illustrations of Direct Writing, Drawing and Painting. London: Rider, (1908), (1) + 64 pp.

An informative treatment of these phenomena. Bennett gives a history of the subject, followed by an account of his witnessing the phenomena produced by David Duguid, Mrs. Everitt, and a few other mediums. [P]

1605. ***Boirac, Emile.***
La psychologie inconnue. Introduction et contribution *à l'étude expérimentale des sciences psychiques.* Paris: Alcan, 1908, (4) + 346 pp. ENGLISH: *Our Hidden Forces ("La Psychologie Inconnue"). An Experimental Study of the Psychic Sciences.* Translated by W. de Kerlor. New York: Frederick A. Stokes Company, 1917.

This book is composed of writings produced between 1893 and 1903. It is Boirac's intention to answer the question: Is it possible to study so-called "psychic" phenomena scientifically? He divides these phenomena into three main categories: hypnoidal phenomena (relating to hypnotism and suggestion), magnetoidal phenomena (natural but unclassified physical forces), and spiritoidal phenomena (resulting from the action of unknown agents). Boirac believes the scientific study of such phenomena is possible and suggests methods for that study. The book covers everything from experiments in mesmerism and suggestion to the investigation of spiritism. [P]

1606. ***Carrington, Hereward.***
The Coming Science. Boston: Small, Maynard & Co., 1908, xii + (1) + 393 pp.

Carrington's second book on psychical research. In it he brings together a great deal of the information available to that time, not to establish the genuineness of the facts, but to examine the various hypotheses which might explain the facts. In the process he covers the whole field of research, from experiments with telepathy to the physical phenomena of mediumship. [P]

1607. *Courtier, Jules.*

Rapport sur les séances d'Eusapia Palladino à l'Institut Général Psychologique en 1905, 1906, 1907 et 1908. Paris: Institut Général Psychologique, 1908, 80 pp.

A valuable report of experiments done with Italian physical medium Eusapia Palladino at the Institut Général Psychologique. The author describes the controls used to prevent fraud, the physical environment of the experiments, and the phenomena observed, such as movements of objects and luminous phenomena that were observed. [P]

1608. *Filiatre, Jean.*

Les sciences utiles. Hypnotisme et magnétisme. Somnambulisme, suggestion et télépathie. Influence personnelle. Partie théorique. Pratique (suite). Historique. Occultisme expérimental. Bourbon l'Archambault: Librairie Genest, (1908), 318 pp.

The second part of a work begun under the title *Hypnotisme et magnétisme* (1905, *see* entry number 1554). [H & P]

1609. *Fournier D'Albe, Edmund Edward.*

New Light on Immortality. London: Longmans, Green and Co., 1908, xix + 334 pp.

Abstruse theories of human survival after death, accompanied by an account of findings in psychical research which the author offers as confirmation of his theories. [P]

1610. *Harris, John William.*

The Pseudo-Occult. Notes on Telepathic Vision and Auditory Messages Proceeding from Hypnotism. London: Wellby, 1908, 30 pp.

[H & P]

1611. *Hyslop, James Hervey.*

Psychical Research and the Resurrection. Boston: Small, Maynard and Company, 1908, xiv + (2) + 409 pp.

Hyslop considered this book to be a supplement to *Science and a Future Life* (1905, entry number 1556) and used it to draw implications from material he presented in *Borderland of Psychical Research* (1906, entry number 1574) and *Engimas of Psychical Research* (1906, entry number 1575). In particular, he discusses the involvement in psychic phenomena of subconscious or subliminal consciousness and, in certain cases, of a "secondary personality." [P]

1612. *Joire, Paul.*

Traité de l'hypnotisme expérimental et thérapeutique. Ses applications à la médecine, à l'éducation, à la psychologie. Paris: Vigot frères, 1908, 456 pp.

A general treatment of hypnotism, its uses and its phenomena. The author's view of hypnotism is influenced by Charcot's three phase schema, but Joire clearly situates himself in the tradition of animal magnetism with the notion that a "nervous force" emanates from the body and produces external effects. Indeed, he even developed a device called the "sthenometer" to measure this "nervous force." The device consisted of a dial with a straw needle balanced on a pivot. Joire believed he obtained clear movements of this needle when a subject's hands were placed in the vicinity of the dial and that these movements were produced by the body's emanations, but other investigators attributed this movement to the action of radiating heat. [H]

1613. *Kotik, Naum.*

Die Emanation der psycho-physischen Energie. Eine experimentelle Untersuchung über die unmittelbare Gedankenübertragung im Zusammenhang mit der Frage über die Radioaktivität der Gehirns. Wiesbaden: J.F. Bergmann, 1908, vi + (2) + 130 pp.

A study of thought transference viewed in relation to the psycho-physical action of the brain. [P]

1614. *Lodge, Oliver Joseph.*

Man and the Universe. A Study of the Influence of the Advance in Scientific Knowledge upon our Understanding of Christianity. London: Methuen & Co., 1908, 356 pp.

Lodge was a physicist and psychical researcher with a long list of credits. As a physicist he made significant experiments in electricity, thermo-electricity, thermal conductivity and radio technology, authored many books, served as President of the British Association, and was knighted in 1902 in recognition of his contributions to science. As a psychical researcher, he was President of the Society for Psychical Research (1901–1904) and the author of major works in the field. Although dealing largely with science and religion, *Man and the Universe*, Lodge's first book in the field of psychical research, contains a long section on the immortality of the soul. Here Lodge cites the findings of psychical research as evidence independent of faith that the human spirit persists after death. [P]

1615. *McComb, Samuel, Coriat, Isador H. and Worcester, Elwood.*

Religion and Medicine: the Moral Control of Nervous Disorders. New York: Moffat, Yard, and Company, 1908, (6) + 427 pp.

Worcester, the rector of Emmanuel Church in Boston, was a co-founder with McComb of the Emmanuel Movement of healing in 1906. This movement undertook the treatment of nervous disorders, combining the moral precepts of Christian teaching with contemporary knowledge of the subconscious mind. *Religion and Medicine* was the foundation volume of the movement. It examines the nature of the subconscious mind and its place in the production of "functional disorders." Emphasis is placed on the value of hypnotism with suggestion as a treatment technique. [P]

1616. ***Morselli, Enrico Agostino.***

Psicologia e "spiritismo", impressioni e note critiche sui fenomeni medianici di Eusapia Palladino. 2 vols. Turin: Fra. Bocca, 1908, xlviii + 463, xviii + 586 pp.

Morselli, a highly respected neurologist and psychologist, describes his impressions of the famous Italian medium Eusapia Palladino. After giving a history of spiritualism in general and of Palladino's mediumship in particular, Morselli describes twenty-eight sittings he had with her between 1901 and 1907. While accepting many of the phenomena that occurred in these séances as genuinely supernormal, he rejects the hypothesis that they are due to the action of discarnate spirits and supports the position that they are produced by unknown forces or influences connected with the human organism. [P]

1617. ***Piddington, John George.***

"A Series of Concordant Automatisms." *Society for Psychical Research. Proceedings* 22 (1908):19–416.

A painstaking study of cross-correspondences, correlations among samples of automatic writing from various mediums which seem to show common elements proceeding from one mind. [P]

1618. ***Podmore, Frank.***

The Naturalization of the Supernatural. New York and London: G.P. Putnam Sons, 1908, viii + 374 pp.

A further development of the thesis of Podmore's earlier works that all the phenomena of psychical research are explicable in terms of telepathy. As usual, Podmore's grasp of the historical and experimental data is excellent and the book is a very useful summary of the material relevant to the issues of psychical research. [P]

1619. ***Quackenbos, John Duncan.***

Hypnotic Therapeutics in Theory and Practice: With Numerous Illustrations of Treatment by Suggestion. New York and London: Harper & Brothers, 1908, iii + (5) + 335 + (1) pp.

Quackenbos describes the use of hypnotism and suggestion to deal with moral and emotional problems. In his discussion of the power of suggestion, he provides a lengthy treatment of auto-suggestion and its place in healing. Quackenbos has his own terminology of hypnotic states, using the term "transliminal" states, or states "across the threshold," to designate the altered consciousness of hypnotism and the realm remote from every day waking consciousness. [H]

1620. ***Vaschide, Nicholas.***
Les hallucinations télépathiques. Paris: Bloud, 1908, x + 97 + (2) pp.

A critique of the psychical research classic *Phantasms of the Living* (1886, entry number 1144) based on experiments personally conducted by the author. [P]

1621. ***Viollet, Marcel.***
Le spiritisme dans ses rapports avec la folie. Paris: Bloud et Cie., 1908, 120 pp.

Viollet examines mental disturbances that he believes can be linked directly or indirectly to a belief in spiritualism or participation in spiritualistic séances. [P]

1909

1622. ***Bormann, Walter.***
Die Nornen, Forschungen über Fernsehen in Raum und Zeit. Leipzig: M. Altmann, 1909, 271 pp.

[P]

1623. ***Bozzano, Ernesto.***
De casi d'identificasione spiritica. Geneva: A. Donath, 1909, 370 pp.

[P]

1624. ***Bramwell, J. Milne.***
Hypnotism and Treatment by Suggestion. London: Cassell and Company, 1909, xii + 216 pp.

Bramwell was both an important historian of hypnotism (*see* his *Hypnotism: its history, practice and theory,*1903, entry number 1517) and a capable practitioner of medical hypnosis. This book reflects his interest in both of these areas. After describing the history of hypnotism and techniques of hypnotic induction and discussing theories about the nature of hypnotism and suggestion, Bramwell devotes the bulk of the book to accounts of his use of hypnotism in the treatment of the ill. [H]

1625. ***Carrington, Hereward.***
Eusapia Palladino and her Phenomena. New York: B.W. Dodge, 1909, (2) + xiv + (2) + 353 pp.

A detailed description of experiments carried out with the famous Italian medium by medical and academic researchers in many places in Europe and Great Britain. Carrington concludes with an examination of the theories put forward to explain the phenomena and a description of his own hypothesis. [P]

1626. *Delanne, Gabriel.*

Les apparitions matérialisées des vivants & des morts. 2 vols. Paris: Librairie spirite, 1909 and 1911, (6) + 527; (2) + 841 pp.

The most important single work on materialized apparitions and related phenomena. Delanne presents a very thorough examination of the subject of apparitions, detailing its history, describing contemporary experiences and experiments, and attempting to explain the causes. Focusing in particular on the phenomenon of materialized apparitions from partial materialization of hands to the whole human form, he discusses the problem of the identity of the materialization, the possibility of fraud, and the processes through which materializations might take place. The work contains many photographs of apparent materialized forms. [P]

1627. *Durville, Hector.*

Le fantôme des vivants: anatomie et physiologie de l'âme. Recherches expérimentales sur le dédoublement des corps de l'homme. Paris: Librairie du magnétisme, 1909, 356 pp.

Durville considers the problem of apparitions from the point of view of the doubling of the human body. He sees them as manifestations of a second non-material body not limited, like the physical body by space and time. The first part of the book is a historical summary of the phenomena. The second describes experiments with individuals who had a special talent in carrying out this doubling. [P]

1628. *Feilding, Everard, Baggally, W. W., and Carrington, Hereward.*

"Report on a series of Sittings with Eusapia Palladino." *Society for Psychical Research. Proceedings* 23 (1909): 309–569.

Detailed description of tests carried out with the famous Italian medium Eusapia Palladino to determine the genuineness of the physical phenomena associated with her mediumship. An important study. [P]

1629. *Janet, Pierre.*

Les Névroses. Paris: Ernest Flammarion, 1909, 397 pp.

A condensation of Janet's ideas about the nature of the neurosis as developed particularly in his *Névroses et idées fixes* (1898, entry number 1441) and *Les obsessions et la psychasthénie* (1903, entry number 1522). [H]

1630. ***Joire, Paul Martial Joseph.***

Les phénomènes psychiques et supernormaux (leur observation, leur expèrimentation). Paris: Vigot frères, 1909, 570 pp. ENGLISH: *Psychical and Supernormal Phenomena, their Observation and Experimentation.* London: W. Rider & Son, 1916.

Joire was a respected French psychical researcher who experimented with hypnotism and the relationship between hypnotic states and psychic phenomena. This book is an impressive compendium of the findings in many areas of psychical research. Most significant is Joire's work in what was called "the externalisation of sensibility" by which subjects in the state of hypnosis seem to extend the range of their senses some distance beyond the body. Joire also includes a great deal of material on experimentation done with the famous Italian physical medium, Eusapia Palladino. [P]

1631. ***Le Goarant de Tromelin, Gustave Pierre Marie.***

Le fluide humain, ses lois et ses propriétés. La science de mouvoir la matière sans être médium. Nombreux appareils et moteurs que l'on peut construire soi-même, mis en mouvement par le fluide humain. L'être psychique, les fantômes, doubles des vivants et images fluidiques. Étude sur la force biolique. Paris: Librairie du magnétisme, (1909), 258 pp.

[P]

1632. ***Lodge, Oliver Joseph.***

The Survival of Man: a Study in Unrecognized Human Faculty. London: Methuen & Co., 1909, xi + 357 pp.

Lodge's second major work on psychical research. It presents a history of the Society for Psychical Research and contains lengthy sections on experimental and spontaneous telepathy. The bulk of the book, however, is taken up with an examination of "automatism and lucidity." In discussing the psychic aspects of automatic writing and trance speaking, Lodge devotes nearly two hundred pages to a discussion of the medium Leonore Piper, long considered to be the best example of a genuine mental medium. His conclusion is that the evidence for the survival of the human spirit after death, which has been accumulating over the centuries, has now been given a powerful boost through the remarkable material obtained from Mrs. Piper and other contemporary mediums. Lodge reveals that he himself believes in survival and the communications of the dead with the living. [P]

1633. ***Lombroso, Cesare.***

Ricerche sui fenomeni ipnotici e spiritici. Turin: Unione tipografico-editrice torinese, 1909, viii + 319 pp. ENGLISH: *After Death—What? Spiritistic Phenomena and Their Interpretation.* Translated by William Sloane Kennedy. Boston: Small, Maynard & Company, 1909.

Lombroso was a renowned psychiatrist and criminal anthropologist interested in problems related to hypnotism and hysteria. In 1891, he turned his attention to psychical research when he was invited by Cavaliere Ercole Chiana of Naples to sit in on séances with the medium Eusapia Palladino. As a result of those sittings, Lombroso felt compelled to accept the facts of spiritualism, although he still had difficulty with the theory. After further investigations of mediums and experimentation with thought transference, Lombroso accepted the hypothesis of spirit communication. This is his most important book in the field of psychical research. The first half is a study of hypnotism as it relates to thought transmission, clairvoyance, precognition, and the physical phenomena of spiritualism. The second half is a study of mediumship, its conditions, and its limitations. The book concludes with an important chapter on unconscious fraud and telepathy. [H & P]

1634. ***Morselli, Enrico Agostino.***
Intorno all'ignoto; fakiri e case infestate in un conflitto sullo spiritismo: prima riposta a Ces. Lombroso. Milan: t. La Compositrice, 50 pp.

[P]

1635. ***Podmore, Frank.***
Mesmerism and Christian Science: a Short History of Mental Healing. London: Methuen, 1909, xv + 306 pp.

Although not as detailed as might be desired, the treatment of the evolution of thought that began with Mesmer's discoveries is readable and accurate. Podmore adeptly traces the early years of animal magnetism, the rise of spiritualistically oriented mesmerism, and the connections between mesmerism and mental healing, culminating in Christian Science. [H & P]

1636. ***Podmore, Frank.***
Telepathic Hallucinations: the New View of Ghosts. London: Milner, (1909), vii + 128 pp.

One of Podmore's last books. Here he reiterates the theory of ghosts and apparitions he first put forth in *Apparitions and Thought-transference* (1894, entry number 1371). [P]

1637. **[*Prince, Morton (ed.)*].**
Psychotherapeutics: A Symposium. By Morton Prince, Frederic H. Gerrish, James J. Putnam, E. W. Taylor, Boris Sidis, George A. Waterman, John Donley, Ernest Jones, Tom A. Williams. Boston: Richard G. Badger, 1909.

Papers originally delivered at an American Therapeutic Society symposium in 1909, published in the *Journal of Abnormal Psychology* and in this collection. [H]

1638. ***Tweedale, Charles Lakeman.***
Man's Survival after Death, or The Other Side of Life in the Light of Human Experience and Modern Research. London: Grant Richards, 1909, 277 pp.

Tweedale was a clergyman of the Church of England who believed that spiritualism had an important message for all Christians. In this book he discusses the evidence for communication from the dead and its significance for the living. He also examines the phenomena of spiritualistic mediumship and compares them to the events recorded in the Bible. [P]

1910–1919

1910

1639. *Atkinson, Willan Walker.*

Telepathy. Its Theory, Facts and Proof. Chicago: New Thought Publishing Co., 1910, 94 pp.

An examination of British experiments in telepathy and a lengthy discussion of a study conducted in the United States in 1907 called the "Weltmer Experiment." [P]

1640. *Congrès international de psychologie expérimentale.*

Compte-rendu des travaux. Edited by Henri Durville. Paris: H. Durville fils, (1910), 245 pp.

The congress, convened in Paris in 1910, gathered together a number of investigators who were interested in human psychology particularly as it related to the more extraordinary phenomena of the psyche. Topics discussed included the divining rod, fluido-magnetic photography, and the doubling of the personality. [H & P]

1641. *Dallas, Helen Alexandria.*

Mors Janua Vitae? A Discussion of Certain Communications Purporting to Come from Frederic W. H. Myers. London: Willam Rider & Son, 1910, xix + 147 pp.

One of the important early women psychical researchers, Dallas was particularly interested in the issue of survival after death. Here she discusses automatic writing scripts produced by Alice Kipling Fleming (who was called Mrs. Holland) and Margaret Verrall that purported to be messages from the late Frederic W. H. Myers. [P]

1642. ***Duchatel, Edmond.***

La vue à distance. Dans le temps et dans l'espace. Enquête sur des cas de psychométrie (Janvier-Décembre 1909). Préface de Joseph Maxwell. Suivie d'une conférence relative à l'influence de l'amour sur l'écriture, par Paul de Fallois. Paris: Leymarie, 1910, xvi + 128 pp.

[P]

1643. ***Hart, Bernard.***

"The Conception of the Subconscious." *Journal of Abnormal Psychology* 4 (1910): 351–71.

Hart examines the evolution of the notion of the subconscious and grapples with the question of whether the subconscious is a "brain fact" or a "mind fact," that is, a subject for physiology or for psychology. He concludes that both lines of investigation yield useful results, but it is important not to confuse the two fields of inquiry. Hart also insists that no matter which line of inquiry the investigator of the subconscious pursues, he must clearly distinguish between the phenomenal (observed facts) and the conceptual (mental constructions). Keeping this in mind, he points out that Janet's subconscious is a phenomenal fact, whereas Freud's unconscious is a conceptual construct. [H]

1644. ***Hollander, Bernard.***

Hypnotism and Suggestion in Daily Life, Education, and Medical Practice. London: Isaac Pitman, 1910, viii + 295 pp.

In addition to discussing hypnotism and suggestion, the author treats thought transference, clairvoyance and apparitions. [H]

1645. ***Mann, G. A.***

La force-pensée: la faculté unique; mécanisme de la télépathie; extériorisation de la volunté; appel et captation des forces cosmiques; théorie nouvelle de l'influence de l'homme sur l'homme. Paris: G. A. Mann, 1910, 234 pp.

[P]

1646. ***Melville, John.***

Crystal-Gazing and the Wonders of Clairvoyance, Embracing Practical Instructions in the Art, History, and Philosophy of This Ancient Science. To Which is Appended an Abridgment of Jacob Dixon's "Hygienic Clairvoyance," with Various Extracts and Original Notes. London: Nichols, 1910, 98 pp.

[P]

1647. *Myers, Gustavus.*

Beyond the Borderline of Life. Boston: Ball, 1910, 249 pp.

A summary of findings in psychical research, including Botazzi's experiments with Eusapia Palladino and the cross correspondences in the writings of the mediums Mrs. Piper and Mrs. Verrall. [P]

1648. *Oesterreich, Traugott Konstantin.*

Die Phänomenologie des Ich in ihren Grundproblemen. Erster Band: Das Ich und das Selbstbewusstsein. Die scheinbare Spaltung des Ich. Leipzig: Johann Ambrosius Barth, 1910, x + 532 pp.

A thorough philosophico-psychological study of the nature of the "I" and the apparent splitting of the "I." Employing the findings of a number of investigators of personality and consciousness, particularly Richet and Janet, Oesterreich attempts to provide a phenomenological analysis of the "I" that can cast light on such puzzling phenomena as multiple personality, apparitions, hallucinations, and the experience of the double. Only one volume was published. [H]

1649. *Podmore, Frank.*

The Newer Spiritualism. London: Fisher Unwin, 1910, 320 pp.

Podmore's last book on psychical research. In it he distinguishes between the "older spiritualism," in which physical phenomena dominate, and the "newer spiritualism," in which mental phenomena dominate. Podmore cannot bring himself to accept any of the physical phenomena as genuine, and sees in them a great deal of self delusion and fraud, along with a bit of telepathic hallucination. Evidence for the genuineness of the mental phenomena of the "newer spiritualism" he finds to be more compelling but he cannot accept the notion that they are communications from the spirits of the dead. Instead he prefers to explain them in terms of telepathy operating through subtle unconscious mechanisms within the medium. [P]

1650. *Revue du psychisme expérimental. Magnétisme. Hypnotisme. Suggestion. Psychologie. Mediumnisme. Mensuelle illustrée.*

One vol. only; 1910–1911.

Published in Paris and edited by Gaston and Henri Durville, this periodical was absorbed by the *Journal du magnétisme*. [H & P]

1651. *Rothe, Georg.*

Die Wünschel-Rute. Historisch-theoretische Studie. Jena: Diederich, 1910, viii + 118 pp.

[P]

1652. ***Samona, Carmelo.***

Psiche misteriosa fenomeni detti spiritici ("metapsichici" del Richet). Palermo: A. Reber, 1910, (6) + 299 + (4) pp.

A study of the nature of spiritualistic mediumship and the psychology of the medium. It includes a description of a sitting with the Italian medium Eusapia Palladino. [P]

1653. ***Stead, Estelle Wilson.***

My Father: Personal and Spiritual Reminiscences. London: Nelson & Sons, (1910), x + 11–378 pp.

A biography of a significant figure in the psychic world, William Thomas Stead (1849–1912). [P]

1654. ***Tanner, Amy Eliza.***

Studies in Spiritism. New York & London: D. Appleton & Co., 1910, xxxviii + (1) + 408 pp.

Amy Tanner was an assistant to G. Stanley Hall (1844–1924), who contributed an introduction to this volume which describes a number of sittings these two investigators had with the famous medium Leonora Piper. Tanner also discusses the material gathered by other investigators of Mrs. Piper and certain related phenomena, such as the cross-correspondences, where messages received by two or more mediums seem to have a common source. The author is highly skeptical about the involvement of any truly paranormal elements. [P]

1655. ***Williams, C.***

Spiritualism and Insanity. An Essay Describing the Disastrous Consequences to the Mental Health Which are Apt to Result from a Pursuit of the Study of Spiritualism. London: The Ambrose Co., (1910), 53 pp.

[P]

1911

1656. ***Barrett, William Fletcher.***

Psychical Research. London: Williams and Norgate, 1911, 255 pp.

Barrett presents a popular account of some of the key issues of psychical research, much of the material being taken from the pages of the *Society for Psychical Research. Proceedings.* [P]

1657. ***Bernheim, Hippolyte.***
De la suggestion. Paris: Aubin Michel, (1911), 267 pp.

[H]

1658. ***Carrington, Hereward and Meader, John.***
Death: Its Causes and Phenomena with Special Reference to Immortality. London: Rider, 1911, ix + 552 pp.

An attempt to examine death from every aspect. Carrington deals with the scientific aspect of death, its causes, its signs, experience at the moment of death, and even premature burial. He also takes up theories of immortality from various cultures and ages and attempts to show the role psychical research can play in enhancing the understanding of death and survival. [P]

1659. ***Coates, James.***
Photographing the Invisible: Practical Studies in Spirit Photography, Spirit Portraiture, and Other Rare but Allied Phenomena. Chicago: Advanced Thought Publishing, 1911, vi + (2) + vii–xxi + 394 pp.

Deals with what Coates calls "spirit produced or supernormal photographs, portraits, and writings." Although not as critical of the data as might be hoped, the book, a later edition, does supply a great deal of information about what had been reported in the field. [P]

1660. ***Constable, Frank Challice.***
Personality and Telepathy. London: Kegan Paul, Trench, Trubner & Co., 1911, xv + 330 pp.

A closely reasoned philosophical examination of the nature of human personality, followed by a discussion of the implications of telepathy and clairvoyance for that subject. In the philosophical discussion, in which he explicitly acknowledges the influence of Kant, Constable holds that human personality in this world is a partial manifestation of a spiritual self (the "intuitive self") existing outside of time and space. In the second part of the book, Constable attempts to show that the findings of psychical research confirming the existence of telepathy support the truth of his theory, since telepathy is in itself evidence of the existence of the intuitive self. [P]

1661. ***Durville, Gaston.***
Le sommeil provoqué et les causes qui le déterminent. Étude étiologique de l'hypnose. Paris: (Librairie du magnétisme), 1911, 65 + (2) pp.

Gaston Durville was a son of Hector Durville (1849–1923), a prolific writer in the field of animal magnetism. The author examines the nature of artificial somnambulism and hypnotism. Concluding that suggestion alone is insufficient to explain the phenomena, he argues for the notion that a fluid is the principle agent. [H]

1662. *Flournoy, Theodore.*
Esprits et mediums: mélanges de métapsychique et de psychologie. Geneva: Librairie Kundig, 1911, viii + 561 pp.

A collection of articles and treatises on psychical research written by a balanced investigator if somewhat skeptical. Flournoy examines the work of some of the principal psychical researchers (such as F.W.H. Myers and Charles Richet), discusses the phenomena of famous mediums, and examines the beliefs of spiritism from a critical point of view. Flournoy distinguishes between "spiritualism," the belief in the immortality of the soul and "spiritism," the belief that the dead communicate with the living. While personally accepting the former belief, he has serious doubts that the departed communicate through mediums. Nevertheless he does hold that evidence supports the reality of such phenomena as telepathy and clairvoyance. [P]

1663. *Hill, John Arthur.*
New Evidences in Psychical Research: A Record of Investigations, with Selected Examples of Recent S.P.R. Results. London: Rider, 1911, xii + 218 pp.

Principally an account of sittings with a medium, "Mr. Watson," who, without going into trance, claimed to see spirit forms and gain information from them. The information obtained in these sittings is remarkably accurate, but little is said about safeguards against the medium obtaining that information by secret means. Hill, who believes the medium is genuine, suggests a theory of mediumship: the mind of the medium is placed in rapport with the source of information (living or dead) by means of the presence of some object (in this case the "object" is the sitter). The last part of the book presents cases of telepathy and apparitions drawn from the investigations of the Society for Psychical Research. [P]

1664. *Kilner, Walter John.*
The Human Atmosphere, or The Aura Made Visible by the Aid of Chemical Screens. London: Rebman, 1911, xiii + 329 pp.

A pioneering discussion of experiments designed to make the field of energy purported to exist around living things visible. [P]

1665. *Klinckowström, Karl Ludwig Friedrich Otto von.*
Bibliographie der Wünschelrute; mit einer Einleitung von dr. Ed. Aigner: Der gegenwärtige Stand der Wünschelruten-Forschung. Munich: Kommissionsverlag von O. Schönhuth, 1911, (2) + 146 pp.

[P]

1666. *McDougall, William.*
Body and Mind: a History and a Defense of Animism. London: Methuen, 1911, xix + 384 pp.

An attempt on the part of a psychologist to come to terms with the problem of "animism," the belief that some non-corporeal principle animates the bodies of human beings. The question has importance for psychical research in that most explanations of human survival of death and the existence of paranormal phenomena are based on some kind of animistic view of human nature. McDougall accepts the notion of animism with qualifications, holding that the acceptance of animism does not imply a belief in specific metaphysical entities such as the soul. He believes that this is the only philosophically sound conclusion and the only way to account for certain paranormal phenomena such as telepathy. [P]

1667. ***[Moberly, Charlott Anne Elizabeth and Jourdain, Eleana Frances].***

An Adventure. London: Macmillan, 1911, vi + (1) + 162 pp.

The book describes a visit by the two authors (given assumed names in the book) to France in 1901. On an excursion to the Petit Trianon they saw and heard people and things that they realized later were hallucinations. They believed these hallucinations were in fact images of the past that somehow intruded on the present. The book created a sensation at the time of its publication and went through a number of editions. [P]

1668. ***Moore, W. Usborne.***

Glimpses of the Next State (The Education of an Agnostic). London: Watts & Co., 1911, xxiv + 642 pp.

A discussion of various kinds of spiritualistic phenomena from several mediums that seem to indicate that human beings survive death. [P]

1669. ***Remy, M.***

Spirites et illusionnistes de France. Paris: A. Leclerc, 1911, 257 pp.

[P]

1670. ***Rochas d'Aiglun, Eugène Auguste Albert de.***

Les vies successives. Documents pour l'étude de cette question. Paris: Chacornac, 1911, 504 pp.

An early attempt to bring about regression to what appears to be "past life" experiences through magnetic (hypnotic) somnambulism. De Rochas also employs the somnambulistic state to explore prevision and uses the material of past life memory and prevision as a basis for hypothesizing an evolution of the soul through successive lives. A remarkable book for its time. [H & P]

1671. ***Tuckett, Ivor Lloyd.***

The Evidence for the Supernatural: A Critical Study Made with "Uncommon Sense." London: Kegan Paul, Trench, Trubner & Co., 1911, (4) + 409 pp.

Written in response to the spread of what the author considered to be a naive and even irresponsible acceptance of the psychic, this volume is an attempt to provide a critical examination of the facts. Although Tuckett does not, by his own admission, bring new material to bear on the question and arrives at a sort of agnostic position with regard to the genuineness of the matter, the book is useful as a readable compendium of relevant material on the subject. [P]

1912

1672. *Bisson, Juliette Alexandre.*

Les phénomènes dits de matérialisation: étude expérimentale. Paris: Félix Alcan, 1912, 311 pp.

A detailed verbal account of sittings with the materialization medium "Eva C" that contains many photographs of materializations at various stages. Bisson includes a chapter on the medium, an examination of the nature of the phenomena, and a discussion of controls for possible fraud. This is an important book for the history of materialization phenomena that helped establish the credibility of this area as worthy of serious consideration by psychical researchers. [P]

1673. *Caillet, Albert Louis.*

Manuel bibliographiques des sciences psychiques ou occultes. Sciences des mages. Hermétique. Astrologie. Kabbale. Franc-Maçonnerie. Médecine ancienne. Mesmérisme. Sorcellerie. Singularités. Aberrations de tout ordre. Curiosités. Sources bibliographiques et documentaires sur ces sujets, etc. 3 vols. Paris: Lucien Dorbon, 1912, lxvii + 531; (4) + 533; (4) + 767 + (folding table) pp.

The best and most comprehensive bibliography of works on subjects related to psychic science. Caillet consulted earlier bibliographies and a number of important collections in France. In the area of mesmerism, it is the best bibliography of French work. Of the more than 11,500 entries, a large number are annotated. [H & P]

1674. *Caillet, Albert Louis.*

Traitement mental et culture spirituelle. La santé et l'harmonie dans la vie humaine. Paris: Vigot frères, 1912, xiii + 399 pp.

A wide ranging work devoted to the application of all the "psychic sciences" to the healing and well-being of mankind. Caillet draws from every relevant field and from a number of cultures to put together a comprehensive manual of healing practices. Subjects dealt with include mesmerism, suggestion, eastern philosophies and psychical research. [P]

1675. ***Carrington, Hereward.***
Death Deferred. New York: Dodd, Mead & Co., 1912.

A book about health and the prolongation of life with some sorties into the area of psychical research and the light it casts on survival after death. [P]

1676. ***Coué, Emile.***
De la suggestion et de ses applications. Conference faite à Nancy et à Chaumont en Janvier 1912. Chaumont: Andriot Moissonnier, 1912.

[H]

1677. ***Crichton-Miller, Hugh.***
Hypnotism and Disease; A Plea for Rational Psychotherapy. London: T. F. Unwin, 1912, (4) + 5–252 pp.

[H]

1678. ***Freud, Sigmund.***
"A Note on the Unconscious in Psycho-Analysis." *Society for Psychical Research. Proceedings* 26 (1912):312–18.

Here Freud speaks of unconscious but active ideas and relates his notion of the unconscious to experiences of post-hypnotic suggestion. [H]

1679. ***Imoda, Enrico.***
Fotografie di fantasmi: Contributo sperimentale alla constatazione dei fenomeni medianici. Turin: Fratelli Bocca, 1912, 254 pp.

Imoda conducted experiments with a remarkable Italian medium in Turin at the house of the Marquise de Ruspoli. The book, illustrated with photographs, describes the strong materialization and telekinetic phenomena that were observed. [P]

1680. ***Schrenck-Notzing, Albert Philibert Franz von.***
Die Phänomene des Mediums Linda Gazerra. Leipzig: Oswald Mutze, 1912, 41 pp.

[P]

1681. ***Staudenmaier, Ludwig.***
Die Magie als experimentelle Naturwissenschaft. Leipzig: Akademische Verlagsgesellschaft, 1912, (4) + 184 pp.

Staudenmaier, chemistry professor at the Freisinger Lyceum, suffered from a kind of doubling of the personality and from hallucinations, which he determined to use as a starting point to study "psychic force." Believing that hallucinations were not always a manifestation of internal disturbance, but could sometimes correspond to an objective reality, he contended that by studying the experience of hallucination, knowledge of "magical" powers might be attained. [H & P]

1913

1682. ***Bergson, Henri.***

"Presidential Address." *Society for Psychical Research. Proceedings* 26 (1913):462–79.

A discussion of the relationship between mind and brain by one of the most eminent philosophers of the day. Bergson was elected President of the Society for Psychical Research in 1913 and used the occasion to discuss this philosophical problem so relevant to the work of psychical research. [P]

1683. ***Bozzano, Ernesto.***

Des phénomènes prémonitoires. (*Pressentiments — rêves prophétiques, clairvoyance dans le future, etc.*) Paris: Annales des sciences psychiques, 1913, 450 pp.

An erudite study of premonitory phenomena by a man who was one of Italy's foremost psychical researchers and a spiritualist. This work seems to have appeared first in French; it was published in Italian in 1914. [P]

1684. ***Carrington, Hereward.***

Personal Experiences in Spiritualism (Including the Official Account and Record of the American Palladino Séances). London: T. Werner Laurie, (1913), xvi + 274 pp.

A description of Carrington's personal experiences with spiritualist mediums and their phenomena. In the preface, he suggests that genuine physical phenomena are very rare indeed and in the first half of the book he writes of experiences which were for the most part fraudulent. However in the second half of the book, he describes his encounters with the Italian medium Eusapia Palladino during her American tour. He states his firm belief that in contrast to other phenomena he had seen, some of her physical phenomena were genuine. Although she was known to resort to trickery and Carrington had himself seen that, he had no doubt that she also produced real psychic feats of materialization and movement of objects at a distance. At the end of the book Carrington emphasizes the need for a psychical laboratory which could study such phenomena in a properly controlled setting. [P]

1685. ***Harris, John William.***

Of Spiritism: i.e. Hypnotic Telepathy and Phantasms—Their Danger. London: F. Griffiths, 1913, (5) + 3–127 pp.

[H & P]

1686. ***Haynes, Edmund Sidney Pollock.***

The Belief in Personal Immortality. London: Watts & Co., 1913, viii + 156 pp.

[P]

1687. ***Hude, Anna.***

The Evidence for Communication with the Dead. London: T. F. Unwin, 1913, viii + 351 pp.

A discussion of telepathy and the automatic writing of Mrs. A. W. Verrall (1859–1916), "Mrs. Holland" (whose real name was Alice Kipling Flemming), and Mrs. Leonora Piper (1859–1950). The automatic writings of these individuals were thought to contain evidence of communication from departed spirits that could not be explained through natural causes or by telepathy between the living. [P]

1688. ***Hyslop, James Hervey.***

Psychical Research and Survival. London: G. Bell & Sons, 1913, x + 208 pp.

Bringing the data of psychical research to bear on the question of human survival of death, Hyslop speaks of his own personal conviction that death does not end human existence. [P]

1689. ***Lancelin, Charles Marie Eugène.***

Méthode de dédoublement personnel (Extérioration de la neuricité. Sorties en astral). Paris: n.p., (1913), 554 pp.

[P]

1690. ***Mysteria. Revue mensuelle illustrée d'études.***

1913.

See *L'initiation* (1888, entry number 1209). [P]

1691. ***Osty, Eugène.***

Lucidité et intuition; étude expérimentale. Paris: Félix Alcan, 1913, xxxix + 477 pp.

A summary of three years' research into the possibility of acquiring knowledge through supernormal means. In this important book, Osty concludes that such acquisition is possible. [P]

1914

1692. *Bérillon, Edgar.*

L'hypnotisme et la psychothérapie dans l'oeuvre de Dumontpallier. Paris: n.p., 1914, 68 pp.

[H]

1693. *Bolton, Gambier.*

Ghosts in Solid Form: an Experimental Investigation of Certain Little-known Phenomena (Materialisations). London: Rider, 1914, vii + 120 pp.

[P]

1694. *Bruce, Henry Addington.*

Adventurings in the Psychical. Boston: Little, Brown, & Co., 1914, vii + 318 pp.

A review of the findings of psychical research by a man who believes that all mental psychic phenomena can be explained by telepathy. [P]

1695. *Carrington, Hereward.*

The Problems of Psychical Research: Experiments and Theories in the Realm of the Supernormal. New York: W. Rickey & Co., 1914, xi + 412 pp.

Carrington takes an approach that is reminiscent of that of his earlier work, *The Coming Science* (1908, *see* entry number 1606) in order to bring together a great many of the facts established by psychical research and examine the various theories that might explain them. He seems, however, to believe that things are a bit more complex than he had originally thought, suggesting that there are many genuine psychic phenomena, mental and physical, but that the spiritualistic explanation is too simplistic and cannot account for some of the facts. He contends that not enough data are available to reach any certain conclusions about what kind of intelligence might be behind the genuine phenomena produced. [P]

1696. *D'aute-Hooper, T.*

Spirit-Psychometry and Trance Communications by Unseen Agencies, Through a Welsh Woman and Dr. T. D'aute-Hooper. London: William Rider & Son, 1914, xii + 13–160 pp.

Following in the tradition of Buchanan and Denton, the Welsh medium studied by D'aute-Hooper provided information of a geological or anthropological type while holding some object from a particular era. Unlike earlier authors, this medium seemed to obtain her information from unseen spirits. The book is an account of those visions and their accuracy and includes a whole section on false or inaccurate material obtained when forgeries of prehistoric objects were used. [P]

1697. *Freimark, Hans.*

Mediumistische Kunst. Mit einem Beitrag über den kunstlerischen Wert mediumistischer Malereien von Eugen Johannes Maecker. Leipzig: W. Heims, 1914, 136 pp.

[P]

1698. *Geley, Gustave.*

Contribution à l'étude des correspondences croisées (documents nouveaux). Conference faite par le docteur Gustave Geley le samedi 20 decembre 1913. Paris: Henri Durville, 1914, 47 pp.

[P]

1699. *Holt, Henry.*

On the Cosmic Relations. 2 vols. Boston and New York: Houghton Mifflin Company, 1914, xi + (1) + 512; (8) + 513–989 pp.

A very ambitious attempt to relate mediumistic phenomena to the philosophy of a cosmic soul. For a philosophical work, the book is surprisingly rich in the data of psychical research. Most notably, it contains about 400 pages of matter relating to the medium Leonore Piper (1859–1950). The author concludes that genuine mediumistic phenomena are the result of possession of the medium or hypnotic influence exerted on the medium by discarnate souls. Holt sees these discarnate personalities, as well as all incarnate personalities, as manifestations of the one cosmic soul which is sometimes called God. [P]

1700. *Kemnitz, Mathilde von.*

Moderne Mediumforschung. Kritische Betrachtungen zu Dr. von Schrenck-Notzing's "Materialisationsphaenomene." Munich: J. F. Lehmann, 1914, 96 pp.

[P]

1701. *Prince, Morton.*

The Unconscious: the Fundamentals of Human Personality Normal and Abnormal. New York: Macmillan, 1914, xiii + 549 pp.

An extremely important work in psychotherapy, dissociation, and the study of the unconscious. After discussing the nature of the subconscious and subconscious processes, Prince describes his conception of the unconscious as a "storehouse of neurographic dispositions or residua." He intends the term to indicate an absence of the attributes of consciousness. Prince points out, however, that such "conserved experiences" may at times be activated and function as subconscious processes. These subconscious (or, in his terminology, "co-conscious") processes do partake of qualities of consciousness, even though they may never emerge into the awareness of our personal consciousness. Prince hoped that through these clear distinctions, the confusions in the use of the terms "unconscious," "subconscious," and "conscious" might be cleared up. [H]

1702. ***Rosacroce, Elia.***

L'ipno-magnetismo alla portata di tutti. Manuale pratico di psichismo illustrato. Naples: Partenopea, 1914, 192 pp.

[H & P]

1703. ***Schrenck-Notzing, Albert Philibert Franz von.***

Der Kampf um die Materialisations-Phänomene. Eine Verteidigungsschrift. Munich: Ernst Reinhardt, 1914, viii + 160 pp.

In this work Schrenck-Notzing answers criticisms leveled against his book on materialization by von Kemnitz, von Gulat-Wellenburg, Morselli, and others. [P]

1704. ***Schrenck-Notzing, Albert Philibert Franz von.***

Materialisations-Phänomene: ein Beitrag zur Erforschung der Mediumistischen Teleplastie. Munich: E. Reinhardt, 1914, xii + 523 pp. ENGLISH: *Phenomena of Materialisation: A Contribution to the Investigation of Mediumistic Teleplastics.* Translated by E. E. Fournier d'Albe. London: Kegan Paul, Trench & Trubner, 1920.

One of the most important scientific studies of the phenomenon of materialization. Schrenck-Notzing was a well-known German psychiatrist who was associated with the Nancy School of hypnotism. He first became interested in the field of psychical research through his experimentation with hypnotism and was drawn further in that direction by Charles Richet, whom he met in 1889. Schrenck-Notzing's interest centered on experimentation with mediums, especially on mediumistic production of physical phenomena. He participated in experiments with Richet, Myers and Sir Oliver Lodge of the Society for Psychical Research, and with such well-known psychiatric figures as C. G. Jung and Eugen Bleuler. *Materialisations-Phänomene* is Schrenck-Notzing's best known work in the field of psychical research. It describes phenomena produced in sittings with two mediums: Eva C. and Stanislava P. Both mediums produced remarkable materializations of figures with human features, along with more amorphous shapes. The book contains over two hundred photographs of these productions at various stages of formation. Schrenck-Notzing gives a detailed account of precautions taken to exclude trickery and bad observation. He also explores possible alternate means of producing the same results. His conclusion is that the phenomena of these two mediums, produced within the stringent conditions which he created, were genuine. The second greatly enlarged edition, published in 1923, contains added data derived from sittings with the same mediums and with Willy Sch., Marie S., Enja Nielsen, Franek Kluski, and others. The first edition in its English translation was also augmented with additional material. [P]

1705. ***Sidis, Boris.***

The Foundations of Normal and Abnormal Psychology. Boston: Richard G. Badger, 1914, 116 pp.

Sidis believes that knowledge of the normal arises out of knowledge of the abnormal. In the first part of the volume, he discusses his solution of the problem of the brain *versus* mind in the study of the psyche, that is, the question of the place of physiological inquiry in psychology. Sidis advocates a form of psycho-physical parallelism, claiming that every psychic change must have its physiological concomitant. The second part of the book contains an analysis of the abnormal mind and an exposition of the author's theory of "moment consciousness" as the basis for the synthetic unity of consciousness, a unity which, according to Sidis, must somehow be explained. [H]

1706. ***Sidis, Boris.***

Symptomatology, Psychognosis, and Diagnosis of Psychopathic Diseases. Boston: Richard G. Badger, 1914, (2) + xvii + (6) + 11–448 pp.

Here Sidis continues to develop his notion of psychological disturbance as a form of the hypnoidal state, an idea touched upon in his earlier writings (see entry number 1450). His elaboration of experiences with hypnosis and suggestion into a comprehensive system is an important evolution of ideas that had first been formulated by Janet in the 1880s. [H]

1915

1707. ***Graves, Lucien Chase.***

The Natural Order of Spirit; a Psychic Study and Experience. Boston: Sherman French & Co., 1915, 365 pp.

Written as a result of the author's experiences with the medium Mrs. Chenoweth and with communications purporting to come from the author's son, who had been killed in a railway accident. After an historical and philosophical discussion of communication with the spirit world, the sittings with Mrs. Chenoweth are recounted. [P]

1708. ***Philpott, Anthony J.***

The Quest for Dean Bridgman Conner. London: W. Heinemann, 1915, 251 pp.

An account of a fruitless search for a young man reported to have died in Mexico City but said to be alive by the famous medium Leonora Piper (1859–1950). After two attempts to find the man, depicted by the medium as abducted and hidden away, it was concluded by the searchers that he had been dead all along and that the medium had produced the supposed clairvoyant information in response to suggestions given her in an initial interview. [P]

1709. ***Prince, Walter Franklin and Hyslop, James Hervey.***

"The Doris Case of Multiple Personality: a Biography of Five Personalities in Connection with One Body and a Daily Record of a Therapeutic Process Ending in the Restoration of the Primary Member to Integrity and Continuity of Consciousness. Part I, by Walter Franklin Prince; The Doris Case of Multiple Personality. Part II, by Walter Franklin Prince; The Doris Case of Multiple Personality. Part III, by James Hervey Hyslop; The Mother of Doris, by Walter Franklin Prince." *American Society for Psychical Research. Proceedings* 9 (1915-1923): 1–700, 10: 701–1419, 11: 5–866, and 17: 1–216.

Walter Franklin Prince, a minister of the Episcopal Church, was the founder of the Boston Society for Psychical Research, president for one term of the Society for Psychical Research in England, and one of America's most important American psychical investigators. In 1909, a twenty-year-old girl, who was enrolled in the Sunday school of the Episcopal Church in Pittsburgh where Prince was then rector, came to the attention of Prince's wife. Mrs. Prince was worried about what appeared to be unusual behavior in the girl, called "Doris Fischer" in the account, and kept an eye on her, finally speaking to her husband about her concern. Prince had read Morton Prince's *The Dissociation of a Personality* (1905, entry number 1559) and thought Doris was suffering from a similar form of dissociation. Desiring to do therapy with the girl and believing that she could only be successfully treated if she were away from her father (her mother having died in 1906), Dr. and Mrs. Prince adopted Doris in 1911. Doris had five distinct personalities: 1) "Real Doris," who knew nothing of the other personalities, 2) "Sick Doris," who knew all about "Real Doris" but nothing of the others, 3) "Margaret," who knew both "Real Doris" and "Sick Doris," and who, although she liked "Real Doris," would play little tricks on her, such as getting her to see hallucinations, 4) "Sleeping Margaret," who, when Margaret was sleeping, would emerge and speak with impressive intelligence and perception, but who ordinarily could not move the limbs of the body, and 5) "Sleeping Real Doris," who acted very much like an automaton, repeating phrases mechanically. Prince conducted his treatment with great skill, encouraging Real Doris to handle as much of daily life as possible, and, while allowing the others some expression, keeping them asleep for longer and longer periods. The abilities of the secondary personalities were gradually absorbed into Real Doris, and, with a single exception, the secondary personalities faded into the background. The exception was Sleeping Margaret, who remained the same throughout this period of integration. Sleeping Margaret was something of a puzzle. She was very mature and detached in her view of things, and unlike the others, she was not subject to suggestion. She eventually depicted herself as a guardian spirit who, when Doris was three, had come as a result of the prayers of Doris's mother. To test the hypothesis that Sleeping Margaret was a spirit who had come in from the outside, James Hyslop, a psychical researcher and Prince's friend took Doris in 1915 to twenty-three sittings with a "Mrs. Chenoweth" (a pseudonym). There had been some evidence of supernormal phenomena (telepathy, clairvoyance) in Doris's life up to this point, but with Hyslop's intervention, that aspect of the case received more attention. Hyslop describes the elaborate measures taken to assure that Mrs. Chenoweth had no knowledge whatsoever about Doris before the sittings. He hoped to gain information from the medium that would provide independent confirmation that she was possessed by a spirit, thus verifying Sleeping Margaret's account of herself. But Mrs. Chenoweth indicated instead that Sleeping Margaret was merely Doris' subconscious self. Hyslop accepted this assertion, but Prince did not, and there the matter stood in 1917 at the end of their account. In 1923, however, Prince wrote another article on the Doris Fischer case, describing communications purporting to come from Doris's deceased mother through Mrs. Chenoweth. It contains a detailed account of the sittings in which those messages occurred. The Doris Fischer case is of considerable interest both as one of the most elaborately described cases of dissociation and multiple personality in the annals of the phenomenon and as an extremely interesting account of an attempt to distinguish multiple personality from possession. [H & P]

1710. ***Sidgwick, Eleanor Mildred Balfour.***
"A Contribution to the Study of the Psychology of Mrs. Piper's Trance Phenomena." *Society for Psychical Research. Proceedings* 28 (1915): 1–652.

A study of the psychology of the phenomena produced by the mental medium Leonora Piper (1859–1950). Sidgwick investigates the nature of the medium's trance, the characteristics of her "controls" or personalities that manifest through her, and the effects of the sitters on the medium. [P]

1916

1711. ***Crawford, William Jackson.***
The Reality of Psychic Phenomena: Raps, Levitations, etc. London: John M. Watkins, 1916, vii + 246 pp.

Crawford was a lecturer in engineering in Queens University, Belfast. Between 1916 and 1920, he used his mechanical skills to investigate the physical mediumship of Kathleen Goligher in her family circle, and this is the first of a series of books in which he reported the results of his investigations. Here he claims to have both witnessed and photographed levitation and the production of ectoplasmic structures. The book caused a sensation and the remarkable photographs were not easily explained away. It was Crawford's contention that phenomena such as raps and levitation of furniture were produced by ectoplasmic "rods" emanating from the medium, and his photographs appeared to confirm that opinion. [P]

1712. ***Driesch, Hans Adolf Eduard.***
Leib und Seele. Eine Untersuchung über das Psycho-Physische Grundproblem. Leipzig: E. Reinicke, 1916, vi + 108 + (2) pp. ENGLISH: *Mind and Body.* Translated by Theodore Besterman. London: Methuen and Co., 1920.

A sophisticated attempt to come to terms with the problem of the relationship between body and mind. Driesch first gives a history of positions taken on the question and then provides his own solution—positing a kind of psychophysical parallelism. He approaches the problem from a phenomenological point of view and states that experience shows that we exist in two separate realms: the realm of mind and that of nature. One cannot be reduced to the other. However, we also perceive a relationship between the two realms, a kind of parallelism. Driesch's solution to the mind-body problem leaves the way open for accepting the phenomena of psychical research as valid objects of scientific study. [P]

1713. ***Hopp, Max.***
Über Hellsehen. Eine dritisch-experimentelle Untersuchung. Inaugural-Dissertation. Berlin: Haussmann, 1916, 127 pp.

[P]

1714. ***Lodge, Oliver Joseph.***
Raymond, or Life and Death: with Examples of the Evidence for Survival of Memory and Affection after Death. London: Methuen, 1916, xi + 403 pp.

Lodge's best known contribution to psychical research, this book focuses on his son Raymond who had been killed in the First World War. Lodge gives the reader an intimate sketch of his son as a young soldier through letters written by him from the front and letters from other army men who knew him. He then presents a mass of mediumistically obtained material which provided what Lodge considered to be evidence of Raymond's survival of death and his communication with his loved ones on earth. The book concludes with a long philosophical discussion of the meaning of life and death in the light of the evidence of human survival. [P]

1715. ***McArthur, Angus [pseudonym].***
Psychic Science in Parliament. A Survey of the Statutes and the Leading Legal Decisions in Regard to Psychic Phenomena. London: n.p., 1916, 18 pp.

[P]

1716. ***Prince, Walter Franklin.***
"The Doris Case of Quintuple Personality." *Journal of Abnormal Psychology* 11 (1916): 73–122.

See entry number 1709. [H & P]

1717. ***Rogers, Louis William.***
Dreams and Premonitions. Los Angeles: Theo Book Concern, 1916, 121 pp.

An attempt to show that dreams may foretell the future and may involve spiritual communication with the departed. [P]

1718. ***Yost, Casper Salathiel.***
Patience Worth: a Psychic Mystery. New York: Henry Holt, 1916, iv + (2) + 290 pp.

The first detailed description of the case of a spirit entity called Patience Worth who purportedly communicated through a St. Louis housewife named Pearl Curran in the early 1900's. "Patience Worth" wrote a number of novels and an extremely long poem that puzzled experts because it was written entirely in Anglo-Saxon words. [P]

1917

1719. ***Baggally, W. W.***
Telepathy, Genuine and Fraudulent. London: Methuen & Co., 1917, vii + (1) + 94 + (2) pp.

An account of telepathic experiments witnessed by the author. Baggally was something of an amateur conjurer and made a point of trying to expose trickery. This book gives both cases in which fraud was discovered and successful cases in which no deceit could be discovered. [P]

1720. ***Bernheim, Hippolyte.***
Automatisme et suggestion. Paris: Félix Alcan, 1917, xv + (1)+ 168 pp.

Bernheim rejects the notion of "unconscious automatic psychism." He asserts that individuals in the state of artificial somnambulism are not automata; they are conscious and know their state. Thus, Bernheim draws a sharp distinction between the hypnotic state and natural sleep. He also denies that people have absolute free will, since that notion is contradictory to the facts of human suggestibility. He concedes, however, that this does not obliterate legal responsibility, since the principle of responsibility is a "social necessity." [H]

1721. *Boirac, Emile.*

L'avenir des sciences psychiques. Paris: Félix Alcan, 1917, (4) + 300 + (2) pp. ENGLISH: *The Psychology of the Future* ("*L'avenir des sciences psychiques*"). Translated by W. de Kerlor. New York: Frederick A. Stokes, 1918.

Boirac attempts to present a complete picture of the status of psychical research at the time. He discusses research methodology and deals with factors that must be taken into account to arrive at sound conclusions. His treatment of hypnotism and animal magnetism, and particularly his discussion of suggestion in psychical research, forms the core of his treatise. [H & P]

1722. *Clodd, Edward.*

The Question: "If a Man Die, Shall He Live Again?" Job. xiv. 14. A Brief History and Examination of Modern Spiritualism. With a Postscript by H. E. Armstrong. London: Richards, 1917, 314 pp.

Clodd examines the evidence from spiritualism and psychical research, and, believing it to be completely explicable in terms of fraud and bad observation finds it insufficient to prove that human beings survive death. [P]

1723. *Coover, John Edgar.*

Experiments in Psychical Research at Leland Stanford Junior University. Stanford University, California: Leland Stanford Junior University Publications, 1917, xxiv + 641 pp.

After Leland Stanford Junior University was given a significant endowment to study psychical research, Coover engaged in experiments investigating thought transference, precognition, subconscious phenomena and other subjects. This book, which contains a long catalogue of books on psychical research held by the University library, was both a result of that endowment and a demonstration of the use to which the money was being put. [P]

1724. *Dessoir, Max.*

Vom Jenseits der Seele. Die Geheimwissenschaften in kritischer Betrachtung. Stuttgart: Ferdínand Enke, 1917, 344 pp.

A critical but friendly study of the principal questions of psychical research. The title may be translated as "On the Beyond of the Soul" where "beyond" refers to the unconscious depths to be found there. [P]

1725. *Doyle, Arthur Conan.*

The New Revelation. New York: Metropolitan Magazine Co., 1917, 40 pp.

Conan Doyle, best known as the originator of Sherlock Holmes, attempts to show the relationship between the revelations of spiritualism and the tenets of conventional religion. The American edition seems to be the first. [P]

1726. ***Hill, John Arthur.***
Psychical Investigations: Some Personally-Observed Proofs of Survival. London: Cassell & Co., 1917, viii + 288 pp.

Although it contains chapters on immortality, the nature of the after-life, and psychical research, *Psychical Investigations* is mainly a verbatim report of mediumistic sittings held in 1914–1916. [P]

1727. ***Kaplan, Leo.***
Hypnotismus, Animismus und Psychoanalyse. Historisch-kritische Versuche. Leipzig and Vienna: Franz Deuticke, 1917, viii + 128 pp.

An historical treatment serves as a backdrop to Kaplan's analysis of the work of Charcot, Benedikt, Freud, and others on the nature of hysteria and the problem of animism as it arises in attempts to give a psychological account of human functioning. [H]

1728. ***Maeterlinck, Maurice.***
L'hôte inconnu. Paris: E. Fasquelle, 1917, vii + 327 pp.

Maeterlinck, winner of the Nobel prize for literature, was deeply interested in the issues of psychical research. This, his major work on the subject, is a well thought out examination of certain central issues, including phantasms of the living and dead, psychometry, and knowledge of the future. Most significant is his long chapter on the "Elberfeld horses." These were the famous "talking" horses which had been previously discussed in *Das Pferd des herrn von Osten* (entry number 1592) by Oskar Pfungst. Maeterlinck himself experimented with the horses and presents material that casts doubt concerning whether the nature of the phenomenon has yet to be understood. [P]

1729. ***Mercier, Charles Arthur.***
Spiritualism and Sir Oliver Lodge. (London): Mental Culture Enterprise, 1917, xi + 132 pp.

[P]

1730. ***Troland, Leonard Thompson.***
A Technique for the Experimental Study of Telepathy and Other Alleged Clairvoyant Processes. A Report on the Work Done in 1916–17 at the Harvard Psychological Laboratory, under the Gift of Mrs. John Wallace Riddle and the Hodgson Fund. (Albany, New York): n.p., (1917) 26 pp.

A brief account of experiments that the author describes as a waste of good technique on a hopeless situation. [P]

1731. *Wright, Dudley.*

The Epworth Phenomena: To Which are Appended Certain Psychic Experiences Recorded by John Wesley in the Pages of his Journal. Collated by Dudley Wright. London: Rider, 1917, xxii + 110 pp.

An historical account of psychic phenomena that occurred in the house of the Reverend Samuel Wesley in the early eighteenth century. The events were recorded in the journal of John Wesley. This volume is mainly a compilation of quotations from that journal and from material related in Southey's biography of Wesley. [P]

1918

1732. *Bond, Frederick Bligh.*

The Gate of Remembrance: the Story of the Psychological Experiment which Resulted in the Discovery of the Edgar Chapel at Glastonbury. Oxford: Blackwell, 1918, x + 176 pp.

Bond describes an experiment in automatic writing aimed at discovering information about buried remnants of Glastonbury Abbey. The experiment was successful in that it produced information relevant to the positioning of the Edgar Chapel. Information about a second chapel, the Loretto Chapel, was also produced. The automatist did not accept a spiritualistic interpretation of the phenomenon but believed that some unknown inner faculty was activated in the process of automatic writing. [P]

1733. *Carrington, Hereward.*

Psychical Phenomena and the War. New York: Dodd, Mead & Co., 1918, ix + (1) + 363 pp.

An unusual study of psychic phenomena that occur in connection with war, and particularly the First World War. The first part of the book deals with normal aspects of the psychology of the war experience such as the soldier's attitude during training and fighting, shell-shock, and the German war mentality. The second part surveys the whole field of psychical research as it pertains to war, covering everything from prophecies of war to the continued survival of the war dead and their communications with the living. [P]

1734. *Coates, James.*

Human Magnetism or How to Hypnotise. A Practical Handbook for Students of Mesmerism. New and rev. ed. London: Fowler & Co., 1918, xvi + 253 pp.

A heavily reworked version of Coates's book *How to Mesmerize* (1893, entry number 1338). It was Coates's belief that "practically, hypnotism is mesmerism." Here he presents the history of animal magnetism and hypnotism, the place of suggestion in hypnotism, and how to use hypnotism. He also brings in the old animal-magnetic themes such as human magnetism, phreno-magnetism, and the "higher phenomena" of somnambulism. [P]

1735. ***Crawford, William Jackson.***

Hints and Observations for Those Investigating the Phenomena of Spiritualism. New York: E. P. Dutton, 1918, (4) + 110 pp.

Drawing from his own experience, Crawford describes the measures to be taken to get good results when sitting in a spiritualist circle in which physical phenomena occur. The emphasis is not on how to avoid fraud, but on how to make dramatic physical phenomena more likely to occur by avoiding things that inhibit the action of psychic forces. [P]

1736. ***Cumberland, Stuart.***

That Other World. Personal Experiences of Mystics and their Mysticism. London: Richards, 1918, 253 pp.

One of a number of books by Cumberland debunking the phenomena associated with spiritualism and the occult. He claims here to believe that all such phenomena are delusions, but he does not hold that conscious deception is always involved. [P]

1737. ***Davis, Abert E.***

Hypnotism and Treatment by Suggestion. London: Simpkin, Marshall, Hamilton, Kent and Co., Ltd., 1918, 124 pp.

[H]

1738. ***Hill, John Arthur.***

Man is a Spirit: A Collection of Spontaneous Cases of Dream, Vision, and Ecstasy. London: Cassell and Company, 1918, (xii) + 13–199 + (1) pp.

Hill starts with the assumption that a human being is not just a collection of matter but has a spirit. He then proceeds to describe incidents that seem to bear out such an assumption. The book contains original material describing occurrences of, among other things, clairvoyance, out-of-the-body experiences, and visions of the dead. [P]

1739. ***Hill, John Arthur.***

Spiritualism: Its History, Phenomena and Doctrine. London: Cassell & Co., 1918, xxii + 270 pp.

A sketchy but useful history of spiritualism and a perceptive discussion of spiritualism as a religion. It is written by a believer. [P]

1740. *Hyslop, James Hervey.*

Life after Death. Problems of the Future Life and Its Nature. New York: E. P. Dutton, 1918, ix + (3) + 346 pp.

Hyslop goes one step beyond his earlier works to examine the nature of existence after death. He describes ancient beliefs and contemporary speculations and evaluates them in terms of the philosophical issues involved. In the process, Hyslop draws heavily from the data of psychical research and from mediumistic communications. [P]

1741. *Jastrow, Joseph.*

The Psychology of Conviction: A Study of Beliefs and Attitudes. Boston and New York: Houghton Mifflin Company, 1918, (2) + (20) + 387 + (3) pp.

A psychological study debunking unfounded belief in psychic phenomena. [P]

1742. *Liljencrants, Johan.*

Spiritism and Religion: "Can You Talk to the Dead?" Including a Study of the Most Remarkable Cases of Spirit Control. New York: The Devin-Adair Co., 1918, (2) + vii + 295 + (1) pp.

A discussion of modern spiritism written from a Catholic point of view. The material on the phenomena of spiritualism is well researched and the author admits the possibility that these phenomena are genuine. However, he sees spiritistic practices as illicit and immoral, since they attribute to men what should be attributed only to God. The author cites two separate condemnations of spiritistic practices by the Vatican's Holy Office, one in 1898 and one in 1917. [P]

1743. *Lodge, Oliver Joseph.*

Christopher: a Study in Human Personality. London: Cassell, 1918, v + 293 pp.

[P]

1744. *Martin, Alfred Wilhelm.*

Psychic Tendencies To-day: an Exposition and Critique of New Thought, Christian Science, Spiritualism, Psychical Research (Sir Oliver Lodge) and Modern Materialism in Relation to Immortality. New York: D. Appleton and Company, 1918, vii + (1) + 161 pp.

Martin sees the movements mentioned in the subtitle of the book as arising from despair provoked by the crass philosophical materialism of the mid-nineteenth century. He believes that these movements have arrived at beliefs and reached conclusions that go beyond the evidence they present. While he himself is convinced that death does not end human existence, this belief derives from philosophical considerations rather than from empirical evidence. [P]

1745. ***Radclyffe-Hall, Miss, and Troubridge, Una.***
"On a Series of Sittings with Mrs. Osborne Leonard." *Society for Psychical Research. Proceedings* 30 (1918): 339–554.

A well-written, detailed study of an important British mental medium, Gladys Osborne Leonard (1882–1968). [P]

1919

1746. ***Baudouin, Charles.***
Suggestion et autosuggestion; étude psychologique et pedagogique d'après les résultats de la nouvelle école de Nancy. Neuchatel: Delachaux et Niestlé, 1919, 246 pp. ENGLISH: *Suggestion and Autosuggestion.* Translated by Eden and Cedar Paul. London: G. Allen and Unwin, 1920.

Influenced by the Nancy School of hypnotism and especially by the "New Nancy School," founded by Emile Coué (1857–1926), Baudouin forcefully separates the concept of suggestion from its common identification with hypnotic suggestion, and puts it forward as the key to a powerful psychotherapeutic and educational system. Baudouin emphasizes the importance of reaching the subconscious through well-constructed suggestions and outlines a technique which he claims to be effective in treating emotional problems and enhancing learning at every level. This enormously popular book went into a number of editions and was translated into many languages. [H]

1747. ***Bond, Frederick Bligh.***
The Hill of Vision: A Forecast of the Great War and of the Social Revolution with the Coming of the New Race. Gathered from Automatic Writings Obtained Between 1909 and 1912, and also, in 1918, Through the Hand of John Alleyne, Under the Supervision of the Author. London: Constable, 1919, xxv + 134 pp.

Describes a series of automatic writings produced by the same individual who authored the scripts that lead to the discovery of the Edgar Chapel at Glastonbury (see *The Gate of Remembrance,* entry number 1732). These writings contain information about the beginning and ending of World War One that is apparently precognitive as well as predictions of events to occur after the war. [P]

1748. ***Bozzano, Ernesto.***
Dei fenomeni d'infestazione. Rome: Casa editrice Luce & Ombra, 1919, 226 pp.

A study of hauntings, telepathy, poltergeists, and related phenomena undertaken by an investigator who accepts the reality of these experiences. [P]

1749. *Carrington, Hereward.*
Modern Psychical Phenomena: Recent Researches and Speculations. New York: Dodd, Mead & Co., 1919, xi + (5) + 331 pp.

A discussion of the implications of the findings of psychical research for life and for science. Carrington also reports on some of his more recent experiments. [P]

1750. *Chowrin, A. N.*
Experimentelle Untersuchungen auf dem Gebiete des raumlichen Hellsehens (der Kryptoskopie und inadaequaten Sinneserregung). Nach dem russischen Original bearbeitet und herausgegeben von Freiherrn von Schrenck-Notzing. Munich: Reinhard, 1919, 79 pp.

[P]

1751. *Crawford, William Jackson.*
Experiments in Psychical Science. Levitation, "Contact," and the "Direct Voice." London: John M. Watkins, 1919, vii + 191 pp.

In this second book on the Goligher Circle, Crawford describes further experiments with the physical phenomena of spiritualism and presents diagrams of the experiments and photographs of his equipment. New to this work are experiments with "direct voice" phenomena. [P]

1752. *Cumberland, Stuart.*
Spiritualism—the Inside Truth. London: Odhams, (1919), 157 pp.

A critical look at spiritualistic phenomena. Cumberland dismisses them all as delusion or deception. [P]

1753. *Doyle, Arthur Conan.*
The Vital Message. London: Hodder & Stoughton, (1919), 228 pp.

Written as a sequel to Doyle's *The New Revelation* (1917, entry number 1725), this volume further develops Doyle's ideas concerning the coming of a new age of knowledge and hope through the revelations of spiritualism. [P]

1754. *Geley, Gustave.*
De l'inconscient au conscient. Paris: Félix Alcan, 1919, xiii + 346 pp.

Geley considered this book to be a continuation of his *L'être subconscient* (1899, entry number 1458) published twenty years earlier. Abandoning his previous notion that the phenomena of mediumship could be attributed entirely to the unconscious operation of a vital energy within the medium, he here posits an outside directing force beyond the medium and those in the immediate vicinity and explains this within the framework of a doctrine of individual and collective evolution. [P]

1755. *Harris, Dean William Richard.*

Essays in Occultism, Spiritism, and Demonology. Toronto: McClelland, Goodchild & Stewart, 1919, (6) + vi + (2) + 181 pp.

Harris includes spiritism among the occult sciences. Writing from a Catholic point of view, he admits the genuineness of many spiritist psychic phenomena, but believes they are produced not by the spirits of the dead, but by the devil. [P]

1756. *Henslow, George.*

The Proofs of the Truths of Spiritualism. London: Kegan Paul, Trench, Trubner & Co., 1919, xii + 255 pp.

Using psychic occurrences witnessed by himself and those known to him, the author attempts to prove the genuineness of mediumistic phenomena. His purpose is to show that the spirits of the departed communicate with the living. Data used include spirit communications, automatic writing, and physical phenomena such as apports, poltergeists, and spirit photography. It is the work of a believer in spiritualism. [P]

1757. *Hyslop, James Hervey.*

Contact with the Other World: the Latest Evidence as to Communication with the Dead. New York: The Century Co., 1919, (8) + 493 pp.

Hyslop's last book on psychical research. In a sense this is a summary of all of his previous writings, concentrating principally on the issue of survival after death and a future life. The intention of the book, and his personal "mission," is summarized in these words from the preface: "If I succeed in leading intelligent people to take scientific interest in the phenomena while they preserve proper cautions in accepting conclusions I shall have accomplished all that can be expected in a work of this kind, and tho I regard the evidence of survival after death as conclusive for most people who have taken the pains to examine the evidence critically, I have endeavored in this work to canvass the subject as tho it had still to be proved." [P]

1758. *Osty, Eugène.*

Le sens de la vie humaine. Paris: La Renaissance du Livre, 1919, xi + 271 pp.

[P]

1759. *Raupert, John Godfrey Ferdinand.*

The New Black Magic and the Truth about the Ouija-Board. New York: Devin-Adair, 1919, vii + (2) + 243 pp.

Further developing the theme of previous books, Raupert expresses his disapproval of spiritualism and its practices. He admits that spontaneous apparitions of the dead may take place, but criticizes the spiritualists' deliberate attempt to communicate with the dead. His main conclusion is that spiritualism has not, as it claimed, discovered a way of tapping hidden but normal powers in man which may be legitimately used and by means of which the dead may communicate with the living. On the contrary, spiritualism is a modern revival of necromancy and black magic universally condemned by the Church. [P]

1760. *Smith, Hester Travers.*
Voices from the Void: Six years' Experience in Automatic Communications. London: Rider & Son, 1919, xv + 108 pp.

After describing her experiences with automatic writing, the author discusses possible sources for the information obtained. She questions whether the subconscious or some external agent is involved, and concludes that her experiences involve communications from the dead. [P]

1920 – 1925

1920

1761. ***Balfour, Arthur James.***
The Ear of Dionysius. Farther Scripts Affording Evidence of Personal Survival. New York: Henry Holt, 1920, (6) + 134 pp.

This book reprints an article which appeared in the *Proceedings* of the Society for Psychical Research (1918). It is a detailed discussion of a series of automatic scripts from different mediums which, when cross referenced, appeared to provide strong evidence that they all came from one discarnate source. [P]

1762. ***Beadnell, Charles Marsh.***
The Reality or Unreality of Spiritualistic Phenomena. Being a Criticism of Dr. W. J. Crawford's Investigations into Levitations and Raps. London: Watts & Co., 1920, 23 pp.

One of a number of attempts to evaluate Crawford's striking experiments and puzzling conclusions. (See entry number 1711.) [P]

1763. ***Bozzano, Ernesto.***
Dei fenomeni di telestesia. Rome: Casa editrice Luce e Ombra, 1920, 55 pp.

[P]

1764. *Bozzano, Ernesto.*

Delle apparizioni di defunti al letto di morte. 2a serie. Rome: Soc. poligr. Italiana, 1920, 49 pp.

[P]

1765. *Bozzano, Ernesto.*

Les phénomènes de hantise. Paris: Félix Alcan, 1920, xii + 310 pp.

[P]

1766. *Carrington, Hereward.*

Your Psychic Powers and How to Develop Them. New York: Dodd, Mead & Co., 1920, xvii + (3) + 358 pp.

Adopting as a working hypothesis the spiritistic view that people survive death, Carrington brings information concerning psychic development together from various sources to guide readers in realizing their spiritual potentials. [P]

1767. *Carrington, Walter Whately, pseudonym: W. Whately Smith.*

A Theory of the Mechanism of Survival: the Fourth Dimension and Its Applications. London: Kegan Paul, Trench, Trubner & Co., 1920, (12) + 195 + (1) pp.

Assuming that human beings do survive death, the author attempts to examine the possible conditions of that continued existence. He begins with an analysis of the dimensions of space and looks at the possibility that existence beyond death takes place in a fourth dimension. [P]

1768. [*Carrington, Walter Whately, pseudonym: W. Whately Smith*].

The Foundations of Spiritualism. London: Kegan Paul, 1920, 134 pp.

Carrington attempts to indicate the kind of empirical evidence that would be necessary to establish human survival of death with reasonable certainty, evidence that some individual personality has undergone death and has yet remained sufficiently intact to warrant being described as the same individual. Contending that the evidence available from mediums is sufficient to make survival the most probable hypothesis, Carrington then investigates the process of mediumistic communication and the various factors that may influence its form. He concludes by deploring the fact that spiritualists are for the most part so uncritical about communications from mediums that they make the dispassionate study of the evidence difficult and give those who undertake such studies a bad name in the minds of thinking people. [P]

1769. *Culpin, Millais.*

Spiritualism and the New Psychology: An Explanation of Spiritualistic Phenomena and Belief in Terms of Modern Knowledge. London: Edward Arnold, 1920, xvi + 159 + (1) pp.

[P]

1770. *Dresser, Horatio Willis.*

The Open Vision: a Study of Psychic Phenomena. London: G. G. Harrap, (1920), xi + 352 pp.

[P]

1771. *Flammarion, Camille.*

La mort et son mystère. 3 vols. Paris: Ernest Flammarion, 1920–1922. ENGLISH: *Death and Its Mystery.* 3 vols. Translated by E. S. Brooks. London: T. Fisher Unwin, 1922–1923.

These three volumes are devoted to the attempt to demonstrate the continuity of human existence after death. The first volume deals with man's spiritual nature and the faculties of the soul that seem to go beyond ordinary material existence, such as telepathy and precognition. The second volume takes up phantasms of the living, that is, apparitions of various kinds, with the emphasis on apparitions of the dying before or at the moment of death. The third volume deals with the return of the dead, describing manifestations of the departed from a few minutes to many years after death. Flammarion drew his material from many sources including, well-attested first hand accounts, the writings of psychical researchers, and the experiences of spiritualist mediums. The work was quite influential. [P]

1772. *Gray, James Martin.*

Spiritism and the Fallen Angels in the Light of the Old and New Testaments. Chicago: The Bible Institute Colportage Association, 1920, 148 pp.

A Christian minister's condemnation of spiritism as the work of the devil. [P]

1773. *Grunewald, Fritz von.*

Physikalisch-mediumistische Untersuchungen. Pfullingen: Johannes Baum, 1920, 112 pp.

After generally discussing the methods of investigating mediums, Grunewald provides a detailed description of the laboratory and its equipment and illustrates the use of the laboratory through practical examples of the author's experiments with mediums. [P]

1774. *Henslow, George.*

The Religion of the Spirit World: Written by the Spirits Themselves. London: Kegan Paul, Trench Trubner, 1920, viii + 223 pp.

Drawing on mediumistic communications gathered from various sources, Henslow attempts to depict the nature of life after death. [P]

1775. *Ince, Richard Basil.*

Franz Anton Mesmer, His Life and Teaching. London: Rider and Son, 1920, 59 + (5) pp.

A readable but sketchy biography of Mesmer and history of his teachings. [H]

1776. *Institut métapsychique international. Bulletin.*

Nos. 1–3; 1920–1921.

Continued as: *Revue métapsychique* (1920). *See* entry number 1768. [P]

1777. *Janet, Pierre Marie Félix.*

Les médications psychologiques; études historiques, psychologique et cliniques sur les méthodes de la psychothérapie. 3 vols in 2. Paris: Félix Alcan, 1920. ENGLISH: *Psychological Healing.* 2 vols. Translated by Eden and Cedar Paul. New York: Macmillan, 1925.

Janet's most exhaustive statement of his theory and technique of psychological healing. In this massive work, Janet presents a system that results from thought evolved through many years of clinical work. In this system, psychological energy plays a central role, providing the means for integrating one's experiences into a functioning whole. The work is divided into four parts: (1) a historical account of the search for mental healing through the ages, (2) a description of the role of human automatism in the understanding and treatment of psychological disturbances, (3) an explanation of the place of mental energy in the formation of an illness and its treatment, and (4) an evaluation of the utility of treatment techniques that have been used by practitioners. Throughout the work, Janet assigns a place of preeminent importance to the animal magnetic and hypnotic traditions in the search to understand and heal mental illness. [H]

1778. *Jung, Carl Gustav.*

"The Psychological Foundation of Belief in Spirits." *Society for Psychical Research. Proceedings* 31 (1920):75–93.

A paper originally read at a general meeting of the Society for Psychical Research on July 4, 1919 and translated into English by H. G. Baynes. Here Jung discusses apparitions, dreams, and nervous disorders as the principle sources for the primitive belief in the existence of spirits. From a psychological viewpoint, he suggests, the experience of spirits arises from the projection of autonomous complexes buried within the psyche. Projection outward is to be expected, because these "spirit-complexes" which belong to the impersonal or collective unconscious, should not ordinarily be associated with the ego. [P]

1779. *Kindborg, Erich.*
Suggestion, Hypnose und Telepathie; ihre Bedeutung für die Erkenntnis gesunden und kranken Geisteslebens. Munich: J. F. Bergmann, 1920, vi + 98 pp.

[H & P]

1780. *King, John Sumpter.*
Dawn of the Awakened Mind. New York: James A. McCann Company, 1920, xxix + (1) + 451 pp.

King was the founder and president of the Canadian Society for Psychical Research, which had a brief eight-year existence in the early 1900s. He was a believer in the reality of spirit communication and was convinced that the evidence provided by good mediums should have a cumulative effect ending with conviction on the part of any impartial investigator. Here King presents descriptions of first hand experience with mediums and letters and notes relevant to those experiences. It is an unusual testament and the only early major book on psychical research written by a Canadian. [P]

1781. *Kingsford, S. M.*
Psychical Research for the Plain Man. London: Kegan Paul, Trench, Trubner & Co., 1920, vii + (1) + 271 pp.

A solid treatment of the subject of psychical research for the layman. [P]

1782. *Lancelin, Charles Marie Eugène.*
L'âme humaine, études expérimentales de psycho-physiologie, par un spiritualiste. Substance de l'âme, formes, biologie organique, dissection, anatomie, éléments matériels, physiologie, propriétés physiques et chimiques. Paris: H. Durville, (1920), (2) + (7)–206 pp.

The year of publication of this work is uncertain. [P]

1783. *McCabe, Joseph.*
Is Spiritualism Based on Fraud? The Evidence of Sir A. Conan Doyle and Others Drastically Examined. London: Watts & Co., 1920, v + 160 pp.

The author writes scornfully of the adherents of spiritualism and criticizes the findings of psychical research. He emphasizes that fraud is commonplace among mediums and takes upon himself the task of keeping the unwary public informed of this fact. [P]

1784. *McCabe, Joseph.*
Spiritualism: a Popular History from 1847. New York: Dodd, Mead and Company, 1920, 243 + (1) pp.

A history of spiritualism written from a rather personal point of view. McCabe attempts to give a "bird's eye view" of a movement which he obviously considers to be questionable. The book contains useful information. [P]

1785. ***Mitchell, Thomas Walker.***
"The Doris Fischer Case of Multiple Personality." *Society for Psychical Research. Proceedings.* 31 (1920):30–74.

[H & P]

1786. ***La psychologie appliquée.***
1920.

See *Revue de l'hypnotisme expérimental et thérapeutique* (1886, entry number 1156). [H]

1787. ***Revue metapsychique.***
1920.

See *Annales des sciences psychiques* (1891, entry number 1283). [P]

1788. ***Schrenck-Notzing, Albert Philibert Franz.***
Physikalische Phaenomene des Mediumismus: Studien zur Erforschung der telekinetischen Vorgange. Munich: Ernst Reinhardt, 1920, x + 201 pp.

Schrenck-Notzing reports on the investigations carried out by a number of researchers on some of the most noteworthy physical mediums. He includes a discussion of his own work and that of J. Ochorowicz with Stanislawa Tomczyk, presents a lengthy summary of Crawford's investigation of the Goligher circle, and a few remarks on Gustav Geley's observation of the medium Eva C. [P]

1789. ***Tischner, Rudolf E.***
Über Telepathie und Hellsehen. Experimentell-theoretische Untersuchungen. Munich and Wiesbaden: J. F. Bergmann, 1920, (8) + 122 pp.

At the time this book appeared it was without question the best work on the subjects of telepathy and clairvoyance yet written in Germany. Tischner carried out a number of well-controlled experiments to study these phenomena and came to the conclusion that, under suitable conditions, some individuals do show themselves to possess faculties which allow them to communicate with other minds and to obtain information through other than ordinary channels. At the end of the book, Tischner discusses a number of theories to explain these faculties but comes to no certain conclusion as to which, if any, of these theories is most suitable. [P]

1790. ***Tweedale, Violet Chambers.***
Ghosts I Have Seen: and Other Psychic Experiences. London: Herbert Jenkins, 1920, 313 pp.

[P]

1791. *Wright, George E.*
Practical Views on Psychic Phenomena. London: Kegan Paul, Trench, Trubner & Co., 1920, viii + 136 pp.

Wright was an advocate of spiritualism within the context of the Christian church. Here he discusses the evidence for both the mental and physical phenomena of spiritualism. [P]

1921

1792. *Bozzano, Ernesto.*
Gli enigmi della psicometria. Rome: Casa editrice Luce e Ombra, 1921, 84 pp.

[P]

1793. [*Carrington, Walter Whately, pseudonym: W. Whately Smith and Patrick, Conrad Vincent*].
The Case against Spirit Photographs. London: Kegan Paul, Trench, Trubner & Co., 1921, 47 pp.

[P]

1794. *Coué, Emile.*
La matrise de soi-même par l'autosuggestion consciente. Nancy and Paris: n.p., 1921, 118 pp. ENGLISH: *Self Mastery through Conscious Autosuggestion.* New York: American Library Service, 1922.

Probably the best known work by this teacher of self suggestion. Coué was a follower of the Nancy School of hypnotism. He helped make principles of suggestion practical by popularizing the notion that one can productively use the power of suggestion on oneself, and that hypnotism is not needed to make suggestion effective. [H]

1795. *Crawford, William Jackson.*
The Psychic Structures at the Goligher Circle. With a Note by David Grow. London: John M. Watkins, 1921, vi + 151 pp.

This last of Crawford's books on the Goligher circle was published posthumously and edited by David Grow, editor of "Light." Again Crawford presents a detailed description of experiments with diagrams and photographs. Among the photographs are some of imprints made by the ectoplasmic "rods" on clay molds. [P]

1796. ***Dr. Beale, or More about the Unseen.***
London: John M. Watkins, 1921, 152 pp.

A description of psychic healing accomplished by a "Dr. Beale" operating through a medium named Miss Rose. "Dr. Beale" claims to be the discarnate spirit of a medical man who continues his work with the ill from the "other side." [P]

1797. ***Doyle, Arthur Conan.***
The Wanderings of a Spiritualist. London: Hodder & Stoughton, (1921), vii + (3) + 9–317 pp.

An account of a long trip to and from Australia undertaken by Conan Doyle and his family and of psychic events that occurred along the way. Conan Doyle offers a detailed description of the psychic milieu of Australia. [P]

1798. ***Freimark, Hans.***
Das Tischrücken. Seine qeschichtliche Entwicklung und seine Bedeutung. Auf Grund der neuesten Forschungsergebnisse dargestellt. 2 ed. Pfullingen: Johannes Baum, 1921, 51 pp.

[P]

1799. ***Grey, Pamela Genevieve Adelaide (Wyndham).***
The Earthen Vessel: a Volume Dealing with Spirit Communications Received in the Form of Book-tests by Pamela Glenconner. London and New York: John Lane, 1921, xxvi + (1) + 29–155 pp.

[P]

1800. ***Heuzé, Paul R.***
Les morts vivent-ils? Enquête sur l'état présent des sciences psychiques. Paris: La renaissance du livre, 1921, 247 pp.

[P]

1801. ***Lambert, Rudolf.***
Geheimsvolle Tatsachen. Stuttgart: Süddeutsches Verlagshaus, 1921, 224 pp.

[P]

1802. ***Ludwig, August Friedrich.***
Okkultismus und Spiritismus im Lichte der Wissenschaft und des katholischen Glaubens. . . . 2 ed. Munich: Verlag Natur und Kultur, 1921, 46 pp.

[P]

1803. *Mitchell, Thomas Walker.*

The Psychology of Medicine. London: Methuen & Co., 1921, vii + 187 pp.

A general treatise on psychology with a section on hypnosis and dissociation. [H & P]

1804. *Nogueira de Faria, ———.*

O trabalho dos mortos. Rio de Janeiro: Livraria da federaçao espirita brasileira, 1921, xvii + 221 pp.

Describes mediumistic phenomena produced by Madame Prado of Belém do Para, Brazil. The phenomena occurred in a spiritualistic circle and included full form and hand and face materializations. [P]

1805. *Oesterreich, Traugott Konstantin.*

Die Besessenheit. Langensalza: Wendt & Klauwell, 1921, 403 pp. ENGLISH: *Possession, Demonical & Other, among Primitive Races in Antiquity, the Middle Ages, and Modern Times.* London: Kegan Paul, Trench & Trubner, 1930.

The single most important work yet written on the subject of possession and exorcism. Oesterreich thoroughly examines material related to possession from ancient to contemporary times and from a number of primitive cultures. Noting that certain themes appear again and again in accounts of possession, no matter where they originate, he points out that possessing entities appear to be of two types: human and non-human spirits. In his classification and comments, Oesterreich brings to bear a solid grasp of the then most recent findings relating to hysteria, multiple personality, and hypnotic trance. He distinguishes between involuntary and voluntary possession, the latter being by far the most common type, observable every day in primitive trance rituals and spiritualistic seances. Oesterreich does not claim to have arrived at a final conclusion concerning the intrinsic nature of possession phenomena, but he clearly leans toward a psychological-parapsychological as opposed to a spiritistic explanation. But regardless of his personal belief about possession, his description of the phenomenon in its various forms remains the most comprehensive available. [P]

1806. *Oesterreich, Traugott Konstantin.*

Grundbegriffe der Parapsychologie. Eine Philosophische Studie. Pfullingen: Johannes Baum, 1921, 55 pp.

[P]

1807. *Oesterreich, Traugott Konstantin.*

Der Okkultismus im modernen Weltbild. Dresden: Sibyllen, 1921, 173 pp. ENGLISH: *Occultism and Modern Science.* Translated from the second German edition. London: Methuen and Co., 1923.

Oesterreich's first work on psychical research ("Okkultismus"). Looking chiefly at the phenomena produced by the mediums Hélène Smith, Leonora Piper, Eusapia Palladino, and Eva C., he concludes that the tenets of spiritualism can be neither proved nor disproved. He comments that many have harbored the fond wish that parapsychology would open the doors to the world beyond the grave and give the race new hope. This, he says, has not happened. He insists, however, that progress has been made and promising directions for research are opening up. [P]

1808. ***Psychica. Revue scientifique du psychisme.***
Vols. 1–5+; 1921–1925+.

Published in Paris and edited by Carita Borderieux. [P]

1809. ***Randall, John Herman.***
The New Light on Immortality, or The Significance of Psychical Research. New York: Macmillin, 1921, vii + (3) + 174 pp.

Randall, a well-known philosopher, appraises the work of psychical research to 1920 and points out the deeper meaning and significance of its findings, particularly in regard to human survival after death. He makes use principally of material written by Maurice Maeterlinck, William James, Oliver Lodge, and James Hyslop. [P]

1810. ***Redgrove, Herbert Stanley and L., I. M.***
Joseph Glanvill and Psychical Research in the Seventeenth Century. London: William Rider, 1921, 94 pp.

With some justification, the authors view Joseph Glanvill (1636–1680) as the first true psychical researcher. [P]

1811. ***Robbins, Anne Manning.***
Past and Present with Mrs. Piper. New York: Henry Holt, 1921, (2) + iv + (2) + 280 pp.

A biographical description of the mediumship of Mrs. Leonore Piper (1859–1950), one of the most important mediums to be investigated by the Society for Psychical Research. Robbins had been acquainted with Mrs. Piper for some 40 years when she wrote the book. [P]

1812. ***Tischner, Rudolf E.***
Einführung in den Okkultismus und Spiritismus. Munich and Wiesbaden: J. F. Bergmann, 1921, viii + 142 pp.

As is so often the case, "occultism" in this title is the equivalent of the English term "psychical research." Writing from a critical and scientific point of view, Tischner describes experiments he conducted in the areas of telepathy and clairvoyance with a significant number of successes but makes no attempt to explain the successes and confesses he does not know how to control conditions to make successes more likely. [P]

1813. ***Tischner, Rudolf E.***
Monismus und Okkultismus. Leipzig: Oswald Mutze, (1921), 104 pp.

In this work Tischner examines various kinds of knowledge, normal and supernormal, and their implications for philosophy. [P]

1814. ***Warcollier, René.***
La télépathie: recherches expérimentales. Paris: Félix Alcan, 1921, xvii + 363 pp.

Warcollier carried out many experiments in telepathy over a number of years before writing this book. In some of the most successful experiments, he was himself the percipient or "receiver" of telepathic information. Here he presents details of circumstances and results of his telepathic experiences and his own view of telepathy as a psycho-physical process involving the transmission of vibrations from one brain to another. Not surprisingly, he does not accept the notion of telepathic communication between the living and the dead. [P]

1815. ***Wasielewski, Waldemar von.***
Telepathie und Hellsehen: Versuche und Betractungen über ungewöhnliche seelische Fähigkeiten. Halle: C. Marhold, 1921, (2) + 276 pp.

An excellent study of telepathy and clairvoyance based on experiments conducted by the author. [P]

1816. ***Wright, George E.***
The Church and Psychical Research: A Layman's View. London: Kegan Paul, Trench, Trubner & Co., 1921, 147 pp.

A favorable presentation of psychical research written for the clergy. [P]

1922

1817. ***Bozzano, Ernesto.***
De fenomeni di telekinesia in rapporto con eventi di morte. Rome: Casa editrice Luce e Ombra, 1922, 46 pp.

[P]

1818. ***Bozzano, Ernesto.***
Musica trascendentale. Rome: Casa editrice Luce e Ombra, 1922, 59 pp.

[P]

1819. ***Doyle, Arthur Conan.***
The Case for Spirit Photography. London: Hutchinson & Co., (1922), x + 110 pp.

[P]

1820. ***Doyle, Arthur Conan.***
The Coming of the Fairies. London: Hodder & Stoughton, 1922, 139 pp.

Contains, among other things, an account of a clairvoyant's experiences with fairies and the famous Cottingley photographs (later shown to be faked) of two young girls with "fairies." [P]

1821. ***Duckworth, J. Herbert.***
Autosuggestion and Its Personal Application. New York: James A. McCann Co., 1922, 180 pp.

An explanation of the work of Emile Coué (1857–1926), whose book *La matrise de soi-même par l'autosuggestion consciente* (1921, entry number 1794) became the bible of self-development through autosuggestion. [H]

1822. **[*Ehrenfreund, Edmund Otto, pseudonym: Ubald Tartaruga*].**
Das Hellseh-Medium Megalis in Schweden. Leipzig: Tausverlag, 1922, 129 pp.

Ehrenfreund was a lawyer and police official in Vienna who had an interest in psychical research. "Megalis" (Karoline Steininger) was a Viennese clairvoyant with whom Ehrenfreund worked in criminal cases (*see* entry number 1823). In 1922 "Megalis" gave performances in various towns in Sweden. This book describes those apparently successful clairvoyant sessions. [P]

1823. **[*Ehrenfreund, Edmund Otto, pseudonym: Ubald Tartaruga*].**
Kriminal-Telepatie und -Retroskopie. Telepathie und Hellsehen im Dienste der Kriminalistik. Leipzig: Max Altmann, 1922, iv + 201 pp.

In this book Ehrenfreund combines his expertise in criminology with his interest in psychical research in describing the work of a Viennese clairvoyant named "Megalis" (Karoline Steininger). When "Megalis" had been placed in a deep hypnotic trance, induced by Dr. Leopold Thoma, she was told to go to the scene of a crime and back to the moment when it was being committed. According to Ehrenfreund, she frequently produced accurate information about the crime and its circumstances. [P]

1824. ***Fournier D'Albe, Edmund Edward.***
(Psychical Research). The Goligher Circle: May to August 1921. With an Appendix Containing Extracts from the Correspondence of the Late W. J. Crawford and Others. London: John Watkins, 1922, 81 pp.

Fournier D'Albe attempted to continue the experiments begun by William Crawford (1880–1920) first made public in Crawford's 1916 work *The Reality of Psychic Phenomena,* (*see* entry number 1711) and further described in two other books. The experiments were conducted with the Goligher family. Fournier D'Albe failed to obtain genuine phenomena and discovered what he considered to be evidence of trickery. The book also contains reports of earlier seances conducted in the Goligher circle by Crawford but never before published. [P]

1825. ***Heredia, C. M. de.***
Spiritism and Common Sense. New York: P. J. Kenedy & Sons, (1922), xv + 220 pp.

An examination of the phenomena of spiritualism by a man who was both a Catholic priest and an amateur conjurer. Heredia, having himself reproduced many mediumistic feats through trickery, believed that most of the phenomena could be attributed to fraud. Admitting that some may not be, however, he counsels Catholics to avoid involvement with spiritualism. [P]

1826. ***Heuzé, Paul R.***
L'ectoplasme. Paris: La renaissance du livre, 1922.

[P]

1827. ***Illig, Johannes.***
Historischen Prophezeiungen mit besonderer Berücksichtigung der Weltkriegsprophezeiungen. Pfullingen; Johannes Baum, (1922), 83 pp.

[P]

1828. ***International Congress for Psychical Research.***
Proceedings. Edited by Carl Vett. Copenhagen: n.p., 1922, 554 pp.

These are the *Proceedings* for the first International Congress for Psychical Research held in Copenhagen in 1921. [P]

1829. ***Klinckowström, Karl Ludwig Friedrich Otto von.***
Die Wünschelrute als wissenschaftliches Problem. Stuttgart: K. Wittwer, 1922, 40 pp.

[P]

1830. ***Ludwig, August Friedrich.***
Geschichte der okkultistischen (metapsychischen) Forschung von der Antike bis zur Gegenwart. 1. Teil. Von der Antike bis zur Mitte der 19. Jahrhunderts. Pfullingen: Johannes Baum, 1922, 152 pp.

An important study of the German occult tradition that led to the rise of psychical research in that country. This is the first part of a two volume series (*see* entry number 1881) and it covers the period from the classical Greek philosphers to the middle of the nineteenth century. A most valuable reference work for the historian of psychical research. [P]

1831. ***Meyer, Adolph F.***

Materialisation und Teleplastie. Munich and Wiesbaden: n.p., 1922, 62 pp. [P]

1832. ***Mitchell, Thomas Walker.***

Medical Psychology and Psychical Research. London: Methuen, 1922, vii + 244 pp.

Made up largely of articles published in the *Proceedings* of the Society for Psychical Research, this book deals with questions of psychology that have some bearing on the problems of psychical research. These are principally hypnotism, hysteria, and multiple personality. The four chapters dealing with the latter subject are very valuable attempts to understand the problem of the "splitting of the mind." [H & P]

1833. ***Muhl, Anita M.***

"Automatic Writing as an Indicator of the Fundamental Factors Underlying the Personality." *Journal of Abnormal Psychology* 17 (1922): 162–83.

A discussion of the psychological factors involved in the phenomenon of automatic writing, elucidating particularly the dynamics of dissociation. [H]

1834. ***Randall, Edward Caleb.***

Frontiers of the After Life. New York: Alfred A. Knopf, 1922, 213 pp.

An examination of mediumship and the nature of the afterlife by a confirmed spiritualist. [P]

1835. ***Revue de psychothérapie et de psychologie appliquée.***

1922.

See *Revue de l'hypnotisme expérimental et thérapeutique* (1886, entry number 1156). [H]

1836. ***Richet, Charles Robert.***

Traité de Metapsychique. Paris: Félix Alcan, 1922, ii + 816 pp.

A valuable, remarkably condensed summary of the psychical research which had been done to date. By the time Richet wrote this book he had accumulated an impressive list of personal honors. He had been professor of physiology at the University of Paris since 1887; he had won the Nobel prize for physiology and medicine in 1913; he was a member of the Académie de Médecine and the Académie des Sciences; and in 1905 he had been elected president of the Society for Psychical Research. The latter position indicates the depth of his involvement in studies of the paranormal. In 1884 Richet began his psychical researches with sittings with the Italian medium Eusapia Palladino and in 1891, he founded the *Annales des Sciences Psychiques,* the first French periodical devoted to psychical research. Richet begins his *Traité* with a definition and discussion of the term "metapsychique," the French word (first introduced by Richet) which is the equivalent of the English "psychical research." After giving a brief history of developments in the area, he proceeds to describe, area by area, the present state of knowledge in research of the paranormal. Richet's book remains to this day one of the most informative and well-balanced treatments of psychical research that is available. [P]

1837. ***Thomas, Charles Drayton.***
Some New Evidence for Human Survival. London: Collins, 1922, xxiv + 261 pp.

Thomas was a British clergyman who interested himself in matters of psychical research. This is a study of book and newspaper tests—experiments in which the medium attempts to tell what is on a certain page of a certain book or newspaper indicated by the experimenter. These tests are designed to exclude telepathy. [P]

1923

1838. ***Bird, Malcolm.***
My Psychic Adventures. London: Allen & Unwin, 1923, vii + 309 pp.

A semi-critical evaluation of sittings with a number of mediums, including Mrs. Gladys Osborne Leonard and William Hope, a spirit photographer. The book includes a chapter on the project undertaken by *The Scientific American* in 1922 to determine the genuineness of psychic phenomena. [P]

1839. ***Bisson, Juliette Alexandre.***
Le médiumisme et la Sorbonne. Paris: Félix Alcan, 1923, 137 + (2) pp.

Presents the report of a group of Sorbonne professors who investigated the materialization medium "Eva C." (Marthe Beraud), along with comments by the author and some supplementary material. Bisson had worked with "Eva C." for ten years before these tests were conducted. In the book she spells out in great detail the conditions of the tests and the procedures used. The results were blank and the report of the investigators was negative. [P]

1840. ***Bozzano, Ernesto.***

Animali e manifestazioni metapsichiche. Rome: Casa editrice Luce e Ombra, 1923, 91 pp.

[P]

1841. ***Doyle, Arthur Conan.***

Our American Adventure. London: Hodder & Stoughton, (1923), 205 pp.

[P]

1842. ***Doyle, Arthur Conan.***

Our Second American Adventure. London: Hodder & Stoughton, (1923), 250 pp.

[P]

1843. ***Flammarion, Camille.***

Les maisons hantées. En marge de la mort et son mystère. Paris: Ernest Flammarion, 1923, v + 435 pp. ENGLISH: *Haunted Houses.* London: T. Fisher Unwin, 1924.

In this, his last major work on psychical research, Flammarion energetically takes up the issue of survival after death. He begins with a discussion of what he considers to be experimental proofs of survival, and moves on to consider the continued presence of the dead in hauntings. Within this general survey, Flammarion scatters a detailed examination of a few cases, including "Calvados castle," the haunted house of La Constantinie, and the "fantastic villa of Comeada, Coimbra." After classifying true haunting phenomena, he discusses those not attributable to the dead (such as poltergeist occurrences) and spurious hauntings. Flammarion concludes with an attempt to specify how haunting phenomena are produced. [P]

1844. ***Flammarion, Camille.***

Discours présidentiel à la Society for Psychical Research (London, 1923), suivi de Essais médiumiques. Uranographie générale. Communication signées Galilée, obtenues à l'âge de vingt ans au cercle d'Allan Kardec, à la Société spirite de Paris (1862–1863). Paris: Éditions de la B.P.S., (1923) 89 + (2) pp.

[P]

1845. ***Forthuny, Pascal.***

Une victoire de la photographie psychique. La romanesque et glorieuse aventure du médium William Hope (de Crewe—Angleterre). Accusé d'être un imposteur. Trainé une année dans la boue. Victime d'une sombre machination. Enfin reconnu parfaitement innocent et incontestable médium photographe. Paris: Librairie des sciences psychiques, 1923, 20 pp.

[P]

1846. ***Janet, Pierre.***
"A propos de la métapsychique." *Revue philosophique* 96 (1923): 5–32.

[P]

1847. ***Janet, Pierre.***
La médecine psychologique. Paris: Ernest Flammarion, 1923, 288 pp.
ENGLISH: *Principles of Psychotherapy.* New York: Macmillan, 1924.

As a preface to describing the principles of psychotherapeutic practice, Janet presents a history of cures accomplished through animal magnetism and religious belief. By so doing, he places himself squarely in those traditions and acknowledges the importance of belief and imagination in therapeutic treatment. [H]

1848. ***Keyserling, Hermann, Hardenberg, Kuno, and Happich, Carl.***
Das Okkulte. Darmstadt: n.p., 1923, 158 pp.

An account of a German mental medium "H. B.," who produced phenomena of medical diagnosis, psychometry, and clairvoyance. [P]

1849. ***Lambert, Rudolf.***
Spuk, Gespenster und Apportphänomene. Berlin: Dr. Schwarz & Co., 1923, 184 pp.

[P]

1850. ***Mackenzie, William.***
Metapsychica moderna: fenomeni medianici e problemi del subcosciente. Rome: Libreria di scienze e lettere, 1923, 450 pp.

Mackenzie intends through this book to give the Italian reader a solid grasp of the principles and potentials of psychical research. While making use of the experiments and investigations of others, Mackenzie contributes his own critical evaluation of the reliability and implications of those findings. He has an unusually lengthy treatment of supernormal phenomena connected with animals (e.g., the Elberfeld horses and the Mannheim dog), and he seems fairly ready to accept them as genuine. [P]

1851. ***Osty, Eugène.***
La connaissance supra-normale: étude expérimentale. Paris: Félix Alcan, 1923, vii + 388 pp.

Working largely from original material drawn from his own investigations, Osty, in a work he considers to supercede his earlier *Lucidité et intuition* (1913), explores some of the most remarkable cases of clairvoyance ever published. These cases are related in great detail and with minute descriptions of the circumstances surrounding each. The phenomena described involve knowledge of unknown events at a distance in both space and time. Any serious criticism of the validity of the findings of psychical research must deal in some way or other with the material of this book. [P]

1852. ***Richet, Charles Robert.***

"Réponse à M. P. Janet: à propos de la métapsychique." *Revue philosophique* 96 (1923): 462–71.

[P]

1853. ***Rutot, Aimé Louis and Schaerer, Maurice.***

Le mécanisme de la survie. Explication scientifique des phénoménes métapsychiques. Paris: Félix Alcan, 1923, 123 pp.

[P]

1854. ***Schwab, Friedrich.***

Teleplasma und Telekinese; ergebnisse meiner zweijahrigen Experimentalsitzungen mit Berliner Medium Maria Volhart, mit 6 textzeichnungen. Berlin: Pyramidenverlag Dr. Schwarz & Co., 1923, 115 + (1) pp.

A general account of a sitting with the Berlin medium Maria Volhart. The phenomena produced were mainly physical, with movement of objects and the production of "ectoplasm." [P]

1855. ***Simon, Gustave Marie Stéphane Charles.***

Chez Victor Hugo: Les tables tournantes de Jersey: procés-verbaux des séances présentés et commentés par Gustave Simon. Paris: L. Conrad, 1923, 393 pp.

[P]

1856. ***Trethewy, A.W.***

The "Controls" of Stainton Moses ("M.A. Oxon"). London: Hurst & Blackett, (1923), 292 pp.

In this unusual book, based on published work and unpublished material made available by the London Spiritualist Alliance, Trethewy attempts an exhaustive study of the question of the nature and identity of the "controls" or "spirits" that communicated through Moses, one of the most prolific and respected spiritual mediums of the nineteenth century. His controls identified themselves to him on various occasions as biblical characters, ancient philosophers, well-known historical personages, and others. Trethewy considers two possible sources for these controls, the subliminal or subconscious self of the medium and spirits. While admitting that it is impossible to prove either alternative, he indicates that the evidence is such that the second cannot be reasonably ruled out. [P]

1924

1857. ***Andry, J.***
"Le mesmér et le somnambulisme à Lyon avant la Revolution." *Mémoires de l'Académie des sciences, belle-lettres et arts de Lyon. Sciences et lettres.* 3rd series, 18 (1924):57–101.

[H]

1858. ***Besterman, Theodore***
Crystal-Gazing: a Study in the History, Distribution, Theory and Practice of Scrying. London: Rider & Son, 1924, xiii + (1) + 183 pp.

A scholarly treatment of scrying: its history, its mechanism, the phenomena associated with it, and its possible explanations. [P]

1859. ***Blackmore, Simon Augustine.***
Spiritism: Facts and Frauds. New York and Cincinnati: Benziger Bros., 1924, 535 pp.

This work, written by an American Jesuit, was meant to be a serious study of the phenomenon of spiritualism, but with an eye to disproving its claim that the spirits of the dead communicate with the living. It is a combination of good scholarship and a heavily partisan viewpoint. [P]

1860. ***Bond, Frederick Bligh.***
The Company of Avalon: A Study of the Script of Brother Symon, Sub-Prior of Winchester Abbey in the Time of King Stephen. Oxford: Blackwell, 1924, xxxiv + 159 pp.

A presentation of automatic writings obtained by Bond after the publication of his *Gate of Remembrance* (1918, entry number 1732) and including the productions of three new automatists, "S.," Mrs. Philip Lloyd, and Mrs. Travers Smith. The book deals with script emanting from a "Brother Symon," who claims to be an earlier incarnation of "S." [P]

1861. ***Bozzano, Ernesto.***
Delle communicazioni medianiche tra viventi. Rome: Casa editrice Luce e Ombra, 1924, 125 pp.

[P]

1862. ***Carrington,* Hereward.**
Death: and Its Problems. Girard, Kansas: Haldeman-Julius Co., 1924, 64 pp.

[P]

1863. ***L'état actuel des recherches psychiques d'après les travaux du IIme Congrès International tenu à Varsovie en 1923 en l'honneur du Dr. Julien Ochorowicz.***
Paris: Les Presses universitaires de France, 1924, vii + (1) + 360 + (1) pp.

A collection of essays on experiments in psychical research by Richet, Geley, Bisson, Brugma Dingwall, Barret, Mrs. Sidgwick, Oesterreich, Schrenck-Notzing, and Grunewald. [P]

1864. ***Geley, Gustave.***
L'ectoplasmie et la clairvoyance. Observations et expériences personelles. Paris: Félix Alcan, 1924, iv + 445 pp. ENGLISH: *Clairvoyance and Materialisation. A Record of Experiments.* Translated by Stanley De Brath. New York: George H. Doran, 1927.

Written shortly before Geley's death in a plane crash and intended to precede a second volume on the meaning of metapsychic phenomena, *L'ectoplasmie et la clairvoyance* is made up largely of experiments in clairvoyance with Stephan Ossowiecki and a Madame B., and experiments in materialization with Eva C., Franek Kluski, and Jean Guzik. All experiments are described in minute detail, presenting an impressive set of data for the psychical researcher to ponder. Failed experiments are reported along with successful ones. The work contains many photographs of targets and results in the clairvoyant experiments and many more photographs relating to the materialization experiments. [P]

1865. ***Hill, John Arthur.***
From Agnosticism to Belief: An Account of Further Evidence for Survival. London: Methuen & Co., 1924, xix + 213 pp.

An account of the author's personal odyssey from agnosticism to belief in a spiritual world and life after death. Hill describes how he was particularly influenced by the mediumship of a Mr. Wilkinson, who gave demonstrations of clairvoyance and spoke from trance. In the process of describing his conversion, Hill discusses issues connected with psychology, religion, and psychical research. [P]

1866. ***International Congress for Psychical Research Proceedings (Warsaw, 1923).***
Paris: n.p., 1924, vii + 360 pp.

[P]

1867. ***Kronfeld, Arthur.***
Hypnose und Suggestion. Berlin: Ullstein, 1924, 158 pp.

[H]

1868. ***Lodge, Oliver Joseph.***

Making of Man: a Study in Evolution. London: Hodder & Stoughton, 1924, 185 pp.

An investigation into human nature which attempts to take into account both the findings of evolutionary science and the data of psychical research. [P]

1869. ***Moll, Albert.***

Der Spiritismus . . . . Nebst einem Beitrag von dr. K. R. Kupffer. Stuttgart: Franckh, (1924), 96 pp.

[P]

1870. ***Northridge, William Lovell.***

Modern Theories of the Unconscious. London: Kegan Paul, Trench, Trubner & Co., 1924, (2) + vii–xv + 193 + (1) pp.

Discussion of various theories of the unconscious and subconscious, especially those of F.W.H. Myers, Sigmund Freud, Morton Prince, W.H.R. Rivers, Boris Sidis and Pierre Janet. [H]

1871. ***Oesterreich, Traugott Konstantin.***

Die philosophische Bedeutung der mediumistischen Phänomene. Erweiterte Fassung des auf dem Zweiten Internationalen Kongress fur parapsychologische Forschung in Warschau Gehaltenen Vortrags. Stuttgart: W. Kohlhammer, 1924, vii + 54 pp.

[P]

1872. ***Prince, Walter Franklin.***

"Five Sittings with Mrs. Saunders." *American Society for Psychical Research. Proceedings* 18 (1924):1–177.

A lengthy description of sittings with a New York City medium and an evaluation of the accuracy of clairvoyant material produced in those settings. It is a very useful account of a well-conducted experiment. [P]

1873. ***Prince, Walter Franklin.***

"Studies in Psychometry." *American Society for Psychical Research. Proceedings* 18 (1924): 178–352.

An account of experiments conducted by Prince and others with mediums practicing a psychic technique called "psychometry." The technique involves the medium gaining "impressions" from an object by which he or she can clairvoyantly produce information about that object or situations connected with the object. The experimenters describe verbatim the words of the medium and comment on the quality of the information given. [P]

1874. *Schiller, Ferdinand Canning Scott.*
Problems of Belief. London: Hodder & Stoughton, (1924) 194 pp.

A tutor at Corpus Christi College at Oxford from 1903 to 1926, a professor of philosophy at the University of Southern California from 1929 to 1936, and a psychical researcher who had close ties with the Society for Psychical Research in England, Schiller's interest was largely in the philosophical problems associated with psychical research. In this work he attempts to classify and analyze the many beliefs that human beings arrive at in the course of a lifetime. He is particularly concerned to establish a philosophical basis for crediting the findings of science, which has to avoid credulity on the one hand and skepticism on the other. One of the beliefs Schiller examines in detail is the belief in survival of death, pointing out the extreme difficulty involved in arriving at anything scientific in this matter. [P]

1875. *Schrenck-Notzing, Albert Philibert Franz von.*
Der Betrug des Mediums Ladislaus Laszlo (Nachahnung von Materialisationsphänomenen). Leipzig: Oswald Mutze, 1924, 32 pp.

An account of a materialization medium who used trickery to produce his effects. [P]

1876. *Schrenck-Notzing, Albert Philibert Franz von.*
Experimente der Fernbewegung (Telekinese) im Psychologischen Institut der Münchener Universität und im Laboratorium des Verfassers. Stuttgart: Union Deutsch Verlag, 1924, xv + 273 pp.

A discussion of telekinetic experiments performed with the medium Willy Schneider. The experiments were successful in producing movement at a distance of various objects without the apparent use of any normal means of force. [P]

1877. *Schröder, Christoph.*
Grundversuche auf dem Gebiete der psychischen Grenzwissenschaften. Berlin: Dr. Schwarz & Co, 1924, 66 pp.

The first publication of the German Society for Scientific Occultism (that is, psychical research), this work contains an account of 68 sittings with 6 "sensitives" held between February 1920 and September 1921. The experiments were constructed to test for clairvoyance as distinguished from telepathy or simply a heightened natural sensitivity. The results seemed to rule out the latter two explanations. [P]

1878. *Stead, Estelle Wilson.*
Faces of the Living Dead: Remembrance Day Messages and Photographs. A Straightforward Statement by Estelle W. Stead. With Supplement Showing

Recognised "Extras" Obtained through the Mediumship of Mrs. Deane. (London): Privately printed, (1924), 87 pp.

[P]

1879. ***Sudre, René.***

La lutte pour la métapsychique. Un chapitre passionnant d'histoire scientifique (1922–24). Les "jugements" de la Sorbonne et la question de la fraude chez les mediums. Paris: Librairie des sciences psychiques, 1924, 66 + (1) pp.

A passionate defense of psychical research in general and a criticism of specific experiments conducted with two mediums at the Sorbonne. [P]

1880. ***Tischner, Rudolph E.***

Fernfühlen und Mesmerismus (Exteriorisation der Sensibilität). Munich: J. F. Bergmann, 1924, (7) + 42 pp.

An excellent study of "exteriorization of sensibility" or sensation at a distance. Tischner begins with an examination of the "magnetic fluid" of Mesmer and the "odic force" of Reichenbach, and traces the notion of an extension of sensibility beyond the organism that arose from those systems. He discusses the recent investigations of de Rochas, Joire, and Boirac, and then describes his own experiments. Asserting that the success of these trials combined with the accumulated evidence of previous experimenters stands strongly in favor of the reality of the phenomenon, Tischner moves on to a general theoretical discussion of what this "exteriorization of sensibility" might be. He concludes that there must be something other than suggestion involved in the phenomenon. There must, he believes, be some physical agent at work that is not as yet well understood. This reopens the question of the validity of mesmerism and "magnetic fluid," in some form or other. Tischner believes that the appeal to suggestion as an explanation has not closed that question, as some had believed. He also suggests that "exteriorization of sensibility" is probably involved in certain phenomena of mediumship, such as telekinesis and materializations. [H & P]

1881. ***Tischner, Rudolph E.***

Geschichte der okkultischen (metapsychischen) Forschung von der Antike bis zur Gegenwart. II. Teil. Von der Mitte des 19 Jahrhunderts bis zur Gegenwart. Pfullingen: Johannes Baum, 1924, 371 pp.

An important and excellent reference source for the history of psychical research. This is the second of two volumes. The first volume (entry number 1830) covers the period from ancient times to the middle of the nineteenth century. This volume covers the period from the middle of the nineteenth century to the time of writing. As is so often the case, the German word "occult" in the title refers to psychical research, not to issues related to esoteric magic. Tischner covers the period in question efficiently, with great accuracy, and with broad coverage that includes the history of psychical research in America, England, France, and Germany. Although the book is, as one might expect, particularly strong in its treatment of psychical research in Germany, its one area of weakness is the brevity of its treatment of work in France. Nevertheless, Tischner gives the reader a clear picture of the main streams of development of psychical research in this important period, and provides a clear framework for understanding the interplay of intellectual currents that influenced its course. [H & P]

1882. [*Weiss, Ehrich, pseudonym: Harry Houdini*].

Houdini Exposes the Tricks Used by the Boston Medium "Margery" to Win $2500 Prize Offered by the Scientific American. Also a Complete Exposure of Argamasilla, the Famous Spaniard Who Baffled Noted Scientists of Europe and America, with His Claim to X-Ray Vision. New York: Adams Press, 1924, 39 + (1) pp.

[P]

1883. [*Weiss, Ehrich, pseudonym: Harry Houdini*].

A Magician Among the Spirits. New York and London: Harper & Brothers, 1924, xxiii + 294 pp.

Here the famous magician who had a particular interest in debunking the phenomena of spiritualistic mediums, discusses the feats of such well-known mediums as the Davenport brothers, D. D. Home, Eusapia Palladino, and Henry Slade. The book contains a very informative chapter on Houdini's long friendship with Arthur Conan Doyle, whose spiritualist beliefs led to many animated debates between the two men. The chapter reproduces some of the correspondence that was exchanged between them. [P]

1884. *Wickland, Carl August.*

Thirty Years among the Dead. In Collaboration with Nelle M. Watts, Celia L. Goerz, Orlando D. Goerz. Los Angeles: National Psychological Institute, 1924, 390 pp.

Wickland, a physician working in collaboration with his wife who was a spiritualist medium, had a unique method for enticing possessing spirits to communicate. Using a static electricity machine, sometimes employed by physicians in those days to treat various illnesses, he would drive the spirit from the victim and into his entranced wife. This enabled him to talk and reason with the spirit through his wife. He believed that in many cases the invading spirit had entered the victim accidentally, being attracted to the light of that person's "aura." After the spirit had become convinced of its true condition (that it was no longer in its own body, that it needed to pass on to another stage of life, etc.), it would usually leave willingly. Sometimes, however, Wickland had to call upon friendly spirits to take the invading entity away forcibly. Wickland claims to have completely cured individuals who had long been institutionalized as insane. In some cases he was able to confirm that the cure still held many years after the treatment. He also spent some time tracking down and verifying information given by the invading spirits about the circumstances of their lives on earth. The book is significant as a rare attempt to give practical application to the spiritualist view of the world through a kind of mediumistic psychotherapy. [P]

1925

1885. ***Baerwald, Richard.***
Die Intellektuellen Phanomene. (Vol. 2 of *Der Okkultismus in Urkunde.* Edited by Max Dessoir) Berlin: Ullstein, 1925, ix + (1) + 382 + (2) pp.

An excellent study of the "intellectual phenomena" of psychical research. These include telepathy, apparitions, clairvoyance, foreknowledge, and communication with the dead. Baerwald's point of departure for the study of psychical phenomena is psychological, an attempt to probe the realms of the unconscious through an examination of some of our most mysterious experiences. Baerwald concentrates on examples of these phenomena drawn from German psychical researchers and, for that reason, the book is an extremely valuable source of information on the history of psychical research in Germany. The book is well researched and copiously footnoted. (For the first volume of this work, see Gulat-Wallenburg, *Der physikalische Mediumismus,* 1925, entry number 1896). [P]

1886. ***Barrett, William Fletcher.***
The Religion of Health: an Examination of Christian Science. Completed by R. M. Barrett. London: Dent, 1925, xiv + 149 pp.

A posthumously published work in which Barrett examines the history of Christian Science and accounts of cures performed under its influence. He concludes that its affects are explained in terms of suggestive therapeutics. [H & P]

1887. ***Bird, James Malcolm.***
"Margery" the Medium. Boston: Small, Maynard & Co., 1925, xi + 518 pp.

A work depicting the mediumship of "Margery," whose actual name was Mrs. Crandon. "Margery," the wife of a well-known Boston surgeon, began producing what appeared to be physical phenomena around 1923. The book attempts to describe those phenomena, but does so in a disorganized and confusing manner. The author believes "Margery's" phenomena to be supernormal, but, although attempts were made by objective persons to test them under laboratory conditions, no concurring conclusions were reached. One of those attempts was made by a committee of the *Scientific*

American set up to investigate "Margery," with Bird as its secretary. That committee's conclusions were negative and the latter half of this book is devoted to a criticism of its findings. [P]

1888. *Broad, Charlie Dunbar.*

The Mind and Its Place in Nature. London: Kegan Paul, Trench, Trubner & Co., 1925, x + 674 pp.

Broad was Professor of Philosophy at Cambridge and a brilliant philosophical lecturer and writer. He developed a particular interest in psychical research and became one of its most distinguished theorists. This work is an expanded version of the Tarner Lectures given at Cambridge in 1923. In the first part of the volume, Broad considers the various common-sense theories of the nature of mind and raises the mind-body problem. He then investigates the ways in which we come to know about things around us, examining sense perception, memory, introspection, and knowledge of other minds. From there, Broad moves on to a discussion of evidence for the existence of an "unconscious." In the fourth section of the book he takes up human survival after death, an issue directly related to psychical research. Pointing out that the Society for Psychical Research has brought forth a mass of facts pertaining to the continued existence of human beings after death, Broad contends that these facts may justly be called "supernormal" in that they cannot be explained through the usual assumptions of science and common sense about the nature and powers of the human mind. Taking these facts into account, Broad discusses first the ethical and then the empirical arguments for survival. He does not believe that the various ethical arguments that have been put forward by philosophers are valid, but he does think that important directions are indicated in the empirical arguments. Human survival of death as a hypothesis, says Broad, goes a long way toward explaining supernormal data. Nonetheless, a hypothesis is only that, and although Broad does not believe it possible to disprove survival, neither does he hold that we have what constitutes convincing proof. Broad believes that any valid philosophical analysis must take into account all the facts as we know them, and in this vein he develops a philosophical psychology that recognizes both the findings of psychical research and those of scientific psychology. [P]

1889. *Bruck, Carl.*

Experimentell Telepathie. Neue Versuche zur telepathischen Ubertrag von Zeichnungen. Stuttgart: J. Püttmann, 1925, ix + 80 pp.

An account of 114 experiments in telepathic transference of drawings and diagrams carried out by the author. Dr. Bruck himself was the "agent" or "sender" in these experiments. He used easily hypnotizable subjects as "receivers" in an attempt to see whether telepathic abilities are in any way related to hypnotic susceptibility. The results of the experiments strongly favored the existence of telepathy. [P]

1890. *Burnett, Charles Theodore.*

Splitting of the Mind: An Experimental Study of Normal Men. Princeton, New Jersey and Albany, New York: Psychological Review Company, 1925, 132 pp. [Psychological Monographs No. 155].

A report of experiments on the dissociation through hypnosis of a normal individual's mental states into two or more groups or systems that function separately but simultaneously. Burnett argues that separate dissociated systems maintain an internal unity through dominating ideas. [H]

1891. *Campbell, Charles MacFie, Langfeld, Herbert Sidney,* *McDougall, William, Roback, Abraham Aaron, and Taylor, E. W. (eds.).*

Problems of Personality. Studies Presented to Dr. Morton Prince, Pioneer in American Psychopathology London: Kegan Paul, Trench, Trubner & Co. Ltd., 1925, xiii + (1) + 434 pp.

An important collection of essays on psychological subjects. Of particular interest are Janet's "On Memories Which are Too Real," Mitchell's "Divisions of the Self and Co-consciousness," and Dunlap's "The Subconscious, the Unconscious, and the Co-conscious." [H]

1892. *Curnow, W. Leslie.*

The Physical Phenomena of Spiritualism: A Historical Survey. Manchester: Two Worlds Publishing Co., 1925, 110 + vi pp.

A collection of articles, originally published in *The Two Worlds* in 1924, that discuss the history of the various physical phenomena of mediumship observed during the rise of modern spiritualism. The work is rather uncritical and does not deal with exposures of fraud. [P]

1893. *De Brath, Stanley.*

Psychical Research, Science and Religion. London: Methuen and Co., (1925) xxiii + 207 pp.

De Brath relates the findings of psychical research to religious teachings and advocates the acceptance of a "higher naturalism" in conjunction with faith. [P]

1894. *Driesch, Hans Adolf Eduard.*

The Crisis in Psychology. Princeton: Princeton University Press, 1925, xvi + 275 pp.

Driesch devotes about one third of general discussion of the contemporary state of psychological investigation to psychical research. He expresses his personal conviction that parapsychological or psychic facts exist, and states his belief that modern psychology must take these human experiences into account if it is to remain relevant. There is no prior German edition. [P]

1895. *Erman, Wilhelm.*

Der tierische Magnetismus in Preussen vor und nach den Freiheitskriegen. Munich and Berlin: R. Oldenbourg, 1925, viii + 124 pp.

After a discussion of the earliest history of animal magnetism in Germany, Erman concentrates on events that took place in Berlin and Bonn, touching on the views of virtually all of the principal German proponents of animal magnetism to the time of Justinus Kerner (1786–1862). [H]

1896. *Gulat-Wallenburg, W. von; Klinckowstroem, Graf Carl von; and Rosenbusch, Hans.*

Der physikalische Mediumismus. (Vol. 1 of *Der Okkultismus in Urkunde.* Edited by Max Dessoir) Berlin: Ullstein, 1925, xiii + (3) + 494 + (2) pp.

The first volume of a scholarly two volume work on psychical research. The second volume, published in the same year, is entitled *Die intellektuellen Phänomene* and is written by Richard Baerwald. There was to have been a third volume, *Suggestion und Hypnose* by Albert Moll, but it never appeared. The present volume, concerned with the physical phenomena of mediumship, begins with a thorough discussion of the methodology to be used in investigating the phenomena and the sources of error against which researchers must be guarded. This is followed by a discussion of the experiments of the London Dialectical Society and of William Crookes in the early 1870s, a brief treatment of the Slade-Zöllner experiments, a lengthy section on the physical phenomena of Eusapia Palladino (1854–1918), a substantial discussion of Stanislawa Tomczyk and Kathleen Goligher, and shorter studies of other well-known mediums, such as "Eva C.," Willy Schneider, and Ladislaus Laszlo. The book is a friendly but highly critical study that concludes that every experiment with physical mediums that had been carried out so far fell short of the scientific standards necessary to establish the reality of the physical phenomena of mediumship once and for all. [P]

1897. *Gruber, Karl.*

Parapsychologische Erkenntnisse. Munich: Drei Masken, 1925, x + (1) + 330 + (1) pp.

Gruber affirms the authenticity of parapsychological phenomena and states his belief that the area is worthy of the attention of scientists. [P]

1898. *Holmes, Archibald Campbell.*

The Facts of Psychic Science and Philosophy Collated and Discussed. London: n.p., 1925, xvi + 512 pp.

A massive compendium of facts related to psychical research. This collection of data is drawn rather uncritically from a great variety of sources. It is footnoted, and therefore useful to the student, but references are not always complete. [P]

1899. ***Hyslop, James Hervey.***
"A Further Record of Mediumistic Experiments." *American Society for Psychical Research. Proceedings* 19 (1925): 1–451.

Hyslop describes sittings with the medium Mrs. Chenoweth to whom he brought a number of individuals with whom she had had no previous contact and whom she was not allowed to see during the sitting—a condition preferred by her in any case. Hyslop found the sittings fruitful not only for the accuracy of much of the material produced by the medium, but also as a study in the progressive psychological development of the trance state in the medium over the course of the sittings. This article is an important study of mental mediumship and the factors that influence its form. [P]

1900. ***Lambert, Rudolph.***
Die okkulten Tatsachen und die neuesten Medienentlarvungen. Eine Entgegnung auf die letzten Verstösse der Verachter der Parapsychologie. Stuttgart, Berlin, and Leipzig: Union Deutsche Verlagsgesellschaft, 1925, 97 pp.

A critical study of mediumship with a valuable section on the Goligher circle in Ireland. [P]

1901. ***Mattiesen, Emil.***
Der Jenseitige Mensch: eine Einführung in die Metapsychologie der Mystischen Erfahrung. Berlin: Walter de Gruyter, 1925, viii + 825 pp.

A learned treatise on the principal phenomena of psychical research by one of Germany's important investigators. The author covers everything from the automatisms and hysteria to mediumistic materializations. [P]

1902. ***Moniz, Egas.***
O Padre Faria na Historia do Hipnotismo. Lisbon: n.p., 1925, 194 + (5) pp.

Along with Dalgado's *Mémoire sur la vie de l'abbé de Faria* (1906, entry number 1571), this book is a rare source of information about José Custodi de Faria, a central figure in the history of hypnotism (see his *De la cause du sommeil lucide,* 1819, entry number 294). After summarizing the history of animal magnetism and somnambulism up to Faria's time, Moniz presents a detailed picture of Faria's life and the development of his theory of hypnotism, which he called "lucid sleep." [H]

1903. ***Price, Harry.***
Stella C.: an Account of Some Original Experiments in Psychical Research. London: Hurst & Blackett, 1925, 106 + (1) pp.

Presents a detailed account of laboratory experiments done with a physical medium given the pseudonym "Stella C." Price, an expert in magic and conjuring, set up his experiments in such a way that, in his opinion, no deception or fraud was possible. Nonetheless, he found that a number of physical effects took place in her presence which could not be reproduced through normal means. Among the phenomena observed were: a notable elevation in temperature around the medium, strong breezes in her vicinity, telekinetic movements of small objects, raps, and occasional flashes of light. [P]

1904. ***Richardson, Mark W.; Hill, Charles S.; Martin, Alfred W.;*** *Harlow, S. Ralph; De Wyckoff, Joseph; and Crandon, L.R.G.*

Margery Harvard Veritas. A Study in Psychics. Boston: Blanchard, 1925, 109 pp.

[P]

1905. ***Tischner, Rudolf E.***

Das Medium D. D. Home. Untersuchung und Beobachtungen (nach Crookes, Butlerow, Varley, Aksakow and Lord Dunraven). Ausegewahlt und herausgegeben von Rudolf Tischner. Leipzig: Oswald Mutze, 1925, 164 pp.

[P]

Indexes

Name Index

This index provides entry numbers for the authors, editors, and other persons associated with the publications included in the Bibliography. Names are followed immediately by entry numbers.

Abbott, David Phelps, 1579
Absonus, Valentine, 195
Acevedo, M. Otero, 1377
Adam, 564
Adamkiewicz, Albert, 1567
Adare, Windham Thomas Wyndham-Quin, viscount, 932
Adler, Alfred, 1146
Agassiz, ———, 963
Aide, Hamilton, 1252
Aigner, Ed., 1665
Aksakov, Alexandr Nikolaevich, 962, 1253, 1293, 1395, 1418, 1430–1431, 1905
Alembert, Jean Le Rond d', 111
Alexander, Patrick Proctor, 935
Alexis. *See* Didier, Alexis
Alibert, ———, 328
Allen, Edward W., 1010
Allen, Thomas Gilchrist, 1501
Alleyne, John, 1747
Allix, Eugene, 701, 722, 761
Almaviva, ———, comte, 44
Almignana, ———, abbé, 557, 702
Amadou, ———, 8
Amouroux, J. A., 506
Andry, Charles Louis François, 26–27, 101
Andry, J., 1857
Angelhuber, J. F., 569
Apis, ———, 306
Archibold, ———, 120
Argamasilla, 1882
Argenti, Francesco, 699
Armitage, Miss, 784
Armstrong, H. E., 1722
Arndt, W., 257
Artois, ———, comte d', 11, 20, 59, 340
Artois, ———, comtesse d', 13
Ashburner, John, 560, 583, 587, 901
Assezat, Jules, 651
Assier, Adolphe d', 1062
Atkinson, Henry George, 605
Atkinson, Willan Walker, 1639
Auguez, Paul, 763, 786
Azais, Pierre Hyacinthe, 271, 403
Azam, Etienne Eugène, 450, 465, 812, 818, 896, 980, 1159, 1311, 1334

B., Madame, 171, 1148, 1864
B—y, Therese von, 431
Baader, Franz von, 270
Babinet, Jacques, 879
Babinski, J., 1223
Bacher, Alexandre André Philippe Frédéric, 28

Bacot, G. F., 906
Baerwald, Richard, 1885, 1895
Baeumler, Christian Gottfried Heinrich, 1035
Baggally, W. W., 1628, 1719
Bahr, Johann Karl, 652
Baillif, Louis Ernest, 907
Bailly, Jean Sylvain, 30, 32–33, 62, 213
Baker, Rachel, 249, 254, 992
Baldinger, Ernst Gottfried, 163
Baldwin, Helen Green, 1311
Baldwin, J. Mark, 1311
Balfour, Arthur James, 1761
Ballou, Aidin, 627
Bapst, F. G., 271
Baraduc, Hippolyte Ferdinand, 1335, 1396–1397, 1531, 1600
Baragnon, P. Petrus, 628
Barbeguière, J. B., 34
Barberin, chevalier de, 49, 133, 158, 175, 399, 451
Baréty, A., 1051, 1160
Barlow, William Frederick, 391
Barré, P. Y., 35
Barreau, Ferdinand, 492
Barret, ———, 1863
Barrett, R. M., 1886
Barrett, William Fletcher, 1244, 1475, 1601, 1656, 1886
Barth, George H., 588, 653
Barth, Henri, 1124
Bates, E. Katharine, 1602
Bathurst, James, 1310
Bauche, Alexandre, 787, 880
Baudot, Louis Antoine, 404
Baudouin, Charles, 1746
Baumann, A.M.F., 570
Bayley, William, 1582
Baylina, Ignacio Ribera, 1549
Baynes, H. G., 1778
Beadnell, Charles Marsh, 1762
Beale, Dr., 1796
Beard, George Miller, 1036, 1052
Beauchamp, Miss, 1495, 1559
Beaunis, Henri Étienne, 896, 1076, 1101, 1125, 1184
Beaux, Jean Jacques, 737
Beecher, Charles, 654
Beecher, William H., 493, 539
Beers, Clifford Whittingham, 1603
Beesel, M., 589
Bekhterev, Vladimir Mikhailovich, 1550
Belden, Lemuel W., 357
Belfiore, Giulio, 1161, 1224, 1432
Bell, John, 121, 178, 201, 380
Bell, Robert, 813
Bellanger, Augustin René, 1077
Belot, Camille, 1078
Benedikt, Moriz, 1357, 1727
Bénézet, E., 703
Bennett, Edward T., 1515, 1532, 1551, 1568, 1604
Bennett, John Hughes, 606
Benoit, Jacques Toussaint, 590
Bentivegni, Adolf von, 1254
Beraud, Marthe ["Eva C."], 1672, 1704, 1787, 1807, 1839, 1864, 1895
Bercham, Berthout van, 180
Bergasse, Nicolas, 15, 36–38, 47, 98, 122–124, 201, 250, 282
Bergson, Henri, 1682
Bérillon, Edgar, 1079, 1225, 1433, 1481, 1499, 1692
Berjon, A., 1126
Berjot, E., 787
Berlancourt de Beauvais, Mademoiselle de, 16
Berna, Didier Jules, 365, 383, 392
Bernard, Prudence, 620
Bernheim, Hippolyte, 294, 450, 896, 947, 1080, 1125, 1127, 1131, 1153, 1228, 1232, 1237, 1284, 1412, 1499, 1657, 1720
Berry, Catherine, 981
Bersot, Ernest, 420, 629
Berthelen, Karl Andreas, 886
Berti, A., 630
Bertrand, Alexandre Jacques François, 294, 313, 324, 726, 1424
Besterman, Theodore, 1712, 1858
Betiero, T. J., 1413
Bettoli, Parmenio, 1285
Bianchi, ———, 1149
Biat, Chrétien, 590
Billot, G. P., 393, 559
Binet, Alfred, 1128, 1162, 1195, 1255, 1270, 1311, 1378, 1434, 1467, 1500
Birchall, James, 1067, 1086
Bird, ———, 1308
Bird, Friedrich, 405
Bird, James Malcolm, 1887
Bird, Malcolm, 1838
Birnstiel, F. H., 163
Birt, William Radcliff, 655
Bisson, Juliette Alexandre, 1672, 1839, 1863
Björnström, Fredrik Johan, 1163
Blackmore, Simon Augustine, 1859
Blakeman, Rufus, 571
Blanc, Hippolyte, 881
Blavatshy, Madame, 1175, 1367, 1394
Bleuler, Eugen, 1506, 1704
Bloch, George, 1, 5, 10
Böckmann, Johann Lorenz, 157
Boehme, ———, 1367

Boin, ———, 242
Boirac, Emile, 1605, 1721, 1880
Boisgelin, Louis de, 138
Bolton, Gambier, 1533, 1693
Bombay, ———, 39
Bompard, Gabrielle, 1296
Bonaventura, Isador, 762
Bonavia, Emmanuel, 919
Bond, Frederick Bligh, 1732, 1747, 1860
Bonjean, Albert, 1256
Bonnaymé, Ernest, 1583
Bonnefoy, Jean Baptiste, 40, 125
Borderieux, Carita, 1808
Boret, ——— de, 865
Boret, Bènigne Ernest, 956
Bormann, Walter, 1452, 1622
Bormes, ———, baron de, 41
Bory, ———, de, 30–31
Bosc, Ernest, 1227, 1595
Bose, Ernest, 1360
Bottey, Fernand, 1081
Bourdin, Claude Etienne, 449
Bourne, Ansel, 1002
Bourneville, Désiré, 947, 982
Bourru, Henri, 1102, 1126, 1164, 1196
Bourzeis, Jacques Aimée de, 19
Bouvier, Marie André Joseph, 42
Bouvignier, L.J.D. de, 435
Bozzano, Ernesto, 1480, 1516, 1623, 1683, 1748, 1763–1765, 1792, 1817–1818, 1840, 1861
Brack, ———, 43–44, 126
Brackett, Edward A., 1129
Brackett, Loraina, 389
Braid, James, 324, 450, 460, 465, 481, 483, 532, 540, 552, 591, 598, 607, 631, 656, 659, 738, 800, 812, 819, 821, 823, 896, 907, 918, 977, 993, 1029, 1035, 1047, 1056, 1159, 1183, 1219, 1275, 1424, 1453, 1504, 1517, 1585
Bramwell, J. Milne, 1517, 1624
Brandis, J. D., 284
Brever, ———, 175
Brierre de Boismont, Alexandre Jacques François, 352, 508
Brigham, A., 381, 386, 389
Brissot de Warville, Jacques Pierre, 45
Bristol Mesmeric Institute, 572, 592
Brittan, S. B., 657, 866, 1014
Britten, Emma Hardinge, 933, 983–984, 1063
Brivasac, ———, 1089
Broad, Charlie Dunbar, 1888
Broca, Paul, 450, 465, 800, 818
Brofferio, Angelo, 1312
Brooks, E. S., 1771
Broussais, ———, 403
Brown, John, 466
Brown-Sequard, ———, 294
Brownson, Orestes Augustus, 704
Bruce, Henry Addington, 1518, 1694
Bruck, Carl, 1889
Brugma Dingwall, ———, 1863
Bruining, Gerbrand, 253
Bruni, Em., 1336
Bruno, ———, de, 127
Buchanan, Joseph Rhodes, 540, 573, 705, 854, 1011, 1014, 1103, 1696
Buckland, Thomas, 608
Bué, Alphonse, 1337
Bué, Hector Joseph, 1053
Burdin, Charles, 436
Burg, Victor, 1003
Burnett, Charles Theodore, 1890
Burot, P., 1102, 1126, 1164, 1196
Burq, Victor Heab Antoine, 658, 1054
Burr, ———, 918
Bush, George, 549
Bute, John, marquess of, 1459
Butlerow, ———, 1905

C., Eva. *See* Beraud, Marthe
C., Stella, 1903
Cadwell, J. W., 1055
Cagliostro, 1470
Cahagnet, Louis Alphonse, 558–559, 577, 593, 609–610, 707–708, 1064
Caille, Claude Antoine, 101
Caillet, Albert Louis, 922, 1673–1674
Caldwell, Charles, 451
Cambry, Jacques, 46
Campbell, Charles MacFie, 1891
Canelle, A., 846
Capern, Thomas, 611
Capron, Elias Wilkinson, 739
Capua, Sara Carlotta Di, 1276
Caritides, ———, 108
Caroli, Giovanni M., 801, 814
Carpenter, J. Estlin, 1197
Carpenter, William Benjamin, 659, 985, 993, 1011, 1025, 1197
Carra, Jean Louis, 128
Carreras, Enrico, 1527
Carrington, Hereward, 1584, 1606, 1625, 1628, 1658, 1675, 1684, 1695, 1733, 1749, 1766, 1862
Carrington, Walter Whately [pseud. W. Whately Smith], 1767, 1768, 1793
Carus, Carl Gustav, 533, 773
Castle, ———, 894
Caullet de Veumorel, ———, 129
Cavaignac, ———, 562
Cavailhon, Edouard, 1056

Cazaintre, ———, 335
Cazotte, Jacques S., 867
Célaphon, ———, 34
Cerise, ———, 778
Chaddock, Charles Gilbert, 1328
Champville, Gustave Fabius de, 1534
Chapelain, ———, 334, 393
Chapman, ———, 1057
Charavet, ———, 777
Charcot, Jean Marie, 1003, 1037, 1048
Charcot, Jean Martin, 450, 818, 896, 947, 977, 982, 1052, 1058, 1080, 1127, 1131, 1148, 1172, 1230, 1234, 1239, 1318, 1386, 1502, 1504, 1517, 1559, 1612, 1727
Chardel, Casimir Marie Marcellin Pierre Célestin, 285, 325, 332, 349
Charles X (king of France), 340
Charpentier, J.B.A., 394
Charpignon, Jules, 437
Charpignon, Louis Joseph Jules, 467–468, 815, 868
Chastellux, ———, marquis de, 36
Chautard, E., 1535
Chazarain, Louis Théodore, 1130
Chenoweth, Mrs., 1707, 1709, 1899
Chevenix, R., 339
Chévillard, Alphonse, 920
Chevreul, Michel Eugène, 709
Chiana, Ercole, 1633
Chowrin, A. N., 1750
Christ. *See* Jesus Christ
Christison, John Sanderson, 1569
Christopher, 1743
Clark, Charles S., 1473
Clarke, Edward Hammond, 1004
Cleveland, William, 1398
Clodd, Edward, 1722
Cloquet, ———, 47
Close, Francis, 683
Coates, James, 1338–1339, 1414, 1552, 1570, 1659, 1734
Cocke, James Richard, 1358
Coddè, Luigi, 612
Cogevina, Angelo, 452
Cohnfeld, Adalbert Salomo, 661
Colas, Albert, 1104
Coleman, Benjamin, 837, 936
Colin, ———, 908
Coll, ———, 273
Colley, Thomas, 994
Collyer, Robert Hanham, 395, 469, 937, 986
Colquhoun, John Campbell, 222, 354, 372, 396, 470, 613, 631
Colsenet, Edmond, 1025
Comet, Charles Jean Baptiste, 816
Concato, L., 614
Condillac, ———, 31
Conner, Dean Bridgman, 1708
Constable, Frank Challice, 1660
Constantin, ———, comte de, 1167
Constantin, Gregor, 1430
Conte, Paolo, 1224
Cook, E. Wake, 1528
Cook, Florence, 1465
Cook, William Wesley, 1468
Cooke, ———, 1001
Coover, John Edgar, 1723
Corbaux, F., 261
Corfe, George, 560
Coriat, Isador H., 1615
Corson, Caroline Rollin, 1318
Coste, Albert, 971, 1379
Coste, Marie Leon, 1226
Coste de Lagrave, ———, 1198
Cottin, Angélique, 548
Cottingly, ———, 1820
Coué, Emile, 1676, 1746, 1794, 1821
Coupland, William C., 924
Court de Gébelin, Antoine, 29, 60, 71, 77
Courtier, Jules, 1607
Cousin, M. V., 363
Cox, Edward William, 938, 951
Cox, Mrs., 1209
Crampon, ———, 333
Crandon, L.R.G., 1904
Crandon, Mrs. ["Margery"], 1882, 1887, 1904
Crawford, William Jackson, 1711, 1735, 1751, 1762, 1787, 1795, 1824
Crichton-Miller, Hugh, 1677
Crocq, Jean, 294, 1359, 1399
Croizet, ———, abbé, 679
Crookes, William, 744, 858, 939, 959, 973, 1406, 1430, 1895, 1905
Crowe, Catherine, 341, 561
Cruikshank, George, 852
Crumpe, G. S., 510
Csanady, Stephan, 817
Cullere, A., 1131
Cullerre, Alexandre, 1340
Culpin, Millais, 1769
Cumberland, Stuart, 1199, 1736, 1752
Curnow, W. Leslie, 1892
Curran, Pearl, 1718

D., Madame, 1145
Dailey, Abram Hoagland, 1361
Dalgado, D. G., 1571, 1585, 1902
Dallas, Helen Alexandria, 1641
Dallmer, Oskar, 1105
Dalloz, A.L.J., 286, 314

Dal Pozzo di Mombello, Enrico, 632, 921
D'Amico, P., 902, 1281
Dammerung, Gottlieb, 886
Dampierre, Antoine Esmonin, marquis de, 49
Darcy, Miss, 1509
Dardeps, ———, 406
Darieux, X., 1283
Darwin, Charles, 900
D'aute-Hooper, T., 1696
Davenport, Ira, 852, 875, 890, 931, 1883
Davenport, Reuben Briggs, 1200
Davenport, William, 852, 875, 890, 931, 1883
Davey, William, 710
David, Pierre, 1132
Davies, Charles Maurice, 972
Davis, Abert E., 1737
Davis, Andrew Jackson, 746, 774, 909
Deane, Mrs., 1878
Debay, A., 471
De Brath, Stanley, 1864, 1893
Debuire, H., 651
Décle, Charles, 1130
Defer, J.B.E., 397
De Gasparin, ———, 1586
Deher, Eugen, 1106
Delaage, Henri, 550, 594, 686, 763
Delandine, Antoine François, 130–131
Delanne, Gabriel, 1107, 1341, 1415, 1454, 1500, 1626
De Laurence, Lauron William, 1469
Delboeuf, Joseph Remi Léopold, 1133–1134, 1165, 1201, 1228, 1259, 1286, 1313
D'Eldir, Alina, 342
Deleuze, Joseph Philippe François, 243, 261, 292, 295, 301, 320, 326, 330, 336, 358, 387, 393, 402, 455, 558, 602, 608, 1089, 1424, 1554
De Mainauduc, John Benoit, 208, 220
Demarquay, Jean Nicolas, 818
De Morgan, Augustus, 853
De Morgan, Sophia Elizabeth, 853
Dendy, Walter Cooper, 438
Denis, Léon, 1287, 1536
Denton, ———, 1696
Denton, Elizabeth M.F., 854
Denton, William, 854
Descartes, René, 62
D'Eslon, Charles, 11–12, 14, 17, 19–21, 23, 31, 35, 52–53, 55, 57, 67, 72, 74, 79, 82, 84, 86, 98, 129, 132, 177, 208, 340, 502
Despine, Antoine, 421
Despine, Prosper, 910, 1026
Dessoir, Max, 1166, 1202, 1229, 1240, 1260, 1473, 1573, 1584, 1724, 1885, 1895
Devillers, Charles Joseph, 54, 133
Dewey, Dellon Marcus, 595
De Wyckoff, Joseph, 1904
Dexter, George T., 664
Dickerson, K.D.D., 472
Didier, Adolphe, 764, 838
Didier, Alexis, 512, 562, 764–765, 806, 895, 1021
Dippel, Joseph, 1038
Dixon, Jacob, 803, 1646
Dobler, Herr. *See* Smith-Buck, George
Dods, John Bovee, 422, 473, 596, 607, 711, 718, 918
Donato. *See* Hont, Albert Edouard d'
Donavan, ———, 1288
Donley, John, 1637
Donné, Alfred, 398
Doppet, François Amédée, 55, 134
Douglas, James S., 453
Doyle, Arthur Conan, 1725, 1753, 1783, 1797, 1819–1820, 1841–1842, 1883
Drake, Daniel, 494
Drayton, H. S., 1230
Dresser, Horatio Willis, 1770
Driesch, Hans Adolf Eduard, 1553, 1584, 1712, 1894
Dubois, Frédéric, 416, 436
Dubois, M.E.F., 383, 392
Dubreuil-Chambardel, ———, 407
Duchatel, Edmond, 1642
Ducie, ———, earl, 572
Duckworth, J. Herbert, 1821
Du Commun, Joseph, 340, 387
Dufau, Julien, 56
Dufay, ———, 1203
Duff, Edward Macomb, 1501
Duguid, David, 1365, 1604
Dujardin, Edouard, 1135
Dulora de la Haye, ———, 1400
Dumas, Alexandre, 562
Dumas, Georges, 1540
Dunand, Tony, 819, 1027
Dunlap, ———, 1891
Dunraven, ———, earl of, 932
Dunraven, ———, Lord, 1905
Dupau, Jean Amédée, 303, 328
Du Potet de Sennevoy, Jules Denis, 302, 334, 346, 359, 367, 373, 399, 423–424, 511, 518, 534, 633, 789, 831, 855, 1089, 1504, 1554
Du Prel, Karl Ludwig August Friedrich Maximillian Alfred, 1082, 1136, 1231, 1261, 1294, 1362, 1455–1456
Dupuy, Antonin, 820
Durand de Gros, Joseph Pierre [pseud.

A.J.P. Philips], 740, 820–821, 911, 1363, 1380
Durant, Charles Ferson, 381, 386
Dureau, Alexis, 922
Duroy de Bruignac, Charles Joseph Albert, 869
Durville, Gaston, 1650, 1661
Durville, Hector, 518, 1137–1138, 1167, 1262, 1381–1383, 1435, 1627, 1661
Durville, Henri, 1640, 1650
Du Vernet, B., 732

Eagle, G. Barnard, 634
Ebhardt, G. F., 287
Eckhartshausen, ———, von, 196
Edard, Guillaume, 1083, 1108
Edmonds, John W., 664, 711, 952
Eduard, Guillaume, 987
Edwards, D., 188
Eeden, F. van, 1247, 1372
Eglinton, William, 1091, 1139, 1300
Ehrenfreund, Edmund Otto [pseud. Ubald Tartaruga], 1822–1823
Elie de la Poterie, Jean Antoine, 135
Elisabetta, 452
Elliotson, John, 302, 339, 346, 372, 396, 399, 424, 464, 474, 490, 512, 535–536, 901, 1517
Ellsworth, Robert G., 1502
Emma, 575
Emmons, Samuel Bullfinch, 804
Encausse, Gérard, 1298, 1416
Ennemoser, Joseph, 293, 316, 454, 635
Ensor, Laura, 1264
Éprémesnil, Jean Jacques Duval d', 35, 57–58, 106, 136, 144
Ermacora, Giovanni Battista, 1314, 1342, 1390, 1436
Erman, Wilhelm, 1895
Erny, Alfred, 1384
Eschenmayer, Carl Adolph von, 258, 269, 343, 348, 361
Esculape, 306
Esdaile, James, 536, 546, 636–637, 1517
Esenbeck, Nees von, 258
Espérance, Madame d'. *See* Hope, Elizabeth
Espinouse, A., 1018
Estelle, 421
Esther, 1178
Evans, Richard, 1569
Everitt, Mrs. Thomas, 981, 1604

Fabre, Pierre Honoré, 741
Fahnestock, William Baker, 923
Fairfield, Francis Gerry, 973
Fajnkind, Stephanie, 1343
Falconer, William, 214
Fallois, Paul de, 1642
Fancher, Mary J., 1361
Fancher, Molly, 1361
Faraday, Michael, 659, 666
Farez, Paul, 1499
Faria, José Custodio de, abbé, 294, 726, 1239, 1424, 1517, 1571, 1585, 1902
Farmer, John Stephen, 1065, 1139
Fauvelle le Gallois, Auguste, 870
Favrye, ———, de la, 137, 164
Fechner, Gustav Theodor, 368, 616, 769, 839, 989
Feilding, Everard, 1628
Feilgenhauer, ———, 1431
Fellner, F. von, 1028
Fenayrou, Gabrielle, 1296
Féré, Charles Samson, 1128, 1162
Ferguson, Jesse Babcock, 883
Ferret, J., 1289
Feytaud, Urbain, 1344
Fiard, Jean Baptiste, 223, 337
Fichte, Immanuel Hermann, 805, 1005
Figaro, 44
Figuier, Guillaume Louis, 822
Figuier, Louis, 940, 1168
Filiatre, Jean, 1554, 1608
Fillassier, Alfred, 353
Finzi, Giorgio, 1390
Fischer, Doris, 1709
Fischer, Engelbert Lorenz, 1066
Fischer, Friedrich, 278, 408
Fishbough, William, 718
Fitzgerald, J., 1184
Flammarion, Camille, 847, 890, 1257, 1457, 1586, 1771, 1843–1844
Flemming, Alice Kipling (Mrs. Holland), 1641, 1687
Fletcher, William, 1300
Flournoy, Theodore, 1263, 1470, 1503, 1525, 1662
Flower, Sidney, 1402
Fludd, Sir Robert, 94, 116, 145, 208
Foissac, Pierre, 321, 329–330, 350, 355
Fontan, Jules, 1169, 1190
Fontenay, Guillaume de, 1437
Fontette Sommery, ———, comte de, 59
Fonvielle, W. de, 1170
Forbes, John, 512
Ford, A. E., 418
Forel, August, 1232, 1240
Forthuny, Pascal, 1845
Fortia de Piles, Alphonse, 138
Fortlage, ———, 769
Fouet, ———, 1509
Foughet, Par, 60

Fournel, Jean François, 139–141, 152
Fournier D'Albe, Edmund Edward, 1609, 1704, 1824
Fournier-Michel, ———, 16
Foveau de Courmelles, François Victor, 1264–1265
Fox, John D., 595
Fox, Kate, 387, 595, 739, 774, 969, 995, 1122, 1200, 1508. *See also* Jencken, Catherine Fox
Fox, Margaret, 387, 595, 739, 774, 969, 995, 1122, 1200, 1508. *See also* Kane, Margaret Fox
Francis, J. G., 638
Franklin, Benjamin, 31, 40, 62, 65, 84, 101, 109, 177, 204, 213
Franklin, Mrs. Benjamin, 30
Frapart, ———, 409
Freimark, Hans, 1697, 1798
Frère, ———, 382
Freud, Sigmund, 175, 910, 1146, 1643, 1678, 1727, 1870
Friedlander, ———, 277
Frisz, ———, 667
Fronda, R., 1290
Frotté, Em., 840
Fugairon, Louis Sophrone, 1364
Fullerton, George S., 1180
Funk, Isaac K., 1537, 1587
Fustier, ———, abbé, 255, 265

Galien, ———, 306
Galilei, Galileo, 177, 613
Gall, ———, 315
Gallert de Montjoie, Christophe Félix Louis, 61–62
Garcin, ———, 742
Gardane, Joseph Jacques de, 63
Gardy, Louis, 1266
Garrett, Julia E., 1315
Gasc-Desfosses, Edouard, 1417
Gasparin, Agénor Étiene, comte de, 702, 712
Gassner, ———, 7, 94, 109, 116, 967
Gathy, August François Servais, 639
Gauthier, Aubin, 425, 455, 495, 498, 501, 513, 537, 1089
Gauthier, Louis Philibert Auguste, 496
Gazerra, Linda, 1680
Geley, Gustave [pseud. Dr. Gyel], 1438, 1458, 1698, 1754, 1787, 1863–1864
Gentil, Joseph Adolphe, 551, 562, 640, 668 713, 743
Gérard, Jules, 788, 872, 891–892, 914
Gérard, ———, 47
Gérardin, Sébastien, 64
Gerrish, Frederic H., 1637
Gessmann, Gustav Wilhelm, 1171
Gibier, Paul, 1140, 1233
Gigot-Suard, Jacques Léon, 823
Gilbert, ———, 1153
Gilbert, ———, 65
Gilbert, Jean Emmanuel, 66
Gilles de la Tourette, George Albert Edouard Brutus, 294, 1172, 1234, 1291
Girardin, Sebastien, 67–68
Giraud-Teulon, M. A., 818
Glanvill, Joseph, 1810
Glenconner, Pamela, 1799
Glendinning, Andrew, 1365
Gley, E., 1141, 1184
Gmelin, Eberhard, 165, 278, 341
Godfrey, Nathaniel Steadman, 669–670
Goerz, Celia L., 1884
Goerz, Orlando D., 1884
Goligher circle, 1751, 1787, 1900
Goligher family, 1824
Goligher, Kathleen, 1711, 1895
Goodhart, Simon P., 1547
Goodrich-Freer, Adela, 1459–1460
Gordon, James Adam, 592
Gorgeret, P.M.E., 439
Gougenot des Mousseaux, Henri Roger, 824, 856
Gould, Tracy, 1395
Goupy, ———, 671
Gragnon, Célestin, 806
Grandvoinet, ———, [pseud. Tedinngarov, J. A.], 475, 477
Grashey, ———, 1392
Grasset, Joseph, 1519–1520, 1538, 1572, 1588
Graves, Lucien Chase, 1707
Gray, James Martin, 1772
Greatrakes, ———, 94, 116, 284
Green, J. H., 1316
Greenhow, Thomas Michael, 514
Gregory, Samuel, 372, 476
Gregory, William, 583, 617, 710
Grellety, J. Lucien, 990
Grey, Pamela Genevieve Adelaide (Wyndham), 1799
Gridley, Josiah A., 714
Grimaldi, A., 1290
Grimes, James Stanley, 515, 540, 607, 775, 974
Gromier, Emile, 597
Grosjean, ———, 733, 758
Grossmann, Jonas, 1292
Grow, David, 1793
Gruber, Karl, 1896
Grunewald, Fritz von, 1773, 1863
Guibert, ———, marquis de, 426
Guidi, Francesco, 618, 715–716, 767,

825–826, 828, 841, 903, 953, 1193, 1281
Guigoud-Pigale, Pierre, 69
Guillard, ———, 672
Gulat-Wallenburg, W. von, 1703, 1885, 1896
Guldenstubbe, Louis, baron de, 776
Guppy, Samuel, 857
Guppy, Mrs. Samuel, 981
Gurney, Edmund, 1084–1085, 1109–1110, 1142–1144, 1173–1174, 1204–1205, 1525
Guthrie, Malcolm, 1067, 1086, 1111
Guyomar de la Roche Derrien, 882
Guyon, Madame, 1367
Guzik, Jean, 1864
Gyel, Dr. *See* Geley, Gustave

H., B., 1848
Haddock, Joseph W., 372, 575, 718
Haldat du Lys, Charles Nicolas Alexandre de, 427, 516, 538, 619, 641
Hall, Caxton, 1471
Hall, G. Stanley, 1654
Hall, G. T., 615
Hall, Spenser Timothy, 483, 517, 522
Halphide, Alvan Cavala, 1482
Hammard, Charles Pierre Guillaume, 369
Hammond, William Alexander, 988
Hanna, Thomas Carson, 1547
Hannapier, C. R., 304
Happich, Carl, 1848
Hardenberg, Kuno, 1848
Hardinge, Emma. *See* Britten, Emma Hardinge
Hare, Robert, 744
Harlow, S. Ralph, 1904
Harnack, Erich, 1555
Harris, Dean William Richard, 1755
Harris, John William, 1610, 1685
Harrison, William H., 975, 1019
Harsu, Jacques de, 22
Hart, Bernard, 1643
Hart, Ernest, 1345
Harte, Richard, 1401, 1504
Hartmann, Eduard von, 533, 924, 1025, 1082, 1112–1113, 1253, 1293
Hartshorn, Thomas C., 320
Harvey, ———, 613
Harvier, Father, 178
Hauffe, Friederike, 341
Haweis, H. R., 1365
Haygarth, John, 214
Haynes, Edmund Sidney Pollock
Hayward, Aaron S. [pseud. A Magnetic Physician], 941
Hébert, ———, 240–241, 245
Hébert de Garnay, L. M., 642
Hébert, Hippolyte Joseph, 842
Heidenhain, Rudolf Peter Heinrich, 1029
Heineken, J., 215
Hell, Maximillian, 2–3, 8–9
Hellenbach von Paczolay, Lazar, 1006, 1020, 1114
Helmont, ——— van, 94, 145
Hélot, Charles, 1461
Hénin de Cuvilliers, Etienne Félix, baron d', 259, 298, 303, 305–306, 309
Hennig, Richard, 1573
Henry, Victor, 1483, 1503
Henslow, George, 1756, 1774
Heredia, C. M. de, 1825
Herholdt, Johan Daniel, 210
Héricourt, J., 1145, 1184
Hering, Charles E., 673
Hermès, 890
Herter, Christian, 1127
Hervier, Charles, 56, 71, 88, 95, 275
Heuzé, Paul R., 1800, 1826
Hewlett, Edgar, 717
Hill, Charles S., 1904
Hill, Ellida, 1389
Hill, John Arthur, 1663, 1726, 1738–1739, 1865
Hippocrates, 306
Hirt, ———, 1392
Hodgson, Richard, 1002, 1175, 1222, 1317, 1419, 1439, 1525
Hoffmann, Rath, 165
Hoffmann von Fallersleben, ———, 684
Hofrath Carus, ———, 769
Hohenheim, Theophrastus Bombastus von [pseud. Philippus Aureolus Paracelsus], 94, 116, 145, 188
Holland, Henry, 410, 505
Holland, Mrs. *See* Flemming, Alice Kipling
Hollander, Bernard, 1644
Holmes, Archibald Campbell, 1898
Holmes, Sherlock, 1725
Holt, Henry, 1699
Home, Alexandrina, 858
Home, Daniel Dunglas, 792, 813, 858, 932, 935, 948, 959, 991, 995, 1206, 1244, 1252, 1268, 1428, 1430, 1883, 1905
Home, Madame Daniel Dunglas, 1206, 1244, 1268
Hont, Alfred Edouard d', 1056, 1150, 1182, 1207, 1259
Hope, Elizabeth [pseud. Madame d'Espérance], 1395, 1418
Hope, William, 1838, 1845
Hopp, Max, 1713
Horne, ———, de, 13–14

Horst, Georg Conrad, 344
Hortensia, 909
Houdini, Harry. *See* Weiss, Ehrich
Houghton, Georgiana, 1040, 1059
Houston, Sam, 596
Hovey, William Alfred, 1115
Howitt, William, 859, 1074
Hubbell, Gabriel G., 1484
Hubbell, Walter, 1208
Hude, Anna, 1687
Hudson, Captain ———, 643
Hudson, Thomson Jay, 1346, 1385, 1473, 1521, 1539
Hue, Charles, 893
Huet, Mademoiselle, 847
Hufeland, Christoph Wilhelm, 260, 268
Hufeland, Friedrich, 234–235, 260, 278
Hugo, Victor, 1855
Hugueny, Charles, 1269
Huguet, Hilarion, 777
Hull, R.F.C., 1506
Husson, Henri Marie, 302, 324, 350, 354, 374, 383
Hutchinson, Horatio Gordon, 1485
Hyslop, James Hervey, 1486, 1556, 1574–1575, 1611, 1688, 1709, 1740, 1757, 1809, 1899

Illig, Johannes, 1827
Imoda, Enrico, 1679
Ince, Richard Basil, 1775
Inchbald, Elizabeth Simpson, 179
Ireland, William W., 1116
Isis, 306

Jackson, J. W., 710
Jage, Gustav, 1462
James, Constantin, 1210
James, John, 1021
James, William, 1002, 1222, 1270–1271, 1274, 1440, 1450, 1487, 1525, 1529, 1603, 1809
Janet, Pierre Marie Félix, 175, 910, 954, 977, 1146–1148, 1153, 1176, 1184, 1211, 1235, 1270, 1318–1319, 1386, 1441, 1496, 1500, 1522, 1540, 1559, 1584, 1589, 1594, 1629, 1643, 1648, 1706, 1777, 1846–1847, 1852, 1870, 1891
Jastrow, Joseph, 1488, 1557, 1594, 1741
Jaubert, Ernest, 1308
Jefferson, Thomas, 205
Jencken, Catherine Fox, 1200. *See also* Fox, Kate
Jenkins, E. Vaughan, 1376
Jesus Christ, 473, 557, 714, 877, 919, 1501
Joan of Arc, 779
Joan, Saint, 1116
Johnson, Alice, 1525
Johnson, Charles P., 497
Johnson, Franklin, 1177
Joire, Paul Martial Joseph, 1320, 1391, 1584, 1612, 1630, 1880
Joly, Henri, 996
Jones, Ernest, 1637
Jones, Henry, 539
Jourdain, Eleana Frances, 1667
Jourgniac de St-Médard, François de, 138
Jozwik, Albert, 360
Judel, ———, 79, 230, 330
Jung, Carl Gustav, 1146, 1506, 1704, 1778
Jung-Stilling, Johann Heinrich, 231–232, 343, 549
Jussieu, Antoine Laurent de, 72, 101, 145

Kachler, Auguste, 527
Kampfmuller, Charles, 210
Kane, Margaret Fox, 1200. *See also* Fox, Margaret
Kant, Immanuel, 616, 1082, 1660
Kaplan, Leo, 1727
Kardec, Allan. *See* Rivail, Hippolyte Léon Dénizard
Kayner, D. P., 1014
Kemnitz, Mathilde von, 1700, 1703
Kennedy, William Sloane, 1633
Kerlor, W. de, 1605, 1721
Kerner, Justinus Andreas Christian, 317, 341, 343, 348, 356, 361, 375–376, 674, 684, 766, 1074, 1136, 1896
Keyserling, Hermann, 1848
Kieser, Dietrich Georg, 258, 310, 351
Kiesewetter, Carl, 1294, 1347, 1348
Kilner, Walter John, 1664
Kindborg, Erich, 1779
King, John Sumpter, 1780
King, Katie, 1465
King, William, 623
Kinger, Johann August, 276
Kingsbury, George Chadwick, 1295
Kircher, Athanasius, 116, 674
Klinckowstroem, Graf Carl von, 1895
Klinckowström, Karl Ludwig Friedrich Otto von, 1665, 1829
Klinkosch, Joseph Thaddaus, 7
Kluge, Karl Alexander Ferdinand, 235
Kluski, Franek, 1704, 1864
Koreff, ———, 320
Kornmann, ———, 36, 47, 201
Kotik, Naum, 1613
Krafft-Ebing, Richard von, 1212

Kramer, Phillipp Walburg, 960
Krause, Ernst Ludwig [pseud. Carus Stern], 860
Krauss, Friedrich, 644, 904
Kronfeld, Arthur, 1867
Kuhnholtz, Henry Marcel, 428, 477
Kupffer, K. R., 1869

L., I. M., 1810
La Breteniere, ———, de, 159
Ladame, Paul Louis, 1041, 1259
Lafarge, ———, 568
Lafayette, ———, 205
Lafisse, ———, 72
Lafont-Gouzi, Gabriel Grégoire, 412
Lafontaine, Charles, 450, 552, 745, 807, 832, 894, 1089, 1504, 1554
La Grezie, Bertrand de, 73
Lallart, C., 873
Lambert, Rudolf, 1801, 1849, 1900
Lambroso, Cesare, 1238
Lancelin, Charles Marie Eugène, 1689, 1782
Landresse, C. de, 74
Laneri, Innocenzo, 145
Lang, Andrew, 1366, 1420, 1443, 1523
Lang, William, 478
Langfeld, Herbert Sidney, 1891
Langsdorff, Georg von, 1236
Lapponi, Giuseppe, 1576
La Salzede, Charles de, 553
Lassaigne, Auguste, 620
Laszlo, Ladislaus, 1875, 1895
Laugier, Esprit Michel, 143
Laurent, Emile, 1296
Laurent, Louis Henri Charles, 1321
Laurent, P., 441, 443, 519, 563
Lausanne, A. de. *See* Sarrazin de Montferrier, Alexandre André Victor
Lavater, Johann Kasper, 284
Lavoisier, Antoine Laurent, 30–31
Lazare, Bernard, 1349
Leaf, Walter, 1274, 1394
Le Cat, Nicolas Claude, 1591
Lee, Edwin, 370, 384, 479, 576, 895
Leger, Theodore, 540
Le Goarant de Tromelin, Gustave Pierre Marie, 1631
Lehmann, Alfred Georg Ludwig, 1272, 1444
Lélut, Louis François, 645, 753, 848
Le Noble, ———, abbé, 27
Leonard, ———, 362
Leonard, Gladys Osborne, 1745, 1838
Léquine, F., 599
Leroux, Auguste [pseud. Rouxel], 1322, 1387
Le Roy, ———, 30
Leroy, Jean Baptiste, 31
Levade, Reynier, 180
Lichtenberg, ———, 284
Liébeault, Ambroise Auguste, 175, 294, 450, 800, 812, 896, 947, 954, 1068, 1080, 1125, 1153, 1228, 1297, 1499, 1504, 1517
Liégeois, Jules, 896, 1125, 1228, 1237, 1499
Liljencrants, Johan, 1742
Lillbopp, C.P.E., 311
Lillie, Arthur, 1367
Lioy, P., 912
Litta Biumi Resta, Carlo Matteo, 202
Lloyd, Mrs. Philip, 1860
Lo-Looz, Robert de, 181
Lodge, Oliver Joseph, 1274, 1584, 1614, 1632, 1704, 1714, 1729, 1743–1744, 1809, 1868
Lodge, Raymond, 1714
Loewenfeld, L., 1472
Loisel, A., 520–521, 541
Loisson de Guinaumont, Claude Marie Louis, 542
Lombard, A., 295
Lombroso, Cesare, 1042, 1149, 1633–1634
London Dialectical Society, 942
London Mesmeric Infirmary, 719
Long, H., 413
Lordat, Jacques, 335
Loubert, Jean Baptiste, 498, 543, 867
Loudun, ———, 822, 982
Louis, Eugène Victor Marie, 1445
Loutherbourg, Mr. and Mrs. ——— de, 189
Lovy, Jules, 646, 827
Lowenfeld, L., 1490
Lowenfeld, Leopold, 1590
Ludwig, August Friedrich, 1802, 1830
Lunt, Edward D., 1524
Luther, ———, 1116
Lutzelbourg, ———, comte de, 160, 167, 182–183
Luys, Jules Bernard, 1178, 1213, 1239, 1273, 1298, 1454
Lyman, D., 1011
Lyon, Jane, 948

Mabru, G., 786, 789
Macario, Maurice Martin Antonin, 778
McArthur, Angus [pseud.], 1715
McCabe, Joseph, 1783–1784
McComb, Samuel, 1615
McDougall, William, 1666, 1891
Mackay, Charles, 647
Mackenzie, William, 1850
MacNab, Donald, 1406

M'Neile, Hugh, 450, 460, 481, 503
Macnish, Robert, 345, 1214
MacWalter, J. G., 720
Macy, Benjamin Franklin, 621
Madden, Richard Robert, 779
Maecker, Eugen Johannes, 1697
Maeterlinck, Maurice, 1728, 1809
Maggiorani, Carlo, 925, 1022, 1030
Magini, Giuseppe, 1179
Maginot, Adèle, 610
A Magnetic Physician. *See* Hayward, Aaron S.
Magnin, Emile, 1558
Magnin, Paul de, 1087
Mahan, Asa, 746
Mahon, Paul Augustin Olivier, 78–79
Maine de Biran, Marie François Pierre Gonthier de Biran dit, 363
Mais, Charles, 249, 254
Maitland, Samuel Roffey, 578, 747
Majewski, Adrien, 1447
Malgras, J., 1577
Mann, G. A., 1645
Marat, Jean Paul, 80, 197
Marcillet, J. B., 551, 562
Margery. *See* Crandon, Mrs.
Maricourt, R. de, 1088
Marie Antoinette (queen of France), 1470
Marne, M . . . de la, 337
Marrin, Paul, 1239
Marryat, Florence, 1300
Martin, Alfred Wilhelm, 1744, 1904
Martin, John, 191
Martineau, Harriet, 522, 555, 605
Mary Jane, 857
Maskelyne, John Nevil, 991, 1001, 1309
Mason, Rufus Osgood, 1421, 1492
Massey, Charles Charleton, 1034, 1082, 1091, 1113, 1180
Masson, Alphonsine, 780
Mattieson, Emil, 1901
Mauduyt de Varenne, Pierre Jean Claude, 101
Maury, Louis Ferdinand Alfred, 843
Maxwell, ———, 83, 116
Maxwell, Joseph, 1526, 1541, 1584, 1642
Maxwell, William, 1391
Mayo, Herbert, 372, 579
Mead, Richard, 1
Meader, John, 1658
Megalis. *See* Steininger, Karoline
Meigs, ———, 209
Meiners, Christoph, 184
Meltier, ———, 169
Melville, John, 1646
Mercier, Charles Arthur, 1729
Méric, Joseph Élie, 1043, 1044
Mesmer, Franz Anton, 1, 3–19, 21–23, 25–26, 28–29, 31, 35–36, 38–39, 41, 43, 46–47, 49–50, 54, 56–58, 62, 66, 71–73, 79, 81–90, 92, 94–96, 98, 105, 109, 111–112, 116–117, 123–124, 126–129, 134, 136, 144–145, 158, 177, 181, 197, 205, 208, 211, 216, 224, 228, 238, 243, 250, 252–253, 256, 260, 268, 275, 294, 296, 299, 307, 315, 331, 342, 399, 411, 450–451, 455, 473, 478, 498, 502, 527, 535, 540, 549, 558, 563, 565–566, 583, 599–600, 629, 633–634, 639, 657, 685, 766, 827, 885, 891, 923, 967, 1051, 1089, 1193, 1234, 1275, 1291, 1330, 1348, 1387, 1445, 1504, 1508, 1517, 1554, 1591, 1635, 1775, 1880
Mesnet, Ernest, 829, 961, 1181, 1368
Metzger, Daniel, 319, 1369, 1493
Meunier, ———, 789
Meyer, Adolph F., 1831
Meyer, J. B., 1061
Meyer, Th.J.A.G., 414
Mialle, Simon, 331, 358, 415
Michal, Victor, 723
Millet, F., 791
Mirville, Jules Eudes, marquis de, 676, 702 704
Mitchell, G. W., 992
Mitchell, John Kearsley, 808
Mitchell, Silas Weir, 1214
Mitchell, Thomas Walker, 1785, 1803, 1832, 1891
Mitchill, Samuel Latham, 254, 262
Moberly, Charlott Anne Elizabeth, 1667
Mocet, ———, 34
Moilin, Tony, 926
Moll, Albert, 450, 1240, 1323, 1507, 1869, 1895
Mongruel, Louis Pierre, 580, 600–601, 622, 706
Mongruel, Madame, 580, 601, 622, 809
Moniz, Egas, 1902
Montegre, Antoine François Jenin de, 239, 263
Montius, E., 461
Moore, W. Usborne, 1668
Morand, J. S., 1241
Moréty, G., 1182
Morin, Alcide, 724, 749, 830
Morin, André Saturnin, 831–832
Morley, Charles, 440
Morogues, Pierre Marie Sébastien Bigot, baron de, 725
Moroni, L., 1188
Morrison, ———, 242, 244
Morselli, Enrico Agostino, 1151, 1242, 1422, 1616, 1634, 1703

Morton, W., 1001
Moses, William Stainton [pseud. M. A. Oxon], 997, 1008, 1023, 1031, 1069, 1243, 1367, 1856
Motet, Auguste, 1045
Mouilleseaux, ———, de, 170
Moulinié, Charles, 89
Moutin, Lucien, 1183, 1403, 1591
Moutinho, An., 1089
Mozzoni, G., 768
Muhammed, 1116
Muhl, Anita M., 1833
Muletier, ———, 90
Mullatera, Giovanni Thommaso, 145
Muller, Catherine Elise, 1470
Müller, Rudolf, 1423
Münsterberg, Hugo, 1594
Myers, A. T., 1280
Myers, Frederic William Henry, 1070, 1090, 1110, 1118, 1144, 1146, 1152–1153, 1184, 1244, 1274, 1301, 1317, 1324–1325, 1350, 1351, 1388, 1439, 1487, 1515, 1525, 1557, 1641, 1661, 1704, 1870
Myers, Gustavus, 1647

N., Mademoiselle, 152, 161–162, 172
Nami, Giacomo D., 602
Nasse, Christian Friedrich, 258, 291
Neal, E. Virgil, 1473
Nees von Esenbeck, Christian Gottfried, 299, 677
Nettesheym, Agrippa von, 1294
Neusser, ———, 684
Newbold, William, 1473
Newman, John B., 564, 718
Newnham, William, 346, 523
Newton, Isaac, 62
Nicholls, ———, 695
Nichols, Thomas Low, 875, 883
Nielsen, Enja, 1704
Nizet, Henri, 1326
Noel, Roden, 1091
Nogueira de Faria, ———, 1804
Noizet, ———, 1424
Noizet, François Joseph, 726
Noizet, Gal, 955
Northridge, William Lovell, 1870

Ochorowicz, Julian, 1146, 1184, 1424, 1787, 1863
Oegger, V.G.E., 315
Oesterreich, Traugott Konstantin, 1648, 1805–1807, 1863, 1871
Olbers, Dr., 222
Olcott, Henry S., 1062
Oppert, C., 277
Ordinaire, ———, 441
Orelut, Pierre, 92
Orioli, Francesco, 452
Orleans, ———, duke of, 13
O'Ryan, ———, 93
Osiris, 306
Ossowiecki, Stephan, 1864
Osten, William von, 1592
Osty, Eugène, 1584, 1691, 1758, 1851
Otto, Bernhard, 751
Ottolenghi, Salvatore, 1238, 1474
Owen, Robert Dale, 833, 943
Oxon, M. A. *See* Moses, William Stainton

P., Stanislava, 1704
Pailloux, Xavier, 861
Palladino, Eusapia, 1406, 1437, 1516, 1586, 1607, 1616, 1625, 1628, 1630, 1633, 1647, 1652, 1684, 1807, 1836, 1883, 1895
Panin, ———, 307
Paracelsus, Philippus Aureolus. *See* Hohenheim, Theophrastus Bombastus von
Parish, Edmund, 1370, 1425
Parker, George M., 1513
Parrot, J. F., 264
Parson, Frederick T., 998
Pasley, T. H., 565
Passavant, Johann Carl, 308
Patrick, Conrad Vincent, 1793
Pattie, Frank, 1, 8
Paul, Cedar, 1746, 1777
Paul, Eden, 1746, 1777
Paulet, Jean Jacques, 8, 14, 22, 94–96, 147
Paz Soldan, Carlos, 1477
Peano, C., 752
Pearson, John, 192
Peaumerelle, C. J. de B. de, 97
Pedro (king of Portugal), 1210
Peebles, James Martin, 944
Pélin, Gabriel, 884, 1245
Pellegrino, G. [pseud. Lisimaco Verati], 524, 699
Pennell, Henry Cholmondeley, 1091
Perdriau, ———, 89
Perkinean Society, 224
Perkins, Benjamin Douglas, 204, 207, 209–210, 212, 217–218
Perkins, Elisha, 204–207, 209–210, 212, 214, 217, 224, 647, 658
Perronnet, Claude, 1092, 1154
Perty, Maximillian, 844, 862, 927, 999, 1046
Perusson, E., 442–443
Petetin, Jacques Henri Désiré, 152, 171,

221, 225, 353, 540, 1424
Pétiau, ———, abbé, 98, 99
Pezzani, André, 753
Pfaff, C. H., 278
Pfungst, Oskar, 1592, 1728
Philip, ———, 20, 23, 28, 100
Philips, A.J.P. *See* Durand de Gros, Joseph Pierre
Philpott, Anthony J., 1708
Pictet, Raoul, 1404
Piddington, John George, 1617
Piérart, Z. J., 792
Piéron, Henri, 1514, 1593
Pigeaire, Jules, 409, 416
Pigeaire, Léonide, 398, 409
Pike, Richard, 1475
Pilati, ———, 1281
Piper, Leonora, 1175, 1270, 1274, 1317, 1419, 1439, 1486, 1511, 1556, 1632, 1647, 1654, 1687, 1699, 1708, 1710, 1807, 1811
Pitres, Jean Albert, 294, 1302, 1499
Plumer, William S., 834
Podmore, Frank, 1144, 1371, 1426, 1508, 1525, 1528, 1618, 1635–1636, 1649
Poe, Edgar Allan, 385
Poincaré, Émile Léon, 885
Poincaré, Henri, 885
Poissonier, Pierre Isaac, 101
Poli, Giuseppe Saverio, 236
Posse, Baron Nils, 1163
Possin, Auguste, 488
Powell, James Henry, 876
Poyen Saint Sauveur, Charles, 374, 386–387, 472, 489
Prado, Madame, 1804
Pratt, Mary, 189
Pressavin, Jean Baptist, 102–103
Pretreaux, J. D., 582
Prevorst, Seeress of, 341, 1136
Preyer, Wilhelm Thierry, 450, 812, 1009, 1047, 1155, 1275, 1389, 1393
Price, Harry, 1584, 1902
Prince, Morton, 1119, 1495, 1559, 1594, 1637, 1701, 1709, 1870, 1891
Prince, Walter Franklin, 1709, 1716, 1872–1873
Prony, Gaspard Clair François Marie Riche, 104
Puel, T., 964
Putnam, Allen, 793, 963
Putnam, James J., 1637
Puységur, Antoine Hyacinte Anne de Chastenet, comte de, 24, 49
Puységur, Armand Marie Jacques de Chastenet, marquis de, 39, 47, 49, 105–106, 121, 139, 148, 152, 158, 160, 169, 175, 178, 193, 201, 211, 228–229, 237, 239–241, 243, 245, 251, 254, 284, 294, 318, 320, 331, 387, 399, 450–451, 455, 558, 789, 956, 1089, 1184, 1424, 1504, 1554
Puységur, Jacques Maxime Paul de Chastenet, comte de, 106, 228

Quackenbos, John Duncan, 1476, 1619

Rabelais, François, 142
Race, Victor, 105, 254, 1184
Radau, Rudolphe, 976
Radclyffe-Hall, Miss, 1745
Radet, J. B., 35
Rafn, ———, 210
Rahn, Carl L., 1592
Rahn, Johan Heinrich, 190
Randall, Edward Caleb, 1834
Randall, Frank H., 1427
Randall, John Herman, 1809
Randolph, Paschal Beverley, 913
Raue, Charles Godlove, 1246
Raupert, John Godfrey Ferdinand, 1497, 1542, 1759
Raymond, F., 1441, 1499, 1522
Rebell, Hughes, 1509
Rebman, F. J., 1507
Redgrove, Herbert Stanley, 1810
Reese, David Meredith, 400
Regnard, Paul, 982, 1185
Regnier, Louis Raoul, 1303
Reichel, Willy, 1304, 1327
Reichenbach, Karl Ludwig von, 532, 583, 648, 655, 657, 711, 727, 740, 746, 769–770, 782, 794, 850, 898–899, 905, 989, 993, 1305–1306, 1555, 1880
Reid, Hiram Alvin, 1498
Remy, M., 1669
Renaud, Calixte, 806
Renterghem, Albert Willem van, 1247, 1372, 1543
Résie, ———, comte de, 679
Résimont, Charles de, 484
Retz, Noel de Rochefort, 25, 109
Reymond, ———, 1149
Reynolds, Mary, 262, 345, 834, 1214
Rhubarbini de Purgandis, ———, 112
Riala, ———, 34
Ribot, Théodule Armand, 1120, 1594
Ricard, Jean Joseph Adolphe, 377, 388, 417, 444, 486–487, 498, 502, 566–567, 755, 1089
Rice, Nathan Lewis, 584
Richardson, Mark W., 1904

Richemont, Eugène Panon Desbassayans, compte de, 680
Richer, Paul Marie Louis Pierre, 1048, 1499
Richet, Charles Robert, 450, 977, 1032, 1071, 1093–1094, 1146, 1157, 1184, 1216–1217, 1283, 1525, 1499, 1560, 1584, 1596, 1648, 1652, 1662, 1704, 1836, 1852, 1863
Richmond, Almon Benson, 1218, 1248
Richmond, B. W., 657
Riddle, Mrs. John Wallace, 1730
Rider, Jane C., 357
Rieger, Conrad, 1095
Riko, A. J., 1544
Riols, J. de. *See* Santini de Riols, Emmanuel Napoléon
Rivail, Hippolyte Léon Dénizard [pseud. Allan Kardec], 781, 796, 810, 845, 847, 877, 887, 915, 1013, 1215, 1407, 1457, 1586, 1844
Rivers, W.H.R., 1870
Roback, Abraham Aaron, 1891
Robbins, Anne Manning, 1811
Robert, ———, 319
Robert, E. W., 712
Robertson, James, 1365
Robertson, R., 1376
Robillard, J.B.P., 888
Robinson, William Elsworth, 1448
Rochas d'Aiglun, Eugène August Albert de, 899, 1187, 1306, 1328, 1352, 1391, 1406, 1428, 1454, 1510, 1670, 1880
Rodgers, W. H., 585
Roff, Mary, 1014
Roger, Madame, 908
Rogers, Edward Coit, 209, 681, 720
Rogers, Louis William, 1717
Rosacroce, Elia, 1702
Rose, Miss, 1796
Rose, W., 1278
Rosenbusch, Hans, 1895
Rosenmüller, Johann Georg, 185–186
Rossi, Pasquale, 1545
Rossi'Pagnoni, Francesco, 1188
Rostan, Louis, 322, 332, 885, 1089
Rothe, Anna, 1541
Rothe, Georg, 1651
Roubaud, Alexandre Félix, 682
Rouby, ———, 1596
Rouget, Ferdinand, 797, 934
Roullier, Auguste, 279
Roux, F., 547, 863, 878
Rouxel. *See* Leroux, Auguste
Rovère, Jules de, 798
Roy, Emile, 430
Royce, Josiah, 1222
Ruspoli, ———, marquise de, 1679
Rutot, Aimé Louis, 1853
Rutter, John Obediah Newell, 623, 728

S., 1860
S., Marie, 1704
Sabatier-Desarnauds, Bernard, 401
Sage, Michel, 1511, 1546
Saint-Dominique, ———, countess de, 966
Saint-Martin, Louis Claude de, 158
Salaville, Jean Baptiste, 110
Sallis, Johann G., 1189
Sally, 1559
Samona, Carmelo, 1652
Samson, George Whitefield, 835, 929
Sandby, George, 503, 738
Santini de Riols, Emmanuel Napoléon [pseud. J. de Riols], 1033, 1072
Sargent, Epes, 930, 978, 1011
Sarrazin de Montferrier, Alexandre André Victor [pseud. A. de Lausanne], 289, 296, 1089
Saunders, Mrs., 1872
Sausse, Henri, 1407
Sauviac, Joseph Alexandre Betbezé Larue de, 194
Savage, Minot Judson, 1353, 1463, 1512
Savile, Bourchier Wrey, 965
Savino, E., 1392
Schade, O., 684
Schaerer, Maurice, 1853
Schauenburg, Carl Hermann, 684
Scheible, J., 624
Schelling, ———, 258, 376
Schiller, Ferdinand Canning Scott, 1529, 1597, 1874
Schindler, Heinrich Bruno, 782
Schleiden, ———, 769
Schneider, Willy, 1704, 1876, 1895
Schofeld, Alfred Taylor, 1449
Schopenhauer, Arthur, 378, 603, 1082
Schrenck-Notzing, Albert Philibert Franz von, 1219, 1305, 1329, 1373, 1393, 1408, 1680, 1700, 1703–1704, 1750, 1788, 1863, 1875–1876
Schröder, Christoph, 1877
Schroeder, H. R. Paul, 1279, 1464
Schwab, Friedrich, 1854
Schwarzschild, Heinrich, 685
Scobardi, ———, 415
Scoresby, W., 586
Seca, ———, 34
Segard, Charles, 1169, 1190
Segouin, A., 756
Semiramide, 981
Seppilli, Giuseppe, 1049
Sérapis, 306

Séré, G. Louis de, 729, 757
Sergi, Giuseppe, 1121
Servan, Joseph Michel Antoine, 111–112, 147
Sexton, George, 957
Sextus, Carl, 1354, 1473
Seybert, Adam, 1191
Seybert Commission, 1191
Seybert, Henry, 1180, 1191, 1218, 1248
Sharp, ———, 466
Sherwood, Henry Hall, 445
Sibly, Ebenezer, 203
Sidgwick, Arthur, 1562
Sidgwick, Eleanor Mildred Balfour, 1562, 1710, 1863
Sidgwick, Henry, 1374, 1394
Sidis, Boris, 1450, 1513, 1547, 1637, 1705–1706, 1870
Sierke, Eugen, 967
Silas, Ferdinand, 686
Simon, Claude Gabriel, 364
Simon, Gustave Marie Stéphane Charles, 1855
Simonin, Amédée H., 1249
Simony, Oskar, 1096
Simrock, K., 684
Sinnett, Alfred Percy, 1330
Sitwell, F., 649–650, 687
Slade, Henry, 986, 997, 1001, 1006, 1015, 1034, 1061, 1073, 1180, 1191, 1406, 1883, 1895
Small, ———, 466
Smith, Gibson, 525
Smith, Hélène, 1470, 1503, 1573, 1807
Smith, Hester Travers, 1760
Smith, Mrs. Travers, 1860
Smith, R. Percy, 1280
Smith, W. Whately. *See* Carrington, Walter Whately
Smith-Buck, George [pseud. Herr Dobler], 931
Snow, Herman, 688
Société de l'harmonie de Guienne, 151
Société Exégétique & Philantropique, Stockholm, 186
Society for Psychical Research, 1060, 1374
Sollier, Paul Auguste, 1530
Soloviev, Vsevolod Sergyeevich, 1394
Sommer, ———, 1149
Souriau, Paul, 1355
Sousselier de la Tour, comte, 113
Southey, ———, 1731
Spicer, Henry, 689–690, 864
Staite, O., 504
Staudenmaier, Ludwig, 1681
Stead, Estelle Wilson, 1653, 1878
Stead, William Thomas, 1307, 1331, 1653
Stearns, Samuel, 198
Stefanoni, Luigi, 1281
Steiglitz, Johann, 252, 260, 268
Steininger, Karoline ["Megalis"], 1822–1823
Stephen (king of England), 860
Stern, Carus. *See* Krause, Ernst Ludwig
Stevens, E. Winchester, 1014
Stoll, Otto, 1375
Stone, G. W., 604
Stone, William Leete, 381, 386, 389
Strombeck, Friedrich Karl von, 246
Sudre, René, 1879
Sue, G.A.T., 338
Sully, James, 1050
Sunderland, LaRoy, 457, 472, 489, 526, 540, 691–692, 916, 918, 923
Suremain de Missery, Antoine, 265
Swedenborg, Emanuel, 186, 231, 418, 549, 559, 616, 664, 746, 1116, 1367
Swinden, Han Hendrik van, 114
Symon, Brother, 1860
Szapary, Ferencz Grof, 431, 527–528, 693, 730–731

Tabor, Heinrich, 190
Tallmadge, Nathaniel P., 664
Tanchou, ———, 548
Tanner, Amy Eliza, 1654
Tanzi, E., 1242
Tarchanoff, Jean de, 1308
Tardy de Montravel, A. A., 152, 161–162, 171–173, 313, 353
Tarkhanov, Ivan Romanovich, 1308
Tartaruga, Ubald. *See* Ehrenfreund, Edmund Otto
Tascher, Paul, 733, 758
Taylor, E. W., 1637, 1891
Taylor, J. Traill, 1365
Taylor, Lemesurier, 1459
Tedinngarov, J. A. *See* Grandvoinet, ———
Teed, Mrs., 1208–1209
Tennent, Rev. Dr., 254
Teresa, Saint, 779
Terzaghi, Giuseppe, 662, 697–699
Teste, Alphonse, 432, 447, 529, 568
Thackeray, William, 813
Themola, ———, 34
Theobald, Morell, 1192
Thoma, Leopold, 1823
Thomas, Charles Drayton, 1837
Thomas d'Onglée, François Louis, 153
Thomas, Northcote Whitridge, 1563–1564
Thouret, Michel Augustin, 26–27, 42, 76, 83, 116, 125, 132, 154–155
Thouvenel, Pierre, 18, 117

Thury, Marc, 759, 1586
Tiffany, Joel, 772
Timmler, Julius Eduardus, 958, 1236
Tinterow, Maurice, 106, 450
Tischner, Rudolf E., 250, 1789, 1812, 1813, 1880–1881, 1905
Tissart de Rouvres, Jacques Louis Noel, marquis de, 118
Tissot, Claude Joseph, 917
Tissot, Honoré, 446
Tizzani, Vincenzo, 463
Tode, ———, 210
Tomczyk, Stanislawa, 1787, 1895
Tommasi, M., 625
Tonna, Charlotte Elizabeth (Browne), 555
Topham, William, 464, 474
Tournier, Anna Marie Valentin, 1478
Touroude, A., 1250
Townshend, Chauncy Hare, 372, 433, 617, 734, 738
Trethewy, A. W., 1856
Trismegiste, Johannes [pseud.], 760
Troland, Leonard Thompson, 1730
Troubridge, Una, 1745
Truesdell, John W., 1073
Trufy, Charles, 1429
Tuckett, Ivor Lloyd, 1671
Tuckey, Charles Lloyd, 1251
Tuke, Daniel Hack, 949, 1098
Turchetti, O., 448
Tuttle, Hudson, 944–945, 1014
Tweedale, Charles Lakeman, 1638
Tweedale, Violet Chambers, 1790
Tweedie, A. C., 784

Ulrich, A., 174
Ulrici, Hermann, 1024
Underhill, Leah, 1122
Underhill, Samuel, 390, 918
Ungher, Efisio, 1276
Unzer, A. M., 5
Unzer, Johann Christoph, 5–6
Usteri, Paulus, 187

Valleton de Boissière, ———, 155
Vallombroso, ———, 1193
Varley, C. F., 1430, 1905
Varnier, Charles Louis, 140
Vaschide, Nicholas, 1514, 1620
Vasseur-Lombard, ———, 811
Vaughan, John, 207
Vay, Adelind (Wurnbrand-Stuppach), 968, 1000
Velpeau, ———, 818
Vélye, ———, abbé de, 227, 247
Vennum, Mary Lurancy, 1014
Ventura de Raulica, ———, 824
Verati, Lisimaco. *See* Pellegrino, G.
Vernet, Jules, 266
Verrall, Margaret de Gaudrion Merrifield (Mrs. A. W.), 1578, 1641, 1647, 1687
Verworn, Max, 1451
Vesme, Cesare Baudi di, 1283, 1390, 1410
Vett, Carl, 1828
Vial, Louis Charles Emile, 1565
Vicq-d'Azyr, ———, 83, 85
Villari, Luigi Antonio, 1479
Villenave, Mathieu Guillaume Thérèse de, 342
Villers, Charles de, 54, 175
Villiot, Jean de, 1509
Vincent, Ralph Harry, 1356
Viollet, Marcel, 1621
Virchow, Hans, 1095
Vires, ———, 294
Virey, Julien Joseph, 290, 292
Vogel, W., 1061
Voisin, A., 1099
Volckman, Mrs. Guppy, 1300
Volhart, Maria, 1854
Vollmer, Carl Gottfried Wilhelm [pseud. W.F.A. Zimmermann], 700, 851
Voltaire, 111
Voltelen, Floris Jacobus, 199

Wagener, Samuel Christoph, 219
Wagner, A., 530
Waite, A. W., 450
Waite, Arthur, 1453
Wallace, Alfred Russel, 900, 969, 979, 1566
Wallace, Mrs. Chandos Leigh (Hunt), 1123
Warcollier, René, 1814
Ward, W. Squire, 464, 474
Washington, George, 205
Wasielewski, Waldemar von, 1815
Waterman, George A., 1637
Watson, Mr., 1663
Watson, Samuel, 950, 970
Watts, Anna Mary (Howitt), 1074
Watts, Nelle M., 1884
Weatherly, Lionel Alexander, 1309
Webb, Arthur L., 1466
Weber, Joseph, 267, 280
Weber, W., 1015
Webster, Daniel, 596
Weinholt, Arnold, 168, 176, 222, 226
Weir, S., 808
Weiss, Ehrich [pseud. Harry Houdini], 1882–1883
Weldon, Georgina, 1001
Wells, David Washburn, 1598

Welton, Thomas, 1100
Wendel-Wurtz, abbé, 281
Werner, Heinrich, 418
Wesermann, H. M., 312
Wesley, John, 1731
Wesley, Samuel, 1731
Wetterstrand, Otto Georg, 1220
White, William A., 1513
Wickland, Carl August, 1884
Widemann, Gustav, 736
Wigan, Arthur Ladbroke, 505
Wilkinson, Mr., 1865
Wilkinson, W. M., 799
Williams, C., 1655
Williams, Tom A., 1637
Wilson, James Victor, 556
Winchester, bishop of, 346
Winter, George, 220
Wirth, Friedrich Moritz, 1061
Wittig, Gregor Constantin, 1253
Wolfart, Karl Christian, 233, 250
Wollny, F., 1221
Wood, Alexander, 626
Woodhead, J. E., 1117
Woodward, ———, 209
Wooldridge, L. C., 1029
Worcester, Elwood, 1615
Worth, Patience, 1718
Wright, Dudley, 1731
Wright, George E., 1791, 1816
Wundt, Wilhelm Maximillian, 1029, 1332
Wurm, Wilhelm, 785
Wyld, George, 1091
Wyndham, Percy, 1091

X, Félida, 1159, 1311

Yerkes, Robert, 1473
Yost, Casper Salathiel, 1718
Young, J. F., 1376
Younger, D., 1194
Yung, Emile, 1075, 1266

Zanardelli, D., 1158, 1281
Zerffi, George Gustavus, 946
Zimmermann, W.F.A. *See* Vollmer, Carl Gottfried Wilhelm
Zöllner, Johann Carl Friedrich, 1015, 1034, 1061, 1180, 1191, 1895
Zschokke, ———, 909
Zuccoli, Ant., 836
Zurbonsen, Friedrich, 1599

Title Index

This index lists the titles of all publications that appear in the Bibliography. All citations refer to entry numbers.

Aberglaube und Zauberei von den ältesten Zeiten an bis in die Gegenwart, 1444
Abrégé de la pratique du magnétisme animal aux dix-huitième et dix neuvième siècles, ou Tableau alphabétique des principales cures opérées depuis Mesmer jusqu' à nos jours, 307
Accès de somnambulisme spontané et provoqué. Prévention d'outrage public à la pudeur; condamnation; irresponsabilité; appel, information et acquittement, 1045
Account of a Case of Successful Amputation of the Thigh, During the Mesmeric State, Without the Knowledge of the Patient: Read to the Royal Medical and Chirurgical Society of London, on Tuesday, the 22nd of November, 1842, 464
An Account of Jane C. Rider, the Springfield Somnambulist: the Substance of which was Delivered as a Lecture Before the Springfield Lyceum, Jan. 22, 1834, 357
An account of some experiments in thought-transference, 1086
Les actes inconscients et la mémoire pendant le somnambulisme, 1146, 1211
Les actes inconscients et le dédoublement de la personnalité pendant le somnambulisme provoqué, 1146
An Adventure, 1667
Adventurings in the Psychical, 1694
L'affaire des magnétiseurs de Braine-le-Chateau. Examen critique du la rapport des médecins experts, 1286
After Death; or Disembodied Man. etc., 913
After Death—What? Spiritistic Phenomena and Their Interpretation, 1633
Agassiz and Spiritualism: Involving the Investigation of Harvard College Professors in 1857, 963
Alcuni saggi di medianita ipnotica, 1188
The Alleged Haunting of B——House, Including a Journal Kept During the Tenancy of Colonel Lemesurier Taylor, 1459
Allgemeine Erläuterungen über den Magnetismus und den Somnambulismus. Als vorläufige Einleitung in das Natursystem. Aus dem Askläpieion abgedruckt, 238, 250
Les altérations de la personnalité, 1311
Alterations of Personality, 1311
L'âme est immortelle, 1454
L'âme humaine, études expérimentales de psycho-physiologie, par un spiritualiste. Substance

de l'âme, formes, biologie organique, dissection, anatomie, éléments matériels, physiologie, propriétés physiques et chimiques, 1782
L'âme humaine, ses mouvements, ses lumières et l'iconographie de l'invisible fluidique, 1396
American Society for Psychical Research. Journal, 1556, 1580, 1601
American Society for Psychical Research. Proceedings, 1222, 1581–1582, 1709, 1872–1873, 1899
American Society for Psychical Research (Old) Proceedings, 1222
L'ami de la nature, ou Manière de traiter les maladies par le prétendu magnétisme animal, 113
Amnésia périodique ou dédoublement de la vie, 980
Analyse apologétique et critique de la brochure du docteur J. A. Tedinngarov, intitulée: Esquisse d'une théorie des phénomènes magnétiques, 477
Analyse du magnétisme de l'homme; manière de l'administrer comme guérison naturelle; des effets et des phénomènes que en résultent, 394
Analyse raisonée de l'ouvrage intitulé: Le magnétisme éclairé, ou Introduction aux Archives du magnétisme animal, 303
Analyse raisonée des rapports des commissaires chargés par le roi de l'examen du magnétisme animal, 40
Analytical Report of a Series of Experiments in Mesmeric Somniloquism, Performed by an Association of Gentlemen: with Speculations on the Production of Its Phenomena, 494
L'anesthésie systématisée et la dissociation des phénomènes psychologiques, 1146
Les anglaises chez elles. Le magnétisme du Fouet, ou les indiscrétions de Miss Darcy, traduit de l'anglais par Jean de Villiot, 1509
Anhung von einigen Briefen und Nachrichten, 5
Animal Electricity and Magnetism, &c. Demonstrated after the Laws of Nature; with New Ideas upon Matter and Motion, 121
Animali e manifestazioni metapsichiche, 1840
Der animalische Magnetismus und die experimentirende Naturwissenschaft, 652
Animalischer Magnetismus und Magie, 378
Animalischer Magnetismus und moderner Rationalismus: eine kultur-historische Betrachtung, 1028
Animal Magnetism, 1162
Animal Magnetism. A Ballad, with Explanatory Notes and Observations: Containing Several Curious Anecdotes of Animal Magnetisers, Ancient as well as Modern, 195
Animal Magnetism: A Farce, in Three Acts, as Performed at the Theatre Royal, Covent-Garden, 179
Animal Magnetism. History of its Origin, Progress and Present State; its Principles and Secrets Displayed, as Delivered by the Late Dr Demainauduc. To which is Added, Dissertations on the Dropsy; Spasms; Epileptic Fits . . . with Upwards of One Hundred Cures and Cases. Also Advice to Those who Visit the Sick . . . a Definition of Sympathy; Antipathy; The Effects of the Imagination on Pregnant Women; Nature; History; and on the Resurrection of the Body, 220
Animal magnetism, or Mesmerism; Being a Brief Account of the Manner of Practicing Animal Magnetism; the Phenomena of the State; Its Applications in Disease, and the Precautions to be Observed in Employing It, Made so Plain that Anyone may Practice it, Experiment upon it and Test Its Effects for Himself, 453
Animal Magnetism; or Psychodunamy, 540
Animal magnetism: Past Fictions—Present Science, 380
Animal Magnetism and Homeopathy; Being the Appendix to Observations on the Principal Medical Institutions and Practice of France, Italy, and Germany, 370
Animal Magnetism and Magnetic Lucid Somnambulism. With Observations and Illustrative Instances of Analogous Phenomena Occurring Spontaneously; and an Appendix of Corroborative and Correlative Observations and Facts, 895
Animal Magnetism and Somnambulism, 764
Animal Magnetism and the Associated Phenomena, Somnambulism, Clairvoyance, etc., 576
Animal Magnetism Examined: in a Letter to a Country Gentleman, 191
Animal Magnetism (Mesmerism) and Artificial Somnambulism: Being a Complete and Practical Treatise on that Science and Its Application to Medical Purposes. Followed by Obser-

vations on the Affinity Existing Between Magnetism and Spiritualism Ancient and Modern, 966
Animal Magnetism Repudiated as Sorcery;—Not . . . Science . . . With an Appendix on Magnetic Phenomena by William H. Beecher, D.D., 539
The Animal Magnetizer: or, History, Phenomena and Curative Effects of Animal Magnetism; with Instructions for Conducting the Magnetic Operation, 434
Animisme et spiritisme, 1253
Animismus und Spiritismus. Versuch einer kritischen prufung der mediumistischen Phanomene mit besonderer Berucksichtigung der Hypothesen der Hallucination und des Unbewussten. Als Entgegnung auf Dr Ed. v. Hartmann's Werk: "Der Spiritismus," 1253, 1293
Anleitung zur mesmerischen Praxis, 635
Annales de la société harmonique des amis réunis de Strasbourg, ou Cures ou des membres de cette société ont opérées par le magnétisme animal, 156
Annales de psychiatrie et d'hypnologie, 1039, 1282
Annales des sciences psychiques. Recueil d'observations et d'expériences paraissant tous les deux mois (et mensuellement consacré aux recherches expérimentales et critiques sur les phénomènes de télépathie, lucidité, prémonition, médiumnité, etc.), 1283, 1378, 1548, 1560, 1786, 1836
Annales du magnétisme animal, 248, 289, 296–298, 309, 340
Annales médico-psychologiques, 980
Annals of Animal Magnetism, 390, 918
Annals of Psychic Science, 1283, 1548
Année psychologique, 1434, 1593
L'antimagnétisme, ou Origine, progrès décadence, renouvellement et réfutation du magnétisme animal, 8, 22, 94
L'antimagnétisme animal, ou Collection de mémoires, dissertations théologiques, physicomédicales, des plus savants théologiens et médecins sur le magnétisme, la magie, les pratiques superstitieuses, etc. . . . Ouvrage utile et nécessaire spécialment aux ecclésiasiques et aux médecins, 446
L'antimagnétisme martiniste ou barbériniste; observations trouvées manuscrites su la marge d'un brochure intitulée: Réflexions impartielles sur le magnétisme animal, faites après la publication du Rapport des commissaires, &c., 133
Aperçu de quelques expériences magnétiques faites à Nimes, 406
Aperçu sur le magnétisme animal ou résultats des observations faites à Lyon sur ce nouvel agent, 66
Aphorismen uber Sensitivitat und Od, 898
Aphorismes de M. Mesmer, dictés à l'assemblée de ses élèves, & dans lesquels ou trouve ses principes, sa théorie & les moyens de magnétiser; le tout forant un corps de doctrine, developpé en trois cents quarant-quatre paragraphes, pour faciliter l'application des commentaires au magnétisme animal. Ouvrage mis au jour par M. C. de V., 129, 502
Apologie de M. Mesmer; ou, Réponse à la brochure intitulée: Mémoire pour servir à l'histoire de la jonglerie dans lequel on démontre les phénomènes du mesmerisme, 81
Apologie der Theorie der Geisterkunde, 232
Apparitions: a Narrative of Facts, 965
Apparitions and Thought-Transference: An Examination of the Evidence for Telepathy, 1371, 1636
Les apparitions matérialisées des vivants & des morts, 1626
Appel au public sur le magnétisme animal, ou Projet d'un journal pour le seul avantage du public, et dont il serait le coopérateur, 170
Appel aux savans observateurs du dix-neuvième siècle, de la décision protée par leurs prédécesseurs contre le magnétisme animal, et fin du traitement du jeune Hébert, 241, 245
Appel de l'Union protectrice à tous les partisans et amis du magnétisme. Protestation en faveur de la libre manifestation des croyances et de la libre application de la science de Mesmer. Rapport collectif et officiel de la commission pour servir à la défense du somnambulisme, 600
Application du somnambulisme magnétique au diagnostic et au traitement des maladies, 729

TITLE INDEX

Application du somnambulisme magnétique au diagnostic et au traitement des maladies, sa nature, ses différences avec le sommeil et les rêves, 757
Après la mort. Exposé de la philosophie des esprits, ses bases scientifiques et expérimentales, ses consequences morales, 1287
A propos de la métapsychique, 1846
A propos d'Eusapia Paladino. Les séances de Montfort-L'Amaury (25–28 Juillet 1897). Compte rendu, photographies, témoignages et commentaires, 1437
A propos d'une observation de sommeil provoqué à distance, 1141
Arcana of Spiritualism: a Manual of Spiritual Science and Philosophy, 945
Archives de la société magnétique de Cambrai, 507
Archives de neurologie, 1223
Archives de psychologie de la Suisse Romande, 1263, 1503
Archives du magnétisme animal, 248, 259, 297–298, 306, 309
Archives générales de médecine, 812
Archiv für den thierischen Magnetismus, 258, 269, 300, 323, 343
Archiv für Magnetismus und Somnambulismus, 157
Argus, 647
Arrêt de la cour suprême touchant le magnétisme animal, 486
L'art de former les somnambules, traité pratique de somnambulisme magnétique, à l'usage des gens du monde et des médecins qui veulent apprendre à magnétiser, 531
L'art de magnétiser ou le magnétisme animal considéré sous le point de vue théorique, pratique et thérapeutique, 552
L'art de se magnétiser ou de se guérir mutuellement, 788
L'art et l'hypnose. Interprétation plastique d'oeuvres littéraires et musicales, 1558
Artificial Somnambulism, Hitherto called Mesmerism; or, Animal Magnetism; Containing a Brief Historical Survey of Mesmer's Operations, and the Examination of the Same by the French Commissioners. Phreno-somnambulism; or, the Exposition of Phreno-magnetism and Neurology. A New View and Division of the Phrenological Organs into Functions, with Descriptions of Their Nature and Qualities, etc., in the Senses and Faculties . . . , 923
Art Magic: or, Mundane, Sub-mundane and Super-mundane Spiritism. A Treatise in Three Parts, and Twenty-three Sections: Descriptive of Art Magic, Spiritism, the Different Orders of Spirits in the Universe Known to be Related to, or in Communication with Man; Together with Directions for Invoking, Controlling, and Discharging spirits and the Uses and Abuses, Dangers and Possibilities of Magical Art . . . Edited by Emma Hardinge, 983
Asklepieion Allgemeines medicinish-chirurgisches Wochenblatt, 233, 238
Astounding Facts from the Spirit World, Witnessed at the House of J. A. Gridley, Southampton, Mass., by a Circle of Friends, Embracing the Extremes of Good and Evil. The Great Doctrines of the Bible such as the Resurrection, Day of Judgment, Christ's Second Coming, Defended, and Philosophically and Beautifully Unfolded by the Spirits . . . , 714
The Athenaeum, 666
Atomic-Consciousness. An Explanation of Ghosts, Spiritualism, Witchcraft, Occult Phenomena, and All Supernormal Manifestations, 1310
An attempt to Explain Some of the Wonders and Mysteries of Mesmerism, Biology, and Clairvoyance, 643
Attestazioni di illustri scienziati ed uomini sommi in favore dell'esistenza del magnetismo animale e della sua efficacia, 697
Attività subconsciente e spiritismo, 1342
Auszug und Anzeig der Schrift des Hernn Leibmedikus Stieglitz über den thierischen Magnetismus, nebst Zusätzen, 260, 268
Automatic Speaking and Writing: A Study, 1551
Automatic Writing, 1118
Automatic Writing. The Slade Prosecution. Vindication of the Truth, 986
Automatic Writing as an Indicator of the Fundamental Factors Underlying the Personality, 1833
Automatisme et suggestion, 1720

Title Index

L'automatisme psychologique: essai de psychologie expérimentale sur les formes inférieures de l'activité humaine, 1146, 1235, 1318, 1441
Autosuggestion and Its Personal Application, 1821
Autour des "Indes à la Planète Mars," 1493
Autres reveries sur le magnétisme animal à un académicien de province, 98
L'autre vie, 1043
L'avenir des sciences psychiques, 1721
L'avenir médical, Journal des intérêts de tous, avant pour but la démonstration pratique du nouvel art de guérir, l'homéopathie et le magnétisme, par la fondation d'un hopital homéopathico-magnétique pour 150 à 200 infants, 491

Le baquet magnétique, comédie en vers et en deux actes, 69
Die Bedeutung der hypnotischen Suggestion als Heilmittel. Gutachten und Heilberichte der hervorragendsten wissenschaftlichen Vertreter des Hypnotismus der Gegenwart, 1292
Die Bedeutung der Suggestion im sozialen Leben, 1550
Behind the Scenes with Mediums, 1579
Beiträge zur Physiologie des Centralnervensystems. Erster Theil. Die sogenannte Hypnose der Thiere, 1451
Beitrag zu den Erfahrungen über den thierischen Magnetismus, 176
Ein Beitrag zur therapeutischen Verwerthung des Hypnotismus, 1219
The Belief in Personal Immortality, 1686
Der Beobachter des thierischen Magnetismus und des Somnambulismus, 174
Beobachtungen und Betrachtungen auf dem Begiete des Lebensmagnetismus oder Vitalismus, 677
Beschreibung eines mit dem kunstlichen Magneten angestellten medicinischen Versuchs, 6
Die Besessenheit, 1805
Der Betrug des Mediums Ladislaus Laszlo (Nachahnung von Materialisationsphänomenen), 1875
Beyond the Borderline of Life, 1647
Beytrage zu den durch animalischen Magnetismus zeither bewirkten Erscheinungen. Aus eigner Erfahrung, 257
Bible, 379, 714, 1638
Bibliographie der Wünschelrute; mit einer Einleitung von dr. Ed. Aigner: Der gegenwärtige Stand der Wünschelruten-Forschung, 1665
Bibliographie des modernen Hypnotismus, 1202, 1260
Bibliographie du magnétisme et des sciences occultes, 1381
Bibliothèque du magnétisme animal, par MM. les membres de la Société du magnétisme, 248, 272, 288
Bien-Boa et Ch. Richet, 1596
Bienfaits du somnambulisme, ouvrage dédié à Mme Roger, aux amis de la verité, et aux personnes amies d'elle-mêmes, 908
Biographie d'Allan Kardec. Discours prononcé à Lyon, le 31 mars 1856, 1407
A Biography of the Brothers Davenport. With Some Account of the Physical and Psychical Phenomena Which Have Occurred in Their Presence, in America and Europe, 875
Blatter aus Prevorst. Originalen und Lebenfruchte fur Freunde des innen Lebens, 348, 429
Blatter für hohere Wahrheit: aus altern und neuern Handschrift und seltenen Buchern; mit besonderer Rucksicht auf Magnetismus, 283
Blicke in das verborgene Leben des Menschengeistes, 927
The Blot upon the Brain: Studies in History and Psychology, 1116
Body and Mind: a History and a Defense of Animism, 1666
The Book of Dreams and Ghosts, 1420
Book of Human Nature: Illustrating the Philosophy (New Theory) of Instinct, Nutrition, Life; with their Correlative and Abnormal Phenomena, Physiological, Mental, Spiritual, 691
Book of Psychology. Pathetism, Historical, Philosophical, Practical; Giving the Technics of Amulets, Charms, Enchantment . . . Mesmerism . . . Hallucination . . . Clairvoyance, Somnambulism, Miracles, Sympathy, etc.: Showing How These Results May Be

Induced . . . and the Benevolent Uses to Which This Knowledge Should be Applied, 692
Borderland, 1307
Borderland of Psychical Research, 1574, 1611
The Bottom Facts Concerning the Science of Spiritualism: Derived from Careful Investigations Covering a Period of Twenty-five Years, 1073
Braidisme et fariisme, ou la doctrine du Dr. Braid sur l'hypnotisme comparée avec celle de l'abbé de Faria sur le sommeil lucide, 1585
Braid on Hypnotism. Neurypnology; or The Rationale of Nervous Sleep Considered in Relation to Animal Magnetism or Mesmerism and Illustrated by Numerous Cases of Its Successful Application in the Relief and Cure of Disease. A New Edition, Edited with an Introduction Biographical and Bibliographical Embodying the Author's Later Views and Further Evidence on the Subject. By Arthur Waite, 1453
Breve saggio sulla calamita e sulle sue virtù medicinale, 236
Briefe über die Phänomene des thierischen Magnetismus und Somnambulismus, 185
"Bringing it to Book:" Facts of Slate-writing through Mr W. Eglinton. Edited by H. Cholmondeley-Pennell. Being Letters Written by the Hon. Roden Noel, Charles Carleton Massey, George Wyld, the Hon. Percy Wyndham, and the Editor, 1091
British and Foreign Medical Review, 512
British Journal of Photography, 1365
British Medical Journal, 1345
British Society of Psychical Research. Proceedings, 1559
Buchanan's Journal of Man, 573
Das Buchlein vom Leben nach dem Tode, 368
Bulletin de l'Institut général de Psychologie, 1514
Bulletin magnétique, Journal des sciences psycho-physiques rédigé par une réunion de magnétistes, de médecines, de savants, sous la direction de M. Mongruel, 706

Can Telepathy Explain? Results of Psychical Research, 1512
Un cas de dématérialisation partielle du corps d'un médium: enquête et commentaires, 1395, 1418
Un cas de somnambulisme à distance, 1145
The Case against Spirit Photographs, 1793
The Case for Spirit Photography, 1819
A Case of Partial Dematerialization of the Body of a Medium, 1395
Cases of Successful Practice with Perkins's Patent Metallic Tractors: Communicated since Jan. 1800, the Date of the Former Publication, by Many Scientific Characters. To Which are Prefixed, Prefatory Remarks . . . , 218
La cause de l'hypnotisme, 1289
Causeries mesmériennes: enseignement élémentaire (histoire, théorie, et pratique) de magnétisme animal, 880
Causeries spirites, 1429
The Celestial Telegraph, 559
Cenni storico-critici sul magnetismo animale, 448
Certificates of the Efficacy of Dr. Perkins's Patent Metallic Instruments, 205–206
La chaine magnétique. Organe des Société Magnétiques de France et de l'Etranger, echo des salons et cabinets de magnétisme et de somnambulisme . . . , 1017
Les charlatans modernes, ou, Lettres sur le Charlatanisme académique, 197
Der Cheiroelektromagnetismus oder die Selbstbewegung und das Tanzen der Tische (Tischrucken). Eine Anweisung in Gesellschaften das werkwürdige Phänomen einer neu entdeckten menschlischen Urkraft hervozubringen. Nach einigen pract. Versuchen u. unter Vergleich aller bisher veröffentlichten Proben mitgeheilt, 660
Chemical News, 959
Cherchons. Réponse aux conférences de M. le Professeur Emile Yung sur le spiritisme, 1266
Chez Victor Hugo: Les tables tournantes de Jersey: procés-verbaux des séances présentés et commentés par Gustave Simon, 1855

Title Index

Christian Science, Medicine, and Occultism, 1507
Christopher: a Study in Human Personality, 1743
Chronicles of the Photographs of Spiritual Beings and Phenomena Invisible to the Material Eye, Interblended with Personal Narrative, 1059
The Church and Psychical Research: A Layman's View, 1816
Le ciel et l'enfer, ou la justice divine selon le spiritisme, contenant l'examen comparé des doctrines sur le passage de la vie corporella à la vie spirituelle, les peines et les récompenses futures, les anges et les démons, les peines éternelles, etc. . . . suivi de nombreux exemples sur la situation réelle de l'âme pendant et après la mort, 887
Clairvoyance and Materialisation. A Record of Experiments, 1864
Clever Hans (The Horse of Mr. Von Osten): a Contribution to Experimental Animal and Human Psychology, 1592
Clinique de psychothérapie suggestive fondée à Amsterdam, 1247
The Clock Struck One, and Christian Spiritualist: Being a Synopsis of the Investigations of Spirit Intercourse by an Episcopal Bishop, Three Ministers, Five Doctors, and Others, at Memphis, Tenn., in 1855; Also, the Opinion of Many Eminent Divines, Living and Dead, on the Subject, and Communications Received from a Number of Persons Recently, 950, 970
The Clock Struck Three, Being a Review of Clock Struck One, and reply to It. Part II. Showing the Harmony Between Christianity, Science, and Spiritualism, 970
Cock Lane and Common Sense, 1366
Le colosse aux pieds d'Argille, 54
The Coming of the Fairies, 1820
The Coming Science, 1606, 1695
Comment l'esprit vient aux tables, par un homme qui n'a pas perdu l'esprit, 724
Communication universelle et instantanée de la pensée à quelque distance que ce soit, à l'aide d'un appareil portatif appelé Boussole Pasilalinique Sympathique, 590
The Company of Avalon: A Study of the Script of Brother Symon, Sub-Prior of Winchester Abbey in the Time of King Stephen, 1860
Compte-rendu du banquet commémoratif de la naissance de Mesmer (ll8ème anniversaire) célébré le 23 Mai 1852 à Paris, 639
Compte rendus et mémoires des sciences de la Société de biologie, 1213
Comptes-rendus hebdomadiares des séances de l'Académie des Sciences, 1058
The Conception of the Subconscious, 1643
Le concile de la libre pensée. Abolition des faux dogmes et des mensonges sacerdotaux . . . , 795
Confession d'un medecin académicien et commissaire d'un rapport sur le magnétisme animal avec les remontrances et avis de son directeur, 48, 122
Les confessions d'un magnétiseur: suivies d'une consultation médico-magnétique sur des cheveux de MM. Lafarge, 568
Confessions of a Magnetiser, being an Exposé of Animal Magnetism, 509, 526
"Confessions of a Magnetizer" Exposed, 526
Confessions of a Medium, 1057
Confessions of a Truth Seeker. A Narrative of Personal Investigations into the Facts and Philosophy of Spirit-intercourse, 802
Congrès International de 1889. Le magnétisme humain appliqué au soulagement et à la guérison des malades, Rapport Général, d'après le compte rendu des séances du Congrès, 1257
Congrès international de l'hypnotisme expérimental et thérapeutique. Comptes rendus, 1225
Congrès international de l'hypnotisme expérimental et thérapeutique. Comptes rendus, 1499
Congrès international de l'hypnotisme expérimental et thérapeutique. Procès-verbaux sommaires, 1481
Congrès international de psychologie expérimentale. Compte-rendu des travaux, 1640
Congrès international de psychologie physiologique. Comptes rendus, 1258
La connaissance supra-normale: étude expérimentale, 1851
Consciousness of Self, 1271
Le conservateur . . . de N. Frannçois (de Neufchateau), 213

Considérations sur le magnétisme animal, ou sur la théorie du monde et des êtres organisés, d'après les principes de M. Mesmer, par M. Bergasse avec des pensées sur le mouvement, par M. le Marquis de Chastellux, de l'Académie française, 36
Considérations sur l'origine, la cause et les effets de la fièvre, sur l'électricité médicale, et sur le magnétisme animal, 230
Contact with the Other World: the Latest Evidence as to Communication with the Dead, 1757
Continuation du traitement magnétique du jeune Hébert (mois de Septembre), 240–241, 245
Contribution à l'étude de la soi-disant télépathie, 1378
Contribution à l'étude de l'hypnotisme chez les hystériques, 1037
Contribution à l'étude des correspondences croisées (documents nouveaux). Conference faite par le docteur Gustave Geley le samedi 20 decembre 1913, 1698
Contribution à l'étude expérimentale des phénomènes télépathiques, 1514
Contributions to Christology, 919
A Contribution to the Study of the Psychology of Mrs. Piper's Trance Phenomena, 1710
Contributo sperimentale alla fisio-psicologia dell'ipnotismo. Richerche sul polso e sul respiro negli stati suggestivi dell'ipnosi, 1242
The "Controls" of Stainton Moses ("M. A. Oxon"), 1856
Copie de la requête à nos seigneurs de Parlement en la grand' Chambre, 82
The Cornhill Magazine, 813
Le corps aromal, ou Réponse en un seul mot à l'Académie des sciences philosophiques à propos du concours proposé par elle sur quelques relations à l'andro magnétisme . . . Explication vraie des tables tournantes et parlantes, 723
Correspondance de M. M. . . . sur les nouvelles découvertes du baquet octogne, de l'homme-baquet et du baquet moral, pour servir de suite aux aphorismes. Recueillie et publiée par MM. de F. . . . ; J. . . . et B. . . . , 138
Coup d'oeil sur le magnétisme, 264
Coup d'oeil sur le magnétisme animal et le somnambulisme considérés sous le rapport médical et religieux, 547
Coup d'oeil sur le magnétisme et examen d'un écrit qui a paru sous ce titre: Letre sur le magnétisme à M. . . . à Paris par M. Morrison de Bourges, 242
Cours de magnétisme animal en douze leçons, 791
Cours de magnétisme animal en 7 leçons, augm. d'un rapport sur les expériences magnétiques faites par la commission de l'Académie de médecine en 1831, 359
Cours théorique et pratique de braidisme ou hypnotisme nerveux: considéré dans ses rapports avec la psychologie, la physiologie et la pathologie et dans ses applications à la médecine, à la chirurgie, à la physiologie expérimentale, à la médecine légale et à l'éducation, par le docteur J. P. Philips, 740, 821
Cours théorique et pratique du magnétisme animal, 417
Le cri de la nature, ou le magnétisme au jour; ouvrage curieux et utile pour les personnes qui cherchent à étudier les causes physiques du magnétisme ainsi que les phénomènes que s'y rapportent, 74
The Crisis in Psychology, 1894
The Critics Criticized, 738
Cronica del magnetismo animale, 662, 698
Crystal-Gazing: a Study in the History, Distribution, Theory and Practice of Scrying, 1858
Crystal Gazing: Its History and Practice, with a Discussion of the Evidence for Telepathic Scrying, 1563
Crystal-Gazing and the Wonders of Clairvoyance, Embracing Practical Instructions in the Art, History, and Philosophy of This Ancient Science. To Which is Appended an Abridgment of Jacob Dixon's "Hygienic Clairvoyance," with Various Extracts and Original Notes, 1646
Curative Mesmerism, 838
Curative Results of Medical Somnambulism, Consisting of Several Authenticated Cases, Including the Somnambule's Own Case and Cure, 570
Cures Effected by Animal Magnetism, 838
Cures faites par M. Le Cte. de L. . . . , 183

Title Index

La curiosité. Journal de l'occultisme scientifique, 1227
La curiosité. Revue des sciences psychiques, 1227, 1360

D. Arnold Wienholt's psycholoquishe Vorlesungen über den naturlichen Somnambulism. Aus den literarischen Nachlass des Verfassers besonders abquedruckt, 226
D. D. Home, His Life and Mission [Myers and Barrett], 1244
D. D. Home: His Life and Mission [Home], 1206, 1268
Les dangers du magnétisme animal, 366
Les dangers du magnétisme animal et l'importance d'en arrêter la propagation vulgaire, 295
The Dangers of Spiritualism, Being Records of Personal Experiences with Notes and Comments. By a Member of the Society for Psychical Research, 1497
La danse des tables, phénomènes physiologiques démontrés par le Dr. Félix Roubaud, 682
La danse des tables dévoilée, expériences de magnétisme animal, manière de fair tourner une bague, un chapeau, une montre, une table, et même jusq'eux têtes des expérientateurs et celles des spectateurs, 663
Dans l'invisible. Spiritisme et médiumnité. Traité de spiritualisme expérimental. Les faits, les lois. Phénomènes spontanés. Typtologie et psychographie. Les fantômes des vivants et les esprits des morts, la médiumnité à travers les âges, 1536
Darstellung der mesmerischen Heilmethode nach naturwissenschaftlichen Grundsätzen; nebst der ersten vollständigen Biographie Mesmer's und einer fasslichen Anleitung zum Magnetisiren, 785
Darstellung und Enthüllung des Somnambulismus, mit besonderer Bezugnahme auf den Somnambulen, Stahlschmiedegesellen Carl Wilhelm Kohn, 589
Dawn of the Awakened Mind, 1780
Death: and Its Problems, 1862
Death: Its Causes and Phenomena with Special Reference to Immortality, 1658
Death and Its Mystery, 1771
Death Bed Visions, 1475
The Death-Blow to Spiritualism: Being the True Story of the Fox Sisters, as Revealed by the Authority of Margaret Fox Kane and Catherine Fox Jencken, 1200
Death-Blow to Spiritualism—Is it? Dr Slade, Messrs Maskelyne and Cooke, and Mr W. Morton, 1001
Death Deferred, 1675
The Debatable Land between This World and the Next. With Illustrative Narratives, 943
Les débris du baquet ou Lettres critiques de la requête de Mesmer, 50
De casi d'identificasione spiritica, 1623
Découverte de la polarité humaine, ou démonstration expérimentale des lois suivant lesquelles l'application des aimants, de l'électricité, et les actions manuelles ou analogues du corps humain déterminent l'état hypnotique et l'ordre de succession de ses trois phases . . . , 1130
Décret de la faculté de médicine de Paris, du 24 août 1784 par lequel est adopté le Rapport des commissaires (Français et Latin), 51
"A Defence of Modern Spiritualism," 969
De fenomeni di telekinesia in rapporto con eventi di morte, 1817
Défense du magnétisme animal contre les attaques dont il est l'objet dans le dictionnaire des science médicales, 292
Défense théologique du magnétisme humain, ou le magnétisme est'il superstition, magie? Est'il condamné à Rome? Les magnétiseurs et les somnambules sont-ils en sûreté de conscience? Peuvent-ils être admis à la participation des sacrements?, 543
Dei fenomeni d'infestazione, 1748
Dei fenomeni di telestesia, 1763
De imperio solis ae lunae, 1
De la baguette divinatoire: du pendule dit explorateur et des tables tournantes, au point de vue de l'histoire, de la critique et de la méthode expérimentale, 709
De la cause du sommeil lucide, ou étude de la nature de l'homme. Tome premier, 294, 1571, 1585, 1902

TITLE INDEX

De la métalloscopie et la métallothérapie, 1003
De la multiplicité des états de conscience chez un hystéro-épileptique, 1102
De la philosophie corpusculaire, ou des connaissances et les procédès magnétiques chez les divers peuples, 130–131
De la phrénologie du magnétisme et de la folie. Ouvrage dédié à la mémoire de Broussais, 403
De la suggestion, 1657
De la suggestion dans l'état hypnotique et dans l'état de veille, 1080, 1127
De la suggestion et de ses applications. Conference faite à Nancy et à Chaumont en Janvier 1912, 1676
De la suggestion et de ses applications à la thérapeutique, 1127
De la suggestion et du somnambulisme dans leurs rapports avec la jurisprudence et la médecine légale, 1237
De la suggestion mentale, 1184
De l'automatisme de la mémoire et du sourvenir dans le somnambulisme pathologique, considérations médico-légales, 961
De l'emploi du magnétisme animal et des eaux minérales dans le traitement des maladies nerveuses suivi d'une observation très curieuse de névropathie, 421
De l'état actuel du magnétisme . . . , 336
De l'inconscient au conscient, 1754
De l'influence de la magnétisation sur le développement de la voix et du goût en musique, 737
Delle apparizioni di defunti al letto di morte. 2a serie, 1764
Delle communicazioni medianiche tra viventi, 1861
Del magnetismo animale, e degli effetti ad esso attribuiti nella cura delle umane infermita, 145
Del magnetismo animale ossia mesmerismo. In ordine alla ragione e alla rivelazione, 801
De l'origine des effets curatifs de l'hypnotisme; étude de psychologie expérimentale, 1165
Del vero spirito scientifico secondo il quale debbono essere esaminate le ragioni dell frenologia e del mesmerismo. Dissertazione di W. G. già pubblicata nel Giornale Frenologico di Edinburgo. Seguita da alcuni esperimenti frenomesmerici de G. T. Hall tratti dallo stesso giornale, 615
De magnetismo animali, 276
Des états seconds: variations pathologiques du champ de la conscience, 1321
Des facultés magnétiques de l'homme des moyens divers par lesquels elles se manifestent; des conditions qu'exige leur emploi; de la responsabilité morale qu'entraine leur exercice; des services qu'on peut en attendre, 906
Des forces naturelles inconnues, à propos des phénomènes produits par les frères Davenport et par les médiums en général. Étude critique par Hermés, 890
Des hallucinations ou histoire raisonnée des apparitions, des visions, des songes, de l'extase, des rêves, du magnétisme et du somnambulisme, 352
Des hallucinations, ou histoire raisonée des apparitions, des visions, des songes, de l'extase, du magnétisme et du somnambulisme, 508
Des Indes à la planète Mars; étude sur un cas de somnambulisme avec glossolalie, 1470, 1493, 1503
Des origines de la métallothérapie. Part qui doit être faite au magnétisme animal dans sa découverte. Le burquisme et le perkinisme, 1054
Des phénomènes prémonitoires. (Pressentiments — rêves prophétiques, clairvoyance dans le future, etc.), 1683
Des principes et des procédés du magnétisme animal, et de leurs rapports avec les lois de la physique et de la physiologie, 296
Des propriétés physiques d'une force particulière du corps humain (force neurique rayonnante) connue vulgairement sous le nom de magnétisme animal. . . . Extrait de la Gazette médicale de Paris, année 1881, 1051
Des tables tournantes, du surnaturel en général et des esprits, 712
Des tables tournantes et du panthéisme, 703
Détail des cures opérées à Buzancy, près Soissons par le magnétisme animal, 47, 87
Détail des cures opérées à Lyon par le magnétisme animal, selon les principes de M. Mesmer. Précédé d'une lettre à M. Mesmer, 92

Deuteroskoppie, oder merkwurdige psychische und physiologische Erscheinungen und Probleme aus dem Gebiete der Pneumatologie. Für Religionsphilosophen, Psychologen und denkende Aerzte. Eine nothige Beilage zur Dämono-magie, wie zur Zauber-Bibliothek, 344
Deux cas de somnambulisme provoqué à distance, 1203
Deuxième note sur le sommeil provoqué à distance et la suggestion mentale pendant l'état somnambulique, 1147
Deux mémoires sur le magnétisme (Recherches sur l'universalité de la force magnétique. Recherches sur l'appréciation de la force magnétique), 538
The Development and Geneology of the Misses Beauchamp: A Preliminary Report of the Case of Multiple Personality, 1495
Devotional Somnium; or A Collection of Prayers and Exhortations, Uttered by Miss Rachel Baker, in the City of New York, in the Winter of 1815, During her Abstracted and Unconscious State; To Which Pious and Unprecedented Exercises is Prefixed, An Account of Her Life, with the Manner in Which She Became Powerful in Praise of God and Addresses to Man; Together with a View of That Faculty of the Human Mind which is Intermediate between Sleeping and Waking. The Facts, Attested by the Most Respectable Divines, Physicians, and Literary Gentlemen; and the Discourses, Correctly Noted by Clerical Stenographers. By Several Medical Gentlemen, 254
Le diable dans l'hypnotisme (Soustraction hypnotique de la conscience. Hypnotisme médical. Evocation du démon. Suggestion, etc. . . .), 1461
Le diagnostic de la suggestibilité, 1403
Dialogue entre un docteur de toutes les universités et académies du monde connu, notamment de la faculté de médecine fondée à Paris dans la rue de la Bucherie, l'an de notre salut 1472 et un homme de bon sens, ancien malade du docteur, 37
Dialogue entre un magnétiseur que cherche les moyens de propager le magnétisme et un incrédule qui croit l'avoir trouvé, 282
Dictionnaire de médecine, 322
Dictionnaire des science médicales, 290, 292
Dieu, l'homme et la nature, Tableau philosophique d'une somnambule, 182
The Direct Phenomena of Spiritualism—Speaking, Writing, Drawing, Music and Painting: a Study. With Facsimile Illustrations of Direct Writing, Drawing and Painting, 1604.
Discours présidentiel à la Society for Psychical Research (London, 1923), suivi de Essais médiumiques. Uranographie générale. Communication signées Galilée, obtenues à l'âge de vingt ans au cercle d'Allan Kardec, à la Société spirite de Paris (1862-1863), 1844
Discours prononcé à l'inauguration de la Société magnétique de France, le 7 octobre 1887, 1167
Discours prononcé en l'assemblée de la Faculté de Médicine de Paris le 18 septembre 1780, 11
Discours prononcé par Mme Alphonsine Masson, au banquet de Mesmer de l'année 1857, 780
Discours sur le magnétisme, 8, 22
Discours sur le magnétisme animal, lu à la séance publique de la Société royale de médecine de Marseille tenue le 11 novembre 1827, 338
Discours sur le magnétisme animal lu dans une assemblée du College des médecins de Lyon le 15 septembre 1784, 93
Discours sur le magnétisme animal prononcé le 13 février 1835 à l'Athénée central, 367
Discours sur les principes généreaux de la théorie végétative et spirituelle de la nature, faisant connaître le premier moteur de la circulation du sang, le principe du magnétisme animal et celui de Sommeil magnétique, dit somnambulisme, 286
Discovery concerning Ghosts: With a Rap at the "Spirit-rappers." To which is Added a Few Parting Raps at the "Rappers," Questions, Suggestions, and Advice to the Davenport Brothers, 852
A Discussion of the Facts and Philosophy of Ancient and Modern Spiritualism, 657
The Diseases of Personality, 1120
Dissertation by F. A. Mesmer, Doctor of Medicine, on His Discoveries, 211
Dissertation on the Discovery of Animal Magnetism, 10

Dissertation sur la médecine et le magnétisme, triomphe du somnambulisme, 327
Dissertation sur le magnétisme animal, thèse soutenue à la Faculté de Paris le 13 aout 1834, 360
Dissertatio physico-medica de planetarum influxu, 1, 5
The Dissociation of a Personality: a Biographical Study in Abnormal Psychology, 1495, 1559, 1709
The Divining Rod: Its History — With Full Instructions for Finding Subterranean Springs, and Other Useful Information. Also an Essay Entitled: Are the Claims and Pretensions of the Divining Rod Valid and True? by E. Vaughan Jenkins, 1376
Divisions of the Self and Co-consciousness, 1891
Les docteurs modernes, comédie-parade en un acte et en vaudeville, suivie du Banquet de santé, divertissement analogue mêlé de couplets représentée pour la première fois à Paris par les comédiens italiens ordinaires du Roy, le mardi 16 Novembre 1784, 35, 57–58
Doctrine du magnétisme humain et du somnambulisme, 377
Documents satiriques sur Mesmer, 1234
Das Doppel-Ich, 1229
The Doris Case of Multiple Personality: a Biography of Five Personalities in Connection with One Body and a Daily Record of a Therapeutic Process Ending in the Restoration of the Primary Member to Integrity and Continuity of Consciousness. Part I, by Walter Franklin Prince; The Doris Case of Multiple Personality. Part II, by Walter Franklin Prince; The Doris Case of Multiple Personality. Part III, by James Hervey Hyslop; The Mother of Doris, by Walter Franklin Prince, 1709
The Doris Case of Quintuple Personality, 1716
Do the Dead Depart?, 1602
A double consciousness, or a duality of person in the same individual, 262
Doutes d'un provincial, proposés à messiers le médecins-commissaires chargés par le roi de l'examen du magnétisme animal, 111
Dr. Beale, or More about the Unseen, 1796
Dreams and Premonitions, 1717
Dreams and Their Meanings. With Many Accounts of Experiences Sent by Correspondents and Two Chapters Contributed Mainly from the Journals of the Psychical Research Society on Telepathic and Premonitory Dreams, 1485
Du fluid-universel, de son activité et de l'utilité de ses modifications par les substances animales dans le traitement des maladies. Aux étudiants qui suivent les cours de toutes les parties de la physique, 227
Du magnétisme; qu'est-ce que le magnétisme ou étude historique et critique des principaux phénomènes que le constituent, suivie de l'explication rationelle qu'il convient d'en donner, 597
Du magnétisme animal, 322
Du magnétisme animal, considéré dans ses rapports avec diverses branches de la physique générale, 228
Du magnétisme animal en France, et des jugements qu'en ont portés les sociétés savantes, avec le texte des divers rapports faits en 1784 par les commissaires de l'Académie des sciences, de la Faculté et de la Société royale de médecine, et une analyse des dernières séances de l'Académie royale de médecine et du rapport de M. Husson; suivi de considérations sur l'apparition de l'extase, dans les traitements magnétiques, 313, 324
Du magnétisme animal et du somnambulisme artificiel, 401
Du magnétisme et de ses partisans; ou, Recueil de pièces importantes sur cet objet, précédé des observations récemment publiées, 239
Du magnétisme et des sciences occultes, 831
Du magnétisme et du somnambulisme artificiel, 428
Du merveilleux, des miracles et des pélerinages, au point de vue medical, 990
Du phréno-mesmérisme, 846
Du sommeil, des rêves, et du somnambulisme dans l'état de santé et de maladie, précédé d'une lettre du Dr Cerise, 778
Du sommeil et des états analogues considérés surtout au point de vue de l'action du moral sur le physique, 896, 1080

Du sommeil magnétique dans l'hystérie. Thèse à la faculté de médecine de Strasbourg et soutenue publiquement le samedi 20 juin 1868, à 3 heures du soir, pour obtenir le grade de docteur en médecine, 907
Du sommeil non naturel: ses diverse formes. Thèse présentée au concours pour l'agregation, 1124
Du somnambulisme, des tables tournantes et des médiums, considérés dans leurs rapports avec la théologie et la physique. Examen des opinions de MM. de Mirville et Gasparin, 702
Du somnambulisme dit naturel (noctambulisme), ses rapports avec l'hystérie et l'attaque hystérique à forme somnambulique, 1343
Du somnambulisme provoqué, 977, 1032
Du spiritualisme et de la nature, 420
Du traitement des maladies, ou Étude sur les propriétés médicinales de 150 plantes les plus usuelles par l'extatique Adèle Maginot, avec une exposition des diverse méthodes de magnétisation, 610
Du traitement et de la guérison de quelques maladie chroniques au moyen de somnambulisme magnétique, et à propos de MM. Calixte Renaud, de Bordeaux et Alexis de Paris, 806
Du traitement externe et psychique des maladies nerveuses. Aimants et couronnes magnétiques. Mirroirs. Traitement diététique. Hypnotisme. Suggestion. Transferts, 1416
Du transfert à distance à l'aide d'une couronne de fer aimanté, d'états névropathiques variés, d'un sujet à l'état de veille sur un sujet à l'état hypnotique, 1298
Du zoomagnétisme, son existence, son utilité en médecine rendues indiscutables par les faits, 1018

The Ear of Dionysius. Farther Scripts Affording Evidence of Personal Survival, 1761
The Earthen Vessel: a Volume Dealing with Spirit Communications Received in the Form of Book-tests by Pamela Glenconner, 1799
Ebauche de psychologie, 954
Eclaircissements sur le magnétisme. Cures magnétiques à Genève, 745
Eclaircissements sur la magnétisme actuel, 63
L'ectoplasme, 1826
L'ectoplasmie et la clairvoyance. Observations et expériences personelles, 1864
The Efficacy of Perkins' Patent Metallic Tractors, in Topical Diseases of the Human Body and Animals; Exemplified by 250 Cases from the First Literary Characters in Europe and America. To Which is Prefixed A Preliminary Discourse in Which the Fallacious Attempts of Dr Haygarth to Detract from the Merits of the Tractors, are Detected, and Fully Confuted, 217
Les effluves odiques: conférences faites en 1866, par le baron de Reichenbach à l'Académie des Sciences de Vienne; précédées d'une notice historique sur les effets mechaniques de l'Od, 899
Einführung in den Okkultismus und Spiritismus, 1812
Electricité animal, prouvée par la découverte des phénomènes physiques et moraux de la catalepsie hystérique, et de ses variétés; et par les bons effets de l'électricité artificielle dans le traitment de des maladies, 225
Electricité naturelle, ou Mesmérisme mis en pratique à l'usage des familles, 582
Electro-Biological Phenomena, and the Physiology of Fascination, 738
Electro-biological Phenomena Considered Physiologically and Psychologically, 607
Electro-biology; or, the Electrical Science of Life, 604
Electro-Biology and Mesmerism, 659, 993
Electro-dynamisme vital ou les ralations physiologiques de l'esprit et de la matière, démontrées par des expériences entièrement nouvelles et par l'histoire raisonnée du système nerveux, 740
Eléments de médecine suggestive, hypnotisme et suggestion: faits cliniques, 1169, 1190
Eléments du magnétisme animal, ou Exposition succinte des procédés, des phénomènes et de l'emploi du magnétisme, 289
Elements of Animal Magnetism, or Pneumatology, 440

The Elements of Hypnotism: the Induction of Hypnosis, Its Phenomena, Its Dangers and Value, 1356
Die Emanation der psycho-physischen Energie. Eine experimentelle Untersuchung über die unmittelbare Gedankenübertragung im Zusammenhang mit der Frage über die Radioaktivität der Gehirns, 1613
Les émotions chez les sujets en état d'hypnotisme. Études de psychologie expérimentale faites à l'aide de substances médicamenteuses ou toxiques impressionnant à distance les réseaux nerveux periphériques, 1178
L'encéphale. Journal des maladies mentales et nerveuses, 1039, 1277, 1282, 1405
Les endormeurs. La verité sur les hypnotisants, les suggestionistes, les magnétiseurs, les Donatistes, les Braidistes, etc., 1170
Enigmas of Psychical Research, 1575, 1611
Gli enigmi della psicometria, 1792
Enquête sur l'authenticité des phénomènes électriques d'Angélique Cottin, 548
L'enseignement du magnétisme, 1382
Die Entdeckung der Seele durch die Geheimwissenschaften, 1362
Die Entdeckung des Hypnotismus, 812, 1047
Entretiens sur le magnétisme animal et le sommeil magnétique dit somnambulisme, dévoilant cette double doctrine et pouvant servir à en porter un jugement raisonné, 314
Entwickelungsgeschichte des magnetischen Schlafs und Traums, 299
Die Entwicklungsgeschichte des Spiritismus von der Urzeit bis zur Gegenwart, 1347
The Epworth Phenomena: To Which are Appended Certain Psychic Experiences Recorded by John Wesley in the Pages of his Journal. Collated by Dudley Wright, 1731
Erinnerungen an die letzten Tage der Odlehre und ihres Urhebers, 989
Die Erklärung des Gedankenlesens, 1155
Erlaüterungen zum Mesmerismus, 256, 260
Les erreurs de la science, 1565
Eine Erscheinung aus dem nachtgebiete der Natur, durch eine Reihe von Zeugen gerichtlich bestatigt und den Naturforschern zum Bedenken mitgetheilt, 375
Erster Nachtrag zur Bibliographie des modernen Hypnotismus, 1260
Los Espiritus, 1377
Esprits et mediums: Mélanges de métapsychique et de psychologie, 1662
Esquisse de la nature humaine expliquée par le magnétisme animal précédée d'un aperçu du système général de l'univers, et contenant l'explication du somnambulisme magnétique et de tous les phénomènes du magnétisme animal, 325
Esquisse de l'histoire du magnétisme humain depuis Mesmer jusqu'en 1848, 566
Esquisse d'une théorie des phénomènes magnétiques, 475
Essai de psychologie physiologique, 349
Essai de revue générale et d'interprétation synthétique du spiritisme, 1438
Essai de spiritisme scientifique, 1369
Essai historique sur le magnétisme et l'universalité de son influence dans la nature, 619
Essai sur la découverte du magnétisme animal, 61
Essai sur la théorie du somnambulisme magnétique, 152, 161
Essai sur le magnétisme animal, thèse présentée et soutenue par H. Long, 413
Essai sur l'enseignement philosophique du magnétisme animal. Par le Baron Du Potet de Sennevoy, 511
Essai sur les phénomènes électriques des êtres vivants comprenant l'explication scientifique des phénomenènes dits spirites, 1364
Essai sur les probabilitiés du somnambulism magnétique: pour servir à l'histoire du magnétisme animal, 139
Essai sur l'hypnotisme, nouvelle découverte précédé d'explications sur le magnétisme et le somnambulisme, 873
An Essay on Somnambulism, or Sleep-walking, produced by Animal Electricity and Magnetism, 178
Essay on Superstition; Being and Inquiry into the Effects of Physical Influence on the Mind, in the Production of Dreams, Visions, Ghosts, and Other Supernatural Appearances, 346, 523

Title Index

Essays Classical, 1070
Essays in Occultism, Spiritism, and Demonology, 1755
Essays in Pyschical Research, by Miss X, 1460
L'état actuel des recherches psychiques d'après les travaux du IIme Congrès International tenu à Varsovie en 1923 en l'honneur du Dr Julien Ochorowicz, 1863
État mental des hystériques. Les accidents mentaux, 1318
État mental des hystériques. Les stigmates mentaux, 1318
Les états profonds de l'hypnose, 1328, 1352
Les états superficiels de l'hypnose, 1352
Etherology; or the Philosophy of Mesmerism and Phrenology: Including a New Philosophy of Sleep and Consciousness, with a Review of the Pretensions of Neurology and Phreno-magnetism, 515
L'être subconscient. Essai de synthèse explicative des phénomènes obscurs de psychologie normale et anormale, 1458, 1754
Étude clinique et expérimentale sur l'hypnotisme; de gg. effets des excitations périphériques ches les hystéro-épileptiques à l'état de veille et d'hypnotisme, 1087
Étude critique du matérialisme et du spiritualisme par la physique expérimentale, 1404
Étude de la vie intérieure ou spirituelle chez l'homme, recherches physiologiques et philosophiques sur le magnétisme, le somnambulisme et le spiritisme; théorie nouvelle de la pensée, de l'extase, de la lucidité somnambulique et médianimique, rôle du coeur et du cerveau, 882
Étude du magnétisme animal sous le point de vue d'une exacte pratique, 628
Étude médico-légale sur le somnambulisme spontané et le somnambulisme provoqué, 1181
Étude physiologique sur le magnétisme animal, 885
Étude raisonnée du magnétisme animal et preuves de l'intervention des puissances infernales dans le phénomènes du somnambulisme magnétique, 337
Étude scientifique sur le somnambulisme, sur les phénomènes qu'il présente et sur son action thérapeutique dans certaines maladies nerveuses du rôle important qu'il joue dans l'épilepsie, dans l'hystérie et dans les névroses dites extraordinaires, 1026
Études cliniques sur l'hystéroépilepsie ou grande hystérie, 1048
Études de psychologie expérimentale. Le fétichisme dans l'amour. La vie psychique des micro-organismes. L'intensité des images mentales. Le problème hypnotique. Note sur l'écriture hystérique, 1195
Études expérimentales sur le fluide nerveux et solution rationelle du problème spirite, 920
Études physiques sur le magnétisme animal soumise à l'Académie des sciences, 467
Études sur la médecine animique et vitaliste, 868
Études sur le somnambulisme envisagé au point de vue pathologique, 829
Étude sur différents attributs de l'âme humaine, et sur la lucidité dans la veille et pendant le sommeil magnétique, 888
Étude sur le zoomagnétisme, 1068, 1297
Étude sur l'hypnotisme et sur les suggestions chez les aliénés, 1099
Eusapia Palladino and her Phenomena, 1625
L'evangile selon le spiritisme contenant l'explication des maximes morales du Christ, leur concordance avec le spiritisme et leur application aux diverses positions de la vie, 877
Evenings at Home in Spiritual Séance, Prefaced and Welded Together by a Species of Autobiography, 1040
Events in the Live of a Seer, 909
The Evidence for Communication with the Dead, 1687
The Evidence for the Supernatural: A Critical Study Made with "Uncommon Sense," 1671
Evidences of the Efficacy of Doctor Perkins's Patent Metallic Instruments, 206
L'évolution animique. Essai de psychologie physiologique suivant le spiritisme, 1415
The Evolution of the Soul and Other Essays, 1539
Examen de la doctrine d'Hippocrate, pour servir à l'histoire du magnétisme animal, 135
Examen de l'ouvrage que a pour titre: "Le mystère des magnétiseurs et des somnambules dévoilé aux droites âmes et virtueuses par un homme du monde," 265
Examen du Compte rendu par M. Thouret, sous le titre de Correspondance de la Société royale de médecins, relativement au magnétisme animal, 125

Examen du magnétisme animal, 382
Examen impartial de la médecine magnétique, de sa doctrine, de ses procédés et de ses cures, 290, 292
Examen physique de magnétisme animal; analyse des éloges & des critiques qu'on en a faits jusqu'à présent; et développement des véritables rapports, sous lesquels on doit en considérer le principe, la théorie, la pratique & le secret, 128
Examen raisonné des prodiges récents d'Europe de d'Amerique notamment des tables tournantes et répondantes, par un philosophe, 665
Examen sérieux et impartial du magnétisme animal, 78
Expériences et considérations à l'appui, du magnétisme animal, thèse présentée et soutenue à la faculté de Paris, 365, 383
Experiences in Spiritualism: A Record of Extraordinary Phenomena Witnessed Through the Most Powerful Mediums, with Some Historical Fragments Relating to Semiramide, Given by the Spirit of an Egyptian Who Lived Contemporary with Her, 981
Experiences in Spiritualism with Mr D. D. Home. With Introductory Remarks by the Earl of Dunraven, 932
Expériences sur le magnétisme animal, 397
Expériences sur le sommeil à distance, 1216
Expérience sur le magnétisme animal, thèse présentée et soutenue à la Faculté de médecine de Paris, 369
Experimental Investigation on the Spirit Manifestations, Demonstrating the Existence of Spirits and Their Communion with Mortals, 744
L'expérimentation en psychologie par le somnambulisme provoqué, 1101
Experimente der Fernbewegung (Telekinese) im Psychologischen Institut der Münchener Universität und im Laboratorium des Verfassers, 1876
Eine experimentelle Studie auf dem Gebiete des Hypnotismus, 1212
Experimentelle Untersuchungen auf dem Gebiete des raumlichen Hellsehens (der Kryptoskopie und inadaequaten Sinneserregung). Nach dem russischen Original bearbeitet und herausgegeben von Freiherrn von Schrenck-Notzing, 1750
Experimentell Telepathie. Neue Versuche zur telepathischen Ubertrag von Zeichnungen, 1889
Experiments in Muscle-reading and Thought-transference, 1166
Experiments in Psychical Research at Leland Stanford Junior University, 1723
Experiments in Psychical Science. Levitation, "Contact," and the "Direct Voice," 1751
Experiments with the Metallic Tractors in Rheumatic and Gouty Affections, Inflammations, and Various Topical Diseases, as Pub. by Surgeons Herholdt and Rafn . . . tr. into German by Professor Tode . . . thence into the English Language by Mr. Charles Kampfmuller: also Reports of about One Hundred and Fifty Cases in England . . . By medical and Other Respectable Characters. Ed. by Benjamin Douglas Perkins, 210
Explicacion et emploi du magnétisme, 271
Exposé des expériences qui one été faites pour l'examen du magnétisme animal. Lu à l'Académie des sciences, par M. Bailly en son nom & aux nom de Mrs. Franklin, Le Roy, de Bory, et Lavoisier, le 4 Septembre 1784, 30
Exposé des expériences sur le magnétisme animal faites à l'Hotel Dieu de Paris pendant les mois d'octobre, novembre et décembre 1820, 302, 534
Exposé of the Davenport Brothers, 931
Exposé par ordre alphabétique des cures opérées en France par le magnétisme animal, depuis Mesmer jusqu'à nos jours (1774–1826), ouvrage où l'on a réuni les attestations de plus de 200 médecins, tant magnétiseurs que témoins, ou guéris par le magnétisme. Suivi d'un catalogue complet des ouvrages français qui ont été publiés pour, sur ou contre le magnétisme, 331
Exposition, or a New Theory of Animal Magnetism with a Key to the Mysteries: Demonstrated by Experiments with the Most Celebrated Somnambulists in America: also, Strictures on "Col. Wm. L. Stone's Letter to Doctor A. Brigham," 381, 386
Exposition critique du système et de la doctrine mystique des magnétistes, 309
Exposition de la doctrine magnétique; ou, Traité philosophique, historique, et critique du magnétisme, 641

Title Index

An Exposition of Views Respecting the Principal Facts, Causes and Peculiarities Involved in Spirit Manifestations: Together with Interesting Phenomenal Statements and Communications, 627
Exposition physiologique des phénomènes du magnétisme animal et du somnambulisme: contenant des observations pratiques sur les avantages et l'emploi d l'un et d l'autre dans le traitement des maladies aigues et chroniques, 279
L'extériorisation de la motricité. Recueil d'expériences et d'observations, 1406
L'extériorisation de la sensibilité. Étude expérimentale et historique, 1391
Extrait de la correspondance de la Société royale de médecine, relativement au magnétisme animal, 154
Extrait des journaux d'un magnétiseur, attaché à la société des amis réunis de Strasbourg; avec des observations sur les crises magnétiques, connues sous la dénomination de somnambulisme magnétique, 160
Extrait du journal de ce que s'est passé concernant le somnambulisme magnétique de Mme, 159
Extrait du journal d'une cure magnétique. Traduit de l'allemand, 167

Faces of the Living Dead: Remembrance Day Messages and Photographs. A Straightforward Statement by Estelle W. Stead. With Supplement Showing Recognised "Extras" Obtained through the Mediumship of Mrs. Deane, 1878
Fact and Fable in Psychology, 1488
Fact and Fancy in Spiritualism, Theosophy and Psychical Research, 1484
Facts and Fallacies of Mesmerism; Demonstrated to Its Friends and Opponents, 530
Facts and Fantasies: a Sequel to Sights and Sounds; the Mystery of the Day, 689
Facts and observations on the mesmeric and magnetic fluids, 587
Facts in Magnetism, Mesmerism, Somnambulism, Fascination, Hypotism, Sycodonamy, Etherology, Pathetism, &c., Explained and Illustrated, 585
Facts in Mesmerism and Thoughts on Its Causes and Uses, 451
Facts in Mesmerism with Reasons for a Dispassionate Inquiry into It, 433, 617
The Facts of Psychic Science and Philosophy Collated and Discussed, 1898
Un fait de somnambulisme à distance, 1157
Faits curieux et intéressants produits par la puissance du magnétisme animal, ou compte-rendus des expériences remarquable opérées en Belgique, 461
The Fallacy of Phreno-magnetism Detected and Exposed, 470
Le fantôme des vivants: anatomie et physiologie de l'âme. Recherches expérimentales sur le dédoublement des corps de l'homme, 1627
Fascination, or The philosophy of Charming, Illustrating the Principles of Life in Connection with Spirit and Matter, 564, 718
La fascination magnétique, précédée d'une préface par Donato et de son protrait photographié, 1056
Fatti relativi a mesmerismo e cure mesmeriche con una prefazione storicocritica, 452
Faut-il étudier le spiritisme?, 1560
Fenomeni rimarchevoli di medianità osservati senza medi di professione, 1314
I fenomeni telepatici e le allucinazioni viridiche. Osservazioni critiche sul neo-misticismo psicologico, 1422
Fernfühlen und Mesmerismus (Exteriorisation der Sensibilität), 1880
Filosofia dello spirito—ovvero del magnetismo animale, 814
Five Essays, 808
Five Sittings with Mrs. Saunders, 1872
Florentii Jacobi Voltelen Oratio de magnetismo animali: publice habita Lugduni Batavorum die VII. Februarii a. CDDCCLXXXXI, quum magistratum academicicum solenniter deponeret, 199
Le fluide des magnétiseurs. Précis des expériences du Baron de Reichenbach sur les propriété physiques et physiologiques, classées et annotées par le Lt-Colonel A. de Rochas, 1306
Le fluide humain, ses lois et ses propriétés. La science de mouvoir la matière sans être médium. Nombreux appareils et moteurs que l'on peut construire soi-même, mis en mouvement

par le fluide humain. L'être psychique, les fantômes, doubles des vivants et images fluidiques. Étude sur la force biolique, 1631
Les fluides, chapitres extraits de la Genèse, 1013
Footfalls on the Boundary of Another World, 833, 943
La force-pensée: la faculté unique; mécanisme de la télépathie; extériorisation de la volonté; appel et captation des forces cosmiques; théorie nouvelle de l'influence de l'homme sur l'homme, 1645
La force psychique, l'agent magnétique et les instruments servant à les mesurer, 1583
Force psychique et suggestion mentale, leur demonstration, leur explication, leurs applications possible, à la thérapeutique et à la médecine légale, 1154
Les forces naturelles inconnues, 1586
Les forces non définies: recherches historiques et expérimentales, 1187
La force vitale. Notre corps vitale fluidique, sa formule biométrique, 1335
Fortnightly Review, 969, 1350, 1525
Fotografie di fantasmi: Contributo sperimentale alla constatazione dei fenomeni medianici, 1679
Foundations of Hypnosis, 106, 450
The Foundations of Normal and Abnormal Psychology, 1705
Foundations of Spiritualism, 1768
Les fous, les insensés, les maniaques et les frénétiques ne seraient-ils que des somnambules désordonnés?, 240–241
La France trompée par les magnéciens, les démonolâtres et les magnétiseurs du dix-huitième siècle, 223
Franz Anton Mesmer, His Life and Teaching, 1775
Franz Anton Mesmer aus Schwaben, Entdecker des thierischen magnetismus. Erinnerungen an denselben, nebst nachrichten von den letzten Jahren seines Lebens zu Meersburg am Bodensee, 766
Franz Anton Mesmer's Leben und Lehre. Nebst einer Vorgeschichte des Mesmerismus, Hypnotismus und Somnambulismus, 1348
Frederic Myers's Service to Psychology, 1487
From Agnosticism to Belief: An Account of Further Evidence for Survival, 1865
From Matter to Spirit. The Result of Ten Years' Experience in Spirit Manifestations. Intended as a Guide to Enquirers. By C. D. With a Preface by A. B. [Augustus De Morgan), 853
Les frontières de la science, 1510
Frontiers of the After Life, 1834
Full and Comprehensive Instructions How to Mesmerize, Ancient and Modern Miracles by Mesmerism, also Is Spiritualism True?, 1055
Further Problems of Hypnotism, 1173
A Further Record of Mediumistic Experiments, 1899
A Further Record of Observations of Certain Phenomena of Trance, 1419, 1439
A Further Record of Observations of Certain Trance Phenomena, 1486
Further Report on experiments in thought-transference at Liverpool, 1111

The Gate of Remembrance: the Story of the Psychological Experiment which Resulted in the Discovery of the Edgar Chapel at Glastonbury, 1732, 1747, 1860
Gazette de Santé, 14
Gazzetta magnetico-scientifico-spiritistica, 871
Geburt und Tod als Wechsel der Anschauungsform, oder die Doppel-Natur des Menschen, 1114
Gedruckte Antwort des Herrn Dr Mesmer vom 19. Januar 1775, 3
Geheimsvolle Tatsachen, 1801
Die Geisterhypothese des Spiritismus und seine Phantome, 1293
The General and Particular Principles of Animal Electricity and Magnetism, &c. in Which Are Found Dr. Bell's Secrets and Practice, As Delivered to His Pupils in Paris, London, Dublin, Bristol, Glocester, Worcester, Birmingham, Wolverhampton, Shrewsbury, Chester, Liverpool, Manchester, &c. &c. Shewing How To Magnetise and Cure Different

Diseases; to Produce Crises, as well as Somnambulism, or Sleepwalking; and in That State of Sleep to Make a Person Eat, Drink, Walk, Sing and Play Upon Any Instruments They Are Used to, &c. To Make Apparatus and Other Accessaries To Produce Magnetical Facts; Also To Magnetise rivers, Rooms, Trees . . . , 201
La genèse, les miracles et les prophéties selon le spiritisme, 915
Gesammelte Acten-Stücke zu Aufdeckung des Geheimnisses des sogenannten thierischen Magnetismus in einigen freundschaftlichen Briefen dem Herrn Ernst Gottfried Baldinger mitgetheilet, 163
Geschichte der okkultischen (metapsychischen) Forschung von der Antike bis zur Gegenwart. II. Teil. Von der Mitte des 19 Jahrhunderts bis zur Gegenwart, 1881
Geschichte der okkultistischen (metapsychischen) Forschung von der Antike bis zur Gegenwart. 1. Teil. Von der Antike bis zur Mitte der 19. Jahrhunderts, 1830
Geschichte des Lebensmagnetismus und des Hypnotismus. Von Urfang bis auf den heutigen Tag, 1464
Geschichte des neureren Occultismus. Geheimwissenschaftliche Systeme von Agrippa von Nettesheym bis zu Carl du Prel, 1294
Geschichte eines allein durch die Natur hervorgebrachten animalischen Magnetismus und der durch denselben bewirkten Genesung; von dem Augenzeugen dieses Phänomens, 246
Geschichten Besessener neuerer Zeit. Beobachtungen aus dem Gebiete kakodämonisch-magnetischer Erscheinungen. Nebst Reflexionen von C. A. Eschenmayer über Bessessenseyn und Zauber, 361
Geschichte zweyer Somnambulen. Nebst einigen andern Denkwudigkeiten aus dem Gebiete der magischen Heilkunde und der Psychologie, 317
Gesundbeten Medizin und Okkultismus, 1507
Ghost Land: Or Researches into the Mysteries of Occultism. Illustrated in a Series of Autobiographical Sketches. Translated and Edited by Emma Hardinge Britten, 984
Ghosts I Have Seen: and Other Psychic Experiences, 1789
Ghosts in Solid Form: an Experimental Investigation of Certain Little-known Phenomena (Materialisations), 1693
The Gift of D. D. Home, 1268
Giornale del magnetismo ed ipnotismo, 1267, 1299
Glimpses of the Next State (The Education of an Agnostic), 1668
Grande belle découverte du magnetisme animal, 28
La grande hystérie chez l'homme. Phénomènes d'inhibition et de dynamo-génie, changements de la personnalité, action des médicaments à distance. D'après les travaux de MM. Bourru et Burot, 1126
Grande isteria ed ipnotismo. Studio medico legale su Paolo Conte imputato di truffa in danno del dr. Fusco, 1224
Grand et petit hypnotisme, 1223
Grandeur et décadence des Rayons N: histoire d'une croyance, 1593
The Great Amherst Mystery. A True Narrative of the Supernatural, 1208
Grosjean à son évèque au sujet des tables parlantes, 733
Grundbegriffe der Parapsychologie. Eine Philosophische Studie, 1806
Grundversuche auf dem Gebiete der psychischen Grenzwissenschaften, 1877
Guardian Spirits, a Case of Vision into the Spiritual World, 418
Guida Elementare dello studente magnetizzatore, 701
Guida teorico-pratica del magnetismo animale, 902
Guide du magnétiseur ou procédés magnétiques d'après Mesmer, de Puységur et Deleuze, mis à la portée de tout le monde suivi des bienfaits et dangers du somnambulisme, 558

Les habitants de l'autre monde. Études d'outre-tombe, 847
Hallucinations, 1109
Hallucinations and Illusions, 1370
Hallucinations or, The Rational History of Apparitions, Visions, Dreams, Ecstasy, Magnetism, and Somnambulism, 508
Les hallucinations télépathiques, 1620

Hand-book of Mesmerism, for the Guidance and Instruction of All Persons who Desire to Practice Mesmerism, 608
Handbuch zur ausübung des Magnetismus, des Hypnotismus, der Suggestion, der Biologie und verwandter Fächer, 1544
Harpers New Monthly Magazine, 834
The Harveian Oration, Delivered before the Royal College of Physicians, London, June 27th, 1846. With an English version and Notes, 535
Haunted Houses, 1843
Die Heilkraft des Lebensmagnetismus und dessen Beweiskraft für die Unsterblichkeit der Seele, 958
Heilkraft des thierischen Magnetismus, 222, 226
Der Heilmagnetismus, 1304
Der heilmagnetismus; seine Theorie und Praxis, 960
Die Heilmethode des Lebensmagnetismus, 1462
Die Heilmethode des Lebensmagnetismus. Theorie und Praxis besprochen und mit eine Nachweise über den wesentlichen Unterschied zwischen Hypnotismus und Heilmagnetismus versehen, 1279
Das Hellseh-Medium Megalis in Schweden, 1822
Henry Sidgwick: a Memoir, 1562
L'Hermés, Journal du magnétisme animal. Publié par une Société des Médecins, 330
Herrn Dr Mesmers Schreiben an die Frankfurter vom 10. Mai 1775, 4
Herrn Professor Zollners Experimente mit dem amerikanischen Medium Herrn Slade und seine Hypothese intelligenter vierdimensionaler Wesen. Mit einer Antwort an die Herren Professoren Herrn. W. Vogel in Berlin und J. B. Meyer in Bonn, 1061
The Hidden Self, 1270
Higher Aspects of Spiritualism, 1031
The Hill of Vision: A Forecast of the Great War and of the Social Revolution with the Coming of the New Race. Gathered from Automatic Writings Obtained Between 1909 and 1912, and also, in 1918, Through the Hand of John Alleyne, Under the Supervision of the Author, 1747
Hints and Observations for Those Investigating the Phenomena of Spiritualism, 1735
Hints on Animal Magnetism, Addressed to the Medical Profession in Great Britain, 396
Histoire académique du magnétisme animal accompagnée de notes et de remarques critiques sur toutes les observations et expériences faites jusqu'à ce jour, 436
Histoire critique du magnétisme animal, 243, 261, 292
Histoire de la médecine et des sciences occultes. Notes bibliographiques pour servir à l'histoire du magnétisme animal. Analyse de tous les livres, brochures, articles de journaux publiés sur le magnétisme animal, en France et à l'étranger, à partir de 1766 jusqu'au 31 décembre 1868. Première partie: livres imprimés en France, 922
Histoire de magnétisme en France, de son régime et de son influence, pour servir à developper l'idée qu'on doit avoir de la médecine universelle, 43
Histoire du magnétisme dont les phénomènes sont rendus sensibles par le mouvement, 516
Histoire du merveilleux dans les temps modernes, 822
Histoire du somnambulisme: chez tous les peuples sous les noms divers d'extases, songes, oracles et visions; examen des doctrines théoriques et philosophiques de l'antiquité et des temps modernes, sur ses causes, ses effets, ses abus, ses avantages, et l'utilité de son concours avec la médecine, 455
Histoire et philosophie du magnétisme chez les anciens et chez les modernes, 1387
Histoire véritable du magnétisme animal, ou nouvelles preuves de la réalités de cet agent tirées de l'ancien ouvrage d'un vieux docteur, 142
Historischen Prophezeiungen mit besonderer Berücksichtigung der Weltkriegsprophezeiungen, 1827
Historisch-psycholoqische Untersuchungen uber den Ursprung und das Wesen der menschlichen Seele überhaupt, und uber die beseelung des Kindes insbesondere, 316
A History of Experimental Spiritualism, 1410
An History of Magic, Witchcraft, and Animal Magnetism, 613
History of the Strange Sounds or Rappings, Heard in Rochester and Western New York, and

Usually Called the Mysterious Noises! Which are Supposed by Many to be Communications from the Spirit World, Together with all the Explanation than Can as yet Be Given of the Matter, 595
The History of the Supernatural in All Ages and Nation, and in All Churches, Christian and Pagan: Demonstrating a Universal Faith, 859
L'homme et l'intelligence. Fragments de physiologie et de psychologie, 1093
Homo duplex: note physiologique sur l'organisme humain presentée aux Facultiés de Médecine et à l'Académie des Sciences, 1245
L'hôte inconnu, 1728
Houdini Exposes the Tricks Used by the Boston Medium "Margery" to Win $2500 Prize Offered by the Scientific American. Also a Complete Exposure of Argamasilla, the Famous Spaniard Who Baffled Noted Scientists of Europe and America, with His Claim to X-Ray Vision, 1882
How to Investigate Spiritualism, 1065
How to Magnetize, or Magnetism and Clairvoyance. A Practical Treatise on the Choice, Management and Capabilities of Subjects, with Instructions on the Method of Procedure, 556
How to Mesmerize, 1338, 1414, 1735
How to Thought-read: a Manual of Instruction in the Strange and Mystic in Daily Life, Psychic Phenomena, Including Hypnotic, Mesmeric, and Psychic States, Mind and Muscle Reading, Thought Transference, Psychometry, Clairvoyance, and Phenomenal Spiritualism, 1339
The Human Atmosphere, or The Aura Made Visible by the Aid of Chemical Screens, 1664
Human Electricity: the Means of Its Development, Illustrated by Experiments. With Additional Notes, 728
Human Immortality: Two Suppossed Objections to the Doctrine, 1440
Humanism: Philosophical Essays, 1529
Human Magnetism: Its Claims to Dispassionate Inquiry: Being an Attempt to Show the Utility of Its Application for the Relief of Human Suffering, 523
Human Magnetism. Its Nature, Physiology and Psychology. Its Uses, as a Remedial Agent, in Moral and Intellectual Improvement, etc., 1230
Human Magnetism or How to Hypnotise. A Practical Handbook for Students of Mesmerism, 1414, 1734
Human Nature, 994
Human Personality and Its Survival of Bodily Death, 1070, 1144, 1525
Human Physiology, 424
Humbugs of New York: Being a Remonstrance Against Popular Delusions; Whether in Science, Philosophy, or Religion, 400
Hygienic Clairvoyance, 803
Hypnologie: du sommeil et des songes au point de vue physiologique: somnambulism, magnétisme, extase, hallucination; exposé d'une théorie du fluide électro sympathique, 471
Die Hypnose und die damit verwandten normalen Zustände. Vorlesungen gehalten auf der Universität Kopenhagen im herbste 1889, 1272
Die Hypnose und ihre civilrechtliche Bedeutung, 1254
Hypnose und Suggestion, 1867
Hypnotic Magazine, 1402, 1409
Hypnotic Suggestion and Its Relation to the Traumatic Neurosis, 1316
Hypnotic Therapeutics, Illustrated by Cases. With an Appendix on Table-Moving and Spirit-Rapping. Reprinted from the Monthly Journal of Medical Science for July 1853, 656
Hypnotic Therapeutics in Theory and Practice: With Numerous Illustrations of Treatment by Suggestion, 1619
Das hypnotische Hellseh-Experiment im Dienste der naturwissenschaftlichen Seelenforschung, 1423
Das hypnotische Verbrechen und seine Entdeckung, 1231
Hypnotism [Moll], 1240
Hypnotism [Foveau de Courmelles], 1264
Hypnotism, a Complete System of Method, Application and Use, 1469

Hypnotism, How It is Done; Its Uses and Dangers, 1358
Hypnotism, Its Facts, Theories and Related Phenomena: with Explanatory Anecdotes, Descriptions and Reminiscences, 1354
Hypnotism: Its History, Practice and Theory, 1517, 1624
Hypnotism: Its History and Present Development, 1163
Hypnotism, Mesmerism and the New Witchcraft, 1345
Hypnotism and Disease; A Plea for Rational Psychotherapy, 1677
Hypnotism and Hypnotic Suggestion. A Scientific Treatise on the Uses and Possibilities of Hypnotism, Suggestion and Allied Phenomena. By Thirty Authors, 1473
Hypnotism and Its Application to Practical Medicine, 1220
Hypnotism and Suggestion in Daily Life, Education, and Medical Practice, 1644
Hypnotism and Suggestions in Therapeutics, Education, and Reform, 1492
Hypnotism and Telepathy, 1204
Hypnotism and the Doctors, 1504
Hypnotism and Treatment by Suggestion [Bramwell], 1624
Hypnotism and Treatment by Suggestion [Davis], 1737
L'hypnotisme, 1264
L'hypnotisme, compte-rendu des conférences du docteur A.J.P. Philips, 820
Hypnotisme, double conscience, et altérations de la personnalité, 1159, 1334
Hypnotisme, états intermédiares entre le sommeil et la veille, 1198
L'hypnotisme: étude critique, 1326
L'hypnotisme, ses phénomènes et ses dangers. Étude, 1250
L'hypnotisme, ses rapports avec le droit et la thérapeutique, la suggestion mentale, 1256
Hypnotisme, suggestion, psychothérapie; étude nouvelles, 1284
Hypnotisme, suggestion et lecture des pensées. Tr. du Russe par Ernest Jaubert, 1308
L'hypnotisme devant les chambres legislative belges, 1313
Hypnotisme et croyances anciennes, 1303
Hypnotisme et double conscience origine de leur étude divers travaux sur des sujets analogues, 1334
L'hypnotisme et la liberté des représentations publiques, 1201
L'hypnotisme et la psychothérapie dans l'oeuvre de Dumontpallier, 1692
L'hypnotisme et la suggestion, 1519
L'hypnotisme et la suggestion dans leurs rapports avec la médecine légale, 1412
L'hypnotisme et la volonté, 1104
L'hypnotisme et la crime, 1359
L'hypnotisme et les états analogues au point de vue médico-légal, les états hypnotiques et les états analogues, les suggestions criminelles, cabinets de somnambules et sociétés de magnétisme et de spiritisme, l'hypnotisme devant la loi, 1172
L'hypnotisme et l'orthopédie mentale, 1433
Hypnotisme et magnétisme. Somnambulisme, suggestion et télépathie. Influence personnelle . . . resumant . . . tous les connaissances humaines sur les possibilités, les usages et la pratique de l'hypnotisme . . . du magnétisme, de la suggestion et de la télépathie . . . , 1554
Hypnotisme expérimental; la dualité cérébrale et l'indépendance fonctionelle des deux hémisphères cérébraux, 1079
L'hypnotisme expliqué dans sa nature et dans ses actes. Mes entretiens avec S. M. l'Empereur Don Pedro sur le Darwinisme, 1210
Hypnotismen, dess utveckling och nuvarande standpunkt. Popular framstallning, 1163
L'hypnotisme scientifique, 1399
L'hypnotisme théorique et pratique, comprenant les procédés d'hypnotisation, 1239
Hypnotism in Mental and Moral Culture, 1476
Der Hypnotism or Animal Magnetism. Physiological Observations, 1029
Der Hypnotismus, 1240
Hypnotismus and Suggestion, 1332
Hypnotismus, Animismus und Psychoanalyse. Historisch-kritische Versuche, 1727
Der Hypnotismus. Handbuch der Lehre von der Hypnose und der Suggestion, mit besonderer Berücksichtigung ihrer Bedeutung für Medicin und Rechtspflege, 1490

Der Hypnotismus. Psychiatrische Beitrage zur Kentniss der sogenannten hypnotischen Zustande. Nebst eieem physiognomischen Beitrag von Dr. Hans Virchow, 1095
Der Hypnotismus: seine Bedeutung und seine Handhabung. In kurzgefasster Darstellung, 1232
Der Hypnotismus, seine Stellung sum Aberglauben und zur Wissenschaft, 1106
Der Hypnotismus: Vorlesungen gehalten an der K. Friedrich-Wilhelms-Universität zu Berlin, 1275
Der Hypnotismus im Münchener Krankenhause (Links der Isar). Eine kritische Studie über die Gefahren der Suggestivbehandlung, 1373
Hypnotismus und Suggestion. Eine klinisch-psychologische Studie, 1357

L'iconographie en anses de la force vitale cosmique et la respiration fluidique de l'âme humaine. Son atmosphére fluidique, 1397
Iconographie photographique de la Salpétrière (service de M. Charcot). I. Hystéro-épilepsie: description des attaques; les possédées du Loudun; du crucifiement. II. Epilepsie partielle et hystéro-épilepsie . . . III. Hystéro-épilepsie: Zones hystérogènes; sommeil; attaque de sommeil; hypnotisme; somnambulisme; magnétisme, catalepsie; procédés de magnétisme, 982
Ideen und Beobachtungen den thierischen Magnetismus und dessen Anwendung betreffend, 215
Idées de physique ou résumé d'une conversation sur la cause des sensations avec la composition de la poudre de sympathie, ouvrage dédié aux dames de Paris, 164
Illusions: A Psychological Study, 1050
Illustrated London News, 647
The Illustrated Practical Mesmerist, Curative and Scientific, 710
Illustrations and Enquiries Relating to Mesmerism. Part I, 578
Illustrations of Modern Mesmerism from Personal Investigation, 512
Illustrations of Political Economy, 522
Illustrations of the Influence of the Mind upon the Body in Health and Disease. Designed to Elucidate the Action of the Imagination, 949
L'imagination, étude psychologique, 996
L'imagination: ses bienfaits et ses égarements sourtout dans le domaine du merveilleux, 917
Incidents in My Life, 858, 948
Incidents in My Life. Second Series, 858, 948
L'inconnu et les problèmes psychiques, 1457
L'inconscient, étude sur l'hypnotisme, 1226
Index [of forbidden books], 716, 825
The Influence of Metallic Tractors on the Human Body, in Removing Various Painful Inflammatory Diseases, etc., by which a New Field of Enquiry is Opened in the Modern Science of Galvanism or Animal Electricity, 204–205
The Influence of Metallic Tractors on the Human Body, in Removing Various Painful Inflammatory Diseases . . . Lately Discovered by Dr. Perkins . . . and Demonstrated in a Series of Experiments and Observations, by Profesors Meigs, Woodward, Rogers, &c, &c. by which the Importance of the Discovery is Fully Ascertained, and a New Field of Enquiry Opened in the Modern Science of Galvanism, or, Animal Electricity . . . , 209
Influenza del magnetismo sul cervelletto: discorso accademico, 1022
Influenza del magnetismo sulla vita animale, 1030
L'initiation, Revue philosophique indépendente des hautes études, hypnotisme, force psychique, théosophie, kabbale gnose, franc maçonnerie, sciences occultes, 1209, 1690
Initiation aux mystères du magnétisme. Théorie du magnétisme. Connaissance des maladies, causes et remèdes. Faits magnétiques. Vision somnambulique. Vision dans l'avenir et dans l'espace, etc. . . . , 550
Initiation aux mystères secrets de la théorie et de la pratique du magnétisme rendue simple et facile quant à la pratique etc., suivie d'expériences inédites faites à Monte Cristo chez Alexandre Dumas, de la biographie de J. B. Marcillet, de la visite faite au somnambule Alexis par le général Cavaignac, 562

Insensibilité produite au moyen du sommeil magnétique. Nouvelle opération chirurgicale faite à Cherbourg, 541
Institut dynamo-thérapique. Du somnambulisme médical, ou esquisse de nososcopie dynamo-thérapique, 777
Institut métapsychique international. Bulletin, 1283, 1776
Instruction explicative des tables tournantes, d'après les publications allemandes, américaines, et les extraits des jounaux allemands, français et américains. Précédée d'une introduction sur l'action motrice du fluide magnétique, par Henri Delaage, 686
Instruction pratique sur le magnétisme animal, suivie d'un lettre écrite à l'auteur par un médecin étranger, 320
Instructions in Mesmerism &c. The Magnetic and Botanic Family Physician, and Domestic Practice of Natural Medicine, with Illustrations Showing Various Phases of Mesmeric Treatment, Including Full and Concise Instructions in Mesmerism, Curative Magnetism, and Massage, 1194
Instructions pratiques sur les manifestations spirites contenant l'exposé complet des conditions nécessaires pour commniquer avec les esprits et les moyens pour développer la faculté médiatrice chez les médiums, 796
Die Intellektuellen Phänomene, 1885, 1895
International Congress for Psychical Research (1st, Copenhagen, 1921) Proceedings, 1828
International Congress for Psychical Research Proceedings (Warsaw, 1923), 1866
Intorno all'ignoto; fakiri e case infestate in un conflitto sullo spiritismo: prima riposta a Ces. Lombroso, 1634
Introduction au magnétisme, examen de son existence depuis les Indiens jusqu'à l'époque actuelle, sa théorie, sa pratique, ses avantages, ses dangers et la nécessité de son concours avec la médicine, 425
Introduction au magnétisme animal par P. Laurent, suivie des principaux aphorismes du Dr Mesmer dictés par lui à l'assemblée de ses élèves, et dans lesquels on trouve ses principes, sa théorie et les moyens de magnétiser, 563
The Introduction of Mesmerism, as an Anaesthetic and Curative Agent, into the Hospitals of India, 636
An Introduction to the Study of Animal Magnetism, 399
Introduzione allo studio del magnetismo animale e del magnetico sonnambulismo, 841
L'ipnotismo, 1267
L'ipnotismo e gli stati affini, 1161
L'ipnotismo e magnetismo svelato e spiegato sulle teorie di Donato Guidi e Mesmer, 1193
Ipnotismo e spiritismo, 1576
L'ipnotismo-magnetismo alla portata di tutti. Manuale pratico de psichismo illustrato, 1702
Ipotesi spiritica e teoriche scientifiche, 1516
Isis Revelata; an Inquiry into the Origin, Progress & Present State of Animal Magnetism, 354, 372, 396
Is Spiritualism Based on Fraud? The Evidence of Sir A. Conan Doyle and Others Drastically Examined, 1783

J. M. Charcot, son oeuvre psychologique, 1386
Der Jenseitige Mensch: eine Einführung in die Metapsychologie der Mystischen Erfahrung, 1901
Der jetzige Spiritualismus und verwandte Erfahrungen der Vergangenheit und Gegenwart. Ein Supplement zu des Verfassers "mystischen Erscheinungen der menschlichen Natur," 999
Joseph Glanvill and Psychical Research in the Seventeenth Century, 1810
Journal de l'anatomie et de la physiologie normales et pathologiques de l'homme et des animaux, 977
Journal de la Société du magnétisme animal à Paris, 288
Journal de Paris, 84–85
Journal de psychologie normal et pathologique, 1540
Journal des débats, 398

Journal du magnétisme, 1012, 1068, 1650
Journal du magnétisme, rédigé par un Société de magnétiseurs et médecins, sous la direction de M. le Baron Du Potet, 518
Journal du magnétisme. Sous la direction de H. Durville, 411, 518
Journal du traitement magnétique de la demoiselle N. Lequel a servi de base à l'Essai sur la Théorie du somnambulism magnétique, 161
Journal du traitement magnétique de Madame B. . . . pour servir de suite au Journal du traitement magnétique de la Dlle. N. . . . & de preuve à la théorie de l'Essai, 172
Journal du traitement magnétique d'un jeune soldat, cavalier dans le régiment des lanciers de la garde royale, attaqué d'un mal à la cheville du pied, dégenéré en ulcère-fistuleux, 318
Journal für Psychologie und Neurologie. Zugleich Zeitschrift für Hypnotismus, 1333, 1505
Journal of Abnormal Psychology, 1559, 1594, 1637, 1643, 1716, 1833
Journal of Magnetism, 1489
Journal of Medical Hypnotism, 1409, 1442
Journal of Mental Science, 1280
Journal of the Phreno-magnetic Society of Cincinnati, 456
Justinus Kerner und die Seherin von Prevorst, 1136

Der Kampf um die Materialisations-Phänomene. Eine Verteidigungsschrift, 1703
Die Kataplexie und der thierische Hypnotismus, 1009
Katechismus des Vital-Magnetismus zur leichteren Direction der Laien-Magnetiseurs. Zusammengetragen während seiner zehnjahrigen magnetischen Laufbahn nach Aussagen von Somnambulen und vieler Autoren, 527, 693
A Key to Hypnotism. A Complete and Authentic Guide to Clairvoyance, Mesmerism and Hypnotism, 1502
A Key to Physic, and the Occult Sciences, 203
A Key to the Science of Electrical Psychology. All its Secrets Explained, with Full and Comprehensive Instructions in the Mode of Operation and its Application to Disease, with Some Useful and Highly Interesting Experiments. Every Person an Operator. By a Professor of the Science, 598
Kort Begrip der psychische Geneeswi jze voodracht gehouden op uitnoodiging van het Bestur van "Ons Huis Buiten de Muiderpoort" . . . , 1543
Kriminal-Telepathie und -Retroskopie. Telepathie und Hellsehen im Dienste der Kriminalistik, 1823

Le langage martien, étude analytique de la genèse d'une lange dans un cas de glossolalie somnambulique, 1483, 1503
Later Phases of Materialisation, with Reflections to Which They Give Rise, 994
The Law of Mental Medicine: the Correlation of the Facts of Psychology and Histology in Their Relation to Mental Therapeutics, 1521
The Law of Psychic Phenomena: A Working Hypothesis for the Systematic Study of Hypnotism, Spiritism, Mental Therapeutics, etc., 1346, 1385
Leçons cliniques sur les principaux phénomènes de l'hypnotisme dans leurs rapports avec la pathologie mentale, 1273
Leçons cliniques sur l'hystérie et hypnotisme: faites à l'Hôpital Saint-André de Bordeaux, 1302
Leçons de clinique médicale faites à l'hôpital Saint-Eloi de Montpellier, 1520
Leçons sur les maladies du système nerveux faites à la Salpêtrière . . . recueillies et publiées par Bourneville, 947, 1058
Lecture on Mysterious Knockings, Mesmerism, &c., with a Brief History of the Old Stone Mill, and a Prediction of Its Fall, Delivered Before the Antiquarian Society of Pappagassett . . . by Benjamin Franklin Macy, D.F., D.D.F., A.S.S., Professor of Hyperflutinated Philosophy, 621
The Lectures of J. B. de Mainauduc, M.D. Part the First, 208
Lectures on Clairmativeness: or, Human Magnetism, 525

Leib und Seele. Eine Untersuchung über das Psycho-Physische Grundproblem, 1712
Le lendemain de la mort; ou la vie future selon la science, 940
Letter from M. Mesmer, Doctor of Medicine at Vienna to A. M. Unzer, Doctor of Medicine, on the Medicinal Usage of the Magnet, 5
A Letter on Animal Magnetism, 493
Letter on the Truths Contained in Popular Superstitions, with an Account of Mesmerism, 579
Letters on Animal Magnetism, 510
Letters on Mesmerism, 522
Letters on the Laws of Man's Nature and Development, 605
Letters to a Candid Inquirer, on Animal Magnetism, 617
A Letter to Col. Wm. L. Stone, of New York, on the Facts Related in his Letter to Dr. Brigham, and a Plain Refutation of Durant's Exposition of Animal Magnetism, &c. by Charles Poyen. With Remarks on the Manner in which the Claims of Animal Magnetism should be Met and Discussed. By a Member of the Massachusetts Bench, 386
Letter to Doctor A. Brigham, on Animal Magnetism: Being an Account of a Remarkable Interview Between the Author and Miss Loraina Brackett While in a State of Somnambulism, 386, 389
Letter to the Royal College of Physicians, London, dated March 28, 1802, 535
Lettre adressée à M. le Marquis de Puységur sur une observation faite à la lune, précédée d'un système nouveau sur le mécanisme de la vue, 169
Lettre adressée par M. d'Eslon aux auteurs du Journal de Paris et voluntairement refusée par eux, concernant l'extrait de la correspondance de la Société royale relativement au magnétisme animal, rédigé par M. Thouret et imprimé au Louvre, 132
Lettre à la Société exqétique et philanthropique de Stockholm . . . , 185
Lettre à M. d'Eslon, médecin ordinaire de Monseigneur le Comte d'Artois, 59
Lettre à M. l'abbé Croizet . . . sur le magnétisme et la danse des tables, 679
Lettre à M. Mesmer, et autre pièces concernant la maladie de mademoiselle de Belancourt de Beauvais, 17
Lettre à Madame la Comtesse de L. . . . contenant une observation magnétique faite par une somnambule sur un enfant de six mois, 166
Lettre à messieurs le membres de l'académie de médecine, sur la marche que convient de suivre pour fixer l'opinion publique relativement à la réalité du magnétisme animal, aux avantages qu'on peut en retirer, et aux dangers qu'il présente lorsqu'on en fait une application inconsidérée, 326
Lettre au docteur Frapart sur le magnétisme, 468
Lettre a un magistrat de province, sur l'existence du magnétisme, 75
Lettre de F. A. Mesmer, docteur en médecine, sur l'origine de la petite vérole et le moyen de la faire cesser, suivie d'une autre lettre du même adressé aux auteurs du Journal de Paris, contenant diverses opinions relatives au système de l'auteur sur le magnétisme animal, 216
Lettre de Figaro au Comte Almaviva sur la crise du magnétisme animal, avec des détails propres à fixer enfin l'opinion sur l'inutilité de cette découverte; nouvelle édition précédé et suivie des réflexions qui ont rapport aux circonstances présentes, traduites de l'espangnol, 44
Lettre de l'auteur de la découverte du magnétisme animal à l'auteur des Réflexions préliminaires. Pour servir de réponse à un imprimé ayant pour titre: Sommes versées entre les mains de M. Mesmer pour acquérir le droit de publiersa découverte, 144
Lettre de l'auteur de l'Examen sérieux et impartial du magnétisme animal à M. Judel, médecin membre de la Société de l'Harmonie, où, en répondant à la critique qu'en a faite ce docteur, et qu'il a insérée dans les affiches du pays chartrain, on fait voir que les disciples de d'Eslon peuvent être aussi instruits de la doctrine du magnétisme animal, que ceux de M. Mesmer et quelquefois mieux, 79
Lettre de l'auteur de monde primitif à messieurs ses souscripteurs sur le magnétisme animal, 29
Lettre de M. A. à M. B., sur le livre intitulé: Recherches & doutes sur le magnétisme animal de M. Thouret, 76

Title Index

Lettre de M. d'Eslon, docteur régent de la Faculté de Paris, et médecine ordinaire de Monseigneur le comte d'Artois, à M. Philip, docteur en médecine, doyen de la Faculté, 20
Lettre de M. L'abbé P. . . . de l'Académie de la Rochelle, à M. . . . de la meme Académie sur le magnétisme animal, 99
Lettre de M. le C de C P a M. le P E de S, 24
Lettre de M. le Marquis de, à un médecine de province, 21
Lettre de M. Mesmer à M. Le Cte de C (31 août 1784), 82
Lettre de M. Mesmer à M (Vicq-d'Azyr), Paris, 16 aout 1784, 83
Lettre de M. Valleton de Boissière, médecin à Bergerac, à M. Thouret, médecin à Paris, pour servir de réfutation à l'Extrait de la correspondance de la Société Royale de Médecine, relativement au magnétisme animal, 155
Lettre d'un Anglais à un Français sur la découverte du magnétisme animal et observations sur cette lettre, 64
Lettre d'un médecin de la faculté de Paris à M. Court de Gébelin, en réponse à celle que ce savant a adressée à ses souscripteurs et dans laquelle il fait un éloge triomphant du magnétisme animal, 60
Lettre d'un médecin de la Faculté de Paris à un médecin du College de Londres; ouvrage dans lequel on prouve contre M. Mesmer que le magnétisme animal n'existe pas, 15
Lettre d'un médecin de Paris à un médecin de province, 86
Lettre du père Hervier aux habitants de Bordeaux, 70
Lettres à M. Lélut sur la question du sommeil, du somnambulisme et des tables tournantes, à propos de son rapport au sujet du dernier concours ouvert à l'Académie des Sciences morales et politiques (section de philosophie), 753
Lettres de M.I.B.d.B., à M.P.L.G.H.D.L.S., à Marseille, sur l'existence du magnétisme animal et l'agent universel de la nature dont le Dr Mesmer se sert pour opérer ses guérisons . . . avec le moyen de se bien porter sans le secours du médecin, 41
Lettres de M. Mesmer, à M. Vicq-d'Azyr, et à messieurs les auteurs du Journal de Paris, 85
Lettres de M. Mesmer à messieurs les auteurs du Journal de Paris, et à M. Franklin, 84
Lettres d'un magnétiseur, 487
Lettres physiologiques et morales sur le magnétisme animal, contenant l'exposé critique des expériences les plus récentes, et une nouvelle théorie sur ses causes, ses phénomènes et ses applications à la médecine; addressées à M. Le Professeur Alibert, 328
Lettres pour servir de suite à l'essai sur la théorie du somnambulism magnétique, 173
Lettres sur le magnétisme, 102
Lettres sur le magnétisme animal, considéré sous le point de vue physiologique et psychologique, à M. le Dr. X, 553
Lettres sur le magnétisme animal; ou l'on discute l'ouvrage de M. Thouret, intitulé: Doutes et recherches . . . et le rapport des commissaires sur l'existence . . . , 42
Lettres sur le magnétisme et le somnambulisme, à l'occasion de Mademoiselle Pigeaire, 409
Lettres sur le magnétisme par le docteur de Boret, publiées pour la première fois par l'Union magnétique, 865
Lettre sur la découverte du magnétisme animal à M. Court de Gébelin, 71, 88, 275
Lettre sur la mort de Court de Gébelin, 77, 87
Lettre sur la seule explication satisfaisante des phénomènes du magnétisme animal et du somnambulisme déduite des vrais principes fondés dans la connaissance du créatur, de l'homme, et de la nature, et confirmée par l'expérience, 185–186
Lettre sur le magnétisme animal, où l'on examine la conformité des opinions des peuples anciens & modernes, des sçavans & notament de M. Bailly avec celles de M. Mesmer: et où l'on compare ces mêmes opinions au rapport des commissaires chargés par le roi de l'examen du magnétisme animal adressé à Monsieur Bailly de l'Académie des Science etc., 62
Lettre sur le magnétisme animal addressée à M. Perdriau, pasteur et professeur d'église et de l'académie de Genève, 89
Lettre sur le magnétisme animal adressée a M. . . . à Paris, 244
Lettre sur le secret de M. Mesmer ou réponse d'un médecin à un autre, qui avait demandé des eclaircissements à ce sujet, 25, 109

Lettre sur un fait relatif à l'histoire du magnétisme animal adressée à M. Philip, doyen de la Faculté de Médecine de Paris, 21, 23
La lévitation, 1428
Library of Mesmerism and Psychology in Two Volumes, Comprising Philosophy of Mesmerism, On Fascination, Electrical Psychology, The Macrocosm, Science of the Soul, 718
Life after Death. Problems of the Future Life and Its Nature, 1740
Life Beyond Death: Begin a Review of the World's Beliefs on the Subject, a Consideration of Present Conditions of Thought and Feeling, Leading to the Question as to Whether It can be Demonstrated as a Fact: to Which is Added An Appendix Containing Some Hints as to Personal Experiences and Opinions, 1463
Life's Borderland and Beyond, 1475
Light, 1795
Lights and Shadows of American Life, 395
Lights and Shadows of Spiritualism, 995
A List of a Few Cures Performed by Mr. and Mrs. De Loutherbourg of Hammersmith Terrace, Without Medicine, 189
Le livre des esprits contenant les principes de la doctrine spirite sur la nature des esprits, leur manifestation et leurs rapports avec les hommes. . . . Écrit sous la dictée et publié par l'ordre d'esprits supérieurs par Allan Kardec, 781, 796, 845
Lois physiques du magnétisme. Polarité humaine. Traité expérimental et thérapeutique de magnétisme, 1137–1138
London Medical and Physical Journal, 339
Luce e ombra. Revista mensile illustrata di scienze spiritualiste, 1491
Luce magnetica, 767
Lucidité et intuition; étude expérimentale, 1691, 1851
Lumière des morts, ou études magnétiques philosophiques et spiritualistes dédiées aux libres penseurs du XIXe siècle, 609
La lumière sur le magnétisme, ses défenses et ses ennemis, 1207
La lutte pour la métapsychique. Un chapitre passionnant d'histoire scientifique (1922–24). Les "jugements" de la Sorbonne et la question de la fraude chez les mediums, 1879

La maçonnerie mesmérienne, ou leçons prononcées par Fr. Mocet, Riala, Themola, Seca et Célaphon, de l'Ordre des F. de l'harmonie, en Loge mesmérienne de Bordeaux, l'an des influences 5.784, et le premier du mesmérisme, 34
The Macrocosm and Microcosm, 718
Madame Piper et la Société Anglo-Américaine pour les Recherches Psychiques, 1511
Mademoiselle Pigeaire, somnmbulisme et magnétisme animal, 398
Magic, Witchcraft, Animal Magnetism, Hypnotism, and Electro-Biology; Being a Digest of the Latest Views of the Author on these Subjects. Third Edition, Greatly Enlarged, Embracing Observations on J. C. Colquhoun's "History of Magic," &c., 631
A Magician Among the Spirits, 1883
The Magic Staff; an Autobiography of Andrew Jackson Davis, 774
Die Magie als experimentelle Naturwissenschaft, 1681
Die Magie als Naturwissenschaft, 1455
La magie aux dix-neuvième siècle, ses agents, ses verités ses mensonges. Precedée d'une letre adressée à l'auteur par le P. Ventura de Raulica, Ancien Générale de l'ordre des Théatins, Consulteur de la Sacrée Congregation des Rites, Examinateur des Evêques et du clergé romain, 824
La magie dévoilée, ou principes de science occulte, 633
Magie du XIXe siècle. Ténèbres. Treize nuits suivies d'un demi-jour sur l'hypnotisme, 830
Magie magnétique ou traité historique et pratique de fascination, miroirs cabalistiques, apports, suspensions, pactes, talismans, possession, envoutements, sortilèges, etc., 707
Magikon: Archiv für Beobachtungen aus dem Gebiete der Geisterkunde und des magnetischen und magischen Lebens. Nebst andern Zugeben fur Freunde des Innern, 348, 429
Das magische Geistesleben. Ein Beitrag zur Psychologie, 782

Magnes sive de arte magnetica, 674
Magnet, 371
The Magnet, 457, 499
La magnete e i nervosi: centuria di osservationi, 925
Magnetic and Cold Water Cure, 544
Magnetic Journal, 1446
Die magnetische Lehre der neuen Schule in Fragen und Antworten nach den Vorlesungen . . . von eim seiner Hörer, 528
Magnetische Magazin fur Nieder-Teutschland, 168, 222
Le magnétiseur, journal du magnétisme animal, 807
Le magnétiseur amoureux, par un membre de la société harmonique du régiment de Metz, 54, 175
Le magnétiseur praticien, 755
Magnétiseurs et médecins, 1259
Les magnétiseurs jugés par eux-mêmes nouvelle enquête sur le magnétisme animal. Ouvrage dédié aux classes letrées, aus médecins, à la magistrature et au clergé, 789
Les magnétiseurs ont-ils tort ou raison? That is the question! Appréciation et solution en deux parties, 798
Le magnétiseur spiritualiste, Journal rédigé par les membres de la société des magnétiseurs spiritualistes de Paris, 577, 708
Les magnétiseurs sont-ils sorciers? La France est-elle hérétique? Les memes hommes l'ont dit, 458
Le magnétiseur univérsel. Echo du monde magnétique, 897
Le magnétiseur universel, recueil des progrès spiritualistes ou études sur les manifestations du spiritualisme moderne, 870
Magnetism and its Healing Power, 838
Le magnétisme, 976
Magnétisme. Arcanes de la vie future devoilés, ou l'existence, la forme, les occupations de l'âme après sa séparation du corps sont prouvées par plusieurs années d'experiences au moyen de huit somnambules extatiques qui ont eu quatre-vingts perceptions de trente-six personnes de diverses conditions décédées à différentes époques, leur signalement, conversations, renseignement preuves irrécusables de leur existence au monde spirituel!, 559, 708
Magnétisme. Encyclopédie magnétique spiritualiste, traitant spécialement de faits psychologiques, magie magnétique, swedenborgianisme, écromancie, magie céleste, etc., 708
Magnétisme. Explication du phénomène du seconde vue et de soustraction de pensée, dont jouissent les somnambules lucides. Du magnétisme au point de vue de la thérapeutique. Marcillet, notice biographique, 551
Magnétisme. Insensibilité produite au moyen du sommeil magnétique. Nouvelle opération chirurgicale faite à Cherbourg, 545
Magnétisme, insensibilité produite au moyen du sommeil magnétique. Trois nouvelles opérations chirurgicales pratiquées à Cherbourg, le 4 juin 1847, en présence de plus de 60 temoins, 554
Le magnétisme. Journal des sciences magnétique, hypnotique et occultes . . . , 849
Magnétisme. L'âme de la terre et des tables parlantes, 713
Magnétisme. Le Christ qualifié de magnétiseur par la synagogue et l'incrédulité modernes et le magnétisme plaidant lui-même la cause du Christ, 557
Le magnétisme, le somnambulisme et le spiritualisme dans l'histoire. Affaire curieuse des possédées de Louviers. (Explications et rapprochements avec les faits actuels, avec les phénomènes produits par M. Home.), 792
Le magnétisme, le spiritisme, et l'eglise, 861
Magnétisme. M. Lafontaine et les sourds muets, 832
Magnétisme, moyen d'affermir le magnétisme dans la voie scientifique, 863
Magnétisme. Moyens magnétiques pour faire tourner les tables, les chapeaux, 675
Le Magnétisme. Revue générale des sciences physio-psychologiques, 1150
Magnétisme. Somnambulisme. Guide des incrédules, 640

Magnétisme, somnambulisme, hypnotisme, considérations nouvelles sur le système nerveux, ses fonctions et ses maladies, 819
Magnétisme, somnambulisme. Manuel élémentaire de l'aspirant magnétiseur, 668
Magnétisme, son histoire, sa théorie, son application au traitement des maladies, mémoire envoyé à l'Académie de Berlin, 362
Le magnétisme. Vérités et chimères de cette science occulte. Un drame dans le somnambulisme, épisode historique, 1077
Le magnétisme à Chateauroux, 506
Le magnétisme animal [Peronnet], 1092
Le magnétisme animal [Binet and Féré], 1162
Le magnétisme animal: à propos d'une visite à l'école de Nancy, 1228
Le "magnétisme animal"; étude critique et expérimental sur l'hypnotisme, ou sommeil nerveux provoqué chez les sujets sains, léthargie, catalepsie, somnambulisme, suggestions, etc., 1081
Le magnétisme animal. Étude historque et critique, 1241
Le magnétisme animal: étudié sous le nom de force neurique: rayonnante et circulante dans ses propriétés physiques, physiologiques et thérapeutiques, 1160
Magnétisme animal. Examen et réfutation du rapport fait par M.E.F. Dubois (d'Amiens) à l'Académie royale de médecine, le 8 aout 1837, sur le magnétisme animal, 392
Magnétisme animal. Mémoires et aphorismes de Mesmer suivis des procédés de D'Eslon. Nouvelle édition avec des notes, 502
Magnétisme animal, refus de l'Académie de médecine de constater le phénomène de la vision à travers les corps opaques, 442
Magnétisme animal, suggestion hypnotique et post-hypnotique. Son emploi comme agent thérapeutique, 1132
Le magnétisme animal à l'usage des gens du monde suivi de quelques lettres en opposition à ce mode de guérison, 333
Le magnétisme animal considéré comme moyen thérapeutique; son application au traitement de deux cas remarquables de névropathique, 484
Le magnétisme animal dévoilé par un zélé citoyen français, 73
Le magnétisme animal expliqué, ou Leçons analytiques sur la nature essentielle du magnétisme, sur ses effets, son histoire, ses applications, les diverses manières de la pratiquer, etc., 529
Le magnétisme animal fantaziéxoussique retrouvé dans l'antiquité . . . , 306
Le magnétisme animal opposé à la médecine, Mémoire pour servir à l'histoire du magnétisme en France et en Angleterre, 423
Le magnétisme animal retrouvé dans l'antiquité ou dissertation historique, étymologique et mythologique sur Esculape, Hippocrate et Galien, sur Apis, Sérapis ou Osisis et sur Isis suivie de recherches sur l'alchimie, 306
Le magnétisme appliqué à la médecine, 872
Le magnétisme catholique; ou, Introduction à la vraie pratique et réfutation des opinions de la médecine sur le magnétisme; ses principes, ses procédés et ses effects, 495
Le magnétisme curatif. Manuel technique, 1337
Le magnétisme devant la loi, 1265
Le magnétisme éclairé, ou introduction aux archives du magnétisme animal, 298, 303, 306
Magnétisme et crédulité ou solution naturelle du problème des tables tournantes, 651
Magnétisme et hypnotisme: Exposé des phénomènes observés pendant le sommeil nerveux provoqué au point de vue clinique, psychologique, therapeutique et médico-légal avec un résumé historique du magnétisme animal, 1131
Le magnétisme et le somnambulism devant les corps savant, le cour de Rome et les théologiens, 498
Le magnétisme et le somnambulisme du docteur Laurent; une somnambule maconnaise, 441
Magnétisme et magnéto-thérapie, 693
Magnétisme et somnambulisme, 646
Magnétisme et somnambulisme. Méthode nouvelle facile et pratique expliquant les principes réels du magnétisme, les moyens infaillables pour arriver promptement à bien magnétiser suivis de documents historiques et de nombreuses anecdotes, 1033

Le magnétisme expliqué par lui-même, ou Nouvelle théorie des phénomènes de l'état magnétique comparés aux phénomènes de l'état ordinaire, 742
Le magnétisme humain, l'hypnotisme, et le spiritualisme moderne considérés aux points de vue théorique et pratique, 1591
Le magnétisme humain considéré comme agent physique, 1262
Le magnétisme humain en cour de Rome, 492
Le magnétisme militant; origine et histoire des luttes, progrès et conquêtes de la science; le somnambulisme aux prises avec les corps savants etc. . . . résumé général des preuves historiques, philosophiques, scientifiques et juridiques, propres à démontrer l'existence du magnétisme et du somnambulisme lucide, etc. . . . , 622
Magnétisme organique. Le magnétisme à la recherche d'une position sociale; sa théorie, sa critique, sa pratique, 892
Le magnétisme triomphant: exposé historique et critique de la question, 1182
Magnétisme vital: contributions expérimentales à l'étude par le galvanomètre de l'électro-magnétisme vital; suivies d'inductions scientifiques et philosophiques, 1417
Il magnetismo, l'ipnotismo e lo spiritismo, ovvero Satana e la moderna magia, 1392
Magnetismo animal. Principios de magnetologia o methodo facil de aprender a magnetisar segundo os systemas de Mesmer, Puysegur, Deleuze, de Lausanne, Rostan, Brivasac, Ricard, Du Potet, Gauthier e Lafontaine, 1089
Il magnetismo animal considerato secondo le leggi della natura, 632
Il magnetismo animale, la fascinazione e qli stati ipnotici, 1151
Il magnetismo animale al cospetto dell-Associazione Midica degli Stati Sardi, 752
Il magnetismo animale considerato secondo le leggi della natura e principalmente diretto alla cura delle malattie. Con note ed un appendice sull ipnotismo, 825
Il magnetismo animale considerato sotto un nuovo punto di vista, 625
Magnetismo animale e sonnambulismo magnetico, 618
Il magnetismo animale svelato ossia teoria e pratica dell'antropo-elettromagnetismo ed una nuova meccanica delle sostanze, 612
Magnetismo ed ipnotismo, 1267, 1299
Magnetismo e ipnotismo, 1432
Magnetismo e ipnotismo svelati. Storia critica, 1281
La magnétismomanie, Comédie folie en un acte, melée de couplets, 266
Il magnetismo smascherato e svelato, 1336
Magnetismus, Somnambulismus, Clairvoyance. Zwolf Vorlesungen für Aerzte und gebildete Nichtärzte, 685
Der Magnetismus des Menschen, 736
Der Magnetismus gegen die Stieglitz-Hufelandische Schrift über den thierischen Magnetismus in seinum wahren Werth behauptet, 268
Der Magnetismus im Verhältnisse zur Natur und Religion, 454
Der Magnetismus nach der allseitiger Beziehung seines Wesens, geschichtlichen Entwickelung von allen Zeiten und bei allen Völkern wissenschaftlich dargestellt, 293
Der Magnetismus und die allgemeine Weltsprache, 312
Magnetismus und Hypnotismus, 1424
Magnetismus und Hypnotismus: eine Darstellung dieses Gebietes mit besonderer Berücksichtigung der Beziehungen zwischen dem mineralischen Magnetismus und dem sogenannten thierischen Magnetismus oder Hypnotismus, 1171
Magnetismus und Mesmerismus, oder, Physische und geistige Kräfte der Natur; der mineralische und thierische Magnetismus sowohl in seiner wirklichen Heilkraft, also in dem Missbrauch, der von Betrugern und Narren damit getrieben worden, im Zusammenhange mit der Geisterklopferei, der Tischruckerei, dem Spiritualismus, 700
Magnetismus und Mesmerismus oder physische und geistige Kräfte der Natur, 851
Der Magnetismus und seine Phänomene, 1327
Magnetizer's Magazine and Annals of Animal Magnetism, 261
Il magnetofilo (Società magnetica di Torino), 721
Magnetoid Currents, Their Forces and Directions; with a Description of the Magnetoscope; a Series of Experiments. To Which is Subjoined a Letter from William King, 623
Il magnetologo, 828

Les maisons hantées. En marge de la mort et son mystère, 1843
The Major Symptoms of Hysteria. Fifteen Lectures Given in the Medical School of Harvard University, 1589
Making of Man: a Study in Evolution, 1868
The Making of Religion, 1443
Les maladies de la personalité, 505, 1120
Les maladies épidemiques de l'esprit. Sorcellerie, magnétisme, morphinisme, délire des grandeurs, 1185
Man and His Relations: Illustrating the Influence of the Mind on the Body; the Relations of the Faculties to the Organs, and to the Elements, Objects and Phenomena of the External World, 866
Man and the Universe. A Study of the Influence of the Advance in Scientific Knowledge upon our Understanding of Christianity, 1614
Man is a Spirit: A Collection of Spontaneous Cases of Dream, Vision, and Ecstasy, 1738
Man's Survival after Death, or The Other Side of Life in the Light of Human Experience and Modern Research, 1638
Manuale del magnetismo animal. Sua storia—sue teorie—modo di magnetizzare—catechismo magnetologico—fenomeni magnetici—applicazione del magnetismo alla medicina—inconvenienti del magnetismo—conclusione—Appendice: Le tavole semoventi ovvero i miracoli del secolo XIX, 874
Manuale del magnetismo animale, desunto dalle più recenti opere magnetiche, 790
Manual of Psychometry: The Dawn of a New Civilization, 1103
Manuel bibliographiques des sciences psychiques ou occultes. Sciences des mages. Hermétique. Astrologie. Kabbale. Franc-Maçonnerie. Médecine ancienne. Mesmérisme. Sorcellerie. Singularités. Aberrations de tout ordre. Curiosités. Sources bibliographiques et documentaires sur ces sujets, etc., 922, 1673
Manuel de l'étudiant magnétiseur, ou Nouvelle instruction practique sur le magnétisme, fondée sur 30 années d'observation; suivi de la 4e éd. des expériences faites en 1820 à l'Hôtel-Dieu de Paris, 534
Manuel du magnétiseur: explication physiologique des phénomènes magnétiques, utilité du somnanbulisme dans l'exercice de la médecine, 741
Manuel historique élémentaire et pratique de magnétisme animal, contenant les principes généraux de l'art magnétique, l'explication des divers phénomènes qui s'y rattachent, la description des symptomes des principales maladies chroniques, leurs causes determinantes et les procédés reconnus les plus convenables à leur guérison au moyen du magnétisme par E. Berjot, membre titulaire de la Société du mesmérisme de Paris, Suivi d'une dissertation sur le fluide magnétique animal par A. Bauche, membre titulaire de la même société, 787
Manuel pratique de magnétisme animal. Exposition méthodique des procédés employés pour produire les phénomènes magnétiques et leur application à l'étude et au traitement des maladies, 432
Le maraveglie dell'ipnotismo. Sommario dei principali fenomeni del sonnambulismo provocato, e metodi di sperimentazione, 1179
Margery Harvard Veritas. A Study in Psychics, 1904
"Margery" the Medium, 1887
Les marveilles du magnétisme et les mystères des tables tournantes et parlantes, 760
Mary Jane; or Spiritualism Chemically Explained with Spirit Drawings: also Essays by, and Ideas (Perhaps Erroneous) of "A Child at School," 857
Mary Reynolds: A Case of Double Consciousness [Plumer], 834
Mary Reynolds: a case of double consciousness [Mitchell], 1214
Materialisations-Phänomene: ein Beitrag zur Erforschung der Mediumistischen Teleplastie, 1700, 1704
Materialisation und Teleplastie, 1831
Materialized Apparitions: If not Beings from Another Life, What are They?, 1129
La matrise de soi-même par l'autosuggestion consciente, 1794, 1821
Le mécanisme de la survie. Explication scientifique des phénomènes métapsychiques, 1853
The Mechanism of Man, 951

Title Index

La médecine psychologique, 1847
Médecine vitale. Réhabilitation du magnétiseur Mesmer, son baquet, sa doctrine, ses luttes et son triomphe, 891
La medianité: rivista mensile di scienze psico-fisiche e morali, 1527
Les médiateurs et les moyens de la magie, les hallucinations et les savants; la fantôme humain et le principe vital, 856
Medical Notes and Reflexions, 410
Medical Psychology and Psychical Research, 1832
Medical Report of the Case of Miss H——— M———, 514
Medical Repository, 262
Medical Times and Gazette, 450
Les médications psychologiques; études historiques, psychologique et cliniques sur les méthodes de la psychothérapie, 1777
Medicinische Philosophie und Mesmerismus, 817
Das medium D. D. Home. Untersuchung und Beobachtungen (nach Crookes, Butlerow, Varley, Aksakow and Lord Dunraven). Ausegewahlt und herausgegeben von Rudolf Tischner, 1905
Le médiumisme et la Sorbonne, 1839
Mediumistische Kunst. Mit einem Beitrag über den kunstlerischen Wert mediumistischer Malereien von Eugen Johannes Maecker, 1697
Médiumnité quérissante par l'application des fluides éléctrique, magnétique et humain, 1447
Mediums Unmasked. An Exposé of Modern Spiritualism. By an Ex-Medium, 1315
Mélanges de philosophie critique, 955
La mémoire chez les hypnotisés, 1133
Mémoire de F. A. Mesmer, docteur en médecine, sur ses découvertes, 211, 502
Mémoire en réponse au rapport de MM. les Commissaires chargés par le roi de l'examen du magnétisme animal, 65
Mémoire physique et médecinal, montrant des rapports évidens entre les phénomènes de la baguette divinatoire, du magnétisme et de l'électricité, avec des éclaircissements sur d'autres objects non moins importants, qui y sont relatifs, 18, 117
Mémoire pour M. Charles Louis Varnier . . . appellant d'un décret de la Faculté; contre les doyen et docteurs de ladite Faculté, intimés, 140
Mémoire pour servir a l'histoire de la jonglerie dans lequel on démontre les phénomènes du mesmerisme. Nouvelle édition précédée d'une letre sur le secret de M. Mesmer . . . . On y a joint une réponse au Mémoire qui paroit ici pour la première fois, 109
Mémoires [of the Société royale de médecine], 1779, 26
Mémoires de l'Académie des sciences, belle-lettres et arts de Lyon. Sciences et lettres, 1857
Mémoires d'un magnétiseur: suivis de l'examen phrénologique de l'auteur, par le Docteur Castle, 894
Mémoires d'un magnétiseur contenant la biographie de la somnambule Prudence Bernard, 620
Mémoires of Extraordinary Popular Delusions and the Madness of Crowds, 647
Mémoires pour servir à l'histoire et à l'établissement du magnétisme animal, 47, 105, 148, 1184
Mémoire sur la découverte des phénomènes que présentent la catalepsie et le somnambulisme, symptomes de l'affection hystérique essentielle, avec des recherches sur la cause physique des ces phénomènes. Premiere partie. Mémoire sur la découverte des phénomènes de l'affection hystérique essentielle, et sur la méthode curative de cette maladie. Second partie, 171, 225
Mémoire sur la découverte du magnétisme animal, 8, 10, 26, 87, 211, 238, 502, 1591
Mémoire sur la faculté de prévision: suivi des notes et pièces justificatives recueillis par M. Mialle, 358
Mémoire sur la vie de l'abbé de Faria. Explication de la charmante légende du chateau d'If dans le roman "Monte-Cristo," 1571, 1902
Mémoire sur l'électricité médicale, 80
Mémoire sur le magnétisme animal, présenté à l'Académie de Berlin, en 1818, 285

Mémoire sur le magnétisme animal adressé à MM. les membres de l'Académie des sciences et de l'Académie de médecine, 321, 329, 355
Mémoire sur le magnétisme animal et sur son application au traitement des maladies mentales, lu au Congrès scientifique de Poitiers, le 11 septembre 1834, 364
Mémoire sur le sommeil, les songes et le somnambulisme, 645, 848
Mémoire sur le somnambulisme et le magnétisme animal adressé en 1820 à l'Académie royale de Berlin, 726
Mémoire sur un cas d'hystérie, traité par le magnétisme animal, 449
Memoranda of Persons, Places, and Events; Embracing Authentic Facts, Visions, Impressions, Discoveries, in Magnetism, Clairvoyance, Spiritualism: also Quotations from the Opposition. With an Appendix, Containing Zschokke's Great Story of "Hortensia," Vividly Portraying the Wide Difference Between the Ordinary State and that of Clairvoyance, 909
Mental Magic. A Rationale of Thought Reading, and Its Attendant Phenomena, and Their Application to the Discovery of New Medicines, Obscure Diseases, Correct Delineations of Character, Lost Persons and Property, Mines and Springs of Water, and All Hidden and Secret Things: To Which is Added the History and Mystery of the Magic Mirror, 1100
The Mental State of Hystericals. A Study of Mental Stigmata and Mental Accidents, 1318
Mental Suggestion, 1184
Ein merkwürdiger Fall von Fascination, 1389
Le merveilleux dans le Jansénisme, le magnétisme, le Méthodisme et le Baptisme américain; l'épidemie de Morzine, le spiritisme. Recherches nouvelles, 881
Le merveilleux et la science, étude sur l'hypnotisme, 1044
Le merveilleux scientifique, 1363
Mesmer & Swedenborg; or, The Relation of the Developments of Mesmerism to the Doctrines and Disclosures of Swedenborg, 549
Mesmer blessé ou réponse à la lettre du P. Hervier sur le magnétisme animal, 95
Mesmer et le magnétisme animal, 629
Le mesmér et le somnambulisme à Lyon avant la Revolution, 1857
Mesmer guéri ou Lettre d'un provincial au R.P.N. . . . , en réponse à sa lettre intitulée Mesmer blessé, 88
La Mesmériade, ou le triomphe du magnétisme animal. Poème en 3 chants dédié à la lune, 100
Mesmeric Experiences, 517
Mesmeric Magazine; or Journal of Animal Magnetism, 459, 469
The Mesmeric Mania of 1851, with a Physiological Explanation of the Phenomena Produced: a Lecture, 606
Mesmerism, 1, 5, 10, 211
Mesmerism, a Letter to Miss Martineau, 555
Mesmerism, Clairvoyance, and Animal Magnetism Explained: Also a Treatise on Mesmerism from the Earliest Ages, including the Life and Death of Mesmer, the Founder of the Above Science; with Instructions to Gentlemen Wishing to Introduce the New Science of Electro-biology. Including G. B. Eagle's Hand-book of magic . . . . One Hundred Beautiful Illusions are Comprehensively Described . . . , 634
Mesmerism, its History, Phenomena, and Practice, with Reports of Cases Developed in Scotland, 478
Mesmerism; Its Pretensions & Effects upon Society Considered, 466
Mesmerism, or, The New School of Arts, with Cases in Point, 504
Mesmerism, or Animal Magnetism, and Its Uses; with Particular Directions for Employing It in Removing Pains and Curing Diseases, in Producing Insensibility to Pain in Surgical and Dental Operations; and in the Examination of Internal Diseases, with Cases of Operations, Examinations and Cures, 476
Mesmerism, Spitirualism, Witchcraft, and Miracle: A Brief Treatise Stating that Mesmerism Is a Key Which Will Unlock Many Miracles and Mysteries, 793
Mesmerism, with Hints for Beginners, 1021
Mesmerism & Hypnotism. An Epitome of all the Best Works on the Hypnotic Phases of Psychology in the Form of Question and Answer. By an Adept, 1471
Mesmerism and Christian Science: a Short History of Mental Healing, 1635

Mesmerism and Its Opponents: with a Narrative of Cases, 503
Mesmerism and its Realities Further Proved by Illustrations of its Curative Powers in Disease as well as by its Development of Some Extraordinary Magnetic Phenomena in the Human Body, 784
Mesmerism and Media, with Full Instructions How to Develop the Alleged Spiritual Rappings in Every Family, 748
Mesmerism and Phreno-Mesmerism. Consisting of Modes of Mesmerizing, 480
Mesmerism and Spiritual Agency, 638
Mesmerism and Spiritualism, &c. Historically and Scientifically Considered. Being Two Lectures Delivered at the London Institution, 993
Mesmerism Considered, 649
Mesmérisme à l'aide d'un bassin et d'un ventilateur inventé par F. Léquine, 599
Mesmerism in India, and Its Practical Application in Surgery and Medicine, 536
Mesmerism in Its Relation to Health and Disease and the Present State of Medicine, 750
Mesmerism or Animal Magnetism, Hypnotism and Thought Reading, 1278
Mesmerism Proved True, and the Quarterly Reviewer Reviewed, 734
Mesmerism the Gift of God: in Reply to "Satanic Agency and Mesmerism," a Sermon Said to Have Been Preached by the Rev. Hugh M'Neile: in a Letter to a Friend by a Beneficed Clergyman, 481
Mesmerism Tried by the Touch-stone of Truth: Being a Reply to Dr. Ashburner's Remarks on Phrenology, Mesmerism, and Clairvoyance, 560
Mesmerism True—Mesmerism False: A Critical Examination of the Facts, Claims, and Pretentions of Animal Magnetism. With an Appendix Containing a Report of Two Exhibitions by Alexis, 512
Mesmerismus. Oder System der Wechselwirkungen, Theorie und Anwendung des thierischen Magnetismus als die allgemeine Heilkunde zur Erhaltung des Menschen. Herausgegeben von D. Karl Christian Wolfart, 250, 256, 260
Mesmerismus und Belletristik in ihren schadlichen Einflussen auf die Psychiatrie, 405
The Mesmerist. A Weekly Journal of Vital Magnetism, 482
Il mesmerista, 722
The Mesmerist's Manual of Phenomena and Practice; with Directions for Applying Mesmerism to the Cure of Diseases, and the Methods of Producing Mesmeric Phenomena. Intended for Domestic Use and the Instruction of Beginners, 588
Mesmer justifié, 96
Mes morts, leurs manifestations, leurs influences, leurs télépathies, 1600
Métallothérapie; traitement des maladies nerveuses, paralysies, rhumatisme chronique, etc. . . . du choléra, etc. . . . , 658
Metapsychica moderna: fenomeni medianici e problemi del subcosciente, 1850
Métapsychique et psychologie, 1263
Méthode de dédoublement personnel (Extérioration de la neuricité. Sorties en astral), 1689
The Mighty Curative Powers of Mesmerism: Proved in Upwards of One Hundred and Fifty Cases of Various Diseases, 611
Mind, 1173
Mind and Body, 1712
The Mind and Its Place in Nature, 1888
Mind and Matter, 1007
Mind in Nature: a Popular Journal of Psychical, Medical and Scientific Information, 1117
Mind-Reading and Beyond, 1115
A mind that Found Itself: An Autobiography. With Introductory letters by William James, 1603
Les miracles de Mesmer, 14
The Missing Link in Modern Spiritualism. Revised and Arranged by a Literary Friend, 1122
I misteri del moderno spiritismo e l'antidoto contro le superstizioni del secolo XIX, 903
Modern American Spiritualism. A Twenty Year's Record of the Communion Between Earth and the World of Spirits, 933, 1063
Moderne Mediumforschung. Kritische Betrachtungen zu Dr. von Schrenck-Notzing's "Materialisationsphaenomene," 1700

Moderne Probleme, 1112
Der moderne Spuk-u. Geisterglaube. Eine Kritik und Erklärung der spiritistischen Phänomene. II. Teil des Werkes "Wunder und Wissenschaft." Mit einem Vorwort von Dr. Max Dessoir, 1573
Modern Mysteries Explained and Exposed. In Four Parts. I. Clairvoyant Revelations of A. J. Davis. II. Phenomena of Spiritualism Explained and Exposed. III. Evidence that the Bible is Given by Inspiration of the Spirit of God, as Compared with Evidence that These manifestations are from the Spirits of Men. IV. Clairvoyant Revelations of Emanuel Swedenborg, 746
The Modern Mystery: or, Table-tapping, Its History, Philosophy, and General Attributes, 720
Modern Mystics and Modern Magic; Containing a Full Biography of the Rev. William Stainton Moses, etc., 1367
A Modern Priestess of Isis, 1394
Modern Psychical Phenomena: Recent Researches and Speculations, 1749
Modern Spiritism: A Critical Examination of Its Phenomena, Character, and Teaching in the Light of Known Facts, 1542
Modern Spiritualism, a History and a Criticism, 1508
Modern Spiritualism: a Short Account of Its Rise and Progress, with Some Exposures of So-called Spirit Media, 991
Modern Spiritualism: Its Facts and Fanaticisms, Its Consistencies and Contradictions, 739
Modern Theories of the Unconscious, 1870
Molly Fancher, the Brooklyn Enigma. An Authentic Statement of Facts in the Life of Mary J. Fancher, the Psychological Marvel of the Nineteenth Century. Unimpeachable Testimony of Many Witnesses, 1361
Le monde occulte, ou mystères du magnétisme et tableau de somnambulisme, 594
Monde primitif, 29
Monismus und Okkultismus, 1813
La moral chrétienne vengée, ou réflexions sur les crimes commis sous les prétextes spécieux de la gloire de Dieu et des intérêts de la religion et observations historiques et philosophiques sur les faux miracles opérés par le magnétisme animal, 305
Le moraliste mesmérien ou lettres philosophiques sur l'influence du magnétisme, 110
Il morbo-cholera curabile col magnetismo: memoria, 715
More Ghost Stories: A Sequel to Real Ghost Stories, 1307, 1331
Mors Janua Vitae? A Discussion of Certain Communications Purporting to Come From Frederic W. H. Myers, 1641
La mort et son mystère, 1771
Les morts vivent-ils? Enquête sur l'état présent des sciences psychiques, 1800
Un mot à l'oreille des académiciens de Paris, 45
Motive Power of Organic Life, and Magnetic Phenomena of Terrestrial and Planetary Motions, with the Application of the Ever-active and All-pervading Agency of Magnetism to the Nature, Symptoms and Treatment of Chronic Diseases, 445
Mr. Slades Aufenthalt in Wien, 1006
Multiple Personality: an Experimental Investigation into the Nature of Human Individuality, 1547
Multiplex Personality, 1152
Musica trascendentale, 1818
My Father: Personal and Spiritual Reminiscences, 1653
My Life: a Record of Events and Opinions, 1566
My Psychic Adventures, 1838
Le mystère de la danse des tables dévoilé par ses rapports avec les manifestations spirituelles d'Amérique par un Catholique, 680
Le mystère des magnétiseurs et des somnambules, dévoilé aux âmes droites et vertueuses, 255, 265
Les mystères de la science, 1168
Les mystères du magnétisme animal et de la magie dévoilés ou la vérité sur le mesmérisme, le somnambulisme dit magnétique et plusieurs phénomènes attribués à l'intervention des esprits démontrés par l'hypnotisme, 823

Les mystères du sommeil et du magnétisme, 471
Mysteria. Revue mensuelle illustrée d'études, 1690
Mysteria. Revue mensuelle industrée d'études initiatiques, 1209
Mysterien des innern Lebens; erläutert aus der Geschichte der Scherin von Prevorst, mit Berücksichtigung der bisher erschienenen Kritiken, 343
Die Mysterien Schlafes und Magnetismus oder Physik un Physiologie des magnetischen Somnambulismus. Eine auf naturwissenschaftliche Principien guestutzte rationelle Erklärung der Phänomene der Schlafes und Traumes, der Ekstase und Gehergabe, der Hallucinationen und Visionen, der electrobiologischen Erscheinungen der Bewegung unbelebter Korper u.v., durch Zuruckfuhrung auf ihre naturlichen Urfachen. Nach Deban, Carpenter v. A., sowie nach eignen Beobachtungen, 762
The Mysteries of Human Nature Explained by a New System of Nervous Physiology: to Which is Added, a Review of the Errors of Spiritualism, and Instructions for developing or Resisting the Influence by Which Subjects and Mediums are made, 775
The Mysteries of the Head and the Heart Explained: Including an Improved System of Phrenology; a New Theory of the Emotions, and an Explanation of the Mysteries of Mesmerism, Trance, Mind-reading, and the Spirit Delusion, 974
Mysteries of the Seance and Tricks and Traps of Bogus Mediums. A Plea for Honest Mediums and Clean Work, 1524
Mysteries of the Vital Element in Connection with Dreams, Somnambulism, Trance, Vital Photography, Faith and Will, Anaesthesia, Nervous Congestion and Creative Function. Modern Spiritualism Explained, 937
Mysterious Psychic Forces, 1586
The Mystery of Animal Magnetism Revealed to the World, Containing Philosophical Reflections on the Publication of a Pamphlet Entitled, A True and Genuine Discovery of Animal Electricity and Magnetism: also, an Exhibition of the Advantages and Disadvantages that may Arise in Consequence of Said Publication, 198
Mystic London: or, Phases of Occult Life in the Metropolis, 972
Die mystischen Erscheinungen der menschlichen Natur, 844

Nachricht von dem Vorkommen des Besessenseyns eines dämonisch-magnetischen Leidens und seiner schon im Alterthum bekannten Heilung durch magisch-magnetisches Einwirken, in einem Sendschreiben an den Herrn Obermedicinalrath Dr. Schelling in Stuttgart, 376
Natural and Mesmeric Clairvoyance with the Practical Application of Mesmerism in Surgery and Medicine, 637
The Naturalization of the Supernatural, 1618
The Natural Order of Spirit; a Psychic Study and Experience, 1707
Natur-Analogien, oder die vornehmsten Erscheinungen des animalischen Magnetismus in ihrem Zusammenhange mit den Ergebnissen der gesammten Naturwissenschaften, mit besonderer Hinsicht auf die Standpunkte und Bedurfnisse heutiger Theologie, 414
Nature and Man, Essays Scientific and Philosophical. With an Introductory Memoire by J. Estlin Carpenter, 1197
The Nature of Mind and Human Automatism, 1119
Die naturgeschichte der Gespenster. Physikalisch-physiologisch-psychologische Studien, 860
Neue Gespenster kurze Erzählungen aus dem Reiche der Wahrheit, 219
Der Neuere Spiritismus, 1038
Der neuere Spiritualismus, sein Werth und seine Tauschungen. Eine anthropologische Studie, 1005
Neuer gelehrter Mercurius, 5–6
Neues Archiv für den thierischen Magnetismus und das Nachtleben überhaupt, 300
Neuro-hypnotism, 450
Neurypnology or The rationale of nervous sleep considered in relation with animal magnetism. Illustrated by numerous cases of its successful application in the relief and cure of disease, 450, 465, 1453, 1585

TITLE INDEX

La névrose hypnotique, ou Le magnétisme dévoilé. Étude de physiologie pathologique sur le système nerveux, 1041
Les Névroses, 1629
Névroses et idées fixes. I. Études expérimentales sur les troubles de la volonté, d'attention, de la mémoire, sur les émotions, les idées obsédantes et leur traitement, par Dr Pierre Janet. II. Fragments des leçons cliniques du mardi sur les névroses, les maladies produits par les émotions, les idées obsédantes et leur traitement. Par Dr F. Raymond et le Dr Pierre Janet, 1441, 1629
The New Black Magic and the Truth about the Ouija-Board, 1759
Newer Spiritualism, 1649
New Evidences in Psychical Research: A Record of Investigations, with Selected Examples of Recent S.P.R. Results, 1663
New Light on Immortality, 1609
The New Light on Immortality, or The Significance of Psychical Research, 1809
The New Psychic Studies in their Relations to Christian Thought, 1177
The New Revelation, 1725, 1753
The New Spiritualism, 1401
New Thought, 1409, 1494
A New View of Insanity. The Duality of the Mind Proved by the Structure, Functions, and Diseases of the Brain, and by the Phenomena of Mental Derangement, and Shown to be Essential to Moral Responsibility, 505
New York Magnet, 457, 499
The Night Side of Nature: or, Ghosts and Ghost Seers, 561
The Nineteenth Century, 1252, 1301, 1350, 1525
Nineteenth Century Miracles; or, Spirits and Their Work in Every Country of the Earth. A Complete Historical Compendium of the Great Movement Known as "Modern Spiritualism," 1063
Das normale Bewusstsein, 1423
Die Nornen, Forschungen über Fernschen in Raum und Zeit, 1622
Note on a Suggested Mode of Psychical Interaction, 1144
A Note on the Unconscious in Psycho-Analysis, 1678
Notes and Studies in the Philosophy of Animal Magnetism and Spiritualism. With Observations upon Catarrh, Bronchitis, Rheumatism, Gout, Scrofula, and Cognate Diseases, 901
Note sur le magnétisme animal et sur les dangers que font courir les magnétiseurs à leurs patients, 263
Note sur le magnétisme et sur l'homéopathie, ou réponse à tout ce qui été imprimé dans les journaux de Nantes contre le magnétisme et contre l'homéopathie, 439
Note sur le sommeil nerveux ou hypnotisme, 812
Note sur quelques phénomènes de somnambulisme, 1148
Nothgedrungene Fortsetzung meines Nothschrei gegen meine Vergiftung mit concentrirtem Lebensäther und gründliche Erklärung der maskirten Einworkungsweise desselben auf Geist und Körper zum Scheinleben, 904
Nothschrei eines Magnetisch-Vergifteten; Thatbestand, erklärt durch ungeschminkte Beschreibung des 36 jährigen Hergangs, belegt mit allen Beweisen und Zeugnissen.Zur Belehrung und Warnung besonders für Familienvater und Geschäftsleute, 644, 904
Notice historique sur les systèmes et les écrits anciens qui se rapportent au magnétisme animal, 131
Notice sur le magnétisme ou manière de se magnétiser soi-meme, 435
Nouveau mécanisme de l'électricité fondé sur les lois de l'équilibre et du mouvement, démontré par des expériences que renversent le système de l'électricité positive et négative et qui etablissent ses rapports avec le mécanisme caché de l'aimant et l'heureuse influence du fluid électrique dans les affections nerveuses, 221
Nouveaux documents satiriques sur Mesmer, 1291
Nouveaux extraits des journaux d'un magnétiseur depuis 1786 jusqu'au mois d'avril 1788, 183
Nouvel almanach du magnétiseur, 756
Le nouvel hypnotisme, 1183

Nouvelle considérations sur le sommeil, les songes et le somnambulisme; mémoire posthume de M. Maine de Biran, lu à l'Acedémie des sciences morales et politiques, le 31 mai 1834, par M. V. Cousin, 363
Nouvelle découverte sur le magnétisme animal; ou, Lettre addressée à un ami de province, par un partisan zélé de la verité, 146
Nouvelles cures opérées par le magnétisme animal [Prony], 104
Nouvelles cures opérées par le magnétisme animal [Tissart de Rouvres], 118
Nouvelles observations sur un cas de somnambulisme avec glossolalie, 1503
Numerous Cases of Surgical Operations without Pain in the Mesmeric State: with Remarks upon the Opposition of Many Members of the Royal Medical and Chirurgical Society and Others to the Reception of the Inestimable Blessings of Mesmerism, 474, 536
Nouvi studi sull'ipnotismo e sulla credulità, 1238

Observation concernant une jeune fille de dix-sept ans amputée d'une jambe à Cherbourg le 2 octobre 1845, pendant le sommeil magnétique, 520
Observation de magnétisme occulte, 430
Observations adressées à messieurs les commissaires de la Société royale de médecine, nommés par le Roi . . . Sur la manière dont ils on procédé, et sur le rapport q'ils en ont fait. Par un médecin de P———. Pour servir de suite à à celles qui ont été adressées sur le même objet à MM. les commissaires tirés de la Faculté de médecine & de l'Académie royale des science de Paris, 68
Observations adressées à Mrs. les commissaires chargés par le Roi de l'examen du magnétisme animal; sur la maniére dont ils y ont procédé, & sur leur rapport. Par un médecin de province, 67
Observations adressées aux médecins qui désireraient établir un traitement magnétique, 301
Observations de l'auteur de l'Esquisse de la nature humaine sur l'article magnétisme animal, inséré dans le 13e volume du dictionnaire de médecine par le dr Rostan, 332
Observations de M. Bergasse sur un écrit du docteur Mesmer, ayant pour titre: Lettre de l'inventeur du magnétisme animal à l'auteur des Reflexions préliminaires, 123
Observations et recherches sur l'usage de l'aimant en médecine; ou Mémoire sur le magnétisme médicinal, 26
Observations of Animal Electricity, In Explanation of the Metallic Operation of Dr. Perkins, 207
Observations on the Principal Medical Institutions and Practice of France, Italy and Germany; with Notices of the Universities, and Cases from Hospital Practice. To Which is Added, An Appendix, on Animal Magnetism and Homeopathy, 384
Observations on Trance: or, Human Hybernation, 591
Observations relatives à la lettre de M. Friedlander, sur l'état actuel du magnétisme en Allemagne, 277
Observations sur le fluide organo-électrique et sur les mouvements électrométriques des baguettes et des pendules, 725
Observations sur le magnétisme animal, 12, 14
Observations sur le rapport des commissaires chargés par le Roi de l'examen du magnétisme animal. Par M.G.C. Membre de diverses académies, 91
Observations sur les deux rapports de MM. les commissaires nommés par sa majesté pour l'examen du magnétisme animal, 52
Observation trés-importante sur les effets du magnétisme animal, 19
Les obsessions et la psychasthénie. I. Études cliniques et expérimentales sur les idées obsédantes, les impulsions, les manies mentales, la folie du doute, les tics, les agitations, les phobies, les délires du contact, les angoisses, les sentiments d'incomplétude, la neurasthénie, les modifications des sentiment du réel, leur pathologie et leur traitement. Par le Dr Pierre Janet. II. Fragments des leçons cliniques du mardi sur les états neurasthéniques, les aboulies, les sentiments d'incomplétude, les agitations et les angoisses diffuses, les algies, les phobies, les délires du contact, les tics, les manies mentales, les folies du doute, les idées obsédantes, les impulsions, leur pathogenie et leur traitement. Par le Dr F. Raymond et le Dr Pierre Janet, 1522, 1629

Occultism and Modern Science, 1807
L'occultisme hier et aujourd'hui, le merveilleux prescientifique, 1588
Odische Begebenheiten zu Berlin in den Jahren 1861 und 1862, 850
Odische Erwiederungen an die Herren Professoren Fortlage, Schleiden, Fechner und Hofrath Carus, 769
Die odische Lohe und einige Bewegungserscheinungen als neuentdeckte Formen des odischen Princips in der Natur, 905
Odisch magnetische Briefe, 648
Of Spiritism: i.e. Hypnotic Telepathy and Phantasms—Their Danger, 1685
Of the Imagination, as a Cause and as a Cure of Disorders of the Body: Exemplified by Fictitious Tractors, and Epidemic Convulsions, 214
Das Okkulte, 1848
Die okkulten Tatsachen und die neuesten Medienentlarvungen. Eine Entgegung auf die letzten Verstösse der Verachter der Parapsychologie, 1900
Der Okkultismus im modernen Weltbild, 1807
Der Okkultismus in Urkunde, 1885, 1895
Okkultismus und Spiritismus im Lichte der Wissenschaft und des katholischen Glaubens . . . , 1802
Om-Hypnotismens anvandande i den praktiska medicinen, 1220
On Animal Magnetism, or Vital Induction, 808
On a Series of Automatic Writings, 1578
On a Series of Sittings with Mrs. Osborne Leonard, 1745
On a Telepathic Explanation of certain so-called Spiritualistic Phenomena, 1118
On a Telepathic Explanation of Some So-called Spiritualistic Phenomena, 1090
On Double Consciousness. Experimental Psychological Studies, 1255, 1311
On Memories Which are Too Real, 1891
On Mesmerism, Improperly Denominated Animal Magnetism, 339
On Miracles and Modern Spiritualism. Three Essays, 979
On Telepathic Hypnotism, and its Relation to Other Forms of Hypnotic Suggestion, 1153
On the Cosmic Relations, 1699
On the Psychology and Pathology of So-called Occult Phenomena, 1506
On the Threshold of a New World of Thought: An Examination of the Phenomena of Spirutualism, 1601
On the Threshold of the Unseen: An Examination of the Phenomena of Spiritualism and the Evidence for Survival after Death, 1601
On the Treatment of Insanity by Hypnotism, 1280
The Open Court, 1255
Open Vision: a Study of Psychic Phenomena, 1770
Opinion prononcée par M. Husson à l'Académie de médecine séance du 22 Aout 1837 sur le rapport de M. Dubois (d'Amiens) relatif au magnétisme animal, 383
Oraison funèbre du célèbre Mesmer, auteur du magnétisme animal et président de la Loge de l'Harmonie, 134
Organ du progrès philosophique et religieux, 795
Organe de l'Institut magnétologique de Paris et New York . . . , 897
L'origine dei fenomeni psichici e loro significazione biologica, 1121
Les origines de la doctrine du magnétisme animal: Mesmer et la Société de l'Harmonie, 1445
Orthodoxie magnétique. Catéchisme raisonné de l'aspirant magnétiseur, suivi d'un simple coup d'oeil sur le triple électro-galvanique et du pilori du magnétisme, 743
Our American Adventure, 1841
Our Hidden Forces ("La Psychologie Inconnue"). An Experimental Study of the Psychic Sciences, 1605
Our Second American Adventure, 1842
Outlines of Lectures on the Neurological System of Anthropology, as Discovered, Demonstrated and Taught in 1841 and 1842, 705
Outrages à la pudeur. Violences sur les organes sexuels de la femme dans le somnambulisme provoqué et la fascination. Étude médico-légale, 1368

O Padre Faria na Historia do Hipnotisma, 1902
Une page nouvelle de magnétisme. Sorcier malgré lui, 987
Pamphlet: I. Des origines de la métallothérapie, part qui doit être faite au magnétisme animal dans sa découverte. Le Burguisme et le Perkinisme, 1016
Parallèle entre le magnétisme animal, l'électricité et les bains médicinaux par distillation, &c. appliqués aux maladies rebelles. On a joint à ce précis l'art de conserver la santé, & de guérir les maladies le plus rebelles . . . , 143
Parapsychologische Erkenntnesse, 1897
Parerga und Paralipomena. Kleine philosophische Schriften, 603
Past and Present with Mrs. Piper, 1811
Pathetism: with Practical Instructions: Demonstrating the Falsity of the Hitherto Prevalent Assumptions in Regard to What has been Called "Mesmerism" and "Neurology," and Illustrating Those Laws which Induce Somnambulism, Second Sight, Sleep, Dreaming, Trance, and Clairvoyance, with Numerous Facts Tending to Show the Pathology of Monomania, Insanity, Witchcraft, and Various Other Mental or Nervous Phenomena, 489
Patience Worth: a Psychic Mystery, 1718
Patologia de las enfermadades epilépticas y mentales, con un estudio del hombre en su modo de ser fisico moral. Tratado de psico-terapia practica y racional, en el que se determinan las leyes que anteceden a los fenomenos del sonambulismo, hipnotismo, sugestion y su relativa afinidad con las enfermedades morales, 1549
Peculiarities of Certain Post-hypnotic States, 1142
Pensée inconsciente et vision de la pensée. Essai d'une explication physiologique du processus de la pensée et de quelques phénomènes surnaturels et psychopathiques, 1567
Per lo spiritismo, 1312
Personal Experiences in Spiritualism (Including the Official Account and Record of the American Palladino Séances) 1684
La personalité et la mémoire dans le somnambulisme, 1071
Personality and Telepathy, 1660
Personal Recollections of the Little Tew Ghost, Reviewed in Connection with the Lancashire Bogie, and the Table-Talking and Spirit-Rapping of the Present Day, 717
Petit catechisme magnétique ou notions élémentaires du mesmérisme, 642
Petit traité sur le magnétisme animal, contenant l. un précis historique sur la matière; 2. une dissertation succincte sur les influences occultes qui dominent l'homme; 3. une courte appréciation de l'avenir du magnétisme; 4. une notice sur la sibylle moderne et sur ses facultés somnambuliques; 5. des preuves positives d'une lucidité prodigieuse et incontestable; 6. des conseils à ceux qui veulent la consulter avec fruit, etc., 601
Das Pferd des herrn von Osten (der kluge Hans), ein experimentellen tier-und Menschen-psychologie, 1592, 1728
Die Pflanzenwelt in ihren Beziehungen zur Sensitivitat und zum Ode. Eine physiologische Skizze, 794
Die Phänomene des Mediums Linda Gazerra, 1680
Die Phänomenologie des Ich in ihren Grundproblemen. Erster Band: Das Ich und das Selbstbewusstsein. Die scheinbare Spaltung des Ich, 1648
Phantasmata or Illusions and Fanaticisms of Protean Forms Productive of Great Evils, 779
Phantasms of the Living, 1144, 1371, 1525, 1621
Les phantises, 1135
Phenomena of Materialisation: A Contribution to the Investigation of Mediumistic Teleplastics, 1704
Les phénomènes d'autoscopie, 1530
Les phénomènes de hantise, 1765
Les phénomènes dits de matérialisation: étude expérimentale, 1672
Le phénomène spirit: témoignages des savants. Étude historique, esposition méthodique de tous les phénomènes, discussion des hypothèses. Conseils aux médiums, la théorie philosophique, 1341
Les phénomènes psychiques: recherches, observations, méthodes, 1526
Les phénomènes psychiques et supernormaux (leur observation, leur expérimentation), 1630

Les phénomènes psychiques occultes. État actuel de la question, 971, 1379
A Philosophical Essay on Credulity and Superstition; and also on Animal Fascination, or Charming, 571
Die Philosophie der Mystik, 1082
Philosophie des Unbewussten, 924
La philosophie des vapeurs ou Correspondance d'une jolie femme, nouvelle édition, augmentée d'un petit traité des crises magnétiques à l'usage des mesmériennes, 97
Philosophie du bon sens. Le spiritisme devant la raison. Le Dieu de la République. L'infaillibilité papale. Qu'était Jésus? Souvenires inédits sur la 32me demi-brigade. Edition posthume, 1478
Philosophie magnétique. Les révolutions du temps, synthèse prophétique du XIXe siècle, 749
La philosophie physiologique et médicale à l'Académie de Médecine, 911
Die philosophische Bedeutung der mediumistischen Phänomene. Erweiterte Fassung des auf dem Zweiten Internationalen Kongress fur parapsychologische Forschung in Warschau Gehaltenen Vortrags, 1871
The Philosophy of Animal Magnetism Together with the System of Manipulating Adopted to Produce Ecstasy and Somnambulism—The Effects and Rationale. By a Gentleman of Philadelphia, 385
Philosophy of Electrical Psychology: In a Course of Twelve Lectures, 596, 718
The Philosophy of Mesmerism, or Animal Magnetism. Being a Compilation of Facts Ascertained by Experience, and Drawn from the Writings of the Most Celebrated Magnetisers in Europe and America. Intended to Facilitate the Honest Inquirer After Truth, and Promote the Happiness of Mankind, By Diffusing the Knowledge of Nature's Wisest Laws and Most Benevolent Institutions, 472
Philosophy of Mysterious Agents, Human and Mundane: or the Dynamic Laws and Relations of Man, embracing the Natural Philosophy of Phenomena styled "Spiritual Manifestations," 681
The Philosophy of Mystery, 438
The Philosophy of Mysticism, 1082
The Philosophy of Sleep, 345, 1214
Philosophy of the Unconscious; Speculative Results According to the Inductive Method of Physical Science, 924, 1112
The Philosophy which Shows the Physiology of Mesmerism, and Explains the Phenomenon of Clairvoyance, 565
La photographie mentale des esprits dévoilée. Connaissance de la cause qui produit les effets naturels et magnétiques du spiritisme depuis l'antiquité jusqu'à nos jours, 934
Photographing the Invisible: Practical Studies in Spirit Photography, Spirit Portraiture, and Other Rare but Allied Phenomena, 1659
Phrenology Examined, and Shown to be Inconsistent with the Principles of Physiology, Mental and Moral Science, and the Doctrines of Christianity: also an Examination of the Claims of Mesmerism, 584
Phreno-Magnet, and Mirror of Nature: a Record of Facts, Experiments and Discoveries in Phrenology, Magnetism, &c., 483
Phreno-magnetic vindicator, 462
Physical Media in Spiritual Manifestations. The Phenomena of Responding Tables and the Planchette, and Their Physical Cause in the Nervous Organism, Illustrated from Ancient and Modern Testimonies, 929
Physical-Medical Treatise on the Influence of the Planets, 1
The Physical Phenomena of Spiritualism: A Historical Survey, 1892
The Physical Phenomena of Spiritualism, Fraudulent and Genuine: Being a Brief Account of the Most Important Historical Phenomena, a Criticism of Their Evidential Value, and a Complete Exposition of the Methods Employed in Fraudulently Reproducing the Same, 1584
The Physical Phenomena Popularly Classed under the Head of Spiritualism, with Facsimile Illustrations of Thought Transference Drawings and Direct Writing, 1568
Der physikalische Mediumismus, 1885, 1895

Title Index

Physikalische Phaenomene des Mediumismus: Studien zur Erforschung der telekinetischen Vorgange, 1787
Physikalisch-mediumistische Untersuchungen, 1773, 1790
Physikalish-physiologische Untersuchungen über die Dynamide des Magnetismus, der Electrizität, der Wärme, des Lichtes, der Krystallisation, des Chemismus in ihren Beziehungen zur Lebenskraft, 583
Physiologie, médecine et métaphysique du magnétisme, 437
Physiologie de la pensée. Recherche critique des rapports du corps avec l'esprit, 848
Physiologie et hygiene du magnétiseur; régime diététique du magnétisé; Mémoires et aphorismes de Mesmer, avec des notes, 502
Physiologie transcendantale. Analyses des choses. Essai sur la science future, son influence certain sur les religions, les philosophies, les sciences et les arts, 1233
The Physiology of Fascination, and The Critics Criticised, 738
The Pioneers of the Spiritual Reformation. Life and Works of Dr Justinus Kerner (Adapted from the German). William Howitt and his Work for Spiritualism. Biographical Sketches, 1074
Les pionniers du spiritisme en France: documents pour la formation d'un livre d'or des sciences psychiques recueillis par J. Malgras, 1577
Une plaie professionelle, ou la médecine exploitée par le somnambulisme, 840
A Plain and Rational Account of the Nature and Effects of Animal Magnetism: in a Series of Letters. With Notes and Appendix by the Editor, 192
Planchette: or, The Despair of Science. Being a Full Account of Modern Spiritualism, Its Phenomena, and the Various Theories Regarding It. With a Survey of French Spiritism, 930
Pneumatologie. Des espirits et de leurs manifestations fluidiques, 676, 704
Pneumatologie positive et expérimentale: la réalité des esprits et le phénomène merveilleux de leur écriture directe démontrés, 776
Polyzoisme ou pluralité animale dans l'homme, 911
Possession, Demonical & Other, among Primitive Races in Antiquity, the Middle Ages, and Modern Times, 1805
The Possibilities of Mal-observation and Lapse of Memory from a Practical Point of View, 1175
Posthumous Humanity: A Study of Phantoms, 1062
Pour transmettre sa pensée; notes et documents sur la télépathie ou transmission de pensée, 1534
The Power of the Mind over the Body: an Experimental Inquiry into the Nature and Cause of the Phenomena Attributed by Baron Reichenbach and Others to a "New Imponderable," 532
A Practical Display of the Philosophical System called Animal Magnetism, in Which is Explained Different Modes of Treating Diseases, 193
Practical Essays on Hypnotism and Mesmerism, 1413
The Practical Hypnotist. Concise Instructions in the Art and Practice of Suggestion; Applied to the Cure of Disease, the Correction of Habits, Development of Will-Power, and Self-Culture, 1552
Practical Instruction in Animal Magnetism, 320
Practical Instruction in Mesmerism, 1427
Practical Instruction in Table-moving, with Physical Demonstrations, 678
Practical Lessons in Hypnotism, 1468
Practical Views on Psychic Phenomena, 1791
The Practice of Hypnotic Suggestion; Being an Elementary Handbook for the Use of the Medical Profession, 1295
Précis historique de magnétisme animal, depuis Mesmer jusqu'à présent, 411
Précis historique des faits relatifs au magnétisme animal jusque en Avril 1781. Tr. de l'allemand, 11, 16
Précis théorique et pratique de neuro-hypnologie; études sur l'hypnotisme et les différents phénomènes nerveux, physiologiques et pathologiques qui s'y rattachent: physiologie, pathologie, thérapeutique, médecin légale, 1320, 1391

Preliminary Report of the Commission Appointed by the University of Pennsylvania to Investigate Modern Spiritualism in Accordance with the Request of the Late Henry Seybert, 1180, 1191
Presidential Address, 1682
Principes universels du magnétisme humain appliqueés au soulagement et à la guérison de tous les êtres malades, 811
Principles of Mental Physiology, with Their Applications to the Training and Discipline of the Mind, and the Study of Its Morbid Conditions, 985
The Principles of Psychology, 1270–1271
Principles of Psychotherapy, 1847
Principles of the Science in Animal Magnetism, 220
Private Instructions in the Science and Art of Organic Magnetism, 1123
Das Problem des Gedankenlesens, 1105
The Problem of Hypnotism, 1084
Problems of Belief, 1874
Problems of Personality. Studies Presented to Dr. Morton Prince, Pioneer in American Psychopathology, 1891
The Problems of Psychical Research: Experiments and Theories in the Realm of the Supernormal, 1695
Procédés du magnétisme animal, 39
Procès verbal du traitement par l'actions magnétique d'une femme malade (par la rapture d'un vaisseau dans la poitrine) de près Soissons, 229
Prodiges et merveilles de l'esprit humain sous l'influence magnétique, 580
Professor Farady on Table Moving, 666
Progress of Animal Magnetism in New England. Being a Collection of Experiments, Reports and Certificates, from the Most Respectable Sources. Preceded by a Dissertation on the Proofs of Animal Magnetism, 387
Prolusione sul magnetismo animale letta al Circolo Popolare di Brera in Milano la sera del 5 maggio 1860, 826
The Proof Palpable of Immortality; Being an Account of the Materialisation Phenomena of Modern Spiritualism: with Remarks on the Relations of the Facts to Theology, Morals, and Religion, 978
The Proofs of the Truths of Spiritualism, 1756
Le propagateur du magnétisme animal. Journal destiné à la publication des faits et des expériences, etc., de l'histoire du magnétisme etc., de la critique des ouvrages etc. etc., par une Société de médecins, 334
Die prophetische Kraft des magnetische Schlafes, oder wunderbare Enthullungen des Zukunft durch Somnambulen psychologische dargestellt und durch zahlreiche Beispiele bestatigt. Nebst Fingerzeigen, die zum Hochschlaf geeigneten Personen in den Zustand der clairvoyance zu versetzen, 569
Prospectus d'un nouveau cours théorique et pratique de magnétisme animal, réduit à des principles simple de physique, de chymie, et de médecine. Dans lequel on démontrera le systéme de M. Mesmer, et ses procédés; on rectifiera quelques unes de ses erreurs; on analysera la cause et le mécanisme par le quel les differents effets magnétiques sont produits; on prouvera enfin l'analogie qu'ils ont avec beaucoup d'autres effets naturels, et pourquoi ils ne présentent rien d'opposé aux connaissances que nous avions jusqu'ici de l'économie animale, 177
Der Prozess Czynski. Thatbestand desselben und Gutachten über Willensbeschränkung durch hypnotisch-suggestiven Einfluss abgegeben vor dem oberbayerischen Schwurgericht zu München, 1393
The Pseudo-Occult. Notes on Telepathic Vision and Auditory Messages Proceeding from Hypnotism, 1610
Psiche. Ipnotismo, magnetismo, spiritismo, 1276
Psiche misteriosa fenomeni detti spiritici ("metapsichici" del Richet), 1652
Psicologia e "spiritismo", impressioni e note critiche sui fenomeni medianici de Eusapia Paladino, 1616

Lo psicologico. Repertorio di magnetismo ad uso di chiunque voglia e debba essere al fatto dell'origine dello scopo e dei frutti di questa scienza, 768
Psyche. Deutsche Zeitschrift fur Odwissenschaft und Geisteskunde, 886
Psyche, zur Entwicklungsgeschichte der Seele, 533
Psychica. Revue scientifique du psychisme, 1808
Psychical and Supernormal Phenomena, their Observation and Experimentation, 1630
Psychical Investigations: Some Personally-Observed Proofs of Survival, 1726
Psychical Phenomena and the War, 1733
Psychical Research, 1656
Psychical Research, Science and Religion, 1893
(Psychical Research). The Goligher Circle: May to August 1921. With an Appendix Containing Extracts from the Correspondence of the Late W. J. Crawford and Others, 1824
Psychical Research and Survival, 1688
Psychical Research and the Resurrection , 1611
Psychical Research for the Plain Man, 1781
The Psychic and Psychism, 1482
Psychic Force, an Experimental Investigation of a Little-known Power, 1533
Psychic Force and Modern Spiritualism: A Reply to the "Quarterly Review" and Other Critics, 939
Psychic Research and Gospel Miracles. A Study of the Evidences of the Gospel's Superphysical Features in the Light of the Established Results of Modern Psychical Research, 1501
The Psychic Riddle, 1587
Psychics: Facts and Theories, 1353
Psychic Science in Parliament. A Survey of the Statues and the Leading Legal Decisions in Regard to Psychic Phenomena, 1715
The Psychic Structures at the Goligher Circle. With a Note by David Grow, 1795
Psychic Tendencies To-day: an Exposition and Critique of New Thought, Christian Science, Spiritualism, Psychical Research (Sir Oliver Lodge) and Modern Materialism in Relation to Immortality, 1744
Psychische Studien. Monatliche Zeitschrift, vorzuglich der Untersuchung der wenig qekannten Phanomene des Seelenlebens, 962, 1253
Le psychisme expérimental, étude sur les phénomènes psychiques, 1384
Le psychisme inférieur. Étude de physiopathologie clinique des centres psychiques, 1572
Psychography: a Treatise on One of the Objective Forms of Psychic or Spiritual Phenomena, 1008
Psychography, or The Embodiment of Thought; with an Analysis of Phreno-magnetism, "Neurology," and Mental Hallucination, Including Rules to Govern and Produce the Magnetic State, 469
The Psychological Foundation of Belief in Spirits, 1778
Psychological Healing, 1777
Psychological Review, 1010
La psychologie appliquée, 1156, 1786
La psychologie du raissonnement. Recherches expérimentales par l'hypnotisme, 1128
La psychologie inconnue. Introduction et contribution à l'étude expérimentale des sciences psychiques, 1605
Psychologie naturelle. Étude sur les facultés intellectuelles et morales dans leur état normal et dans leurs manifestations anomales chez les aliénés et chez les criminels. Tome premier, 910
Psychology; or the Science of the Soul, Considered Physiologically and Philosophically, 718
Psychology and the Occult, 1506
Psychology Applied to Medicine. Introductory Studies, 1598
Psychology as a Natural Science Applied to the Solution of Occult Psychic Phenomena, 1246
The Psychology of Conviction: A Study of Beliefs and Attitudes, 1741
The Psychology of Medicine, 1803
The Psychology of Suggestion: A Research into the Subconscious Nature of Man and Society, 1450

The Psychology of the Future ("L'avenir des sciences psychiques"), 1721
Psychopathological Researches: Studies in Mental Dissociation, 1513
Psycho-physiological Researches on the Dynamides or Imponderables, Magnetism, Electricity, Heat, Light, Crystallization, and Chemical Attraction, in Their Relation to the Vital Force, 583
The Psycho-Physiological Sciences and Their Assailants. Being a Response by A.R.W. . . . J. R. Buchanan, . . . D. Lyman, . . . E. Sargent . . .; to the Attacks of W. B. Carpenter, . . . and Others, 1011
Psychotherapeutics: A Symposium. By Morton Prince, Frederic H. Gerrish, James J. Putnam, E. W. Taylor, Boris Sidis, George A. Waterman, John Donley, Ernest Jones, Tom A. Williams, 1637
Psycho-Therapeutics; or Treatment by Sleep and Suggestion, 1251
Psycho-thérapie: communications statistiques, observations cliniques nouvelles. Compte rendu des résultats obtenus dans la clinique de psycho-thérapie suggestive d'Amsterdam, pendant la deuxième période, 1372
Puissance de l'électricité animale, ou, du magnétisme vital et de ses rapports avec la physique, la physilologie et la médecine, 416
Les Puységur. Leurs oeuvres de littérature, l'économie politique et de science. Étude, 956

Quaere et invenies, 671
Quarterly Journal of Science, 939, 959
Quarterly Review, 659, 993
Quatrième congrès international de psychologie. Compte rendu, 1496
Quelques faits et considerations pour servir à l'histoire du magnétisme animal. Thèse No. 243, 353
Quelques mots sur le magnétisme animal, suivis d'une observation de variole congénitale, 404
Quelques réflexions sur le magnétisme animal, thèse, 407
Qu'est-ce-que le spiritisme; introduction à la connaissance du monde invisible ou des esprits, contenant les principes fondamentaux de la doctrine spirite et la réponse aux quelques objections préjudicielles, 810
The Quest for Dean Bridgman Conner, 1708
The Question: "If a Man Die, Shall He Live Again?" Job. xiv. 14. A Brief History and Examination of Modern Spiritualism. With a Postscript by H. E. Armstrong, 1722
Questions du jeune Docteur Rhubarbini de Purgandis, adressés à Messieurs les docteurs-regents, de toutes les facultés de médecine de l'universe, au sujet de M. Mesmer & du magnétisme animal, 112

The Railway Age and Northwestern Railroader, 1316
Rapport au public de quelques abus en médecin; avec des réflexions & notes historiques, critiques & médicales, 153
Rapport confidentiel sur le magnétisme animal et sur la conduite récente de l'Académie royale de médecine adressé à la congrégations de l'Index, et traduit de l'Italien du R. P. Scobardi, 415
Rapport de la Société royale de médecine sur l'ouvrage intiulée Recherches et doutes sur le magnétisme animal, etc., 107
Rapport de l'un des commissaires chargés par le Roi, de l'examen du magnétisme animal, 72
Rapport des commissaires chargés par le roi de l'examen du magnetisme animal, 31
Rapport des commissaires de la Société royale de médecine nommés par le Roi pour fair l'examen du magnétisme animal, imprimé par ordre du Roi, 101
Rapport des cures opérées à Bayonne par le magnétisme animal, adressé à M. l'abbé de Poulouzat, conseiller clerc au Parlement de Bordeaux, par le comte de Puységur, avec notes de M. Duval d'Espremenil, conseiller au Parlement de Paris, 106
Rapport du magnétisme avec la jurisprudence et la médecine légale, 815
Rapport du rapport de MM. les Commissaires nommés par le Roi, etc. par un amateur de la vérité excité par l'imagination, l'attouchement et magnétisé par le bon sens et la raison.

Adressé à M. Caritides, fils de cet illustre savant qui avait conçu l'ingénieux projet de mettre toutes les côtes du royaume en port de mer, actuellement résident au Monomotapa, 108
Rapport fait à la Société des Sciences Physiques de Lausanne, sur un somnambule naturel, 180
Der Rapport in der Hypnose. Untersuchungen über den thierischen Magnetismus, 1323
Rapporto della Commissione nominata dalla sezione medica della Società di Incoraggiamento de Scienze, Lettere ed Arti in Milano per l'esame delle memorie di concorso al premio proposto pel 1855 sopra un argomento de magnetismo animale, 754
Rapports du magnétisme et du spiritisme, 1322
Rapport secret présenté au ministre et signé par la commission précédente, 32, 213
Rapport secret sur le mesmérisme, 213
Rapports et discussions de l'académie royale de médecine sur le magnétisme animal, recueillis par un sténographe, et publiés, avec des notes explicatives, 350, 355
Rapport sur les aimans presentés par M. l'abbé Le Noble, lu dans la séance tenus au Louvre, le mardi premier avril 1983, 27
Rapport sur les epériences magnétiques faites par la commission de l'Académie royale de médecin, lu dans les séances des 21 et 28 de Juin, par M. Husson, rapporteur, 350, 354, 374, 383
Rapport sur les séances d'Eusapia Palladino à l'Institut Général Psychologique en 1905, 1906, 1907 et 1908, 1607
The Rationale of Mesmerism, 1330
Raymond, or Life and Death: with Examples of the Evidence for Survival of Memory and Affection after Death, 1714
Real Ghost Stories: a Record of Authentic Apparitions. Being the Christmas Number of the Review of Reviews. Collated and Edited by W. T. Stead, 1307, 1331
Die Realität magischer Kräfte und Wirkungen des Menschen gegen die Widersacher vertheidigt. Ein Supplement zu des Verfassers "Mystischen Erscheinungen der menschlichen Natur," 862
The Reality of Psychic Phenomena: Raps, Levitations, etc., 1711, 1824
The Reality or Unreality of Spiritualistic Phenomena. Being a Criticism of Dr W. J. Crawford's Investigations into Levitations and Raps, 1762
Recent Experiments in Hypnotism, 1205
Un récent procès spirite. L'aventure du médium aux fleurs, 1541
Recherche des bases qui la constituent et des lois que gouvernent l'univers physique et moral et l'homme en particulier, par un étudiant de 90 ans, 1269
Recherches, expériences et observations physiologiques: sur l'homme dans l'état de somnambulisme naturel, et dans le somnambulisme provoqué par l'acte magnétique, 237
Recherches et considérations critiques sur le magnétisme animal avec un programme relatif au somnambulisme aritificiel ou magnétique traduit du latin du docteur Metzger accompangné de notes et suivi de réflexions morales ou pensées détachées applicables au sujet, 319
Recherches et doutes sur le magnétisme animal, 26, 83, 116
Recherches expérimentales sur les conditions de l'activité cérébrale et sur la physiologie des nerfs; études physiologiques et psychologiques sur le somnambulsiem provoqué, 1076
Recherches historiques sur l'exercice de la médecine dans les temples, chez les peuples de l'antiquité, suivies de considérations sur les rapports qui peuvent exister entre les quérisons qu'on obtenait dans les anciens temples, à l'aide des songes, et le magnétisme animal, et sur l'origine des hopitaux, 496
Recherches physiques et métaphysiques sur les influences célestes, sur le magnétisme universel et sur le magnétisme animal dont on trouve la pratique de temps immémorial chez les Chinois, 181
Recherches physiques sur le magnétisme; insérées dans le Journal des Savans, en l'année 1790, 194
Recherches psychologiques sur la cause des phénomènes extraordinaires observés chez les modernes voyans improprement dits somnambules magnétiques ou Correspondance sur le magnétisme vital entre un salitaire et M. Deleuze, bibliothécaire du Muséum à Paris, 393

Recherches sur la direction du fluide magnétique, 127
Recherches sur la médiumnité. Étude de travaux des savants, l'écriture automatique des hystériques, l'écriture mécanique des médiums, preuves absolues de nos communications avec le monde des esprits, 1500
Recherches sur l'hypnotisme et ses causes, suivies d'un discours prononcé dans l'assemblée des chirurgiens-dentistes du mois de décembre 1860, 842
Recherches sur l'hypnotisme ou sommeil nerveux, comprenant une série d'expériences instituées à la maison municipale de santé, 818
Recherches sur quelques phénomènes du magnétisme, le fantome magnétique, et sur la diffraction complexe, 427
Recit de l'avocat-général de—, aux chambres assemblées du public, sur le magnétisme animal, 149
Record of Experiments in Thought-transference, 1067
A Record of Observation of Certain Phenomena of Trance [Myers], 1274
A Record of Observations of Certain Phenomena of Trance [Hodgson], 1317
Recueil de mémoires sur l'analogie de l'électricité et du magnétisme: couronnés & publiés par l'Académie de Baviere; traduits du Latin & de L'Allemand, augmentés de notes, & de quelques dissertations nouvelles, 114
Recueil des effets salutaires de l'aimant dans les maladies, 8, 22
Recueil des pièces le plus intéressantes sur le magnétisme animal, 83, 86
Recueil d'observations et de faits relatifs au magnétisme animal, présenté à l'auteur de cette découverte, et publié par la société de Guienne, 120
Recueil d'observations et des faits relatifs au magnétisme animal présenté à l'auteur de cette découverte et publié par la Société de l'harmonie de Guienne, 151
Recueil d'opérations chirurgicales pratiquées sur des sujets magnétisés, 521
Réflexions impartiales sur le magnétisme animal, faites après la publication du rapport des commissaires chargés par le roi de l'examen de cette découverte, 49
Réflexions préliminaires à l'occasion de la pièce intitulée Les docteurs modernes jouée sur le Théatre Italien, le seize, Novembre 1784, 57
Réflexions sur le magnétisme animal, d'après lesquelles on cherche à établir le degré de croyance que peut mériter justqu'ici le système de M. Mesmer, 90
Reflexions sur le magnétisme animal, et sur le système de M. Mesmer, 114
Réforme médicale: compérage magnétique réprimé, questions et observations d'ordre public sur la pratique du magnétisme, du mesmérisme et du somnambulisme, considérée comme exercice de la médecine, etc. . . . , 537
Relation de diverses experiences sur la transmission mentale, la lucidité, et autres phenomenes non explicables par les données scientifiques actuelles, 1217
Relazione di una interessantissima cura magnetica fatta in Berlino, 953
Religion, magnétisme, philosophie. Les éleus de l'avenir ou le progrés réalisé par le Christianisme, 763
Religion and Medicine: the Moral Control of Nervous Disorders, 1615
The Religion of Health: an Examination of Christian Science. Completed by R. M. Barrett, 1886
The Religion of Modern Spiritualism, and Its Phenomena Compared with the Christian Religion and Its Miracles, 1398
The Religion of the Spirit World: Written by the Spirits Themselves, 1774
Remarks on animal magnetism, 391
Remarques sur la conduite de sieur Mesmer et de son commis le P. Hervier, et de ses autres adhérents; où l'on tache de venger la médecine de leurs outrages, 56
Remonstrances des malades aux médecins de la Faculté Paris, 141
Réponse à la lettre de M. le Docteur Cazaintre sur un cas de transposition des sens, 335
Réponse à l'auteur des Doutes d'un provincial, proposés à MM. les médecins commissaires chargés par le Roi de l'examen du mangétisme animal, 147
Réponse à M. P. Janet: à propos de la métapsychique, 1852
Réponse aux articles du Journal des Débats, contre le magnétisme animal, 259
Réponse aux objections contre le magnétisme, 274

Réponse d'un médicin de Paris à un médicin de province, sur le prétendu magnétisme animal de M. Mesmer, 13–14
Report of cures by animal magnetism occurring at Bayonne with verifications, 106
Report of the Annual Meeting, May 1850; James Adam Gordon, Esq., Vice-President, in the Chair, 592
Report of the Annual Meetings [of the London Mesmeric Infirmary], 719
Report of the Committee Appointed by Government to Observe and Report upon Surgical Operations by Dr. J. Esdaile, upon Patients under the Influence of Alleged Mesmeric Agency, 546
Report of the Experiments on Animal Magnetism, Made by a Committee of the Medical Section of the French Royal Academy of Sciences: Read at the Meetings of the 21st and 28th of June, 1831, Translated and Now for the First Time Published; with an Hisotrical and Explanatory Introduction, and an Appendix, 354, 372
Report of the First Public Meeting; the Rt. Hon. Earl Ducie, President, in the Chair, 572
Report on a series of Sittings with Eusapia Palladino, 1628
Report on Spiritualism of the Committee of the London Dialectical Society, Together with Evidence Oral and Written, and a Selection from the Correspondence, 942
Report on the Census of Hallucinations, 1374, 1425
Report on the Magnetical Experiments Made by the Commission of the Royal Academy of Medicine, of Paris, Read in the Metings of June 21 and 28, 1831. by Mr. Husson, the Reporter, Translated from the French, and Preceded with an Introduction, by Charles Poyen St. Sauveur, 374
Reports of about One Hundred and Fifty Cases in England Demonstrating the Efficacy of the Metallic Practice in a Variety of Complaints Both upon the Human Body and on Horses, 212
Report upon the Phenomena of Clairvoyance or Lucid Somnambulism (From Personal Observation). With Additional Remarks. An Appendix to the Third Edition of "Animal Magnetism," 479
Requête burlesque et arrêt de la cour du parlement, concernant la suppression du magnétisme animal, 150
Researches in the Phenomena of Spiritualism (Reprinted from the Quarterly Journal of Science), 939, 942, 959
Résultat des opérations magnétiques de M. le Marquis de Guibert à Fontchâteau commune de Tarascon, 426
A Return of Departed Spirits of the Highest Character of Distinction, as well as the Indiscriminate of All Nations, into the Bodies of the "Shakers," or "United Society of Believers in the Second Advent of the Messiah," By an Associate of Said Society, 485
La révélateur. Journal de magnétisme animal publié par une société de magnétiseurs, 388
Les révélations d'un magnétiseur, trucs ingénieux employés au théatre pour obtenir les phénomènes de la transmission de pensée, du magnétisme et de l'hypnotisme, 1535
Revelations of a Spirit Medium; or Spiritualistic Mysteries Exposed. A Detailed Explanation of the Methods Used by Fraudulent Mediums, 1288
Revenants et Fantomes. Essais sur l'humanité posthume et le spiritisme, 1062
Les rêves d'une femme de province sur le magnétisme animal. ou Essai théoretique & pratique sur la doctrine à la mode, 137
Review of Reviews, 1307, 1331
A Review of the "Spiritual Manifestations". Read Before the Congregational Association of New York and Brooklyn, 654
Une révolution en philosophie, résultant de l'observation des phénomènes du magnétisme animal. Étude physiologique et psychologique de l'homme, 1027
Revue de l'hypnotisme et de la psychologie physiologique, 1039, 1079
Revue de l'hypnotisme expérimental et thérapeutique, 1156, 1297, 1785, 1835
Revue de psychiatrie, de neurologie et d'hypnologie. Recueil des travaux publiés en France et à l'etranger, 1039, 1405
Revue de psychiatrie. Médecin mentale, neurologie, psychologie, 1039
Revue de psychiatrie et de psychologie expérimentale, 1039

Revue de psychologie expérimentale. Études sur le sommeil, le somnambulisme, l'hypnotisme et le spiritualisme, 964
Revue de psychothérapie et de psychologie appliquée, 1039, 1156, 1835
Revue des deux mondes, 1093
Revue des études psychiques, 1283
Revue des sciences hypnotiques, 1186
Revue d'hypnologie théorique et pratique dans ses rapports avec la psychologie, les maladies mentales et nerveuses, 1039, 1277
Revue du psychisme expérimental. Magnétisme. Hypnotisme. Suggestion. Psychologie. Mediumnisme. Mensuelle illustrée, 1650
Revue générale des sciences psychiques, 1595
Revue magnétique. Journal des faits et des cures magnétiques et somnambuliques, des théories, recherches historiques, discussions scientifiques et progrès généreaux du magnétisme en France et dans les pays étrangers, 501
La revue magnétique, journal des malades, 788, 891, 914, 928
La revue magnétique. Organe du Cercle électro-magnétique de Paris, 1012
Revue médical historique et philosophique, 303
Revue métapsychique, 1283, 1776, 1787
Revue philosophique, 1032, 1071, 1093, 1101–1102, 1133, 1141, 1145–1148, 1157, 1176, 1184, 1196, 1211, 1216, 1235, 1318–1319, 1386, 1846, 1852
Revue Scientifique, 1203
Revue Spirite. Journal d'études psychologiques, 1215, 1465
Revue spiritualiste. Journal principalement consecré à l'étude des facultés de l'âme et à la démonstration de son immoralité par l'examen raisonné des divers genres de manifestations médianimiques et de phénomènes psychiques tels que le somnambulisme, l'extase, la vue à distance, etc., 795
Ricerca psichica . . . , 1491
Ricerche sui fenomeni ipnotici e spiritici, 1633
The Riddle of Personality, 1518
Riflessioni sul magnetismo animale; fatte ad oggetto di illuminare i suio cittadini aveudodo trovato salutare in molti mali, 202
The Rise and Progress of Spiritualism in England, 936
Rivelazioni ed insegnamento del givoco col simulare i fenomeni magnetici e ipnotici, 1285
Rivista delle riviste di studi psichici. Rassegna mensile di 300 reviste de studi psichici, 1561
Rivista di studi psichici. Periodico mensile dedicato alle richerche sperimentali e critiche sui fenomeni di telepatia, chiaroveggenza, premonizione, medianità, etc., 1390
Russky Vyestnik (Russian Messenger), 1394

La salute. Giornale d'igiene populare e di altre cognizioni utili, 889
Salute (Società magnetica d'Italia residente in Bologna), 871
Sammlung der neuesten gedruckten und geschriebenen Nachrichten von Magnet-Curen, vorzüglich der Mesmerischen, 3–4, 9
Sammlung von Actenstücken, als da sind: Eingaben und Adressen in Sachen der gemeingefährlichen Einwirkungen durch Magnetisation auf telepathischen Wege, an verschiedene Behörden, Vereine &c. gerichtet, 1221
Sanctuaire du spiritualisme. Etude de l'âme humaine, ses rapports avec l'univers, d'après le somnambulisme et l'extase, 593
Satan et la magie de nos jours, réflexions sur la magnétisme, le spiritisme et la magie, 869
Satanic Agency and Mesmerism. A Sermon, 460, 503
Satanic Agency and Mesmerism Reviewed, 450, 460, 465, 481
Satanic Agency and Table Turning. A Letter to the Rev. Francis Close, in Reply to His Pamphlet, "Table-turning not Diabolical," 683
Schediasma, de Mesmerismo ante Mesmerum, 253
Der Schotte Home, ein physiopsychischer Zeuge der Transscendenten im 19 Jahrhundert, 1452
Schreiben den Thier. Magnetismus u. die sich selbst weider ersetzende Kraft Betreffend, 7

Schreiben über die Magnetkur von Herrn. A. Mesmer, Doktor der Arzneygelahrtheit, an einen auswartigen Arzt, 3, 5
Schriften der Gesellschaft für psychologische Forschung, 1323
Die Schutzgeister, oder merkwurdige Blicke zweier Scherinnen in die Geisterwelt, nebst der wunderbaren Heilung einer zehn Jahre stumm Gewesenen durch den Lebensmagnetismus, und einer vergleichenden Uebersicht aller bis jetzt beobachteten Erscheinungen desselben, 418
Schwärmer und Schwindler zu Ende des 18. Jahrhunderts, 967
Ein Schwerer sensitiv-somnambuler Krankheitsfall geheilt ausschliesslich mittelst einfacher Anwendung der Gesetze des Odes. Für Physiker, Physiologen, Psycholgen, Psychiatriker und insbesondere für Ärzte. Hrgb. von A. Freiherr von Schrenck-Notzing, 1305
Science and a Future Life [Myers], 1301, 1350
Science and a Future Life [Hyslop], 1556, 1575, 1611
Science and a Future Life: With Other Essays, 1350
Les Sciences Occultes au 19e siècle: les tables tournantes et les manifestations prétendues surnaturelles, considérées au point du vue des principes qui servent de guide dans les sciences d'observation.—Des tables tournantes au point de vue de la mécanique et de la physiologie, 879
Les sciences utiles. Hypnotisme et magnétisme. Somnambulisme, suggestion et télépathie. Influence personnelle. Partie théorique. Pratique (suite). Historique. Occultisme expérimentale, 1608
Science vs. Modern Spiritualism. A Treatise on Turning Tables, the Supernatural Agent in General, and Spirits, 712
Scientific American, 1838, 1882, 1887
The Scientific Aspect of the Supernatural: Indicating the Desirableness of an Experimental Enquiry by Men of Science into the Alleged Powers of Clairvoyants and Mediums, 900
A Scientific Demonstration of the Future Life, 1385
Le scienze occulte. Magnetismo, elletrobiologia, spiritualismo e negromanzia ossia la duplice scienza d'una levatrice . . . , 836
Scribner's Magazine, 1270
Second lettre de Grosjean à son évèque, 758
Second Mémoire physique et médicinal, montrant des rapports évidents entre les phénomènes de la baguette divinatoire, du magnétisme, et de l'électricité, avec des éclaircissements sur d'autres objets non moins importants, qui y sont relatifs, 117
Second mémoire sur le magnétisme animal. Observations particulières sur une somnambule présentée à la commission nommée par l'Académie royale de médecine pour l'examen du magnétisme animal, 329
Second Sight: Problems Connected with Prophetic Vision and Records Illustrative of the Gift, Especially Derived from an Old Work not now Available for General Use, 1243
Les secrets du magnétisme, 1078
Seeing the Invisible: Practical Studies in Psychometry, Thought-transference, Telepathy, and Allied Phenomena, 1570
Seeress of Prevorst. Being Revelations Concerning the Inner-Life of Man, and the Inter-Diffusion of a World of Spirits in the One We Inhabit, 341, 343
Die Seherin von Prevorst; Eröffnungen über das innere Leben des Menschen und über das Hereinragen einer Geisterwelt in die unsere, 341, 343
Self Mastery through Conscious Autosuggestion, 1794
Le sens de la vie humaine, 1758
Der sensitive Mensch und sein Verhalten zum Ode. Eine Reihe experimentellen Untersuchungen über ihre gegenseitigen Kräfte und Eigenschaften mit Rücksicht auf die praktische Bedeutung, welche sie für Physik, Chemie, Mineralogie, Botanik, Physiologie, Heilkunde, gerichtliche Medicin, Rechtskunde, Kriegswesen, Erziehung, Psychologie, Theologie, Irrenwesen, Kunst, Gewerbe, häusliche Zustände, Menschenkentniss und das gesellschaftliche leben im weitesten Anfange haben, 727
A Series of Concordant Automatisms, 1617
Seven Lectures on Somnambulism, 222, 226
Sfinge, 1276

Shadowland, or Light from the Other Side, 1418
Die sichtbare und die unsichtbare Welt. Dieseits und Jenseits, 1046
Sights and Sounds: the Mystery of the Day: Comprising an Entire History of the American "Spirit" Manifestations, 689–690
Six Lectures on the Philosophy of Mesmerism, 718
Six Lectures on the Philosophy of Mesmerism, Delivered in the Marlboro Chapel, Boston. Reported by a hearer, 473
The Slade Case; Its Facts and Its Lessons. A Record and a Warning, 997
Sleep-walking and Hypnotism, 1098
The Society for Psychical Research: Its Rise and Progress and a Sketch of its Work. With Facsimile Illustrations of Three Pairs of the Thought-Transference Drawings, 1515
Society for Psychical Research. Proceedings, 1060, 1067, 1084–1086, 1090, 1109–1111, 1118, 1142–1143, 1152–1153, 1166, 1175, 1204–1205, 1217, 1274, 1317, 1324–1325, 1366, 1374, 1388, 1394, 1419, 1439, 1486–1487, 1495, 1525, 1578, 1601, 1617, 1628, 1656, 1678, 1682, 1710, 1745, 1761, 1778, 1832
Society for Psychical Research Journal, 1097, 1244, 1366
Der sogenannte animalische Magnetismus oder Hypnotismus: unter Zugrundelegung eines fuer die Akademische Gesellschaft zu Freiburg i. B. gehaltenen popoulären Vorträges, 1035
Der sogenannte Lebensmagnetismus oder Hypnotismus, 1066
Soirées magnétiques de Monsieur Laurent à Clalon sur Saone, 443
El Sol. Revista de historia magnetismo, estudios psiquicos, 1477
Solution du problème de la suggestion hypnotique. La Salpêtrière et l'hypnotisme. La Suggestion criminelle. La loi doit intervenir, 1249
Some Facts in Mesmerism, 1582
Some Higher Aspects of Mesmerism, 1110
Some New Evidence for Human Survival, 1837
Le sommeil et les rêves, 843
Le sommeil magnétique expliqué par le somnambule Alexis en état de lucidité, 765
Le sommeil naturel et l'hypnose. Leur nature, leurs phases. A qu'ils nous disent en faveur de l'immortalité de l'âme, 1546
Le sommeil normal et le sommeil pathologique, magnétisme animal, hypnotisme, névrose hystérique, 1075
Le sommeil provoqué et les causes qui le déterminent. Étude étiologique de l'hypnose, 1661
Le sommeil provoqué et les états analogues, 896
Sommes versées entre les mains de monsieur Mesmer pour acquérir le droit de publier sa découverte, 136, 144
Le somnambule. Journal de magnétisme, 488
Die Somnambulen Tische. Zur Geschichte und Erklärung deiser Erscheinung, 674
Somnambulism, 1466
Somnambulisme et magie, 1400
Somnambulisme magnétique, 878
Somnambulisme ou supplément aux journaux dans lesquels il a été question de ces phénomènes physiologiques, 247
Le somnambulisme provoqué: études physiologiques et psychologiques, 1125
Der Somnambulismus, 408
Somnambulismus und Spiritismus, 1472, 1590
Somnolism and Psycheism, Otherwise Vital Magnetism, or Mesmerism: Considered Physiologically and Philosophically. With an Appendix Containing Notes of Mesmeric and Psychical Experience, 575
Somnologie magnétique, ou recueil de faits et opinions somnambuliques pour servir à l'histoire du magnétisme humain, 542
Der soqenannte thierische magnetismus. Physologische Beobachtungen, 1029
Soul of Things: or Psychometric Researches and Discoveries, 854
Souvenires d'un magnétiseur, 1088
Souvenirs des banquets de Mesmer, toastes et chansons, 827

Specimen bibliothecae criticae magnetismi sic dicti animalis, 187
Sphinx: Neues Archiv für den thierischen Magnetismus und das Nachtleben überhaupt, 269, 323
Spirit Drawings: A Personal Narrative, 799
Spirites et illusionnistes de France, 1669
Spirit-Identity, 1023
Spirit-intercourse: Containing Incidents of Personal Experience while Investigating the new Phenomena of Spirit Thought and Action; with Various Spirit Communications Through Himself as Medium, 688
Spiritism, 1113
Spiritism: Facts and Frauds, 1859
Spiritism and Common Sense, 1825
Spiritism and Religion: "Can You Talk to the Dead?" Including a Study of the Most Remarkable Cases of Spirit Control, 1742
Spiritism and the Fallen Angels in the Light of the Old and New Testaments, 1772
Le spiritisme, la démonologie et la folie; explication de tous les faits magnétiques, 884
Le spiritisme contemporain, 1319
Le spiritisme dans ses rapports avec la folie, 1621
Le spiritisme devant la conscience, 1344
Spiritisme devant la science, 1107
Le spiritisme devant la science, 1538
Spiritisme et tables tournantes. Nouvelle méthode facile et complète, expliquant les principes réels du spiritisme, les moyens infaillilbes pour arriver promptement à évoquer les esprits et se mettre en rapport avec eux, suivie de la démonstration théorique et pratique du pendule-explorateur et de la baquette divinatoire, 1072
Spiritisme expérimental. Le livre des médiums ou guide des médiums et des évocateurs, contenant l'enseignement spécial des esprits sur la théorie de tous les genres de manifestations, les moyens de communiquer avec le monde invisible, le développement de la médiumnité, les difficultés et les écueils que l'on peut rencontrer dans la pratique du spiritisme. Pour faire suite au Livre des Esprits, 845
Le spiritisme (fakirisme occidental): étude historique, critique et expérimentale, 1140
Spiritisme moderne. Katie King. Histoire de ses apparitions d'après les documents anglais avec illustrations par Un Adepte, 1465
Lo spiritismo di fronte alla scienza: note polemiche, 1480
Spiritismo e magnetismo. Letture fatte a Vincenza, 912
Spiritismo e magnetismo. Note e confronti polemici, 1479
Der Spiritismus, 1113, 1253, 1293
Der Spiritismus. Eine sogenannte wissenschaftliche Frage. Offener Brief an Herrn Prof Hermann Ulrici in Halle, 1024
Der Spiritismus . . . . Nebst einem Beitrag von dr. K. R. Kupffer, 1869
The Spirit Land, 804
Spirit Manifestations Examined and Explained. Judge Edmonds refuted; or, an Exposition of the Involuntary Powers and Instincts of the Human Mind, 711
Spirit Mediums and Conjurers. An Oration Delivered in the Cavendish Rooms, London, on Sunday Evening, June 15th, 1873. To Which is Appended Rules to be Observed at the Spirit Circle, 957
Spirit People: a Scientifically Accurate Description of Manifestations Recently Produced by Spirits, and Simultaneously Witnessed by the Author and Other Observers in London, 975
Spirit-Psychometry and Trance Communications by Unseen Agencies, Through a Welsh Woman and Dr. T. D'aute-Hooper, 1696
Spirit-rapper; an Autobiography, 704
Spirits Before our Eyes, 1019
Spirit Slate Writing and Kindred Phenomena, 1448
Spirit Teachings, 1069
Spiritualism: A Narrative with a Discussion, 935

Spiritualism: a Popular History from 1847, 1784
Spiritualism: Is Communication with the Other World an Established Fact? Pro—E. Wake Cook . . . Con—Frank Podmore, 1528
Spiritualism; Its Facts and Phases. Illustrated with Personal Experiences, 876
Spiritualism: Its History, Phenomena and Doctrine, 1739
Spiritualism—the Inside Truth, 1752
Spiritualism. With an Appendix by Nathaniel P. Tallmadge, 664
Spiritualism and Allied Causes and Conditions of Nervous Derangement, 988
Spiritualism and Animal Magnetism. A Treatise on Dreams, Second Sight, Somnambulism, Magnetic Sleep, Spiritual Manifestations, Hallucinations, and Spectral Visions, 946
Spiritualism and Insanity. An Essay Describing the Disastrous Consequences to the Mental Health Which are Apt to Result from a Pursuit of the Study of Spiritualism, 1655
Spiritualism and Sir Oliver Lodge, 1729
Spiritualism and the New Psychology: An Explanation of Spiritualistic Phenomena and Belief in Terms of Modern Knowledge, 1769
Spiritualism Answered by Science, 938
Spiritualisme, faits curieux précédés d'une lettre à M. G. Mabru suivis de l'extrait d'un compte rendu de la fête Mesmérienne du 23 Mai 1858, et d'une relation américaine des plus extraordinaire publiés, 786
Spiritualism Explained: Being a Series of Twelve Lectures delivered Before the New York Conference of Spiritualists, 772
Spiritualism in America, 837
Spiritualism Tested; or, —the Facts of Its History Classified and Their Cause in Nature Verified from Ancient and Modern Testimonies, 835
Der Spiritualismus und die Wissenschaft. Experimentelle Untersuchungen über die psychische Kraft, von William Crookes, nebst bestätigenden Zeugnissen des Physikers C. F. Varley. Prüfungs-Sitzungen des Mr. D. D. Home mit den Gelehrten zu St. Petersburg und London, 1430
The Spiritualist: Being a Short Exposition of Psychology Based upon Material Truths and of the Faith to Which It Leads, 783
Spiritual Tracts, No. 1–No. 13, 952
Spirit Workers in the Home Circle. An Autobiographic Narrative of Psychic Phenomena in Family Daily Life Extending over a Period of Twenty Years, 1192
Splitting of the Mind: An Experimental Study of Normal Men, 1890
Die Sprache der Verstorbenen oder das Geisterklopfen. Stimmen aus dem Jenseits und enthullte Geheimnisse des Grabes. Ein unumstösslicher Beweis fur die Fortdauer der Seele nach dem Tode und deren Wiedervereinigung mit ihren Lieben. Noch gesammelten authentischen Thatsachen dargestellt, 751
Spuk, Gespenster und Apportphänomene, 1849
Stages of Hypnotic Memory, 1143
The Stages of Hypnotism, 1085
Stella C.: an Account of Some Original Experiments in Psychical Research, 1903
Storia, teoria e practica del magnetismo animale, 524
Storia dello spiritismo, 1410
Stranger Than Fiction, 813
Strange Things Among Us, 864
Studien aus dem Gebiete der Geheimwissenschaften, 1261
Studien über die Geisterwelt, 968
Studien über Hautelektrizität und Hautmagnetismus des Menschen. Nach eigenen Versuchen und Beobachtungen, 1555
Studies in Humanism, 1597
Studies in Physchical Research, 1426
Studies in Psychometry, 1873
Studies in Spiritism, 1654
Gli studi recenti sul cosi detto magnetismo animale, 1049
Studi sull 'ipnotismo, con richerche oftalmoscopiche del Reymond e dei prof. Bianchi e Sommer sulla polarizzazione psichica, 1149

The Study of Trance, Muscle-reading and Allied Nervous Phenomena in Europe and America, with a Letter on the Moral Character of Trance Subjects, and a Defence of Dr. Charcot, 1052
The Subconscious, 1557
The Subconscious, the Unconscious, and the Co-conscious, 1891
The Subliminal Consciousness, 1324–1325, 1351
The Subliminal Self, 1388
Les suggesteurs et la foule. Psychologie des meneurs: artistes, ovateurs, mystiques, guerriers, criminels, écrivains, enfants, etc. . . . , 1545
La suggestibilité, 1467
La suggestibilité au point de vue de la psychologie individuelle, 1434
Suggestion, Hypnose und Telepathie; ihre Bedeutung für die Erkenntnis gesunden und kranken Geisteslebens, 1779
Suggestion and Autosuggestion, 1746
La suggestion dans l'art, 1355
La suggestione e le facoltà psichiche occulte in rapporto alla pratica legale e medico-forense, 1474
Suggestion et autosuggestion; étude psychologique et pedagogique d'après les résultats de la nouvelle école de Nancy, 1746
La suggestion mentale et l'action à distance des substances toxigues et médicamenteuses, 1164
La suggestion mentale et le calcul des probabilités, 1094
Les suggestions criminelles. Viols. Faux et captations. Viols moraux, les suggestions en emour. Gabrielle Fenayrou et Gabrielle Bompard, 1296
Suggestions hypnotiques criminelles, 1380
Die Suggestions-Therapie bei Krankhaften Erscheinungen des Geschlectssinnes, mit besonderer berücksichtigung der conträren Sexualempfindung, 1329
Suggestion und Hypnose, 1895
Suggestion und Hypnotismus in der Völkerpsychologie, 1375
Suggestive Therapeutics, 1127, 1409, 1442, 1489, 1494
Suite de la correspondance de Monsieur Pressavin gradué etc., avec les magnétiseurs de la même ville, 103
Suite des mémoires pour servir à l'histoire et à l'établissement du magnétisme animal, 105, 148, 241
Suite des réflexions préliminaires à l'occasion des Docteurs modernes, 58
Suite du traitement magnétique de la demoiselle N., lequel a servi de base à l'Essair sur la théorie du somnambulisme magnétique, 162
Sulla causa dei Fenomeni Mesmerici, 771
Sulla potenza motrice trasfusa dall'uomo nella materia bruta. Fenomeno comunemente noto sotto il titolo di Tavola Girante. Esperimenti ed osservazioni. Aggiunge alcune dell'illustre magnetologo prof Lisimaco Verati et altre del dottor Francesco Argenti, giá decano della facoltà medica dell'Università di Padova e membro di varie accademie, 699
Sulla trasmissione del pensiero, 1042
Sull'azione del magnetismo animale nell'umano organismo, 614
Sul magnetismo animale, discorso istorico-critico. Letto all 'Accademia di Religione Cattolica il di 21 Luglio 1842, 463
Sul magnetismo animale e sul metodo per istudiarlo, 630
Supermudane Facts in the Life of Rev. Jesse Babcock Ferguson, A.M., LL.D., Including Twenty Years' Observation of Preternatural Phenomena, 883
The Supernatural? With Chapter on Oriental Magic, Spiritualism, and Theosophy, by J. N. Maskelyne, 1309
Superstition and Science: and Essay, 747
Superstitions et prestiges des philosophes, ou les demonolatres du siècle des luminières par l'Auteur des Precurseurs de l'Ante-Christ, 281
Supplément aux deux rapports de MM. les commissaires de l'Académie & de la Faculté de Médecine, & de la Société Royale de Médecine [D'Eslon], 53
Supplément aux deux rapports de MM. les commissaires de l'Académie et de la Faculté de médecine et de la Société royale de médecine [Bailly], 33

Supplément aux Observations de M. Bergasse, ou Règlamens des sociétés de l'harmonie universelle, adoptés par la société de l'harmonie de France dans l'assemblée générale tenue à Paris, le 12 Mai 1785; avec des notes pour servir à l'intelligence du texte, 124
Sur l'anesthésie chirurgicale hypnotique. Note présentée à l'Académie des sciences le 5 décembre 1859 suivie d'une lettre addressée au rédacteur en chef du Moniteur des sciences médicales, 800
Sur les diverse états nerveux déterminés par l'hypnotisation chez les hystériques, 1058
Sur les faits que semblent prouver une communication des somnambules avec let êtres spirituels et sur les consequences qu'on peut tirer de ces faits. (Extrait d'une lettre d M., à M. Deleuze), 402
Sur l'état de fascination déterminé chez l'homme à l'aide de surfaces brillantes en rotation (action somnifère des miroirs à alouettes), 1213
The Surprising Case of Rachel Baker, Who Prays and Preaches in her Sleep: with Specimens of her Extraordinary Performances Taken Down Accurately in Short Hand at the Time; and Showing the Unparalleled Powers She Possesses to Pray, Exhort, and Answer Questions, During Her Unconscious State. The Whole Authenticated by the Most Respectable Testimony of Living Witnesses, 249, 254
The Survival of Man: a Study in Unrecognized Human Faculty, 1632
Die sympathetisch-magnetische heilkunde in ihrem ganzen Umfange: oder die Lehre von der Transplantation der Krankheiten, die Amulete, die Signaturen u.s.w. zum ersten Male ausführlich nach den Schriften der Paracelsisten erläutert und mit einer reichhaltigen Sammlung von Vorschriften sympathetischen Kuren ausgestattet . . . die sogenannten zauberischen Krankheiten und die magisch-magnetischen Heilungen bei den alten Hebräern, 624
A Symposium on the Subconscious, 1594
Symptomatology, Psychognosis, and Diagnosis of Psychopathic Diseases, 1706
A Synopsis of Mesmerism; or, Animal Magnetism, Pathetism, Electrical Psychology; or the Philosophy of Impressions, 574
System des Tellurismus oder thierischen Magnetismus. Ein Handbuch fur Naturforscher und Aerzte, 310
Le systéme de la rose magnétique, 115
Système des passes magnétiques, ou ensemble des procédés de magnétisation, 519
Système raisonné du magnétisme universel. D'après les principes de M. Mesmer, ouvrage auquel on a joint l'explication des procédés du magnétisme animal accomodés au cures des différentes maladies, tant par M. Mesmer que par M. le Chevalier de Barberin et par M. de Puységur, relativement au somnambulisme ainsi qu'une notice de la constitution des Sociétés dites de l'harmonie . . . Par la Société d l'harmonie d'Ostende, 158

Table Moving, its causes and phenomena: with directions how to experiment. Profusely illustrated, on plate paper, by Nicholls, 695
Table-Moving. Somnambulish-Magnetische Traumbedeutung, 731
Table Moving by Animal Magnetism demonstrated: with directions how to perform the Experiment. Also, A full and detailed account of the experiments already performed, 694
Table-Moving Popularly Explained; with an Inquiry into Reichenbach's Theory of Od Force; Also an Investigation into the Spiritual Manifestations known as Spirit-rappings, 655
Table-Moving Tested and Proved to be the Result of Satanic Agency, 669
La table parlante. Journal des faits merveilleux, tables tournantes, mesmérisme, somnambulisme magnétique . . . , 732
Table qui danse et table qui répond, expériences à la portée de tout le mond, 672
Les tables et les têtes qui tournent, ou la fièvre de rotation en 1853. Cent et un croquis, 667
Les tables tournantes considérées au point de vue de la question de physique générale, qui s'y rattache, 759
Table-Turning, the Devil's Modern Master-piece. Being the Result of a Course of Experiments, 670
Table Turning and Table Talking, Containing Detailed Reports of an Infinite Variety of Experiments Performed in England, France, and Germany, with Most Marvellous Results; also

Minute Directions to Enable Every One to Practise Them, and the Various Explanations Given of the Phenomena by the Most Distinguished Scientific Men of Europe. Second Edition with Professor Faraday's Experiments and Explanation, 696
A Technique for the Experimental Study of Telepathy and Other Alleged Clairvoyant Processes. A Report on the Work Done in 1916–17 at the Harvard Psychological Laboratory, under the Gift of Mrs. John Wallace Riddle and the Hodgson Fund, 1730
Telepathic Hallucinations: the New View of Ghosts, 1636
La télépathie: recherches expérimentales, 1814
La télépathie et le néo-spiritualisme, 1349
Telepathie und Hellsehen: Versuche und Betractungen über ungewöhnliche seelische Fähigkeiten, 1815
Telepathy, Genuine and Fraudulent, 1719
Telepathy. Its Theory, Facts and Proof, 1639
Telepathy and the Subliminal Self, 1421
La telepatia, 1436
Teleplasma und Telekinese; ergebnisse meiner zweijahrigen Experimentalsitzungen mit Berliner Medium Maria Volhart, mit 6 textzeichungen, 1854
Témoignage spiritualiste d'outre tombe sur le magnétisme humain, fruit d'un long pèlerinage par J.S.C. publié et annoté par l'abbé Loubert, 867
Ten Years with Spiritual mediums: an Inquiry Concerning the Etiology of Certain Phenomena Called Spiritual, 973
Tératoscopie du fluide vital de la mensambulance, ou démonstration physiologique et psychologique de la possibilité d'une infinité de prodiges réputés fabuleux, ou attribués par l'ignorance des philosophes et par la superstition des ignorants à des causes fausses et imaginaires, 304
Tertium Quid: Chapters on Various Disputed Questions, 1174
Testament politique de M. Mesmer, ou la précaucion d'un sage, avec le dénombrement des adeptes; le tout traduit de l'Allamand par un Bostonien, 126
That Other World. Personal Experiences of Mystics and their Mysticism, 1736
Theologische und philosophische raisonnements in Bezug auf den animalischen Magnetismus nebst einer Beleuchtung uber Realität und Irrealität. Oder: Blicke auf Gott, Natur und den Menschen, 287
Theorie der Geister-Kunde, in einer Natur-, Vernunft- und Bibelmässigen Beantwortung der Frage: Was von Ahnungen, Gesichten und Geisterrerscheinungen geglaubt und nicht geglaubt werden müsse, 231–232
Theorie des Somnambulismus oder des thierischen Magnetismus. Ein Versuch, die Mysterien des magnetischen Lebens, den Rapport der Somnambulen mit dem Magnetiseur, ihre Ferngesichte und Ahnungen, und ihren Verkehr mit der Geisterwelt vom Standpunkte vorurtheilsfreier Kritik aus zu erhellen und erklären für Gebildete überhaupt, und für Mediciner und Theologe insbesondere, 379
Théorie du mesmérisme, par un ancien ami du Mesmer, où l'on explique aux dames ses principes naturels, pour le salut de leurs familles; et aux sages de tous les pays, ses causes et ses effets, comme un bienfait de la nature qu'ils sont invités à répandre avec les précautions convenable, et d'après lesquelles plusieurs rois de l'Europe en ont encouragé l'usage dans leurs états, 275
Théorie du monde et des êtres organisés suivant les principes de M. . . . , 38
A Theory of the Mechanism of Survival: the Fourth Dimension and Its Applications, 1767
Therapeutic Suggestion in Psychopathia Sexualis, 1329
Thérapeutique du magnétisme et due somnambulisme appropriée aux maladies les plus communes; aidée par l'emploi des plantes les plus usuelles en médecine. Renseignements sur la compositions et sur l'application des remèdes conseillés Planches anatomiques avec explication philosophique, 1064
Thérapeutique magnétique, règles de l'application dit magnétisme à l'expérimentation pure et au traitement des maladies. Spiritualisme, son principe et ses phénomènes. Par M. Le Baron Du Potet, 855
Thérapeutique suggestive: son mécanisme. Propriétés diverses du sommeil provoqué et des états analogues, 1297

La thérapeutique suggestive et ses applications aux maladies nerveuses et mentales, à la chirurgie, à l'obstétrique et à la pédagogie, 1340
There is No Death, 1300
Thirty Short Sermons on Various Important Subjects, Both Doctrinal and Practical, 422
Thirty Years among the Dead. In Collaboration with Nelle M. Watts, Celia L. Goerz, Orlando D. Goerz, 1884
A Thought-Reader's Thoughts: Being the Impressions and Confession of Stuart Cumberland, 1199
Thought Transference: a Critrical and Historical Review of the Evidence for Telepathy, with a Record of New Experiments 1902–1903, 1564
Three Lectures on Animal Magnetism, as Delivered in New York, at the Hall of Science, on the 26th of July, 2d and 9th of August, 340, 918
Der tierische Magnetismus, oder das Geheimnis des menschlichen Lebens aus dynamisch-psychischen Kraften gemacht, 267
Der tierische Magnetismus in Preussen vor und nach den Freiheitskriegen, 1896
Der tierische Magnetismus und seine Genese. Ein Beitrag zur Aufklärung und eine Mahnung an die Sanitätsbehorden, 1189
Times (London), 666
Das Tischrucken, 673
Das Tischrücken. (Fortsetzung.) Geistige Agapen. Plychographische Mittheilungen der Pariser Deutsch-Magnetischen Schule, 730
Das Trischrücken. Seine qeschichtliche Entwicklung und seine Bedeutung. Auf Grund der neuesten Forschungsergebnisse dargestellt, 1798
Tischrucken und Tischklopfen eine Thatsache. Mit Documenten von den Herren: K. Simrock, Hoffmann von Fallersleben, O. Schade und Neusser, 684
Der Tod, das Jenseits und das Leben im Jenseits, 1456
O trabalho dos mortos, 1804
Traces du magnétisme, 46
The Tragedy of Chicago. A Study in Hypnotism. How an Innocent Young Man, Richard Evans, was hypnotised to the Gallows, 1569
Traité complet de magnétisme animal, 359
Traité de l'hypnotisme expérimental et thérapeutique. Ses applications à la médicine, à l'éducation, à la psychologie, 1612
Traité de magnétisme, suivi des paroles d'un somnambule, et d'un recueil de traitements magnétiques, 581
Traité de Metapsychique, 1836
Traité des sensations et des passions, 1591
Traité du magnétisme animal considéré sous le rapport de l'hygiène, de la médecine légale et de la thérapeutique, 412
Traité du somnambulisme, et des différentes modifications qu'il presente, 313, 324
Traité élémentaire, théorique et pratique du magnétisme contenant toutes les indications mécessaires pour traiter soi-même, à l'aide du magnétisme animal, les maladies les plus communes, 926
Traité expérimental de magnétisme avec figures dans le teste. Cours professé à l'école pratique de magnétisme et de massage. Théories et procédés, 1383, 1435
Traité expérimental de magnétisme avec portrait de l'auteur et figures dans le texte. Cours professé à l'école pratique de magnétisme et de massage. Physique magnétique, 1383, 1435
Traité expérimental et thérapeutique de magnétisme, 1137–1138
Traitement magnétique suivi d'une guérison remarquable opérée par M. Coll, archiprêtre du canton de Dangé, près Chatellerault, département de la Vienne, 273
Traitement mental et culture spirituelle. La santé et l'harmonie dans la vie humaine, 1674
Traité philosophique sur la nature de l'âme et des facultés, où l'on examine le rapport qu'ont avec la morale, le magnétisme de M. Mesmer et le système de Gall, 315
Traité pratique de magnétisme humain, ou résume de tous les procédés du magnétisme humain, pour rétablir et développer les fonctions physiques et les facultés intellectuelles dans l'état de maladie, 797

Traité pratique du magnétisme et du somnambulisme: ou, résumé de tous les principes et procédés du magnétisme, avec la théorie et la définition du somnambulisme, le description du caractère et des facultés des somnambules, et les règles de leur direction, 513
Traité théorique et pratique du magnétisme animal [Doppet], 55
Traité théorique et pratique du magnétisme animal, ou Méthode facile pour apprendre à magnétiser [Ricard], 444
The Trance and Correlative Phenomena, 916
Trance and Trancoidal States in the Lower Animals, 1036
Transactions du magnétisme animal, 447
Transactions of the College of Physicians of Philadelphia, 1214
Transactions of the Perkinean Society, Consisting of a Report on the Practice with the Metallic Tractors, at the Institution in Frith-Street, and Experiments Communicated by Several Correspondents, 224
Transcendental Physics: An Account of Experimental Investigations. From the Scientific Treatises of Johann Carl Friedrich Zöllner. Translated from the German, with an Preface and Appendices, by Charles Charleton Massey, 1034
Transmissione del pensiero e suggestione mentale. Studio sperimentale e critico . . . e seguito da alcune indagini fatte sullo stesso soggetto a richiesta, 1290
Trattato pratico del magnetismo animale, 921
Trattato teorico-pratico di magnetismo animale considerato sotto il punto di vista fisiologico e psicologico, con note illustrative e appendice, 716, 825
Trattato teorico-pratico sul magnetismo animale, 602
Traume eines Geistersehers, 616
A Treatise on Animal Magnetism [Johnson], 497
Treatise on Animal Magnetism; Discovering the Method of Making the Said Magnets, for the Cure of Most Diseases Incident to the Human Body, From the Writing of Paracelsus [Edwards], 188
Trials of Animal Magnetism on the Brute Creation, 419
Twenty Years of Psychical Research: 1882–1901. With Facsimile Illustrations of Thought-Transference Drawings, 1532
'Twixt Two Worlds: a Narrative of the Life and Work of William Eglinton, 1139 Two Worlds, 1892

Über das Besessenseyn oder das Daseyn und den Einfluss de boesen Geisterreichs in der alten Zeit. Mit Berücksichtungung dämonischen Besitzungen der neuen Zeit, 356
Über den Hypnotismus, 1047
Über den thierischen Magnetismus [Meiners], 184
Über den thierischen Magnetismus [Steiglitz], 252
Über den Willen in der Natur. Eine Erorterung der Bestatigungen, welche die Philosophie des Verfassers, seit ihrem Auftreten, durch die empirischen Wissenschaften erhalten hat, 378
Über die eigenthumliche Seelenstorung der sogenannten "Scherin von Prevorst," 351
Über die Extase oder das Verzucktseyn der Magnetischen Schlafredner, 270
Über die Seelenfrage. Ein Gang durch die sichtbare Welt, um die unsichtbare zu finden, 839
Über die Trugwahrnehmung (Hallucination und Illusion) mit besonderer Berücksichtigung der internationalen Enquete über Wachhallucinationen bei Gesunden, 1370
Über Hellsehen. Eine dritisch-experimentelle Untersuchung. Inaugural-Dissertation, 1713
Über Legensmagnetismus und über die magischen Wirkungen überhaupt, 773
Über Naturerklärung überhaupt und über die Erklärung der thierisch magnetischen Erscheinungen aus dynamish psychischen Kraften inbesondere. Ein ergänzender Beitrag zum Archiv den thierischen Magnetismus, 280
Über psychische Heilmittel und Magnetismus, 284
Über Spaltung der Personlichkeit (Sogennantes Doppel-Ich), 1408
Über spiritistische Manifestationen vom naturwissenschaftlichen Standpunkt, 1096
Über Sympathie, 234
Über Sympathie und Magnetismus. Aus den Lateinischen übersazt und mit anmerkungen begleitet von Heinrich Tabor, 190

Über Telepathie und Hellsehen. Experimentell-theoretische Untersuchungen, 1788
Über thierischen Magnetismus. In einem Brief an Herrn Geheimer Rath Hoffmann in Mainz, 165
Über und gegen den thierischen Magnetismus und die jetzt vorherrschende Tendenz auf dem Gebiete desselben, 278
The Unconscious: the Fundamentals of Human Personality Normal and Abnormal, 1701
The Unconscious Mind, 1449
Underhill on Mesmerism, with Criticisms on Its Opposers, and a Review of Humbugs and Humbuggers, with Practical Instructions for Experiments in the Science, Full Directions for Using it as a Remedy in Disease . . . the Philosophy of Its Curative Powers; How to Develop a Good Clairvoyant; the Philosophy of Seeing Without Eyes. The Proofs of Immortality Derived from the Unfoldings of Mesmerism . . . , 918
L'union magnétique. Journal de la Société philanthropico-magnétique de Paris, 735
L'Université de Montpellier et le magnétisme animal, ou une vérité nouvelle en présence de vieilles erreurs, 373
Unpartheyischer Bericht der allhier gemachten Entdeckungen der sonderbaren Würkung der kunstlichen Stahlmagneten in verschiedenen Nervenkrankheiten, 2
Unseen Faces Photographed. A Condensed Report of Facts and Findings During a Year of Most Critical Research into the Phenomena Called Spirit Photography, 1498
Untersuchungen über den Lebensmagnetismus und das Hellsehen, 308

Vade mecum du magnétiseur, 567
The Valet's Tragedy and Other Studies, 1523
Variations de la personnalité, 1196
Varieties of Religious Experience, 1270
The Veil Lifted. Modern Developments of Spirit Photography. A Paper by J. Traill Taylor Describing Experiments in Psychic Photography, Letter by the Rev. H. R. Haweis, Addresses by James Robertson, Glasgow, and Miscellanea by the Editor, Andrew Glendinning, 1365
Der Veränderungsgesetz, 1423
La verità sull'ipnotismo. Rivelazioni, 1158
La verità sul magnetismo animale, 761
La verité aux médecins et aux gens du monde sur le diagnostic et la thérapeutique des maladies, éclairés par le somnambulisme naturel lucide. De la catalepsie, observations de facultés surnaturelles de clairvoyance, d'intuition, de prévision et d'extase. Du magnétisme animal et de ses effets. Instruction pratique sur son application au traitement des maladies, 816
La vérité du magnétisme prouvée par les faits; extraits des notes et des papiers de Mme Alina D'Eldir, née dans l'Hindoustan, par un ami de la vérité; suivie d'une notice inédite sur Mesmer, qui avait été composée et mise en page pour la "Biographie Universelle," 342
Les verités cheminent, tot ou tard elles arrivent, 251
Verschiedenes zum Unterricht und zur Unterhaltung für Liebhaber der Gauckeltasche, des Magnetismus, und anderer Seltenheiten, Gesammelt und herausgegeben von dem Hofrath von Eckhartshaufen), 196
Versuch die scheinbare Magie des thierischen Magnetismus aus physiologischen und psychischen Gesetzen zu erklären, 258
Versuch einer Darstellung des animalischen Magnetismus, als Heilmittel, 235
Versuch über das Geistersehn und was damit zusammenhängt, 603
Les vibrations de la vitalité humaine. Méthode biométrique appliquée aux sensitifs et aux névroses, 1531
Une victoire de la photographie psychique. La romanesque et glorieuse aventure du médium William Hope (de Crewe—Angleterre). Accusé d'être un imposteur. Trainé une année dans la boue. Victime d'une sombre machination. Enfin reconnu parfaitement innocent et incontestable médium photographe, 1845
La vie et la santé, ou la médecine est-elle une science?, 1053
La vie inconsciente de l'esprit, 1025

La vie par la magnétisme et l'électricité, 1083
Les vies successives. Documents pour l'étude de cette question, 1670
La vision contenant l'explication de l'écrit intitulé: Traces du magnétisme, et la théorie des vrais sages, 119
Visionen im Wasserglase, 1000
Le visionnaire, ou la victime imaginaire du magnétisme. Histoire véritable, contenant la description d'une monomanie sans exemple et dans laquelle sont consignés les lettres autographes, ainsi que les réflexions, traits de démence et récits du monomane; le tout précédé d'une esquisse sur sa vie et ses actions jusqu'à ce jour, 347
Visions: A Study of False Sight (Pseudopia), 1004
Une visite à la Salpêtrière, 1134
Vitalisme curatif par les appareils électro-magnétiques, 1108
Der Vitalismus als Geschichte und also Lehre, 1553
Vital Magnetic Cure: an Exposition of Vital Magnetism and Its Application to the Treatment of Mental and Physical Disease. By a Magnetic Physician, 941
Vital Magnetism: A Remedy, 500
Vital Magnetism: Its power over Disease. A Statement of the Facts Developed by Men who Have Employed This Agent under Various Names, as Animal Magnetism, Mesmerism, Hypnotism, etc., from the Earliest Times down to the Present, 998
The Vital Message, 1753
Voices from the Void: Six Years' Experience in Automatic Communications, 1760
Les voix de l'avenir dans le présent et dans le passé, ou les oracles et les somnambules comparés, 809
Vom Jenseits der Seele. Die Geheimwissenshaften in kritischer Betrachtung, 1724
Vorläufer des Spiritismus. Hervorragende Falle Willkürlicher mediumistischer Erscheinungen aus den letzten drei Jahrhunderten, 1431
Die Vorurteile der Menschheit, 1020
Le vrai et le faux magnétisme, ses partisans ses ennemis, thèse présentée à la Société de magnétisme de Paris, précédée d'un avant propos sur le fluide magnétique suivie d'aphorismes ou opinions de 60 docteurs, médecins, praticiens, prêtres, et du Pape, sur le magnétisme et le somnambulisme, et de notions sur l'origine du magnétisme, sur la Société de magnétisme et sur un project de dispensaire, 893
La vue à distance. Dans le temps et dans l'espace. Enquête sur des cas de psychométrie (Janvier-Decémbre 1909). Préface de Joseph Maxwell. Suivie d'une conférence relative à l'influence de l'amour sur l'écriture, par Paul de Fallois, 1642
The Wanderings of a Spiritualist, 1797
Was I Hypnotized?, 1252
The Watseka Wonder: a Startling and Instructive Psychological Study, and Well Authenticated Instance of Angelic Visitation. A Narrative of the Leading Phenomena Occurring in the Case of Mary Lurancy Vennum. With Comments by Joseph Rodes Buchanan, D. P. Kayner, S. B. Brittan, and Hudson Tuttle, 1014
Wer ist sensitiv, wer nicht? Oder kurze Anleitung, sensitive Menschen mit Leichtigkeit zu finden, 770
What Am I? Popular Introduction to the Study of Psychology, 951
What I saw at Cassadaga Lake: 1888. Addendum to a Review in 1887 of the Seybert Commissioners' Report, 1248
What I saw at Cassadaga Lake: A Review of the Seybert Commissioners' Report, 1218, 1248
What is Mesmerism?, 687
What is Mesmerism? An Attempt to Explain its Phenomena on the Admitted Principles of Physiological and Psychical Science, 626
What is Mesmerism? and What its Concomitants Clairvoyance and Necromancy?, 650
What is Mesmerism? The Question Answered by a Mesmeric Practitioner, or, Mesmerism not Miracle: An Attempt to Show that Mesmeric Phenomena and Mesmeric Cures are not Supernatural; to Which is Appended Useful Remarks and Hints for Sufferers Who are Trying Mesmerism for a Cure, 653
The Widow's Mite and Other Psychic Phenomena, 1537, 1587
Willie; Hypnose; Zweck, 1423

Wissenschaftlichen Abhandlungen, 1015, 1034
Wonderful Works of God. A Narrative of the Wonderful Facts in the Case of Ansel Bourne, of West Shelby, Orleans Co., N. Y., Who, in the Midst of Opposition to the Christian Religion, was Suddenly Struck Blind, Dumb and Deaf; and After Eighteen Days was Suddenly and Completely Restored in the Presence of Hundreds of Persons, in the Christian Chapel, at Westerly on the 15th of November, 1857. Written under His Direction, 1002
Wonders and Mysteries of Animal Magnetism Displayed, or the History, Art, Practice, and Progress of that Useful Science, from Its First Rise in the City of Paris, to the Present Time. With Several Curious Cases and New Anecdotes of the Principal Professors, 200
Ein Wort über animalischen Magnetismus, Seelenkorper und Lebensessenz; nebst Beschreibung des ideo-somnambulen Zustandes des Fräuleins Therese von B—y zu Vasarhely im Jahre 1838, und einem Anhang, 431, 527
Die Wunder des Christentums und deren Verhältnis zum tierischen Magnetismus, mit Berücksichtigung der neuesten Wunderheilungen nach römisch-Kathol. Principien, 311
Die Wunder-Erscheinungen des Vitalismus; Tischdrehen, Tischklopfen, Tischsprechen u. nebst ihrer rationellen Erklärung in Briefen an eine Dame, 661
Die Wünschel-Rute. Historisch-theoretische Studie, 1651
Die Wünschelrute als wissenschaftliches Problem, 1829

X + Y = Z; or The Sleeping Preacher of North Alabama. Containing an Account of Most Wonderful Mysterious Mental Phenomena, Fully Authenticated by Living Witnesses, 992

The Year Book of Spiritualism for 1871; Presenting the Status of Spiritualism for the Current Year Throughout the World; Philosophical, Scientific, and Religious Essays; Review of Its Literature; History of American Associations; State and Local Societies; Progressive Lyceums; Lecturers; Mediums; and other Matters Relating to the Momentous Subject, 944
Your Psychic Powers and How to Develop Them, 1766

Zeitschrift für die Gesammte Strafrechtswissenchaft, 1232
Zeitschrift für Hypnotismus, Psychotherapie, sowie andere psycho-physiologische und psychopathologische Forschungen, 1333, 1411
Zeitschrift für Hypnotismus, Suggestionstherapie, Suggestionslehre und verwandte psychologische Forschungen, 1333, 1411, 1505
Zeitschrift für Parapsychologie, 962, 1253
Zeitschrift für psychische Aerzte, 291
Zeitschrift für psychische Aerzte, mit besonderer Berücksichtigung des Magnetismus, 291
Zend-Avesta oder über die Dinge des Himmels und des Jenseits, von Standpunkt der Naturbetrachtung, 616
The Zoist. A Journal of Cerebral Physiology & Mesmerism, and their Applications to Human Welfare, 424, 490, 901
Zoistic Magnetism: Being the Substance of Two Lectures, Descriptive of Original Views and Investigations Respecting This Mysterious Agency; Delivered, by Request, at Torquay, on the 24th of April and lst of May, 1849, 586
Zöllner. An Open Letter to Professor George S. Fullerton of the University of Pennsylvania, Member and Secretary of the Seybert Commission for Investigating Modern Spiritualism, 1180
Zur Einführung in das Studium des Magnetismus, Hypnotismus, Spiritualismus nebst Kritik von drei Broschuren und eines Buches des magnetischen Heilers Dr. Jul. Ed. Timmler in Altenburg, 1236
Zur Kritik des telepathischen beweis Materiels, 1425
Zur Psychologies und Pathologie sogenannter occulter Phänomene, 1506
Zur seelen lehre. Ein philosophische Confession, 805
Das zweite Gesicht (Die "Vorgesichten") nach Wirklichkeit und Wesen, 1599

Subject Index

Alterations of personality, 105
Alternating selves, 1271
American Society for Psychical Research, 1556, 1559, 1601
"old," 1222, 1317
Amnesia, 910, 980, 1071, 1284, 1318, 1441
Anesthesia
animal magnetism, 359, 464, 473, 474, 520, 521, 536, 541, 545, 546, 554, 636, 637, 653, 1032
chemical, 536
hypnotism, 818
hystical, 1318, 1589
Animal electricity, 121, 194, 198, 207, 211, 225, 238, 540
Animal gravitation, 1
Animal magnetism
abuse of, 152, 213, 274, 295, 506, 509, 815
dangers of, 237, 263, 295, 366, 412, 503, 506, 644, 904
history of, 17, 46, 83, 94, 116, 131, 135, 145, 175, 200, 222, 228, 243, 253, 296, 298, 309, 310, 372, 399, 425, 434, 436, 444, 451, 454, 455, 476, 478, 496, 524, 529, 566, 613, 629, 685, 716, 766, 787, 789, 822, 967, 1018, 1080, 1241, 1348, 1387, 1421, 1464, 1504, 1508, 1775, 1857, 1896
legal implications, 1172
the occult and, 34, 38, 49, 133, 158, 185, 186, 211, 223, 228, 238, 258, 343, 348, 430, 454, 458, 539, 586, 594, 609, 613, 633, 831, 1077
paranormal phenomena, 313, 322, 350, 365, 379, 391, 436, 549, 575, 1151, 1183, 1281
procedures for, 39, 49, 182, 183, 222, 228, 558
psychotherapy and, 160, 175, 240, 241, 245, 246
scientific study of, 170, 174
social implications, 45, 49, 250
theory of, 1, 3, 4, 5, 7, 10, 11, 23, 25, 31, 38, 45, 55, 62, 94, 116, 128, 175, 186, 228, 250, 260, 264, 268, 275, 296, 312, 417, 425, 450, 513, 523, 524, 529, 534, 550, 552, 602, 606, 612, 641, 771
Animals: animal magnetism and, 106, 419, 478, 513
Animism, 1666
Apparitions, 508, 561, 603, 676, 915, 965, 1019, 1062, 1144, 1177, 1307, 1341, 1369, 1371, 1422, 1443, 1475, 1525, 1537, 1626, 1627, 1644, 1648, 1663, 1771, 1778, 1885
Apports, 1756
Astral projection, 1307, 1689
Aura, 1664
Automatic writing, 664, 799, 853, 968, 986, 997, 1069, 1090, 1114, 1118, 1229, 1369, 1525, 1551, 1578, 1687, 1732, 1747, 1756, 1760, 1833, 1860
Automatisms, 910, 977, 1026, 1048, 1119, 1197, 1205, 1229, 1235, 1255, 1467, 1500, 1525, 1632, 1777

Baquet, magnetic, 35, 39, 50, 69, 599, 827
Book tests, 1837
Boston Society for Psychical Research, 1709
Braidism, 820, 821, 949, 1186, 1363, 1585
Brain, dual hemispheres, 410, 505, 1079

Canadian Society for Psychical Research, 1780
Christian science, 1635, 1886
Clairvoyance, 139, 152, 160, 211, 228, 238, 308, 317, 341, 344, 357, 381, 387, 389, 408, 409, 416, 473, 479, 508, 515, 550, 560, 565, 569, 576, 578, 637, 659, 685, 692, 747, 750, 764, 771, 774, 775, 782, 784, 803, 824, 858, 866, 909, 1021, 1034, 1110, 1177, 1217, 1243, 1307, 1361, 1457, 1472, 1500, 1525, 1527, 1537, 1575, 1599, 1633, 1642, 1644, 1662, 1691, 1708, 1713, 1738, 1750, 1788, 1812, 1815, 1822, 1823, 1848, 1851, 1864, 1865, 1873, 1885
  diagnostic, 152, 158, 211, 238, 387, 476, 885
Co-consciousness, 1559, 1701, 1891
Commission: Royal Academy of Medicine, 302, 321, 324, 329, 340, 350, 354, 355, 416, 436, 455, 540, 1504
Commission: Royal Academy of Sciences, 30–33, 40, 52, 53, 62, 65, 67, 68, 82, 86, 101, 109, 111, 122, 145, 177, 205, 213, 239, 324, 436, 498
Commission: Royal Society of Medicine, 31, 50, 53, 67, 68, 72, 101, 111, 145, 239, 324, 436, 498
Contraction of field of consciousness, 1318, 1589
Controls (spirit), 1856
Critical sleep, 211, 238
Cross-correspondences, 1578, 1617, 1654, 1698
Cryptomnesia, 1506
Crystal gazing, 1420, 1525, 1563, 1575, 1646, 1858

Death-bed experiences, 1019
Direct voice, 1587
Disaggregation, 1235
Dissociation, 910, 911, 1002, 1176, 1229, 1513, 1547, 1574, 1594, 1597, 1701, 1709, 1803, 1833, 1890
Distance: animal magnetism at a, 578
Distraction, 1211, 1235
Divided consciousness, 105. *See also* Double consciousness
Divination, 455
Divining rod, 18, 117, 579, 725, 1376, 1640, 1651, 1665, 1829
Double consciousness, 105, 357, 478, 505, 1076, 1098, 1114, 1229, 1255, 1334, 1506
Double ego, 1082
Double memory. *See* Double consciousness
Doubling of the personality, 977, 1080, 1146, 1159, 1176, 1178, 1211, 1355, 1408, 1640, 1681
Dreams, 308
Dual personality, 1120, 1172, 1589

Ecstacy, 313, 317, 352, 508, 559, 577, 593, 676, 855, 1186
Ectoplasm, 1711, 1795, 1826, 1854
Elberfeld horses, 1592, 1728, 1850
Electrical psychology, 574, 596, 598, 718
Electro-biology, 604, 607, 738, 836
Elongation of the body, 858, 931
Etherology, 515, 540, 585
Exegetic and Philanthropic Society of Stockholm, 185, 186
Exteriorization of motricity, 1406
Exteriorization of sensibility, 1328, 1391, 1630, 1880

Fairies, 1820
Fascination, 564, 585, 708, 718, 738, 866, 1151, 1182, 1186, 1213, 1389
Fire-immunity, 858, 931
Fixation of the eyes, 90, 1032
Fixed ideas, 954, 1235, 1318, 1324, 1441
Franklin Commission. *See* Commission: Royal Academy of Sciences
Free love, 772

German Society for Scientific Occultism, 1877
Ghosts, 438, 571, 579, 717, 852, 1062, 1208, 1331, 1426, 1636, 1790, 1849
  animal, 1420
Goligher circle, 1711, 1751, 1788, 1795, 1824, 1900
Gyro-magnetism, 693

Hallucinations, 352, 408, 508, 692, 831, 843, 866, 977, 1019, 1032, 1047, 1080, 1109, 1116, 1128, 1144, 1235, 1273, 1284, 1309, 1311, 1369, 1370, 1374, 1422, 1525, 1574, 1648, 1667, 1681

Subject Index

Hauntings, 348, 561, 1177, 1307, 1420, 1426, 1459, 1460, 1520, 1532, 1748, 1765, 1843
Healing
  by animal magnetism, 10, 12, 16, 29, 31, 33, 42, 47, 49, 56, 60, 66, 71, 73, 74, 89, 95, 104–106, 111, 113, 116, 145, 158, 159, 167, 175, 177, 182, 189, 208, 211, 220, 222, 227, 229, 284, 301, 307, 310, 320, 331, 333, 348, 364, 394, 396, 404, 416, 418, 421, 437, 444, 461, 478, 514, 517, 522, 527, 550, 552, 564, 570, 577, 598, 611, 653, 693, 729, 750, 757, 785, 806, 811, 868, 949, 987, 1110, 1257, 1337, 1383, 1435, 1502
  by hypnotism, 450, 656, 818, 896, 907, 1183, 1251, 1552, 1624
  by spiritualism, 968, 1796
  by suggestion, 1284, 1521
Hibernation, 591
Higher phenomena of somnambulism (or animal magnetism), 328, 579, 617, 637
Human electricity, 548, 728, 1447
Hypnoid states, 1547
Hypnotic rapport, 896, 910, 1323, 1328, 1391
Hypnotic sleep, 450, 896
Hypnotic trance, 47
Hypnotism
  abuse of, 1172, 1368
  animals and, 1473
  dangers of, 1250, 1356
  history of, 1240, 1264, 1354, 1424, 1517, 1624
  legal implications of, 1172, 1237, 1358, 1368, 1412
  memory and, 1143
  procedures for, 465, 598, 1125, 1163, 1239, 1358, 1624
  sex and, 1368
  stages of, 947, 1080, 1087, 1162, 1163, 1178, 1239, 1328, 1352, 1386, 1391, 1612
  theory of, 450, 465, 540, 812, 1058, 1080, 1084, 1085, 1240, 1490
Hysteria, 449, 779, 818, 829, 907, 925, 947, 982, 1037, 1048, 1079, 1080, 1126–1128, 1224, 1229, 1235, 1255, 1284, 1318, 1343, 1386, 1441, 1520, 1589, 1633, 1727, 1805, 1832

Illuminati, the, 1367
Imagination, 25, 31, 54, 62, 72, 214, 220, 228, 328, 391, 438, 450, 523, 779, 804, 917, 996, 1170, 1847
Inner sense, 238, 250, 299
Insanity, 240, 241, 245, 505, 1280, 1415, 1476, 1513, 1655
Institut métapsychique international, 1283

Levitation
  of the body, 858, 931, 1341, 1428
  of objects, 712, 1428, 1711
London Dialectical Society, 942, 1406, 1895
Lower phenomena of somnambulism (or animal magnetism), 617, 637
Lucid sleep, 294, 1902
Luminous phenomena, 858, 997, 1607

Magic, 308, 378, 391, 782, 823, 831
Magicians (stage), 1027
Magnetic crisis, 49, 158, 160, 170, 201, 211, 213, 228
Magnetic fluid, 10, 13, 31, 54, 62, 66, 72, 94, 121, 127, 152, 158, 175, 211, 238, 290, 294, 313, 320, 324, 349, 399, 437, 444, 465, 489, 532, 583, 587, 596, 599, 633, 656, 674, 676, 712, 726, 738, 742, 781, 789, 824, 885, 915, 1027, 1029, 1068, 1240, 1323, 1591, 1661, 1880
Magnetic passes, 105, 158, 1068
Magnetic rapport, 105, 148, 152, 178, 234, 237, 238, 241, 414, 431, 494, 596, 644, 771
  and love, 152, 175, 1056
  and sex, 213, 274
  and transference, 241
Magnetic sleep, 47, 105, 201, 345, 450, 653, 685, 818, 1160
Magnetic somnambulism, 47, 105, 121, 139, 148, 152, 158, 161, 177, 178, 186, 211, 227, 237, 238, 241, 302, 357, 385, 389, 433, 513, 559, 580, 867, 895
Magnetized trees, 39
Magnetized water, 39
Magnetizing at a distance, 62, 158, 201
Mannheim dog, 1850
Materializations, 858, 978, 994, 1113, 1129, 1341, 1369, 1395, 1418, 1626, 1672, 1679, 1693, 1700, 1703, 1704, 1804, 1831, 1864, 1880, 1901
Medical electricity, 80, 203, 221
Mediums
  and fraud, 845, 875, 991, 995, 997, 1034, 1073, 1139, 1199, 1200, 1274, 1285, 1288, 1315, 1406, 1484, 1524, 1579, 1584, 1586, 1607, 1626, 1628, 1672, 1684, 1783, 1875, 1887, 1903
  mental, 1175, 1511
  physical, 744, 928

Mediumship, 693, 748, 772, 831, 844, 845, 847, 858, 973, 997, 1090, 1139, 1271, 1470, 1500, 1503, 1508, 1526, 1652, 1663, 1697, 1699, 1700, 1726, 1884, 1900
  dangers of, 845
  phenomena of, 781, 813, 853, 883, 900, 957, 1244, 1246, 1472, 1526, 1536, 1575, 1606, 1774, 1788, 1856, 1895
  scientific investigation of, 1470, 1788, 1903
  trance and, 1458, 1899
Mesmeric sleep. *See* Magnetic sleep
Mesmerism. *See* Animal magnetism
Metallotherapy, 658, 1003, 1016, 1054, 1186
Mineral magnetism, 2, 3, 6, 10, 26, 27, 236, 310, 467, 925
Multiple personality, 254, 261, 345, 505, 579, 834, 1102, 1159, 1196, 1214, 1311, 1334, 1361, 1495, 1525, 1547, 1559, 1648, 1709, 1785, 1805, 1832
Muscle-reading, 1052, 1166, 1308

Nancy School of hypnotism, 896, 947, 1080, 1127, 1228, 1232, 1237, 1323, 1517, 1794
  "New," 1746
Negative hallucinations, 1176
Neurasthenia, 1052
Neuric force, 1051, 1160
Neuro-hypnotism. *See* Hypnotism

Od (odic, odyle) emanations, 532, 583, 648, 655, 666, 681, 746, 769, 794, 898, 899, 905, 988, 1880

Pain: immunity to, 160, 586
Paracelsus, 94, 116, 145, 188
Paranormal phenomena, 346, 379, 508, 747, 844, 855, 890, 895, 951, 955, 1090, 1118, 1144, 1244, 1301, 1374, 1394, 1410, 1420, 1497, 1666, 1850, 1851, 1885, 1888
  and animals, 1592, 1728, 1840, 1850
  and somnambulism, 416, 437
  scientific investigation of, 1144, 1374, 1704, 1730, 1734, 1741, 1744
Parapsychology, 1807, 1894, 1897
Pathetism, 489, 515, 540, 574, 585, 692, 918, 923
Pendulum, 623, 725
Perkinism, 204–207, 209, 210, 212, 214, 217, 218, 224, 658, 1016, 1054
Personality, 1071
Photography: spirit, 933
Phreno-hypnotism, 1047
Phrenology, 395, 403, 478, 483, 489, 515, 517, 560, 584, 789
Phreno-magnetism, 470, 478, 480, 489, 523, 1734
Pitié school of hypnotism, 1079, 1087
Poltergeists, 561, 1208, 1420, 1426, 1748, 1756, 1843
Possession (spirit), 356, 361, 376, 408, 676, 792, 822, 915, 1014, 1048, 1120, 1235, 1271, 1355, 1415, 1443, 1525, 1537, 1709, 1805, 1884
Powder of sympathy, 109, 1391
Precognition, 341, 358, 508, 561, 858, 915, 1307, 1426, 1457, 1500, 1527, 1539, 1575, 1633, 1683, 1717, 1723
Premonition. *See* Precognition
Presentification, 1522
Presentiments. *See* Precognition
Prevision. *See* Precognition
Prophecy, 308, 455, 569, 782
Psychasthenia, 1441, 1522
Psychic force, 948, 959, 973, 1444, 1533, 1583, 1681
Psychic phenomena, 833, 864, 890, 907, 983, 1482, 1554, 1601, 1605, 1611, 1630, 1731, 1733
Psychical research, 561, 666, 799, 822, 883, 899, 937, 951, 969, 1144, 1253, 1263, 1270, 1294, 1301, 1341, 1353, 1366, 1377, 1384, 1415, 1418, 1421, 1422, 1426, 1438, 1443, 1459, 1460, 1488, 1496, 1500, 1501, 1510, 1520, 1525, 1529, 1537, 1553, 1556, 1561, 1566, 1574, 1586, 1587, 1597, 1601, 1606, 1614, 1618, 1620, 1630, 1632, 1633, 1647, 1649, 1656, 1658, 1660, 1662, 1672, 1674, 1675, 1682, 1688, 1694, 1695, 1704, 1712, 1714, 1721–1724, 1726, 1728, 1740, 1749, 1757, 1761, 1771, 1781, 1809, 1810, 1812, 1816, 1830, 1832, 1836, 1850, 1851, 1863, 1865, 1868, 1874, 1879, 1881, 1885, 1888, 1893–1895, 1898, 1901
Psychodunamy, 540
Psychometry, 854, 1103, 1570, 1696, 1792, 1848, 1873
Psychophysics, 616, 988
Psychotherapy, 1372, 1522, 1637, 1677, 1692, 1701, 1746, 1777, 1884

Reflexology, 1550
Reincarnation, 781, 1574, 1670

Retrocognition, 1667

Salpetrière School of hypnotism, 818, 947, 1080, 1087, 1127, 1134, 1162, 1172, 1178, 1210, 1228, 1303, 1318
Secondary consciousness, 1270
Secondary personalities, 1146, 1270, 1443, 1537, 1574, 1611, 1709
Secondary states, 1321
Second sight. *See* Clairvoyance
Secret report of the King's Commission, 32, 213, 239, 436, 506
Seeress of Prevorst, 341, 343, 351
Seybert Commission, 1180, 1191, 1218, 1248
Shakers, 485, 883
Sixth sense, 152, 161, 178, 201
Slate writing, 1448
Sleep-talking, 249, 345
Sleep-walking, 105, 345, 433, 438, 586, 653, 1098
Société des amis réunis de Strasbourg, 156, 185, 186, 1269
Society for Psychical Research, 666, 676, 937, 959, 1060, 1084, 1090, 1097, 1115, 1144, 1175, 1222, 1274, 1317, 1366, 1371, 1374, 1394, 1460, 1485, 1496, 1515, 1525, 1532, 1559, 1562, 1601, 1614, 1632, 1663, 1682, 1761, 1778, 1811, 1836, 1844, 1874, 1888
Society of Harmony
  other, 43, 120, 122, 151, 228
  Paris, 36, 43, 122, 201
Somnambulism
  artificial, 105, 222, 231, 237, 250, 254, 294, 308, 327, 401, 408, 444, 726, 789, 805, 866, 977, 1076, 1077, 1093, 1098, 1101, 1141, 1159, 1160, 1186, 1255, 1337, 1415, 1661, 1670
  higher phenomena of, 1338, 1345, 1734
  induced at a distance, 1141, 1145, 1147, 1148, 1153, 1157, 1203, 1216
  legal implications of, 1181
  natural, 105, 139, 237, 238, 250, 308, 408, 444, 726, 1124, 1186, 1343
  paranormal phenomena, 350
Spirit photography, 1365, 1369, 1498, 1537, 1659, 1756, 1793, 1819, 1845, 1877
Spiritism, 231, 485, 822, 1072, 1319, 1377
Spirits
  communication with, 231, 393, 402, 418, 549, 558, 559, 595, 627, 633, 666, 746, 781, 799, 802, 804, 901, 1023, 1090, 1346, 1385, 1410, 1439, 1463, 1500, 1512, 1515, 1537, 1542, 1556, 1632, 1638, 1662, 1687, 1733, 1756, 1760, 1768, 1780, 1799, 1885
  evil (the devil, Satan), 281, 460, 481, 665, 669, 670, 676, 683, 869, 1392, 1461
Spiritual force, 744
Spiritualism, 231, 393, 444, 473, 559, 595, 603, 616, 666, 689, 690, 700, 712, 772, 774, 776, 783, 786, 792, 793, 802, 810, 833, 857, 859, 883, 900, 901, 907, 929, 943, 944, 945, 946, 950, 952, 955, 957, 964, 970, 972, 979, 983, 995, 997, 1001, 1005, 1031, 1038, 1073, 1113, 1140, 1245, 1528
  American, 387, 629, 714, 739, 837, 932, 1063, 1122
  British, 935, 1063
  dangers of, 1369, 1759, 1772, 1859
  fraud and, 438, 941, 1366, 1825
  French, 781, 796, 1063, 1140
  general history of, 1347, 1366, 1410, 1431, 1508, 1739, 1784
  German, 1063
  other countries, 1063
  phenomena of, 393, 654, 670, 681, 688, 712, 775, 813, 835, 876, 942, 968, 990, 993, 1023, 1024, 1113, 1118, 1175, 1199, 1235, 1248, 1253, 1293, 1384, 1385, 1426, 1516, 1532, 1576, 1584, 1591, 1601, 1604, 1633, 1638, 1649, 1668, 1742, 1754, 1755, 1773, 1791, 1892; *see also* Mediumship, phenomena of
  scientific investigation of, 666, 744, 746, 942, 959, 969, 1191, 1824
  self-delusion and, 666, 1346, 1736, 1752
Splitting of the "I," 1648
Splitting of the mind, 1270, 1832, 1890
Subconscious, 105, 1146, 1324, 1441, 1513, 1551, 1594, 1611, 1615, 1643, 1891
Subconscious acts, 1211, 1235, 1318, 1346, 1500, 1547, 1557, 1701, 1723
Subconscious self, 1421
Subconscious states, 1318, 1441
Subliminal consciousness, 105, 1152, 1324, 1325, 1351, 1388, 1487, 1525, 1611
Subliminal self, 1421, 1515, 1525, 1557
Subwaking self, 1450
Suggestibility, 653, 961, 974
  in animal magnetism, 105, 175, 278, 294, 324, 381, 386, 438, 1032, 1068
  auto, 1375, 1746, 1794, 1821
  history of, 1284
  hysteria and, 1146
  in hypnotism, 607, 896, 1048, 1052, 1125, 1127, 1132, 1162, 1169, 1172, 1187, 1190, 1232, 1235, 1237, 1244,

1249, 1251, 1273, 1292, 1295, 1316, 1352, 1357, 1412, 1450, 1476, 1492, 1519, 1598, 1624
legal implications, 1296, 1380
mental, 1154, 1184, 1204
movement and, 1197
od emanations and, 583
post-hypnotic, 170, 1080, 1132, 1142, 1211, 1235, 1237, 1240, 1678
in psychical research, 1721
sociology of, 1375, 1450, 1545, 1550, 1720
in spiritualism, 993
in table-moving, 659
therapeutic, 1297, 1494, 1615, 1619, 1657, 1676, 1737, 1886
waking, 1080, 1127, 1237, 1355, 1389, 1467
Supernormal faculty, 1324, 1388
Supernormal phenomena. *See* Paranormal phenomena
Supraliminal consciousness, 1324, 1525
Supraliminal self, 1525
Survival of death, 231, 368, 616, 751, 1090, 1301, 1350, 1439, 1457, 1463, 1518, 1529, 1556, 1609, 1614, 1632, 1641, 1658, 1668, 1675, 1688, 1714, 1722, 1733, 1757, 1766–1768, 1771, 1809, 1837, 1843, 1853, 1874
Sympathy, 450, 596, 1035, 1391
diagnosis by, 158
healing by, 109, 130, 190, 685, 1391
rapport and, 234, 308, 489

Table-knocking, 684
Table-moving, 653, 655, 656, 661, 666, 669, 672, 674, 678, 679, 681, 682, 693–695, 730, 853, 928
Table-talking, 629, 659, 665, 671, 696, 713, 717, 732, 733, 751
Table-tapping, 720, 781
Table-turning, 628, 651, 653, 659, 660, 662, 667, 670, 673, 675, 676, 683, 684, 686, 696, 700, 702, 712, 723, 747, 753, 759, 760, 781, 789, 822, 831, 880, 899, 900, 1072, 1366, 1798, 1855
Telekinesis, 1418, 1527, 1679, 1817, 1854, 1876, 1880, 1903
Telepathy, 588, 1086, 1090, 1094, 1118, 1144, 1153, 1166, 1301, 1341, 1349, 1371, 1378, 1421, 1422, 1425, 1426, 1436, 1457, 1472, 1500, 1508, 1512, 1514, 1515, 1525, 1527, 1532, 1534, 1537, 1563, 1565, 1570, 1610, 1618, 1633, 1636, 1639, 1649, 1660, 1662, 1663, 1666, 1687, 1694, 1719, 1730, 1748, 1789, 1812, 1814, 1815, 1889; *see also* Thought transference
Thought transference (thought reading), 105, 620, 1042, 1067, 1086, 1100, 1105, 1111, 1144, 1155, 1204, 1217, 1308, 1371, 1436, 1613, 1633, 1644, 1723; *see also* Telepathy
Tongues, speaking with, 730
Transposition of senses, 152, 171, 353, 885

Unconscious, the, 105, 533, 773, 924, 1025, 1082, 1226, 1324, 1449, 1678, 1701, 1870, 1888, 1891
Unconscious acts, 740, 746, 993, 1113, 1146, 1211, 1503, 1649
Unconscious cerebration, 985
Unconscious, collective, 1778
Unconscious fraud, 1633
Unconscious thought, 1567

Veridical phenomena, 1090
Visions, 559, 593, 603, 1004
Vital electricity, 804
Vital fluid, 726
Vital force, 1335, 1396, 1397
Vitalism, 1553
Vital magnetism, 482, 500, 940, 998, 1417

Waking self, 1450
Will
in animal magnetism, 121, 148, 158, 193, 201, 378, 451, 771
in hypnotism, 1104, 1198, 1289
the unconscious and, 924
Willing: and the paranormal, 1110

Zoistic magnetism, 586
Zoomagnetism, 1018, 1068, 1297